POWER GENERATION, OPERATION, AND CONTROL

POWER GENERATION, OPERATION, AND CONTROL

SECOND EDITION

Allen J. Wood
*Power Technologies, Inc. and
Rensselaer Polytechnic Institute*

Bruce F. Wollenberg
University of Minnesota

A WILEY-INTERSCIENCE PUBLICATION

JOHN WILEY & SONS, INC.

New York ● Chichester ● Brisbane ● Toronto ● Singapore

Copyright © 1984, 1996 by John Wiley & Sons, Inc.

Library of Congress Cataloging in Publication Data:
Wood, Allen J.
    Power generation, operation, and control / Allen J. Wood, Bruce F.
Wollenberg. — 2nd ed.
        p.   cm.
    Includes index.
    ISBN 0-471-58699-4 (cloth : alk. paper)
    1. Electric power systems.   I. Wollenberg, Bruce F.   II. Title.
TK1001.W64   1996
621.31—dc20   95-10876

Printed in the United States of America

20 19 18 17 16

CONTENTS

PREFACE TO THE SECOND EDITION

It has been 11 years since the first edition was published. Many developments have taken place in the area covered by this text and new techniques have been developed that have been applied to solve old problems. Computing power has increased dramatically, permitting the solution of problems that were previously left as being too expensive to tackle. Perhaps the most important development is the changes that are taking place in the electric power industry with new, nonutility participants playing a larger role in the operating decisions.

It is still the intent of the authors to provide an introduction to this field for senior or first-year graduate engineering students. The authors have used the text material in a one-semester (or two-quarter) program for many years. The same difficulties and required compromises keep occurring. Engineering students are very comfortable with computers but still do not usually have an appreciation of the interaction of human and economic factors in the decisions to be made to develop "optimal" schedules; whatever that may mean. In 1995, most of these students are concurrently being exposed to courses in advanced calculus and courses that explore methods for solving power flow equations. This requires some coordination. We have also found that very few of our students have been exposed to the techniques and concepts of operations research, necessitating a continuing effort to make them comfortable with the application of optimization methods. The subject area of this book is an excellent example of optimization applied in an important industrial system.

The topic areas and depth of coverage in this second edition are about the same as in the first, with one major change. Loss formulae are given less space and supplemented by a more complete treatment of the power-flow-based techniques in a new chapter that treats the optimal power flow (OPF). This chapter has been put at the end of the text. Various instructors may find it useful to introduce parts of this material earlier in the sequence; it is a matter of taste, plus the requirement to coordinate with other course coverage. (It is difficult to discuss the OPF when the students do not know the standard treatment for solving the power flow equations.)

The treatment of unit commitment has been expanded to include the Lagrange relaxation technique. The chapter on production costing has been revised to change the emphasis and introduce new methods. The market structures for bulk power transactions have undergone important changes throughout the world. The chapter on interchange transactions is a "progress report" intended to give the students an appreciation of the complications that may accompany a competitive market for the generation of electric energy. The sections on security analysis have been updated to incorporate an introduction to the use of bounding techniques and other contingency selection methods. Chapter 13 on the OPF includes a brief coverage of the security-constrained OPF and its use in security control.

The authors appreciate the suggestions and help offered by professors who have used the first edition, and our students. (Many of these suggestions have been incorporated; some have not, because of a lack of time, space or knowledge.) Many of our students at Rensselaer Polytechnic Institute (RPI) and the University of Minnesota have contributed to the correction of the first edition and undertaken hours of calculations for home-work solutions, checked old examples, and developed data for new examples for the second edition. The 1994 class at RPI deserves special and honorable mention. They were subjected to an early draft of the revision of Chapter 8 and required to proofread it as part of a tedious assignment. They did an outstanding job and found errors of 10 to 15 years standing. (A note of caution to any of you professors that think of trying this; it requires more work than you might believe. How would you like 20 critical editors for your lastest, glorious tome?)

Our thanks to Kuo Chang, of Power Technologies, Inc., who ran the computations for the bus marginal wheeling cost examples in Chapter 10. We would also like to thank Brian Stott, of Power Computer Applications, Corp., for running the OPF examples in Chapter 13.

ALLEN J. WOOD
BRUCE F. WOLLENBERG

PREFACE TO THE FIRST EDITION

The fundamental purpose of this text is to introduce and explore a number of engineering and economic matters involved in planning, operating, and controlling power generation and transmission systems in electric utilities. It is intended for first-year graduate students in electric power engineering. We believe that it will also serve as a suitable self-study text for anyone with an undergraduate electrical engineering education and an understanding of steady-state power circuit analysis.

This text brings together material that has evolved since 1966 in teaching a graduate-level course in the electric power engineering department at Rensselaer Polytechnic Institute (RPI). The topics included serve as an effective means to introduce graduate students to advanced mathematical and operations research methods applied to practical electric power engineering problems. Some areas of the text cover methods that are currently being applied in the control and operation of electric power generation systems. The overall selection of topics, undoubtedly, reflects the interests of the authors.

In a one-semester course it is, of course, impossible to consider all the problems and "current practices" in this field. We can only introduce the types of problems that arise, illustrate theoretical and practical computational approaches, and point the student in the direction of seeking more information and developing advanced skills as they are required.

The material has regularly been taught in the second semester of a first-year graduate course. Some acquaintance with both advanced calculus methods (e.g., Lagrange multipliers) and basic undergraduate control theory is needed. Optimization methods are introduced as they are needed to solve practical problems and used without recourse to extensive mathematical proofs. This material is intended for an engineering course: mathematical rigor is important but is more properly the province of an applied or theoretical mathematics course. With the exception of Chapter 12, the text is self-contained in the sense that the various applied mathematical techniques are presented and developed as they are utilized. Chapter 12, dealing with state estimation, may require more understanding of statistical and probabilistic methods than is provided in the text.

The first seven chapters of the text follow a natural sequence, with each succeeding chapter introducing further complications to the generation scheduling problem and new solution techniques. Chapter 8 treats methods used in generation system planning and introduces probabilistic techniques in the computation of fuel consumption and energy production costs. Chapter 8 stands alone and might be used in any position after the first seven chapters. Chapter 9 introduces generation control and discusses practices in modern U.S. utilities and pools. We have attempted to provide the "big picture" in this chapter to illustrate how the various pieces fit together in an electric power control system.

The topics of energy and power interchange between utilities and the economic and scheduling problems that may arise in coordinating the economic operation of interconnected utilities are discussed in Chapter 10. Chapters 11 and 12 are a unit. Chapter 11 is concerned with power system security and develops the analytical framework used to control bulk power systems in such a fashion that security is enhanced. Everything, including power systems, seems to have a propensity to fail. Power system security practices try to control and operate power systems in a defensive posture so that the effects of these inevitable failures are minimized. Finally, Chapter 12 is an introduction to the use of state estimation in electric power systems. We have chosen to use a maximum likelihood formulation since the quantitative measurement–weighting functions arise in a natural sense in the course of the development.

Each chapter is provided with a set of problems and an annotated reference list for further reading. Many (if not most) of these problems should be solved using a digital computer. At RPI we are able to provide the students with some fundamental programs (e.g., a load flow, a routine for scheduling of thermal units). The engineering students of today are well prepared to utilize the computer effectively when access to one is provided. Real bulk power systems have problems that usually call forth Dr. Bellman's curse of dimensionality—computers help and are essential to solve practical-sized problems.

The authors wish to express their appreciation to K. A. Clements, H. H. Happ, H. M. Merrill, C. K. Pang, M. A. Sager, and J. C. Westcott, who each reviewed portions of this text in draft form and offered suggestions. In addition, Dr. Clements used earlier versions of this text in graduate courses taught at Worcester Polytechnic Institute and in a course for utility engineers taught in Boston, Massachusetts.

Much of the material in this text originated from work done by our past and current associates at Power Technologies, Inc., the General Electric Company, and Leeds and Northrup Company. A number of IEEE papers have been used as primary sources and are cited where appropriate. It is not possible to avoid omitting, references and sources that are considered to be significant by one group or another. We make no apology for omissions and only ask for indulgence from those readers whose favorites have been left out. Those interested may easily trace the references back to original sources.

We would like to express our appreciation for the fine typing job done on the original manuscript by Liane Brown and Bonnalyne MacLean.

This book is dedicated in general to all of our teachers, both professors and associates, and in particular to Dr. E. T. B. Gross.

ALLEN J. WOOD
BRUCE F. WOLLENBERG

1 Introduction

1.1 PURPOSE OF THE COURSE

The objectives of a first-year, one-semester graduate course in electric power generation, operation, and control include the desire to:

1. Acquaint electric power engineering students with power generation systems, their operation in an economic mode, and their control.
2. Introduce students to the important "terminal" characteristics for thermal and hydroelectric power generation systems.
3. Introduce mathematical optimization methods and apply them to practical operating problems.
4. Introduce methods for solving complicated problems involving both economic analysis and network analysis and illustrate these techniques with relatively simple problems.
5. Introduce methods that are used in modern control systems for power generation systems.
6. Introduce "current topics": power system operation areas that are undergoing significant, evolutionary changes. This includes the discussion of new techniques for attacking old problems and new problem areas that are arising from changes in the system development patterns, regulatory structures, and economics.

1.2 COURSE SCOPE

Topics to be addressed include:

1. Power generation characteristics.
2. Economic dispatch and the general economic dispatch problem.
3. Thermal unit economic dispatch and methods of solution.
4. Optimization with constraints.
5. Using dynamic programming for solving economic dispatch and other optimization problems.

1

    6. Transmission system effects:
      a. power flow equations and solutions,
      b. transmission losses,
      c. effects on scheduling.
    7. The unit commitment problem and solution methods:
      a. dynamic programming,
      b. the Lagrange relaxation method.
    8. Generation scheduling in systems with limited energy supplies.
    9. The hydrothermal coordination problem and examples of solution techniques.
  10. Production cost models:
      a. probabilistic models,
      b. generation system reliability concepts.
  11. Automatic generation control.
  12. Interchange of power and energy:
      a. interchange pricing,
      b. centrally dispatched power pools,
      c. transmission effects and wheeling,
      d. transactions involving nonutility parties.
  13. Power system security techniques.
  14. An introduction to least-squares techniques for power system state estimation.
  15. Optimal power flow techniques and illustrative applications.

In many cases, we can only provide an introduction to the topic area. Many additional problems and topics that represent important, practical problems would require more time and space than is available. Still others, such as light-water moderated reactors and cogeneration plants, could each require several chapters to lay a firm foundation. We can offer only a brief overview and introduce just enough information to discuss system problems.

1.3 ECONOMIC IMPORTANCE

The efficient and optimum economic operation and planning of electric power generation systems have always occupied an important position in the electric power industry. Prior to 1973 and the oil embargo that signaled the rapid escalation in fuel prices, electric utilities in the United States spent about 20% of their total revenues on fuel for the production of electrical energy. By 1980, that figure had risen to more than 40% of total revenues. In the 5 years after 1973, U.S. electric utility fuel costs escalated at a rate that averaged 25%

compounded on an annual basis, The efficient use of the available fuel is growing in importance, both monetarily and because most of the fuel used represents irreplaceable natural resources.

An idea of the magnitude of the amounts of money under consideration can be obtained by considering the annual operating expenses of a large utility for purchasing fuel. Assume the following parameters for a moderately large system.

Annual peak load: 10,000 MW

Annual load factor: 60%

Average annual heat rate for converting fuel to electric energy: 10,500 Btu/kWh

Average fuel cost: $3.00 per million Btu (MBtu), corresponding to oil priced at 18 $/bbl

With these assumptions, the total annual fuel cost for this system is as follows.

Annual energy produced: 10^7 kW $\times$ 8760 h/yr $\times$ 0.60 = 5.256 $\times$ 10^{10} kWh

Annual fuel consumption: 10,500 Btu/kWh $\times$ 5.256 $\times$ 10^{10} kWh
    = 55.188 $\times$ 10^{13} Btu

Annual fuel cost: 55.188 $\times$ 10^{13} Btu $\times$ 3 $\times$ 10^{-6} $/Btu = $1.66 billion

To put this cost in perspective, it represents a direct requirement for revenues from the average customer of this system of 3.15 cents per kWh just to recover the expense for fuel.

A savings in the operation of this system of a small percent represents a significant reduction in operating cost, as well as in the quantities of fuel consumed. It is no wonder that this area has warranted a great deal of attention from engineers through the years.

Periodic changes in basic fuel price levels serve to accentuate the problem and increase its economic significance. Inflation also causes problems in developing and presenting methods, techniques, and examples of the economic operation of electric power generating systems. Recent fuel costs always seem to be ancient history and entirely inappropriate to current conditions. To avoid leaving false impressions about the actual value of the methods to be discussed, all the examples and problems that are in the text are expressed in a nameless, fictional monetary unit to be designated as an "Ŗ."

1.4 PROBLEMS: NEW AND OLD

This text represents a progress report in an engineering area that has been and is still undergoing rapid change. It concerns established engineering problem areas (i.e., economic dispatch and control of interconnected systems) that have taken on new importance in recent years. The original problem of economic dispatch for thermal systems was solved by numerous methods years ago. Recently there has been a rapid growth in applied mathematical methods and the availability of computational capability for solving problems of this nature so that more involved problems have been successfully solved.

The classic problem is the economic dispatch of fossil-fired generation systems to achieve minimum operating cost. This problem area has taken on a subtle twist as the public has become increasingly concerned with environmental matters, so that "economic dispatch" now includes the dispatch of systems to minimize pollutants and conserve various forms of fuel, as well as to achieve minimum costs. In addition, there is a need to expand the limited economic optimization problem to incorporate constraints on system operation to ensure the "security" of the system, thereby preventing the collapse of the system due to unforeseen conditions. The hydrothermal coordination problem is another optimum operating problem area that has received a great deal of attention. Even so, there are difficult problems involving hydrothermal coordination that cannot be solved in a theoretically satisfying fashion in a rapid and efficient computational manner.

The post World War II period saw the increasing installation of pumped-storage hydroelectric plants in the United States and a great deal of interest in energy storage systems. These storage systems involve another difficult aspect of the optimum economic operating problem. Methods are available for solving coordination of hydroelectric, thermal, and pumped-storage electric systems. However, closely associated with this economic dispatch problem is the problem of the proper commitment of an array of units out of a total array of units to serve the expected load demands in an "optimal" manner.

A great deal of progress and change has occurred in the 1985–1995 decade. Both the unit commitment and optimal economic maintenance scheduling problems have seen new methodologies and computer programs developed. Transmission losses and constraints are integrated with scheduling using methods based on the incorporation of power flow equations in the economic dispatch process. This permits the development of optimal economic dispatch conditions that do not result in overloading system elements or voltage magnitudes that are intolerable. These "optimal power flow" techniques are applied to scheduling both real and reactive power sources, as well as establishing tap positions for transformers and phase shifters.

In recent years the political climate in many countries has changed, resulting in the introduction of more privately owned electric power facilities and a reduction or elimination of governmentally sponsored generation and transmission organizations. In some countries, previously nationwide systems have been privatized. In both these countries and in countries such as the United States, where electric utilities have been owned by a variety of bodies (e.g., consumers, shareholders, as well as government agencies), there has been a movement to introduce both privately owned generation companies and larger cogeneration plants that may provide energy to utility customers. These two groups are referred to as independent power producers (IPPs). This trend is coupled with a movement to provide access to the transmission system for these nonutility power generators, as well as to other interconnected utilities. The growth of an IPP industry brings with it a number of interesting operational problems. One example is the large cogeneration plant that provides steam to an industrial plant and electric energy to the power system. The industrial-plant steam demand schedule sets the operating pattern for the generating plant, and it may be necessary for a utility to modify its economic schedule to facilitate the industrial generation pattern.

Transmission access for nonutility entities (consumers as well as generators) sets the stage for the creation of new market structures and patterns for the interchange of electric energy. Previously, the major participants in the interchange markets in North America were electric utilities. Where nonutility, generation entities or large consumers of power were involved, local electric utilities acted as their agents in the marketplace. This pattern is changing. With the growth of nonutility participants and the increasing requirement for access to transmission has come a desire to introduce a degree of economic competition into the market for electric energy. Surely this is not a universally shared desire; many parties would prefer the status quo. On the other hand, some electric utility managements have actively supported the construction, financing, and operation of new generation plants by nonutility organizations and the introduction of less-restrictive market practices.

The introduction of nonutility generation can complicate the scheduling-dispatch problem. With only a single, integrated electric utility operating both the generation and transmission systems, the local utility could establish schedules that minimized its own operating costs while observing all of the necessary physical, reliability, security, and economic constraints. With multiple parties in the bulk power system (i.e., the generation and transmission system), new arrangements are required. The economic objectives of all of the parties are not identical, and, in fact, may even be in direct (economic) opposition. As this situation evolves, different patterns of operation may result in different regions. Some areas may see a continuation of past patterns where the local utility is the dominant participant and continues to make arrangements and schedules on the basis of minimization of the operating cost that is paid by its own customers. Centrally dispatched power pools could evolve that include nonutility generators, some of whom may be engaged in direct sales to large consumers. Other areas may have open market structures that permit and facilitate competition with local utilities. Both local and remote nonutility entities, as well as remote utilities, may compete with the local electric utility to supply large industrial electric energy consumers or distribution utilities. The transmission system may be combined with a regional control center in a separate entity. Transmission networks could have the legal status of "common carriers," where any qualified party would be allowed access to the transmission system to deliver energy to its own customers, wherever they might be located. This very nearly describes the current situation in Great Britain.

What does this have to do with the problems discussed in this text? A *great deal.* In the extreme cases mentioned above, many of the dispatch and scheduling methods we are going to discuss will need to be rethought and perhaps drastically revised. Current practices in automatic generation control are based on tacit assumptions that the electric energy market is slow moving with only a few, more-or-less fixed, interchange contracts that are arranged *between interconnected utilities.* Current techniques for establishing optimal economic generation schedules are really based on the assumption of a single utility serving the electric energy needs of its own customers at minimum cost. Interconnected operations and energy interchange agreements are presently the result of interutility arrangements: all of the parties share common interests. In a world with a transmission-operation entity required to provide access to many parties, both utility and nonutility organizations, this entity has the task of developing operating schedules to accomplish the deliveries scheduled in some (as yet to be defined) "optimal" fashion within the physical constraints of the system, while maintaining system reliability and security. If all (or any) of this develops, it should be a fascinating time to be active in this field.

FURTHER READING

The books below are suggested as sources of information for the general area covered by this text. The first four are "classics;" the next seven are specialized or else are collections of articles or chapters on various topics involved in generation operation and control. Reference 12 has proven particularly helpful in reviewing various thermal cycles. The last two may be useful supplements in a classroom environment.

1. Steinberg, M. J., Smith, T. H., *Economy Loading of Power Plants and Electric Systems*, Wiley, New York, 1943.
2. Kirchmayer, L. K., *Economic Operation of Power Systems*, Wiley, New York, 1958.
3. Kirchmayer, L. K., *Economic Control of Interconnected Systems*, Wiley, New York, 1959.
4. Cohn, N., *Control of Generation and Power Flow on Interconnected Systems*, Wiley, New York, 1961.
5. Hano, I., *Operating Characteristics of Electric Power Systems*, Denki Shoin, Tokyo, 1967.
6. Handschin, E. (ed.), *Real-Time Control of Electric Power Systems*, Elsevier, Amsterdam, 1972.
7. Savulescu, S. C. (ed.), *Computerized Operation of Power Systems*, Elsevier, Amsterdam, 1976.
8. Sterling, M. J. H., *Power System Control*, Peregrinus, London, 1978.
9. El-Hawary, M. E., Christensen, G. S., *Optimal Economic Operation of Electric Power Systems*, Academic, New York, 1979.
10. Cochran, R. G., Tsoulfanidis, N. M. I., *The Nuclear Fuel Cycle: Analysis and Management*, American Nuclear Society, La Grange Park, IL, 1990.
11. Stoll, H. G. (ed.), *Least-Cost Electric Utility Planning*, Wiley, New York, 1989.
12. El-Wakil, M. M., *Power Plant Technology*, McGraw-Hill, New York, 1984.

13. Debs, A. S., *Modern Power Systems Control and Operation*, Kluwer, Norwell, MA, 1988.

14. Strang, G., *An Introduction to Applied Mathematics*, Wellesley-Cambridge Press, Wellesley, MA, 1986.

15. Miller, R. H., Malinowski, J. H., *Power System Operation*, Third Edition, McGraw-Hill, New York, 1994.

16. Handschin, E., Petroianu, A., *Energy Management Systems*, Springer-Verlag, Berlin, 1991.

2 Characteristics of Power Generation Units

2.1 CHARACTERISTICS OF STEAM UNITS

In analyzing the problems associated with the controlled operation of power systems, there are many possible parameters of interest. Fundamental to the economic operating problem is the set of input–output characteristics of a thermal power generation unit. A typical boiler–turbine–generator unit is sketched in Figure 2.1. This unit consists of a single boiler that generates steam to drive a single turbine–generator set. The electrical output of this set is connected not only to the electric power system, but also to the auxiliary power system in the power plant. A typical steam turbine unit may require 2–6% of the gross output of the unit for the auxiliary power requirements necessary to drive boiler feed pumps, fans, condenser circulating water pumps, and so on. In defining the unit characteristics, we will talk about *gross* input versus *net* output. That is, gross input to the plant represents the total input, whether measured in terms of dollars per hour or tons of coal per hour or millions of cubic feet of gas per hour, or any other units. The net output of the plant is the electrical power output available to the electric utility system. Occasionally engineers will develop gross input–gross output characteristics. In such situations, the data should be converted to net output to be more useful in scheduling the generation.

In defining the characteristics of steam turbine units, the following terms will be used

H = Btu per hour heat input to the unit (or MBtu/h)

F = Fuel cost times H is the ₽ per hour (₽/h) input to the unit for fuel

Occasionally the ₽ per hour operating cost rate of a unit will include prorated operation and maintenance costs. That is, the labor cost for the operating crew will be included as part of the operating cost if this cost can be expressed directly as a function of the output of the unit. The output of the generation unit will be designated by P, the megawatt net output of the unit.

Figure 2.2 shows the input–output characteristic of a steam unit in idealized form. The input to the unit shown on the ordinate may be either in terms of heat energy requirements [millions of Btu per hour (MBtu/h)] or in terms of

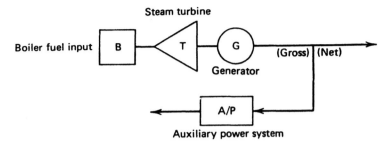

FIG. 2.1   Boiler–turbine–generator unit.

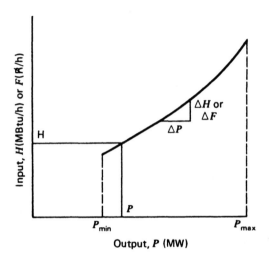

FIG. 2.2   Input–output curve of a steam turbine generator.

total cost per hour (R per hour). The output is normally the net electrical output of the unit. The characteristic shown is idealized in that it is presented as a smooth, convex curve.

These data may be obtained from design calculations or from heat rate tests. When heat rate test data are used, it will usually be found that the data points do not fall on a smooth curve. Steam turbine generating units have several critical operating constraints. Generally, the minimum load at which a unit can operate is influenced more by the steam generator and the regenerative cycle than by the turbine. The only critical parameters for the turbine are shell and rotor metal differential temperatures, exhaust hood temperature, and rotor and shell expansion. Minimum load limitations are generally caused by fuel combustion stability and inherent steam generator design constraints. For example, most supercritical units cannot operate below 30% of design capability. A minimum flow of 30% is required to cool the tubes in the furnace of the steam generator adequately. Turbines do not have any inherent overload capability, so that the data shown on these curves normally do not extend much beyond 5% of the manufacturer's stated valve-wide-open capability.

The incremental heat rate characteristic for a unit of this type is shown in Figure 2.3. This incremental heat rate characteristic is the slope (the derivative) of the input–output characteristic ($\Delta H/\Delta P$ or $\Delta F/\Delta P$). The data shown on this curve are in terms of Btu per kilowatt hour (or ₽ per kilowatt hour) versus the net power output of the unit in megawatts. This characteristic is widely used in economic dispatching of the unit. It is converted to an incremental fuel cost characteristic by multiplying the incremental heat rate in Btu per kilowatt hour by the equivalent fuel cost in terms of ₽ per Btu. Frequently this characteristic is approximated by a sequence of straight-line segments.

The last important characteristic of a steam unit is the unit (net) heat rate characteristic shown in Figure 2.4. This characteristic is H/P versus P. It is proportional to the reciprocal of the usual efficiency characteristic developed for machinery. The unit heat rate characteristic shows the heat input per kilowatt hour of output versus the megawatt output of the unit. Typical conventional steam turbine units are between 30 and 35% efficient, so that their unit heat rates range between approximately 11,400 Btu/kWh and 9800 Btu/kWh. (A kilowatt hour has a thermal equivalent of approximately 3412 Btu.) Unit heat rate characteristics are a function of unit design parameters such as initial steam conditions, stages of reheat and the reheat temperatures, condenser pressure, and the complexity of the regenerative feed-water cycle. These are important considerations in the establishment of the unit's efficiency. For purposes of estimation, a typical heat rate of 10,500 Btu/kWh may be used occasionally to approximate actual unit heat rate characteristics.

Many different formats are used to represent the input–output characteristic shown in Figure 2.2. The data obtained from heat rate tests or from the plant design engineers may be fitted by a polynomial curve. In many cases, quadratic

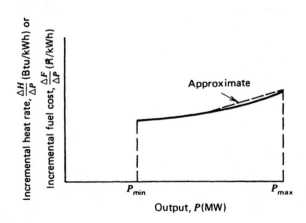

FIG. 2.3   Incremental heat (cost) rate characteristic.

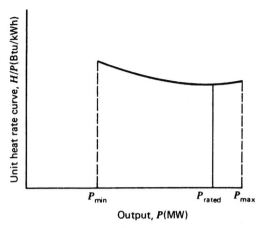

FIG. 2.4 Net heat rate characteristic of a steam turbine generator unit.

characteristics have been fit to these data. A series of straight-line segments may also be used to represent the input–output characteristics. The different representations will, of course, result in different incremental heat rate characteristics. Figure 2.5 shows two such variations. The solid line shows the incremental heat rate characteristic that results when the input versus output characteristic is a quadratic curve or some other continuous, smooth, convex function. This incremental heat rate characteristic is monotonically increasing as a function of the power output of the unit. The dashed lines in Figure 2.5 show a stepped incremental characteristic at results when a series of straight-line segments are used to represent the input–output characteristics of the unit. The use of these different representations may require that different scheduling methods be used for establishing the optimum economic operation of a power

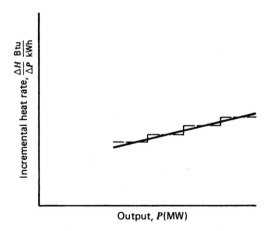

FIG. 2.5 Approximate representations of the incremental heat rate curve.

system. Both formats are useful, and both may be represented by tables of data. Only the first, the solid line, may be represented by a continuous analytic function, and only the first has a derivative that is nonzero. (That is, d^2F/dP^2 equals zero if dF/dP is constant.)

At this point, it is necessary to take a brief detour to discuss the heating value of the fossil fuels used in power generation plants. Fuel heating values for coal, oil, and gas are expressed in terms of Btu/lb, or joules per kilogram of fuel. The determination is made under standard, specified conditions using a *bomb calorimeter*. This is all to the good except that there are *two* standard determinations specified.

1. The higher heating value of the fuel (HHV) assumes that the water vapor in the combustion process products condenses and therefore includes the latent heat of vaporization in the products.
2. The lower heating value of the fuel (LHV) does not include this latent heat of vaporization.

The difference between the HHV and LHV for a fuel depends on the hydrogen content of the fuel. Coal fuels have a low hydrogen content with the result that the difference between the HHV and LHV for a fuel is fairly small. (A typical value of the difference for a bituminous coal would be of the order of 3%. The HHV might be 14,800 Btu/lb and the LHV 14,400 Btu/lb.) Gas and oil fuels have a much higher hydrogen content, with the result that the relative difference between the HHV and LHV is higher; typically in the order of 10 and 6%, respectively. This gives rise to the possibility of some confusion when considering unit efficiencies and cycle energy balances. (A more detailed discussion is contained in the book by El-Wakil: Chapter 1, reference 12.)

A uniform standard must be adopted so that everyone uses the same heating value standard. In the USA, the standard is to use the HHV *except that engineers and manufacturers that are dealing with combustion turbines (i.e., gas turbines) normally use <u>LHVs</u> when quoting heat rates or efficiencies.* In European practice, LHVs are used for all specifications of fuel consumption and unit efficiency. In this text, HHVs are used throughout the book to develop unit characteristics. Where combustion turbine data have been converted by the authors from LHVs to HHVs, a difference of 10% was normally used. When in doubt about which standard for the fuel heating value has been used to develop unit characteristics—*ask!*

2.2   VARIATIONS IN STEAM UNIT CHARACTERISTICS

A number of different steam unit characteristics exist. For large steam turbine generators the input–output characteristics shown in Figure 2.2 are not always as smooth as indicated there. Large steam turbine generators will have a number of steam admission valves that are opened in sequence to obtain ever-increasing output of the unit. Figure 2.6 shows both an input–output and an incremental heat rate characteristic for a unit with four valves. As the unit loading increases, the input to the unit increases and the incremental heat rate decreases between the opening points for any two valves. However, when a valve is first opened, the throttling losses increase rapidly and the incremental heat rate rises suddenly. This gives rise to the discontinuous type of incremental heat rate characteristic shown in Figure 2.6. It is possible to use this type of characteristic in order to schedule steam units, although it is usually not done. This type of input–output characteristic is nonconvex; hence, optimization techniques that require convex characteristics may not be used with impunity.

Another type of steam unit that may be encountered is the *common-header plant*, which contains a number of different boilers connected to a common steam line (called a common header). Figure 2.7 is a sketch of a rather complex

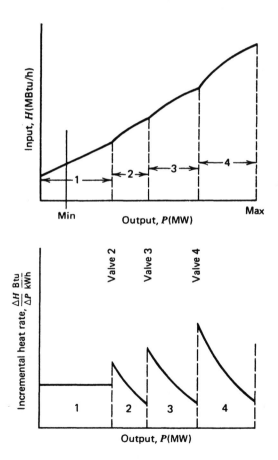

FIG. 2.6   Characteristics of a steam turbine generator with four steam admission valves.

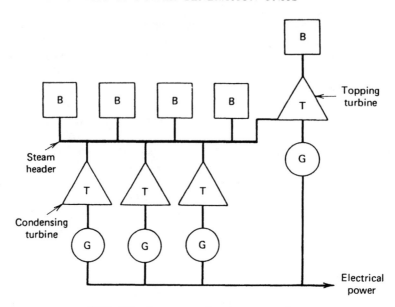

FIG. 2.7 A common-header steam plant.

common-header plant. In this plant there are not only a number of boilers and turbines, each connected to the common header, but also a "topping turbine" connected to the common header. A *topping turbine* is one in which steam is exhausted from the turbine and fed not to a condenser but to the common steam header.

A common-header plant will have a number of different input–output characteristics that result from different combinations of boilers and turbines connected to the header. Steinberg and Smith (Chapter 1, reference 1) treat this type of plant quite extensively. Common-header plants were constructed originally not only to provide a large electrical output from a single plant, but also to provide steam sendout for the heating and cooling of buildings in dense urban areas. After World War II, a number of these plants were modernized by the installation of the type of topping turbine shown in Figure 2.7. For a period of time during the 1960s, these common-header plants were being dismantled and replaced by modern, efficient plants. However, as urban areas began to reconstruct, a number of metropolitan utilities found that their steam loads were growing and that the common-header plants could not be dismantled but had to be expected to provide steam supplies to new buildings.

Combustion turbines (gas turbines) are also used to drive electric generating units. Some types of power generation units have been derived from aircraft gas turbine units and others from industrial gas turbines that have been developed for applications like driving pipeline pumps. In their original applications, these two types of combustion turbines had dramatically different duty cycles. Aircraft engines see relatively short duty cycles where power requirements vary considerably over a flight profile. Gas turbines in pumping duty on pipelines would be expected to operate almost continuously throughout the year. Service in power generation may require both types of duty cycle.

Gas turbines are applied in both a simple cycle and in combined cycles. In the simple cycle, inlet air is compressed in a rotating compressor (typically by a factor of 10 to 12 or more) and then mixed and burned with fuel oil or gas in a combustion chamber. The expansion of the high-temperature gaseous products in the turbine drives the compressor, turbine, and generator. Some designs use a single shaft for the turbine and compressor, with the generator being driven through a suitable set of gears. In larger units the generators are driven directly, without any gears. Exhaust gases are discharged to the atmosphere in the simple cycle units. In combined cycles the exhaust gases are used to make steam in a heat-recovery steam generator before being discharged.

The early utility applications of simple cycle gas turbines for power generation after World War II through about the 1970s were generally to supply power for peak load periods. They were fairly low efficiency units that were intended to be available for emergency needs and to insure adequate generation reserves in case of unexpected load peaks or generation outages. Net full-load heat rates were typically 13,600 Btu/kWh (HHV). In the 1980s and 1990s, new, large, simple cycle units with much improved heat rates were used for power generation. Figure 2.8 shows the approximate, reported range of heat rates

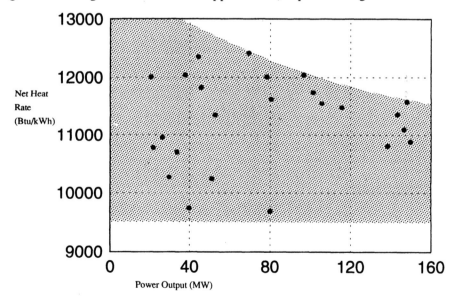

FIG. 2.8 Approximate net heat rates for a range of simple cycle gas turbine units. Units are fired by natural gas and represent performance at standard conditions of an ambient temperature of 15°C at sea level. (Heat rate data from reference 1 were adjusted by 13% to represent HHVs and auxiliary power needs.)

for simple cycle units. These data were taken from a 1990 publication (reference 1) and were adjusted to allow for the difference between lower and higher heating values for natural gas and the power required by plant auxiliaries. The data illustrate the remarkable improvement in gas turbine efficiencies achieved by the modern designs.

Combined cycle plants use the high-temperature exhaust gases from one or more gas turbines to generate steam in heat-recovery steam generators (HRSGs) that are then used to drive a steam turbine generator. There are many different arrangements of combined cycle plants; some may use supplementary boilers that may be fired to provide additional steam. The advantage of a combined cycle is its higher efficiency. Plant efficiencies have been reported in the range between 6600 and 9000 Btu/kWh for the most efficient plants. Both figures are for HHVs of the fuel (see reference 2). A 50% efficiency would correspond to a net heat rate of 6825 Btu/kWh. Performance data vary with specific cycle and plant designs. Reference 2 gives an indication of the many configurations that have been proposed.

Part-load heat rate data for combined cycle plants are difficult to ascertain

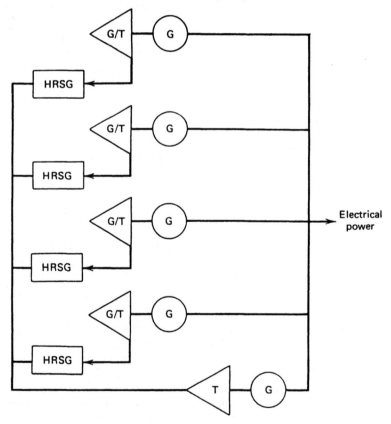

FIG. 2.9   A combined cycle plant with four gas turbines and a steam turbine generator.

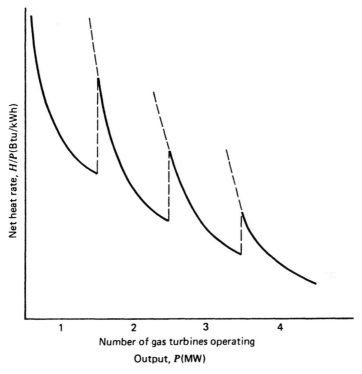

FIG. 2.10   Combined cycle plant heat rate characteristic.

from available information. Figure 2.9 shows the configuration of a combined cycle plant with four gas turbines and HRSGs and a steam turbine generator. The plant efficiency characteristics depend on the number of gas turbines in operation. The shape of the net heat rate curve shown in Figure 2.10 illustrates this. Incremental heat rate characteristics tend to be flatter than those normally seen for steam turbine units.

2.3   COGENERATION PLANTS

Cogeneration plants are similar to the common-header steam plants discussed previously in that they are designed to produce both steam and electricity. The term "cogeneration" has usually referred to a plant that produces steam for an industrial process like an oil refining process. It is also used to refer to district heating plants. In the United States, "district heating" implies the supply of steam to heat buildings in downtown (usually business) areas. In Europe, the term also includes the supply of heat in the form of hot water or steam for residential complexes, usually large apartments.

For a variety of economic and political reasons, cogeneration is assuming a larger role in the power systems in the United States. The economic incentive is due to the high efficiency electric power generation "topping cycles" that can generate power at heat rates as low as 4000 Btu/kWh. Depending on specific plant requirements for heat and power, an industrial firm may have large amounts of excess power available for sale at very competitive efficiencies. The recent and current political, regulatory, and economic climate encourages the supply of electric power to the interconnected systems by nonutility entities such as large industrial firms. The need for process heat and steam exists in many industries. Refineries and chemical plants may have a need for process steam on a continuous basis. Food processing may require a steady supply of heat. Many industrial plants use cogeneration units that extract steam from a simple or complex (i.e., combined) cycle and simultaneously produce electrical energy.

Prior to World War II, cogeneration units were usually small sized and used extraction steam turbines to drive a generator. The unit was typically sized to supply sufficient steam for the process and electric power for the load internal to the plant. Backup steam may have been supplied by a boiler, and an interconnection to the local utility provided an emergency source of electricity. The largest industrial plants would usually make arrangements to supply an excess electric energy to the utility. Figure 2.11 shows the input–output characteristics for a 50-MW single extraction unit. The data show the heat

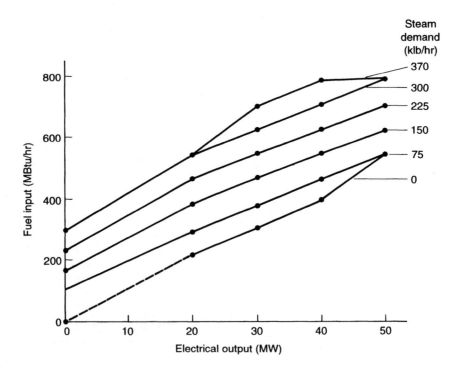

FIG. 2.11 Fuel input required for steam demand and electrical output for a single extraction steam turbine generator.

input required for given combinations of process steam demand and electric output. This particular example is for a unit that can supply up to 370,000 lbs/h of steam.

Modern cogeneration plants are designed around combined cycles that may incorporate separately fired steam boilers. Cycle designs can be complex and are tailored to the industrial plant's requirements for heat energy (see reference 2). In areas where there is a market for electric energy generated by an IPP, that is a nonutility-owned generating plant, there may be strong economic incentives for the industrial firm to develop a plant that can deliver energy to the power system. This has occurred in the United States after various regulatory bodies began efforts to encourage competition in the production of electric energy. This can, and has, raised interesting and important problems in the scheduling of generation and transmission system use. The industrial firm may have a steam demand cycle that is level, resulting in a more-or-less constant level of electrical output that must be absorbed. On the other hand, the local utility's load may be very cyclical. With a small component of nonutility generation this may not represent a problem. However, if the IPP total generation supplies an appreciable portion of the utility load demand, the utility may have a complex scheduling situation.

2.4 LIGHT-WATER MODERATED NUCLEAR REACTOR UNITS

U.S. utilities have adopted the light-water moderated reactor as the "standard" type of nuclear steam supply system. These reactors are either pressurized water reactors (PWRs) or boiling water reactors (BWRs) and use slightly enriched uranium as the basic energy supply source. The uranium that occurs in nature contains approximately seven-tenths of 1% by weight of ^{235}U. This natural uranium must be enriched so that the content of ^{235}U is in the range of 2–4% for use in either a PWR or a BWR.

The enriched uranium must be fabricated into fuel assemblies by various manufacturing processes. At the time the fuel assemblies are loaded into the nuclear reactor core there has been a considerable investment made in this fuel. During the period of time in which fuel is in the reactor and is generating heat and steam, and electrical power is being obtained from the generator, the amount of usable fissionable material in the core is decreasing. At some point, the reactor core is no longer able to maintain a critical state at a proper power level, so the core must be removed and new fuel reloaded into the reactor. Commercial power reactors are normally designed to replace one-third to one-fifth of the fuel in the core during reloading.

At this point, the nuclear fuel assemblies that have been removed are highly radioactive and must be treated in some fashion. Originally, it was intended that these assemblies would be reprocessed in commercial plants and that valuable materials would be obtained from the reprocessed core assemblies. It is questionable if the U.S. reactor industry will develop an economically viable reprocessing system that is acceptable to the public in general. If this is not done, either these radioactive cores will need to be stored for some indeterminate period of time or the U.S. government will have to take over these fuel assemblies for storage and eventual reprocessing. In any case, an additional amount of money will need to be invested, either in reprocessing the fuel or in storing it for some period of time.

The calculation of "fuel cost" in a situation such as this involves economic and accounting considerations and is really an investment analysis. Simply speaking, there will be a total dollar investment in a given core assembly. This dollar investment includes the cost of mining the uranium, milling the uranium core, converting it into a gaseous product that may be enriched, fabricating fuel assemblies, and delivering them to the reactor, plus the cost of removing the fuel assemblies after they have been irradiated and either reprocessing them or storing them. Each of these fuel assemblies will have generated a given amount of electrical energy. A pseudo-fuel cost may be obtained by dividing the total net investment in dollars by the total amount of electrical energy generated by the assembly. Of course, there are refinements that may be made in this simple computation. For example, it is possible by using nuclear physics calculations to compute more precisely the amount of energy generated by a specific fuel assembly in the core in a given stage of operation of a reactor.

In the remainder of this text, nuclear units will be treated as if they are ordinary thermal-generating units fueled by a fossil fuel. The considerations and computations of exact fuel reloading schedules and enrichment levels in the various fuel assemblies are beyond the scope of a one-semester graduate course because they require a background in nuclear engineering, as well as detailed understanding of the fuel cycle and its economic aspects (see Chapter 1, reference 10).

2.5 HYDROELECTRIC UNITS

Hydroelectric units have input–output characteristics similar to steam turbine units. The input is in terms of volume of water per unit time; the output is in terms of electrical power. Figure 2.12 shows a typical input–output curve for hydroelectric plant where the net hydraulic head is constant. This characteristic shows an almost linear curve of input water volume requirements per unit time as a function of power output as the power output increases from minimum to rated load. Above this point, the volume requirements increase as the efficiency of the unit falls off. The incremental water rate characteristics are shown in Figure 2.13. The units shown on both these curves are English units. That is, volume is shown as acre-feet (an acre of water a foot deep). If necessary, net hydraulic heads are shown in feet. Metric units are also used, as are thousands of cubic feet per second (kft^3/sec) for the water rate.

Figure 2.14 shows the input–output characteristics of a hydroelectric plant

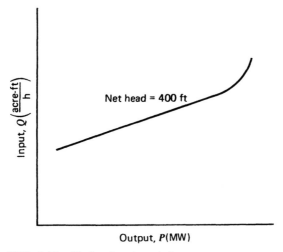

FIG. 2.12   Hydroelectric unit input–output curve.

with variable head. This type of characteristic occurs whenever the variation in the storage pond (i.e., forebay) and/or afterbay elevations is a fairly large percentage of the overall net hydraulic head. Scheduling hydroelectric plants with variable head characteristics is more difficult than scheduling hydroelectric plants with fixed heads. This is true not only because of the multiplicity of input–output curves that must be considered, but also because the maximum capability of the plant will also tend to vary with the hydraulic head. In Figure 2.14, the volume of water required for a given power output decreases as the head increases. (That is, $\partial Q/\partial \text{head}$ or $\partial Q/\partial \text{volume}$ are negative for a fixed power.) In a later section, methods are discussed that have been proposed

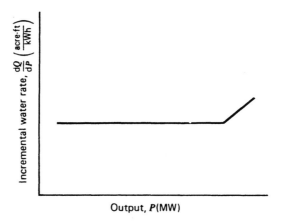

FIG. 2.13   Incremental water rate curve for hydroelectric plant.

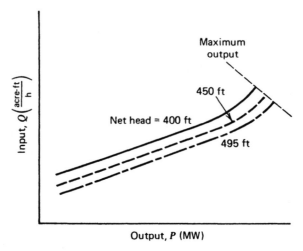

FIG. 2.14   Input–output curves for hydroelectric plant with a variable head.

for the optimum scheduling of hydrothermal power systems where the hydro-electric systems exhibit variable head characteristics.

Figure 2.15 shows the type of characteristics exhibited by pumped-storage hydroelectric plants. These plants are designed so that water may be stored by pumping it against a net hydraulic head for discharge at a more propitious time. This type of plant was originally installed with separate hydraulic turbines and electric-motor-driven pumps. In recent years, reversible, hydraulic pump turbines have been utilized. These reversible pump turbines exhibit normal input–output characteristics when utilized as turbines. In the pumping mode,

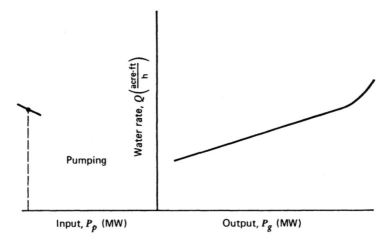

FIG. 2.15   Input–output characteristics for a pumped storage hydroplant with a fixed, net hydraulic head.

however, the efficiency of operation tends to fall off when the pump is operated away from the rating of the unit. For this reason, most plant operators will only operate these units in the pumping mode at a fixed pumping load. The incremental water characteristics when operating as a turbine are, of course, similar to the conventional units illustrated previously.

The scheduling of pumped-storage hydroelectric plants may also be complicated by the necessity of recognizing the variable-head effects. These effects may be most pronounced in the variation of the maximum capability of the plant rather than in the presence of multiple input–output curves. This variable maximum capability may have a significant effect on the requirements for selecting capacity to run on the system, since these pumped-storage hydroplants may usually be considered as spinning-reserve capability. That is, they will be used only during periods of highest cost generation on the thermal units; at other times they may be considered as readily available ("spinning reserve"). That is, during periods when they would normally be pumping, they may be shut off to reduce the demand. When idle, they may be started rapidly. In this case, the maximum capacity available will have a significant impact on the requirements for having other units available to meet the system's total spinning-reserve requirements.

These hydroelectric plants and their characteristics (both the characteristics for the pumped-storage and the conventional-storage hydroelectric plants) are affected greatly by the hydraulic configuration that exists where the plant is installed and by the requirements for water flows that may have nothing to do with power production. The characteristics just illustrated are for single, isolated plants. In many river systems, plants are connected in both series and in parallel (hydraulically speaking). In this case, the release of an upstream plant contributes to the inflow of downstream plants. There may be tributaries between plants that contribute to the water stored behind a downstream dam. The situation becomes even more complex when pumped-storage plants are constructed in conjunction with conventional hydroelectric plants. The problem of the optimum utilization of these resources involves the complicated problems associated with the scheduling of water, as well as the optimum operation of the electric power system to minimize production cost. We can only touch on these matters in this text and introduce the subject. Because of the importance of the hydraulic coupling between plants, it is safe to assert that no two hydroelectric systems are exactly the same.

APPENDIX
Typical Generation Data

Up until the early 1950s, most U.S. utilities installed units of less than 100 MW. These units were relatively inefficient (about 950 psi steam and no reheat cycles). During the early 1950s, the economics of reheat cycles and advances in materials

TABLE 2.1 Typical Fossil Generation Unit Heat Rates

Fossil Unit—Description	Unit Rating (MW)	100% Output (Btu/kWh)	80% Output (Btu/kWh)	60% Output (Btu/kWh)	40% Output (Btu/kWh)	25% Output (Btu/kWh)
Steam—coal	50	11000	11088	11429	12166	13409[a]
Steam—oil	50	11500	11592	11949	12719	14019[a]
Steam—gas	50	11700	11794	12156	12940	14262[a]
Steam—coal	200	9500	9576	9871	10507	11581[a]
Steam—oil	200	9900	9979	10286	10949	12068[a]
Steam—gas	200	10050	10130	10442	11115	12251[a]
Steam—coal	400	9000	9045	9252	9783	10674[a]
Steam—oil	400	9400	9447	9663	10218	11148[a]
Steam—gas	400	9500	9548	9766	10327	11267[a]
Steam—coil	600	8900	8989	9265	9843	10814[a]
Steam—oil	600	9300	9393	9681	10286	11300[a]
Steam—gas	600	9400	9494	9785	10396	11421[a]
Steam—coal	800–1200	8750	8803	9048	9625[a]	
Steam—oil	800–1200	9100	9155	9409	10010[a]	
Steam—gas	800–1200	9200	9255	9513	10120[a]	

[a] For study purposes, units should not be loaded below the points shown.

TABLE 2.2   Approximate Unit Heat Rate Increase Over
Valve-Best-Point Turbine Heat Rate

Unit Size (MW)	Coal (%)	Oil (%)	Gas (%)
50	22	28	30
200	20	25	27
400	16	21	22
600	16	21	22
800–1200	16	21	22

technology encouraged the installation of reheat units having steam temperatures of 1000°F and pressures in the range of 1450 to 2150 psi. Unit
sizes for the new design reheat units ranged up to 225 MW. In the late
1950s and early 1960s, U.S. utilities began installing larger units ranging
up to 300 MW in size. In the late 1960s, U.S. utilities began installing even
larger, more efficient units (about 2400 psi with single reheat) ranging in size
up to 700 MW. In addition, in the late 1960s, some U.S. utilities began installing
more efficient supercritical units (about 3500 psi, some with double reheat)
ranging in size up to 1300 MW. The bulk of these supercritical units ranged
in size from 500 to 900 MW. However, many of the newest supercritical
units range in size from 1150 to 1300 MW. Maximum unit sizes have remained
in this range because of economic, financial, and system reliability considerations.

Typical heat rate data for these classes of fossil generation are shown in
Table 2.1. These data are based on U.S. federal government reports and
other design data for U.S. utilities (see *Heat Rates for General Electric Steam
Turbine-Generators* 100,000 *kW and Larger*, Large Steam Turbine Generator
Department, G.E.).

The shape of the heat rate curves is based on the locus of design "valve-
best-points" for the various sizes of turbines. The magnitude of the turbine heat
rate curve has been increased to obtain the unit heat rate, adjusting for the
mean of the valve loops, boiler efficiency, and auxiliary power requirements.
The resulting approximate increase from design turbine heat rate to obtain the
generation heat rate in Table 2.1 is summarized in Table 2.2 for the various
types and sizes of fossil units.

Typical heat rate data for light-water moderated nuclear units are:

Output (%)	Net Heat Rate (Btu/kWh)
100	10400
75	10442
50	10951

These typical values for both PWR and BWR units were estimated using design valve-best-point data that were increased by 8% to obtain the net heat rates. The 8% accounts for auxiliary power requirements and heat losses in the auxiliaries.

Typical heat rate data for newer and larger gas turbines are discussed above. Older units based on industrial gas turbine designs had heat rates of about 13,600 Btu/kWh. Older units based on aircraft jet engines were less efficient, with typical values of full-load net heat rates being about 16,000 Btu/kWh.

Unit Statistics

In North America, the utilities participate in an organization known as the North American Electric Reliability Council (NERC) with its headquarters in Princeton, New Jersey. NERC undertakes the task of supporting the interutility operating organization which publishes an operating guide and collects, processes, and publishes statistics on generating units. NERC maintains the *Generating Availability Data System* (GADS) that contains over 25 years of data on the historical performance of generating units and related equipments. This information is made available to the industry through special reports done by the NERC staff for specific organizations and is also issued in an annual report, the *Generating Availability Report*. These data are extremely useful in tracking unit performance, detecting trends in maintenance needs, and in

TABLE 2.3 Typical Maintenance and Forced Outage Data

Unit Type	Size Range (MW)	Scheduled Maintenance Requirement (days/yr)	Equivalent Forced Rate (%)	Availability Factor (%)
Nuclear	All	67	18.3	72
Gas turbines	All	22	—	91
Fossil-fueled steam	1–99	31	7.2	88
	100–199	42	8.0	85
	200–299	43	7.2	85
	300–399	52	9.5	82
	400–599	47	8.8	82
	600–799	45	7.6	84
	800–999	40	5.8	88
	≥ 1000	44	9.0	82

From *Generating Unit Statistics* 1988–1992 issued by NERC, Princeton, NJ.

planning capacity additions to maintain adequate system generation reserves. The GADS structure provides standard definitions that are used by the industry in recording unit performance. This is of vital importance if collected statistics are to be used in reliability and adequacy analyses. Any useful reliability analysis and prediction structure requires three essential elements

1. Analytical (statistical and probability) methods and models,
2. Performance measures and acceptable standards,
3. Statistical data in a form that is useful in the analysis and prediction of performance measures.

In the generation field, GADS performs the last two in an excellent fashion. Its reputation is such that similar schemes have been established in other countries based on GADS.

Table 2.3 contains typical generating unit data on scheduled maintenance requirements, the "equivalent forced outage rate" and the "availability factor" that were taken from a NERC summary of generating unit statistics for the period 1988-1992. For any given, specified interval (say a year), the NERC definitions of the data are:

Equivalent forced outage rate = (forced outage hours + equivalent forced
                                 derated hours ÷ (forced outage hours + hours
                                 in service + equivalent forced derated hours
                                 during reserve shutdown)
          Availability factor (AF) = available hours ÷ period hours

Scheduled maintenance requirements were estimated from the NERC data using the reported "scheduled outage factor," the portion of the period representing scheduled outages.

The reported, standard equivalent forced outage rate for gas turbines has been omitted since the low duty cycle of gas turbines in peaking service biases the value of effective forced outage rate (EFOR). Using the standard definition above, the reported EFOR for all sizes of gas turbine units was 58.9%. This compares with 8.4% for all fossil-fired units. Instead of the above definition of EFOR, let us use a different rate (call it the EFOR') that includes reserve shutdown hours and neglects all derated hours to simplify the comparison with the standard definition:

EFOR = forced outage hours ÷ (forced outage hours + hours in service)

or

EFOR' = forced outage hours ÷ (forced outage hours + available hours)

where the available hours are the sum of the reserve shutdown and service hours. The effect of the short duty cycle may be illustrated using the NERC data:

	Effective Outage Rates (%)		Service Factor = (service hours) ÷ (period hours) (%)
	EFOR	EFOR'	
All fossil units	5.7	4.1	60.5
All gas turbines	55.5	3.4	2.6

The significance is not that the NERC definition is "wrong;" for some analytical models it may not be suitable for the purpose at hand. Further, and much more important, the NERC reports provide sufficient data and detail to adjust the historical statistics for use in many different analytical models.

REFERENCES

1. 1990 Performance Specs, *Gas Turbine World*, Oct. 1990, Vol. 11, Pequot Publications, Inc., Fairfield, CT.
2. Foster-Pegg, R. W., *Cogeneration—Interactions of Gas Turbine, Boiler and Steam Turbine*, ASMS paper 84-JPGC-GT-12, 1984 Joint Power Generation Conference.

3 Economic Dispatch of Thermal Units and Methods of Solution

This chapter introduces techniques of power system optimization. For a complete understanding of how optimization problems are carried out, first read the appendix to this chapter where the concepts of the Lagrange multiplier and the Kuhn–Tucker conditions are introduced.

3.1 THE ECONOMIC DISPATCH PROBLEM

Figure 3.1 shows the configuration that will be studied in this section. This system consists of N thermal-generating units connected to a single bus-bar serving a received electrical load P_{load}. The input to each unit, shown as F_i, represents the cost rate* of the unit. The output of each unit, P_i, is the electrical power generated by that particular unit. The total cost rate of this system is, of course, the sum of the costs of each of the individual units. The essential constraint on the operation of this system is that the sum of the output powers must equal the load demand.

Mathematically speaking, the problem may be stated very concisely. That is, an objective function, F_T, is equal to the total cost for supplying the indicated load. The problem is to minimize F_T subject to the constraint that the sum of the powers generated must equal the received load. Note that any transmission losses are neglected and any operating limits are not explicitly stated when formulating this problem. That is,

$$F_T = F_1 + F_2 + F_3 + \cdots + F_N$$

$$= \sum_{i=1}^{N} F_i(P_i) \tag{3.1}$$

$$\phi = 0 = P_{\text{load}} - \sum_{i=1}^{N} P_i \tag{3.2}$$

---

* Generating units consume fuel at a specific rate (e.g., MBtu/h), which as noted in Chapter 2 can be converted to ℝ/h, which represents a cost rate. Starting in this chapter and throughout the remainder of the text, we will simply use the term generating unit "cost" to refer to ℝ/h.

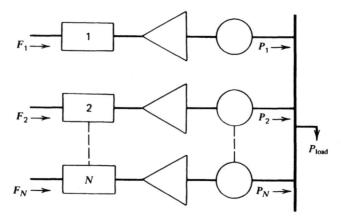

FIG. 3.1 *N* thermal units committed to serve a load of P_{load}.

This is a constrained optimization problem that may be attacked formally using advanced calculus methods that involve the Lagrange function.

In order to establish the necessary conditions for an extreme value of the objective function, add the constraint function to the objective function after the constraint function has been multiplied by an undetermined multiplier. This is known as the *Lagrange function* and is shown in Eq. 3.3.

$$\mathscr{L} = F_T + \lambda\phi \tag{3.3}$$

The necessary conditions for an extreme value of the objective function result when we take the first derivative of the Lagrange function with respect to each of the independent variables and set the derivatives equal to zero. In this case, there are $N + 1$ variables, the N values of power output, P_i, plus the undetermined Lagrange multiplier, λ. The derivative of the Lagrange function with respect to the undetermined multiplier merely gives back the constraint equation. On the other hand, the N equations that result when we take the partial derivative of the Lagrange function with respect to the power output values one at a time give the set of equations shown as Eq. 3.4.

$$\frac{\partial \mathscr{L}}{\partial P_i} = \frac{dF_i(P_i)}{dP_i} - \lambda = 0$$

or

$$0 = \frac{dF_i}{dP_i} - \lambda \tag{3.4}$$

That is, the necessary condition for the existence of a minimum cost-operating condition for the thermal power system is that the incremental cost rates of all the units be equal to some undetermined value, λ. Of course, to this necessary condition we must add the constraint equation that the sum of the power outputs must be equal to the power demanded by the load. In addition, there are two inequalities that must be satisfied for each of the units. That is, the power output of each unit must be greater than or equal to the minimum power permitted and must also be less than or equal to the maximum power permitted on that particular unit.

These conditions and inequalities may be summarized as shown in the set of equations making up Eq. 3.5.

$$\frac{dF_i}{dP_i} = \lambda \qquad N \text{ equations}$$

$$P_{i,\min} \le P_i \le P_{i,\max} \qquad 2N \text{ inequalities} \qquad (3.5)$$

$$\sum_{i=1}^{N} P_i = P_{\text{load}} \qquad 1 \text{ constraint}$$

When we recognize the inequality constraints, then the necessary conditions may be expanded slightly as shown in the set of equations making up Eq. 3.6.

$$\frac{dF_i}{dP_i} = \lambda \qquad \text{for } P_{i,\min} < P_i < P_{i,\max}$$

$$\frac{dF_i}{dP_i} \le \lambda \qquad \text{for } P_i = P_{i,\max} \qquad (3.6)$$

$$\frac{dF_i}{dP_i} \ge \lambda \qquad \text{for } P_i = P_{i,\min}$$

Several of the examples in this chapter use the following three generator units.

Unit 1: Coal-fired steam unit: Max output = 600 MW
    Min output = 150 MW

Input–output curve:

$$H_1\left(\frac{\text{MBtu}}{\text{h}}\right) = 510.0 + 7.2P_1 + 0.00142P_1^2$$

Unit 2: Oil-fired steam unit: Max output = 400 MW
    Min output = 100 MW

Input–output curve:

$$H_2\left(\frac{\text{MBtu}}{\text{h}}\right) = 310.0 + 7.85P_2 + 0.00194P_2^2$$

Unit 3: Oil-fired steam unit: Max output = 200 MW

Min output = 50 MW

Input–output curve:

$$H_3\left(\frac{\text{MBtu}}{\text{h}}\right) = 78.0 + 7.97P_3 + 0.00482P_3^2$$

EXAMPLE 3A

Suppose that we wish to determine the economic operating point for these three units when delivering a total of 850 MW. Before this problem can be solved, the fuel cost of each unit must be specified. Let the following fuel costs be in effect.

Unit 1:   fuel cost = 1.1 ₽/MBtu

Unit 2:   fuel cost = 1.0 ₽/MBtu

Unit 3:   fuel cost = 1.0 ₽/MBtu

Then

$$F_1(P_1) = H_1(P_1) \times 1.1 = 561 + 7.92P_1 + 0.001562P_1^2 \text{ ₽/h}$$
$$F_2(P_2) = H_2(P_2) \times 1.0 = 310 + 7.85P_2 + 0.00194P_2^2 \text{ ₽/h}$$
$$F_3(P_3) = H_3(P_3) \times 1.0 = 78 + 7.97P_3 + 0.00482P_3^2 \text{ ₽/h}$$

Using Eq. 3.5, the conditions for an optimum dispatch are

$$\frac{dF_1}{dP_1} = 7.92 + 0.003124P_1 = \lambda$$

$$\frac{dF_2}{dP_2} = 7.85 + 0.00388P_2 = \lambda$$

$$\frac{dF_3}{dP_3} = 7.97 + 0.00964P_3 = \lambda$$

and

$$P_1 + P_2 + P_3 = 850 \text{ MW}$$

Solving for λ, one obtains

$$\lambda = 9.148 \text{ ₽/MWh}$$

and then solving for P_1, P_2, and P_3,

$$P_1 = 393.2 \text{ MW}$$
$$P_2 = 334.6 \text{ MW}$$
$$P_3 = 122.2 \text{ MW}$$

Note that all constraints are met; that is, each unit is within its high and low limit and the total output when summed over all three units meets the desired 850 MW total.

EXAMPLE 3B

Suppose the price of coal decreased to 0.9 ₽/MBtu. The fuel cost function for unit 1 becomes

$$F_1(P_1) = 459 + 6.48P_1 + 0.00128P_1^2$$

If one goes about the solution exactly as done here, the results are

$$\lambda = 8.284 \text{ ₽/MWh}$$

and

$$P_1 = 704.6 \text{ MW}$$
$$P_2 = 111.8 \text{ MW}$$
$$P_3 = 32.6 \text{ MW}$$

This solution meets the constraint requiring total generation to equal 850 MW, but units 1 and 3 are not within limit. To solve for the most economic dispatch while meeting unit limits, use Eq. 3.6.

Suppose unit 1 is set to its maximum output and unit 3 to its minimum output. The dispatch becomes

$$P_1 = 600 \text{ MW}$$
$$P_2 = 200 \text{ MW}$$
$$P_3 = 50 \text{ MW}$$

From Eq. 3.6, we see that λ must equal the incremental cost of unit 2 since it is not at either limit. Then

$$\lambda = \left. \frac{dF_2}{dP_2} \right|_{P_2 = 200} = 8.626 \text{ ₽/MWh}$$

Next, calculate the incremental cost for units 1 and 3 to see if they meet the conditions of Eq. 3.6.

$$\left.\frac{dF_1}{dP_1}\right|_{P_1 = 600} = 8.016 \text{ R/MWh}$$

$$\left.\frac{dF_3}{dP_3}\right|_{P_3 = 50} = 8.452 \text{ R/MWh}$$

Note that the incremental cost for unit 1 is less than λ, so unit 1 should be at its maximum. However, the incremental cost for unit 3 is not greater than λ, so unit 3 should not be forced to its minimum. Thus, to find the optimal dispatch, allow the incremental cost at units 2 and 3 to equal λ as follows.

$$P_1 = 600 \text{ MW}$$

$$\frac{dF_2}{dP_2} = 7.85 + 0.00388P_2 = \lambda$$

$$\frac{dF_3}{dP_3} = 7.97 + 0.00964P_3 = \lambda$$

$$P_2 + P_3 = 850 - P_1 = 250 \text{ MW}$$

which results in

$$\lambda = 8.576 \text{ R/MWh}$$

and

$$P_2 = 187.1 \text{ MW}$$

$$P_3 = 62.9 \text{ MW}$$

Note that this dispatch meets the conditions of Eq. 3.6 since

$$\left.\frac{dF_1}{dP_1}\right|_{P_1 = 600 \text{ MW}} = 8.016 \text{ R/MWh}$$

which is less than λ, while

$$\frac{dF_2}{dP_2} \quad \text{and} \quad \frac{dF_3}{dP_3}$$

both equal λ.

3.2 THERMAL SYSTEM DISPATCHING WITH NETWORK LOSSES CONSIDERED

Figure 3.2 shows symbolically an all-thermal power generation system connected to an equivalent load bus through a transmission network. The economic-dispatching problem associated with this particular configuration is slightly more complicated to set up than the previous case. This is because the constraint equation is now one that must include the network losses. The objective function, F_T, is the same as that defined for Eq. 3.1. However, the constraint equation previously shown in Eq. 3.2 must now be expanded to the one shown in Eq. 3.7.

$$P_{\text{load}} + P_{\text{loss}} - \sum_{i=1}^{N} P_i = \phi = 0 \tag{3.7}$$

The same procedure is followed in the formal sense to establish the necessary conditions for a minimum-cost operating solution, The Lagrange function is shown in Eq. 3.8. In taking the derivative of the Lagrange function with respect to each of the individual power outputs, P_i, it must be recognized that the loss in the transmission network, P_{loss}, is a function of the network impedances and the currents flowing in the network. For our purposes, the currents will be considered only as a function of the independent variables P_i and the load P_{load}. Taking the derivative of the Lagrange function with respect to any one of the N values of P_i results in Eq. 3.9. There are N equations of this type to be satisfied along with the constraint equation shown in Eq. 3.7. This collection, Eq. 3.9 plus Eq. 3.7, is known collectively as the *coordination equations.*

$$\mathscr{L} = F_T + \lambda\phi \tag{3.8}$$

$$\frac{\partial \mathscr{L}}{\partial P_i} = \frac{dF_i}{dP_i} - \lambda\left(1 - \frac{\partial P_{\text{loss}}}{\partial P_i}\right) = 0 \tag{3.9}$$

or

$$\frac{dF_i}{dP_i} + \lambda \frac{\partial P_{\text{loss}}}{\partial P_i} = \lambda$$

$$P_{\text{load}} + P_{\text{loss}} - \sum_{i=1}^{N} P_i = 0$$

It is much more difficult to solve this set of equations than the previous set with no losses since this second set involves the computation of the network loss in order to establish the validity of the solution in satisfying the constraint equation. There have been two general approaches to the solution of this problem. The first is the development of a mathematical expression for the losses in the network solely as a function of the power output of each of the units. This is the loss-formula method discussed at some length in Kirchmayer's *Economic Operation of Power Systems* (see Chapter 1, reference 2). The other

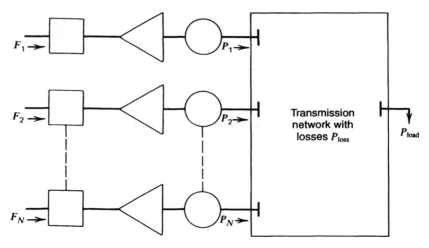

FIG. 3.2 N thermal units serving load through transmission network.

basic approach to the solution of this problem is to incorporate the power flow equations as essential constraints in the formal establishment of the optimization problem. This general approach is known as the *optimal power flow*.

EXAMPLE 3C

Starting with the same units and fuel costs as in Example 3A, we will include a simplified loss expression.

$$P_{\text{loss}} = 0.00003P_1^2 + 0.00009P_2^2 + 0.00012P_3^2$$

This simplified loss formula will suffice to show the difficulties in calculating a dispatch for which losses are accounted. Note that real-world loss formulas are more complicated than the one used in this example.

Applying Eqs. 3.8 and 3.9,

$$\frac{dF_1}{dP_1} = \lambda\left(1 - \frac{\partial P_{\text{loss}}}{\partial P_i}\right)$$

becomes

$$7.92 + 0.003124P_1 = \lambda[1 - 2(0.00003)P_1]$$

Similarly for P_2 and P_3,

$$7.85 + 0.00388P_2 = \lambda[1 - 2(0.00009)P_2]$$

$$7.97 + 0.00964P_3 = \lambda[1 - 2(0.00012)P_3]$$

and

$$P_1 + P_2 + P_3 - 850 - P_{\text{loss}} = 0$$

We no longer have a set of linear equations as in Example 3A. This necessitates a more complex solution procedure as follows.

Step 1   Pick a set of starting values for P_1, P_2, and P_3 that sum to the load.
Step 2   Calculate the incremental losses $\partial P_{loss}/\partial P_i$ as well as the total losses P_{loss}. The incremental losses and total losses will be considered constant until we return to step 2.
Step 3   Calculate the value of λ that causes P_1, P_2, and P_3 to sum to the total load plus losses. This is now as simple as the calculations in Example 3A since the equations are again linear.
Step 4   Compare the P_1, P_2, and P_3 from step 3 to the values used at the start of step 2. If there is no significant change in any one of the values, go to step 5, otherwise go back to step 2.
Step 5   Done.

Using this procedure, we obtain

Step 1   Pick the P_1, P_2, and P_3 starting values as

$$P_1 = 400.0 \text{ MW}$$

$$P_2 = 300.0 \text{ MW}$$

$$P_3 = 150.0 \text{ MW}$$

Step 2   Incremental losses are

$$\frac{\partial P_{loss}}{\partial P_1} = 2(0.00003)400 = 0.0240$$

$$\frac{\partial P_{loss}}{\partial P_2} = 2(0.00009)300 = 0.0540$$

$$\frac{\partial P_{loss}}{\partial P_3} = 2(0.00012)150 = 0.0360$$

Total losses are 15.6 MW.
Step 3   We can now solve for λ using the following:

$$7.92 + 0.003124P_1 = \lambda(1 - 0.0240) = \lambda(0.9760)$$

$$7.85 + 0.00388P_2 = \lambda(1 - 0.0540) = \lambda(0.9460)$$

$$7.97 + 0.00964P_3 = \lambda(1 - 0.0360) = \lambda(0.9640)$$

and

$$P_1 + P_2 + P_3 - 850 - 15.6 = P_1 + P_2 + P_3 - 865.6 = 0$$

These equations are now linear, so we can solve for λ directly. The results are

$$\lambda = 9.5252 \, \text{R/MWh}$$

and the resulting generator outputs are

$$P_1 = 440.68$$
$$P_2 = 299.12$$
$$P_3 = 125.77$$

Step 4   Since these values for P_1, P_2, and P_3 are quite different from the starting values, we will return to step 2.

Step 2   The incremental losses are recalculated with the new generation values.

$$\frac{\partial P_{\text{loss}}}{\partial P_1} = 2(0.00003)440.68 = 0.0264$$

$$\frac{\partial P_{\text{loss}}}{\partial P_2} = 2(0.00009)299.12 = 0.0538$$

$$\frac{\partial P_{\text{loss}}}{\partial P_3} = 2(0.00012)125.77 = 0.0301$$

Total losses are 15.78 MW.

Step 3   The new incremental losses and total losses are incorporated into the equations, and a new value of λ and P_1, P_2, and P_3 are solved for

$$7.92 + 0.003124P_1 = \lambda(1 - 0.0264) = \lambda(0.9736)$$
$$7.85 + 0.00388P_2 = \lambda(1 - 0.0538) = \lambda(0.9462)$$
$$7.97 + 0.00964P_2 = \lambda(1 - 0.0301) = \lambda(0.9699)$$
$$P_1 + P_2 + P_3 - 850 - 15.78 = P_1 + P_2 + P_3 - 865.78 = 0$$

resulting in $\lambda = 9.5275 \, \text{R/MWh}$ and

$$P_1 = 433.94 \, \text{MW}$$
$$P_2 = 300.11 \, \text{MW}$$
$$P_3 = 131.74 \, \text{MW}$$
$$\vdots$$

Table 3.1 summarizes the iterative process used to solve this problem.

TABLE 3.1 Iterative Process Used to Solve Example 3

Iteration	P_1 (MW)	P_2 (MW)	P_3 (MW)	Losses (MW)	λ (R/MWh)
Start	400.00	300.00	150.00	15.60	9.5252
1	440.68	299.12	125.77	15.78	9.5275
2	433.94	300.11	131.74	15.84	9.5285
3	435.87	299.94	130.42	15.83	9.5283
4	434.13	299.99	130.71	15.83	9.5284

3.3 THE LAMBDA-ITERATION METHOD

Figure 3.3 is a block diagram of the lambda-iteration method of solution for the all-thermal, dispatching problem-neglecting losses. We can approach the solution to this problem by considering a graphical technique for solving the problem and then extending this into the area of computer algorithms.

Suppose we have a three-machine system and wish to find the optimum

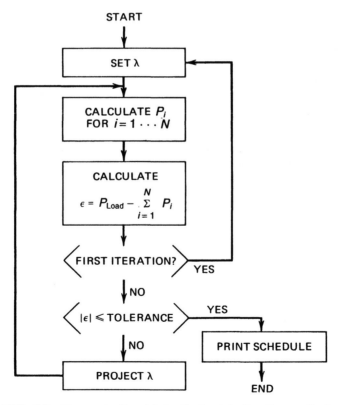

FIG. 3.3 Economic dispatch by the lambda-iteration method.

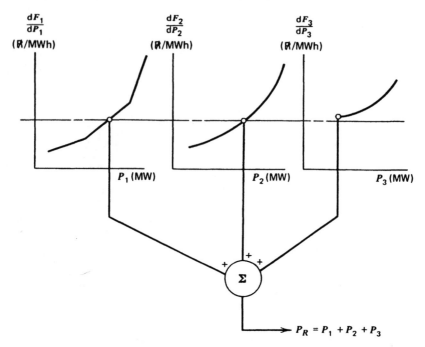

FIG. 3.4   Graphical solution to economic dispatch.

economic operating point. One approach would be to plot the incremental cost characteristics for each of these three units on the same graph, such as sketched in Figure 3.4. In order to establish the operating points of each of these three units such that we have minimum cost and at the same time satisfy the specified demand, we could use this sketch and a ruler to find the solution. That is, we could assume an incremental cost rate (λ) and find the power outputs of each of the three units for this value of incremental cost.

Of course, our first estimate will be incorrect. If we have assumed the value of incremental cost such that the total power output is too low, we must increase the λ value and try another solution. With two solutions, we can extrapolate (or interpolate) the two solutions to get closer to the desired value of total received power (see Figure 3.5).

By keeping track of the total demand versus the incremental cost, we can rapidly find the desired operating point. If we wished, we could manufacture a whole series of tables that would show the total power supplied for different incremental cost levels and combinations of units.

This same procedure can be adopted for a computer implementation as shown in Figure 3.3. That is, we will now establish a set of logical rules that would enable us to accomplish the same objective as we have just done with ruler and graph paper. The actual details of how the power output is established as a function of the incremental cost rate are of very little importance. We

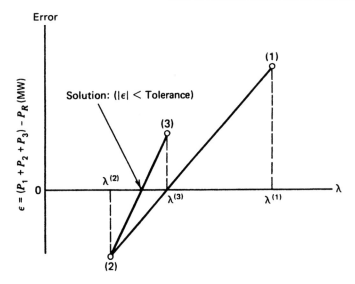

FIG. 3.5   Lambda projections.

could, for example, store tables of data within the computer and interpolate between the stored power points to find exact power output for a specified value of incremental cost rate. Another approach would be to develop an analytical function for the power output as a function of the incremental cost rate, store this function (or its coefficients) in the computer, and use this to establish the output of each of the individual units.

This procedure is an iterative type of computation, and we must establish stopping rules. Two general forms of stopping rules seem appropriate for this application. The first is shown in Figure 3.3 and is essentially a rule based on finding the proper operating point within a specified tolerance. The other, not shown in Figure 3.3, involves counting the number of times through the iterative loop and stopping when a maximum number is exceeded.

The lambda-iteration procedure converges very rapidly for this particular type of optimization problem. The actual computational procedure is slightly more complex than that indicated in Figure 3.3, since it is necessary to observe the operating limits on each of the units during the course of the computation. The well-known Newton–Raphson method may be used to project the incremental cost value to drive the error between the computed and desired generation to zero.

EXAMPLE 3D

Assume that one wishes to use cubic functions to represent the input–output characteristics of generating plants as follows.

$$H \text{ (MBtu/h)} = A + BP + CP^2 + DP^3 \qquad (P \text{ in MW})$$

For the three units, find the optimum schedule using the lambda-iteration method.

	A	B	C	D
Unit 1	749.55	6.95	9.68×10^{-4}	1.27×10^{-7}
Unit 2	1285.0	7.051	7.375×10^{-4}	6.453×10^{-8}
Unit 3	1531.0	6.531	1.04×10^{-3}	9.98×10^{-8}

Assume the fuel cost to be 1.0 R/MBtu for each unit and unit limits as follows.

$$320 \text{ MW} \leq P_1 \leq \ 800 \text{ MW}$$

$$300 \text{ MW} \leq P_2 \leq 1200 \text{ MW}$$

$$275 \text{ MW} \leq P_3 \leq 1100 \text{ MW}$$

Two sample calculations are shown, both using the flowchart in Figure 3.3. In this calculation, the value for λ on the second iteration is always set at 10% above or below the starting value depending on the sign of the error; for the remaining iterations, lambda is projected as in Figure 3.5.

The first example shows the advantage of starting λ near the optimum value.

$$P_{\text{load}} = 2500 \text{ MW}$$

$$\lambda_{\text{start}} = 8.0 \text{ R/MWh}$$

The second example shows the oscillatory problems that can be encountered with a lambda-iteration approach.

$$P_{\text{load}} = 2500 \text{ MW}$$

$$\lambda_{\text{start}} = 10.0 \text{ R/MWh}$$

Iteration	λ	Total Generation (MW)	P_1	P_2	P_3
1	8.0000	1731.6	494.3	596.7	640.6
2	8.8000	2795.0	800.0	1043.0	952.0
3	8.5781	2526.0	734.7	923.4	867.9
4	8.5566	2497.5	726.1	911.7	859.7
5	8.5586	2500.0	726.9	912.7	860.4

Iteration	λ	Total Generation (MW)	P_1	P_2	P_3
1	10.0000	3100.0	800.0	1200.0	1100.0
2	9.0000	2974.8	800.0	1148.3	1026.5
3	5.2068	895.0	320.0	300.0	275.0
4	8.1340	1920.6	551.7	674.5	694.4
5	9.7878	3100.0	800.0	1200.0	1100.0
6	8.9465	2927.0	800.0	1120.3	1006.7
7	6.8692	895.0	320.0	300.0	275.0
8	8.5099	2435.0	707.3	886.1	841.7
9	8.5791	2527.4	735.1	924.0	868.3
10	8.5586	2500.1	726.9	912.8	860.4

3.4 GRADIENT METHODS OF ECONOMIC DISPATCH

Note that the lambda search technique always requires that one be able to find the power output of a generator, given an incremental cost for that generator. In the case of a quadratic function for the cost function, or in the case where the incremental cost function is represented by a piecewise linear function, this is possible. However, it is often the case that the cost function is much more complex, such as the one below:

$$F(P) = A + BP + CP^2 + D \exp\left[\frac{(P - E)}{F}\right]$$

In this case, we shall propose that a more basic method of solution for the optimum be found.

3.4.1 Gradient Search

This method works on the principle that the minimum of a function, f(x), can be found by a series of steps that always take us in a downward direction. From any starting point, x^0, we may find the direction of "steepest descent" by noting that the gradient of f, i.e.,

$$\mathbf{Vf} = \begin{bmatrix} \dfrac{\partial f}{\partial x_1} \\ \vdots \\ \dfrac{\partial f}{\partial x_n} \end{bmatrix} \tag{3.10}$$

always points in the direction of maximum ascent. Therefore, if we want to move in the direction of maximum descent, we negate the gradient. Then we should go from x^0 to x^1 using:

$$\mathbf{x}^1 = \mathbf{x}^0 - \nabla \mathbf{f}\, \alpha \tag{3.11}$$

Where α is a scalar to allow us to guarantee that the process converges. The best value of α must be determined by experiment.

3.4.2   Economic Dispatch by Gradient Search

In the case of power system economic dispatch this becomes:

$$f = \sum_{i=1}^{N} F_i(P_i) \tag{3.12}$$

and the object is to drive the function to its minimum. However, we have to be concerned with the constraint function:

$$\Phi = \left(P_{\text{load}} - \sum_{i=1}^{N} P_i \right) \tag{3.13}$$

To solve the economic dispatch problem which involves minimizing the objective function and keeping the equality constraint, we must apply the gradient technique directly to the Lagrange function itself.

The Lagrange function is:

$$\mathcal{L} = \sum_{i=1}^{N} F_i(P_i) + \lambda \left(P_{\text{load}} - \sum_{i=1}^{N} P_i \right) \tag{3.14}$$

and the gradient of this function is:

$$\nabla \mathcal{L} = \begin{bmatrix} \dfrac{\partial \mathcal{L}}{\partial P_1} \\[2ex] \dfrac{\partial \mathcal{L}}{\partial P_2} \\[2ex] \dfrac{\partial \mathcal{L}}{\partial P_3} \\[2ex] \dfrac{\partial \mathcal{L}}{\partial \lambda} \end{bmatrix} = \begin{bmatrix} \dfrac{d}{dP_1} F_1(P_1) - \lambda \\[2ex] \dfrac{d}{dP_2} F_2(P_2) - \lambda \\[2ex] \dfrac{d}{dP_3} F_3(P_3) - \lambda \\[2ex] P_{\text{load}} - \sum_{i=1}^{N} P_i \end{bmatrix} \tag{3.15}$$

The problem with this formulation is the lack of a guarantee that the new points generated each step will lie on the surface Φ. We shall see that this can be overcome by a simple variation of the gradient method.

The economic dispatch algorithm requires a starting λ value and starting values for P_1, P_2, and P_3. The gradient for $\mathscr{L}$ is calculated as above and the new values of λ, P_1, P_2, and P_3, etc., are found from:

$$\mathbf{x}^1 = \mathbf{x}^0 - (\nabla\mathscr{L})\alpha \tag{3.16}$$

where the vector $\mathbf{x}$ is:

$$\mathbf{x} = \begin{bmatrix} P_1 \\ P_2 \\ P_3 \\ \vdots \\ \lambda \end{bmatrix} \tag{3.17}$$

EXAMPLE 3E

Given the generator cost functions found in Example 3A, solve for the economic dispatch of generation with a total load of 800 MW.

Using $\alpha = 100$ and starting from $P_1^0 = 300$ MW, $P_2^0 = 200$ MW, and $P_3^0 = 300$ MW, we set the initial value of λ equal to the average of the incremental costs of the generators at their starting generation values. That is:

$$\lambda^0 = \frac{1}{3} \sum_{i=1}^{3} \left[\frac{\mathrm{d}}{\mathrm{d}P_i} F_i(P_i^0) \right]$$

This value is 9.4484.

The progress of the gradient search is shown in Table 3.2. The table shows that the iterations have led to no solution at all. Attempts to use this formulation

TABLE 3.2 Economic Dispatch by Gradient Method

Iteration	P_1	P_2	P_3	P_{total}	λ	Cost
1	300	200	300	800	9.4484	7938.0
2	300.59	200.82	298.59	800	9.4484	7935
3	301.18	201.64	297.19	800.0086	9.4484	7932
4	301.76	202.45	295.8	800.025	9.4570	7929.3
5	302.36	203.28	294.43	800.077	9.4826	7926.9
⋮						
10	309.16	211.19	291.65	811.99	16.36	8025.6

will result in difficulty as the gradient cannot guarantee that the adjustment to the generators will result in a schedule that meets the correct total load of 800 MW.

A simple variation of this technique is to realize that one of the generators is always a dependent variable and remove it from the problem. In this case, we pick P_3 and use the following:

$$P_3 = 800 - P_1 - P_2$$

Then the total cost, which is to be minimized, is:

$$\text{Cost} = F_1(P_1) + F_2(P_2) + F_3(P_3) = F_1(P_1) + F_2(P_2) + F_3(800 - P_1 - P_2)$$

Note that this function stands by itself as a function of two variables with no load-generation balance constraint (and no λ). The cost can be minimized by a gradient method and in this case the gradient is:

$$\nabla \text{Cost} = \begin{bmatrix} \dfrac{d}{dP_1}\text{Cost} \\[2ex] \dfrac{d}{dP_2}\text{Cost} \end{bmatrix} = \begin{bmatrix} \dfrac{dF_1}{dP_1} - \dfrac{dF_3}{dP_1} \\[2ex] \dfrac{dF_2}{dP_2} - \dfrac{dF_3}{dP_2} \end{bmatrix}$$

Note that this gradient goes to the zero vector when the incremental cost at generator 3 is equal to that at generators 1 and 2. The gradient steps are performed in the same manner as previously, where:

$$x^1 = x^0 - \nabla \text{Cost} \times \alpha$$

and

$$x = \begin{bmatrix} P_1 \\ P_2 \end{bmatrix}$$

Each time a gradient step is made, the generation at generator 3 is set to 800 minus the sum of the generation at generators 1 and 2. This method is often called the "reduced gradient" because of the smaller number of variables.

EXAMPLE 3F

Reworking example 3E with the reduced gradient we obtain the results shown in Table 3.3. This solution is much more stable and is converging on the optimum solution.

TABLE 3.3   Reduced Gradient Results ($\alpha = 10$)

Iteration	P_1	P_2	P_3	P_{total}	Cost
1	300	200	300	800	7938.0
2	320.04	222.36	257.59	800	7858.1
3	335.38	239.76	224.85	800	7810.4
4	347.08	253.33	199.58	800	7781.9
5	355.97	263.94	180.07	800	7764.9
⋮					
10	380.00	304.43	115.56	800	7739.2

3.5   NEWTON'S METHOD

We may wish to go a further step beyond the simple gradient method and try to solve the economic dispatch by observing that the aim is to always drive

$$\nabla \mathscr{L}_x = 0 \tag{3.18}$$

Since this is a vector function, we can formulate the problem as one of finding the correction that exactly drives the gradient to zero (i.e., to a vector, all of whose elements are zero). We know how to find this, however, since we can use Newton's method. Newton's method for a function of more than one variable is developed as follows.

Suppose we wish to drive the function $g(x)$ to zero. The function $\mathbf{g}$ is a vector and the unknowns, $\mathbf{x}$, are also vectors. Then, to use Newton's method, we observe:

$$\mathbf{g}(\mathbf{x} + \Delta\mathbf{x}) = \mathbf{g}(x) + [g'(x)]\Delta\mathbf{x} = 0 \tag{3.19}$$

If we let the function be defined as:

$$\mathbf{g}(x) = \begin{bmatrix} g_1(x_1, x_2, x_3) \\ g_2(x_1, x_2, x_3) \\ g_3(x_1, x_2, x_3) \end{bmatrix} \tag{3.20}$$

then

$$g'(x) = \begin{bmatrix} \dfrac{\partial g_1}{\partial x_1} & \dfrac{\partial g_1}{\partial x_2} & \dfrac{\partial g_1}{\partial x_3} \\ \dfrac{\partial g_2}{\partial x_1} & & \end{bmatrix} \tag{3.21}$$

which is the familiar Jacobian matrix. The adjustment at each step is then:

$$\Delta\mathbf{x} = -[g'(x)]^{-1}\mathbf{g}(\mathbf{x}) \tag{3.22}$$

Now, if we let the g function be the gradient vector $\nabla \mathscr{L}_x$ we get:

$$\Delta \mathbf{x} = -\left[\frac{\partial}{\partial \mathbf{x}} \nabla \mathscr{L}_x\right]^{-1} \Delta \mathscr{L} \qquad (3.23)$$

For our economic dispatch problem this takes the form:

$$\mathscr{L} = \sum_{i=1}^{N} F_i(P_i) + \lambda \left(P_{\text{load}} - \sum_{i=1}^{N} P_i\right) \qquad (3.24)$$

and $\nabla \mathscr{L}$ is as it was defined before. The Jacobian matrix now becomes one made up of second derivatives and is called the Hessian matrix:

$$\left[\frac{\partial}{\partial \mathbf{x}} \nabla \mathscr{L}_x\right] = \begin{bmatrix} \dfrac{d^2 \mathscr{L}}{dx_1^2} & \dfrac{d^2 \mathscr{L}}{dx_1 \, dx_2} & \cdots \\[2ex] \dfrac{d^2 \mathscr{L}}{dx_2 \, dx_1} & \cdots & \\[2ex] \vdots & \vdots & \\[2ex] \dfrac{d^2 \mathscr{L}}{d\lambda \, dx_1} & \cdots & \end{bmatrix} \qquad (3.25)$$

Generally, Newton's method will solve for the correction that is much closer to the minimum generation cost in one step than would the gradient method.

EXAMPLE 3G

In this example we shall use Newton's method to solve the same economic dispatch as used in Examples 3E and 3F.

The gradient is the same as in Example 3E, the Hessian matrix is:

$$[H] = \begin{bmatrix} \dfrac{d^2 F_1}{dP_1^2} & 0 & 0 & -1 \\[2ex] 0 & \dfrac{d^2 F_2}{dP_2^2} & 0 & -1 \\[2ex] 0 & 0 & \dfrac{d^2 F_3}{dP_3^2} & -1 \\[2ex] -1 & -1 & -1 & 0 \end{bmatrix}$$

In this example, we shall simply set the initial λ equal to 0, and the initial generation values will be the same as in Example 3E as well. The gradient of the Lagrange function is:

$$\nabla\mathscr{L} = \begin{bmatrix} 8.8572 \\ 8.6260 \\ 10.8620 \\ 0 \end{bmatrix}$$

The Hessian matrix is:

$$[H] = \begin{bmatrix} 0.0031 & 0 & 0 & -1 \\ 0 & 0.0039 & 0 & -1 \\ 0 & 0 & 0.0096 & -1 \\ -1 & -1 & -1 & 0 \end{bmatrix}$$

Solving for the correction to the **x** vector and making the correction, we obtain

$$\mathbf{x} = \begin{bmatrix} P_1 \\ P_2 \\ P_3 \\ \lambda \end{bmatrix} = \begin{bmatrix} 369.6871 \\ 315.6965 \\ 114.6164 \\ 9.0749 \end{bmatrix}$$

and a total generation cost of 7738.8. Note that no further steps are necessary as the Newton's method has solved in one step. When the system of equations making up the generation cost functions are quadratic, and no generation limits are reached, the Newton's method will solve in one step.

We have introduced the gradient, reduced gradient and Newton's method here mainly as a way to show the variations of solution of the generation economic dispatch problem. For many applications, the lambda search technique is the preferred choice. However, in later chapters, when we introduce the optimal power flow, the gradient and Newton formulations become necessary.

3.6 ECONOMIC DISPATCH WITH PIECEWISE LINEAR COST FUNCTIONS

Many electric utilities prefer to represent their generator cost functions as single or multiple segment linear cost functions. The curves shown in Figure 3.6 are representative of such functions. Note that were we to attempt to use the lambda-iteration search method on the single segment cost function, we would always land on P_{min} or P_{max} unless λ exactly matched the incremental cost at

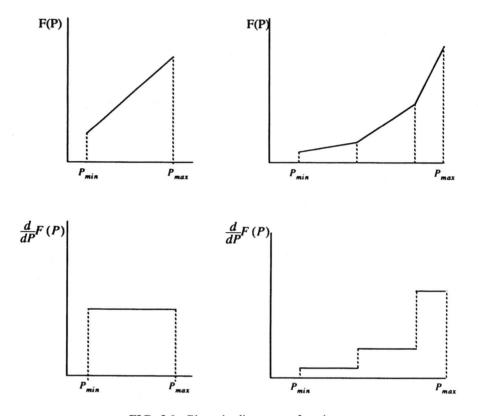

FIG. 3.6   Piecewise linear cost functions.

which point the value of P would be undetermined. To resolve this problem, we perform the dispatch differently.

For all units running, we start with all of them at $P_{\min}$, then begin to raise the output of the unit with the lowest incremental cost segment. If this unit hits the right-hand end of a segment, or if it hits $P_{\max}$, we then find the unit with the next lowest incremental cost segment and raise its output. Eventually, we will reach a point where a unit's output is being raised and the total of all unit outputs equals the total load, or load plus losses. At that point, we assign the last unit being adjusted to have a generation which is partially loaded for one segment. Note, that if there are two units with exactly the same incremental cost, we simply load them equally.

To make this procedure very fast, we can create a table giving each segment of each unit its MW contribution (the right-hand end MW minus the left-hand end MW). Then we order this table by ascending order of incremental cost. By searching from the top down in this table we do not have to go and look for the next segment each time a new segment is to be chosen. This is an extremely fast form of economic dispatch.

3.7 ECONOMIC DISPATCH USING DYNAMIC PROGRAMMING

As we saw in Chapter 2 when we considered the valve points in the input–output curve (for example, Figure 2.6), the possibility of nonconvex curves must be accounted for if extreme accuracy is desired. If nonconvex input–output curves are to be used, we cannot use an equal incremental cost methodology since there are multiple values of MW output for any given value of incremental cost.

Under such circumstances, there is a way to find an optimum dispatch which uses dynamic programming (DP). If the reader has no background in DP, Appendix 3B of this chapter should be read at this time.

The dynamic programming solution to economic dispatch is done as an allocation problem, as given in Appendix 3B. Using this approach, we do not calculate a single optimum set of generator MW outputs for a specific total load supplied—rather we generate a set of outputs, at discrete points, for an entire set of load values.

EXAMPLE 3H

There are three units in the system; all are on-line. Their input–output characteristics are *not* smooth *nor* convex. Data are as follows.

Power Levels (MW)	Costs (R/hour)		
$P_1 = P_2 = P_3$	F_1	F_2	F_3
0	∞	∞	∞
50	810	750	806
75	1355	1155	1108.5
100	1460	1360	1411
125	1772.5	1655	11704.5
150	2085	1950	1998
175	2427.5	∞	2358
200	2760	∞	∞
225	∞	∞	∞

The total demand is $D = 310$ MW. This does not fit the data exactly, so that we need to interpolate between the closest values that are available from the data, 300 and 325 MW.

Scheduling units 1 and 2, we find the minimum cost for the function

$$f_2 = F_1(D - P_2) + F_2(P_2)$$

over the allowable range of P_2 and for $100 \leq D \leq 350$ MW. The search data are given in the table below. We need to save the cost for serving each value of D that is minimal and the load level on unit 2 for each demand level.

	$P_2 = 0$	50	75	100	125	150 (MW)		
	$F_2(P_2) = \infty$	750	1155	1360	1655	1950 (ℝ/h)		
D	$F_1(D)$						f_2	P_2^*
(MW)	(ℝ/h)						(ℝ/h)	(MW)
0	∞	∞	∞	∞	∞	∞	∞	∞	
50	810	∞	∞	∞	∞	∞	∞	∞	
75	1355	∞	∞	∞	∞	∞	∞	∞	
100	1460	∞	1560	∞	∞	∞	∞	1560	50
125	1772.5	∞	2105	1965	∞	∞	∞	1965	75
150	2085	∞	2210	2510	2170	∞	∞	2170	100
175	2427.5	∞	3177.5	2615	2715	2465	∞	2465	125
200	2760	∞	2834	2927.5	2820	3010	2760	2760	150
225	∞	∞	3177.5	3240	3125	3115	3305	3115	125
250	∞	∞	3510	3582.5	3445	3427	3410	3410	150
275	∞	∞	∞	3915	3787.5	3740	3722.5	3722.5	150
300	∞	∞	∞	∞	4120	4082.5	4025	4035	150
325	∞	∞	∞	∞	∞	4415	4377.5	4377.5	150
350	∞	∞	∞	∞	∞	∞	4710	4710	150

This results in:

D	f_2	$P_2{}^a$
50	∞	
100	1560	50
125	1965	75
150	2170	100
175	2465	125
200	2760	150
225	3115	125
250	3410	150
275	3722.5	150
300	4035	150
325	4377.5	150
350	4710	150
375	∞	

^{a} Loading of unit 2 at minimal cost level.

Next we minimize

$$f_3 = f_2(D - P_3) + F_3(P_3)$$

for $50 \leq P_3 \leq 175\,\text{MW}$ and $D = 300$ and $325\,\text{MW}$. Scheduling the third unit for the two different demand levels only requires two rows of the next table.

	$P_3 = 0$	50	75	100	125	150	175	(MW)		
	$F_3(P_3) = \infty$	806	1108.5	1411	1704.5	1998	2358	(R/h)		
D	f_2									
(MW)	(R/h)								f_3	P_3^*
.		.								
.			.							
.										
300	4035	∞	4216	4223.5	4171	4169.5	<u>4168</u>	4323	4168	150
325	4377.5	∞	4528.5	4518.5	4526	4464	<u>4463</u>	4528	4463	150
.		.								
.			.							
.										

The results show:

D	Cost	P_3^*	P_2^*	P_1^*
300	4168	150	100	50
325	4463	150	125	50

so that between the 300 and 325 MW demand levels, the marginal unit is unit 2. (That is, it is picking up all of the additional demand increase between 300 and 325 MW.) We can, therefore, interpolate to find the cost at a load level of 310 MW, or an output level on unit 2 of 110 MW. The results for a demand level of 310 MW are:

$$P_1 = 50, \ P_2 = 110, \text{ and } P_3 = 150 \text{ for a total cost of } 4286\ \text{R/h}$$

One problem that is common to economic dispatch with dynamic programming is the poor control performance of the generators. We shall deal with the control of generators in Chapter 9 when we discuss automatic generation control (AGC). When a generator is under AGC and a small increment of load is added to the power system, the AGC must raise the output of the appropriate units so that the new generation output meets the load and the generators are at economic dispatch. In addition, the generators must be able to move to the new generation value within a short period of time. However, if the generators are large steam generator units, they will not be allowed to change generation output above a prescribed "maximum rate limit" of so many megawatts per minute. When this is the case, the AGC must allocate the change in generation to many other units, so that the load change can be accommodated quickly enough.

When the economic dispatch is to be done with dynamic programming and the cost curves are nonconvex, we encounter a difficult problem whenever a small increment in load results in a new dispatch that calls for one or more generators to drop their output a great deal and others to increase a large amount. The resulting dispatch may be at the most economic values as determined by the DP, but the control action is not acceptable and will probably violate the ramp rates for several of the units.

The only way to produce a dispatch that is acceptable to the control system, as well as being the optimum economically, is to add the ramp rate limits to the economic dispatch formulation itself. This requires a short-range load forecast to determine the most likely load and load-ramping requirements of the units. This problem can be stated as follows.

Given a load to be supplied at time increments $t = 1 \ldots t_{max}$, with load levels of P^t_{load}, and N generators on-line to supply the load:

$$\sum_{i=1}^{N} P^t_i = P^t_{load} \tag{3.26}$$

Each unit must obey a rate limit such that:

$$P^{t+1}_i = P^t_i + \Delta P_i \tag{3.27}$$

and

$$-\Delta P^{max}_i \leq \Delta P_i \leq \Delta P^{max}_i \tag{3.28}$$

Then we must schedule the units to minimize the cost to deliver power over the time period as:

$$F^{total} = \sum_{t=1}^{T_{max}} \sum_{i=1}^{N} F_i(P^t_i) \tag{3.29}$$

subject to:

$$\sum_{i=1}^{N} P^t_i = P^t_{load} \quad \text{for} \quad t = 1 \ldots t_{max} \tag{3.30}$$

and

$$P^{t+1}_i = P^t_i + \Delta P_i \tag{3.31}$$

with

$$-\Delta P^{max}_i \leq \Delta P_i \leq \Delta P^{max}_i \tag{3.32}$$

This optimization problem can be solved with dynamic programming and the "control performance" of the dispatch will be considerably better than that using dynamic programming and no ramp limit constraints (see Chapter 9, reference 19).

3.8   BASE POINT AND PARTICIPATION FACTORS

This method assumes that the economic dispatch problem has to be solved repeatedly by moving the generators from one economically optimum schedule to another as the load changes by a reasonably small amount. We start from a given schedule—the *base point*. Next, the scheduler assumes a load change and investigates how much each generating unit needs to be moved (i.e., "participate" in the load change) in order that the new load be served at the most economic operating point.

Assume that both the first and second derivatives in the cost versus power output function are available (i.e., both F_i' and F_i'' exist). The incremental cost curve of the ith unit is given in Figure 3.7. As the unit load is changed by an amount ΔP_i, the system incremental cost moves from λ^0 to $\lambda^0 + \Delta\lambda$. For a small change in power output on this single unit,

$$\Delta\lambda_i = \Delta\lambda \cong F_i''(_i^0)\Delta P_i \qquad (3.33)$$

This is true for each of the N units on the system, so that

$$\Delta P_1 = \frac{\Delta\lambda}{F_1''}$$

$$\Delta P_2 = \frac{\Delta\lambda}{F_2''}$$

$$\vdots$$

$$\Delta P_N = \frac{\Delta\lambda}{F_N''}$$

The total change in generation (= change in total system demand) is, of course,

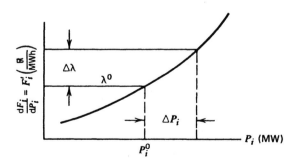

FIG. 3.7   Relationship of $\Delta\lambda$ and ΔP_i.

the sum of the individual unit changes. Let P_D be the total demand on the generators (where $P_D = P_{\text{load}} + P_{\text{loss}}$), then

$$\Delta P_D = \Delta P_1 + \Delta P_2 + \cdots + \Delta P_N$$

$$= \Delta \lambda \sum_i \left(\frac{1}{F_i''} \right) \tag{3.34}$$

The earlier equation, 3.33, can be used to find the *participation factor* for each unit as follows.

$$\left(\frac{\Delta P_i}{\Delta P_D} \right) = \frac{(1/F_i'')}{\sum_i \left(\dfrac{1}{F_i''} \right)} \tag{3.35}$$

The computer implementation of such a scheme of economic dispatch is straightforward. It might be done by provision of tables of the values of F_i'' as a function of the load levels and devising a simple scheme to take the existing load plus the projected increase to look up these data and compute the factors.

A somewhat less elegant scheme to provide participation factors would involve a repeat economic dispatch calculation at $P_D^0 + \Delta P_D$. The base-point economic generation values are then subtracted from the new economic generation values and the difference divided by ΔP_D to provide the participation factors. This scheme works well in computer implementations where the execution time for the economic dispatch is short and will always give consistent answers when units reach limits, pass through break points on piecewise linear incremental cost functions, or have nonconvex cost curves.

EXAMPLE 3I

Starting from the optimal economic solution found in Example 3A, use the participation factor method to calculate the dispatch for a total load of 900 MW.

Using Eq. 3.24,

$$\frac{\Delta P_1}{\Delta P_D} = \frac{(0.003124)^{-1}}{(0.003124)^{-1} + (0.00388)^{-1} + (0.00964)^{-1}} = \frac{320.10}{681.57} = 0.47$$

Similarly,

$$\frac{\Delta P_2}{\Delta P_D} = \frac{(0.00388)^{-1}}{681.57} = 0.38$$

$$\frac{\Delta P_3}{\Delta P_D} = \frac{103.73}{681.57} = 0.15$$

$$\Delta P_D = 900 - 850 = 50$$

The new value of generation is calculated using

$$P_{new_i} = P_{base_i} + \left(\frac{\Delta P_i}{\Delta P_D}\right)\Delta P_D \qquad \text{for } i = 1, 2, 3$$

Then for each unit

$$P_{new_1} = 393.2 + (0.47)(50) = 416.7$$
$$P_{new_2} = 334.6 + (0.38)(50) = 353.6$$
$$P_{new_3} = 122.2 + (0.15)(50) = 129.7$$

3.9 ECONOMIC DISPATCH VERSUS UNIT COMMITMENT

At this point, it may be as well to emphasize the essential difference between the unit commitment and economic dispatch problem. The economic dispatch problem *assumes* that there are N units already connected to the system. The purpose of the economic dispatch problem is to find the optimum operating policy for these N units. This is the problem that we have been investigating so far in this text.

On the other hand, the unit commitment problem is more complex. We may assume that we have N units available to us and that we have a forecast of the demand to be served. The question that is asked in the unit commitment problem area is approximately as follows.

Given that there are a number of subsets of the complete set of N generating units that would satisfy the expected demand, which of these subsets should be used in order to provide the minimum operating cost?

This unit commitment problem may be extended over some period of time, such as the 24 h of a day or the 168 h of a week. The unit commitment problem is a much more difficult problem to solve. The solution procedures involve the economic dispatch problem as a subproblem. That is, for each of the subsets of the total number of units that are to be tested, for any given set of them connected to the load, the particular subset should be operated in optimum economic fashion. This will permit finding the minimum operating cost for that subset, but it does not establish which of the subsets is in fact the one that will give minimum cost over a period of time.

A later chapter will consider the unit commitment problem in some detail. The problem is more difficult to solve mathematically since it involves integer variables. That is, generating units must be either all on or all off. (How can you turn a switch half on?)

APPENDIX 3A
Optimization within Constraints

Suppose you are trying to maximize or minimize a function of several variables. It is relatively straightforward to find the maximum or minimum using rules of calculus. First, of course, you must find a set of values for the variables where the first derivative of the function with respect to each variable is zero. In addition, the second derivatives should be used to determine whether the solution found is a maximum, minimum, or a saddle point.

In optimizing a real-life problem, one is usually confronted with a function to be maximized or minimized, as well as numerous constraints that must be met. The constraints, sometimes called *side conditions*, can be other functions with conditions that must be met or they can be simple conditions such as limits on the variables themselves.

Before we begin this discussion on constrained optimization, we will put down some definitions. Since the objective is to maximize or minimize a mathematical function, we will call this function the *objective function*. The constraint functions and simple variable limits will be lumped under the term *constraints*. The region defined by the constraints is said to be the *feasible region* for the independent variables. If the constraints are such that no such region exists, that is, there are no values for the independent variables that satisfy all the constraints, then the problem is said to have an *infeasible* solution. When an optimum solution to a constrained optimization problem occurs at the boundary of the feasible region defined by a constraint, we say the constraint is *binding*. If the optimum solution lies away from the boundary, the constraint is *nonbinding*.

To begin, let us look at a simple elliptical objective function.

$$f(x_1, x_2) = 0.25x_1^2 + x_2^2 \tag{3A.1}$$

This is shown in Figure 3.8 for various values of f.

Note that the minimum value f can attain is zero, but that it has no finite maximum value. The following is an example of a constrained optimization problem.

Minimize: $f(x_1, x_2) = 0.25x_1^2 + x_2^2$

Subject to the constraint: $\omega(x_1, x_2) = 0$ $\tag{3A.2}$

Where: $\omega(x_1, x_2) = 5 - x_1 - x_2$

This optimization problem can be pictured as in Figure 3.9.

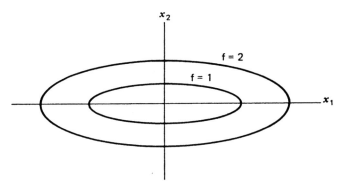

FIG. 3.8 Elliptical objective function.

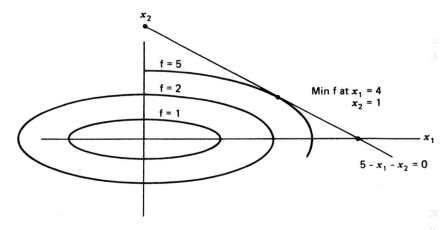

FIG. 3.9 Elliptical objective function with equality constraint.

We need to observe that the optimum as pictured, gives the minimum value for our objective function, f, while also meeting the constraint function, ω. This optimum point occurs where the function f is exactly tangent to the function ω. Indeed, this observation can be made more rigorous and will form the basis for our development of Lagrange multipliers.

First, redraw the function f for several values of f around the optimum point. At the point (x'_1, x'_2), calculate the gradient vector of f. This is pictured in Figure 3.10 as $\nabla f(x'_1, x'_2)$. Note that the gradient at (x'_1, x'_2) is perpendicular to f but not to ω, and therefore has a nonzero component along ω. Similarly, at the point (x''_1, x''_2) the gradient of f has a nonzero component along ω. The nonzero component of the gradient along ω tells us that a small move along ω in the direction of this component will increase the objective function. Therefore, to minimize f we should go along ω in the opposite direction to the

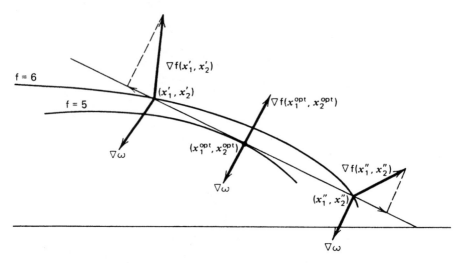

FIG. 3.10   Gradients near a constrained optimum.

component of the gradient projected onto ω. At the optimum point, the gradient of f is perpendicular (mathematicians say "normal") to ω and therefore there can be no improvement in f by moving off this point. We can solve for this optimum point mathematically by using this "normal" property at the optimum. To guarantee that the gradient of f (i.e., ∇f) is normal to ω, we simply require that ∇f and the gradient of ω, $\nabla\omega$, be linearly dependent vectors. Vectors that are linearly dependent must "line up" with each other (i.e., they point in exactly the same or exactly the opposite direction), although they may be different in magnitude. Mathematically, we can then set up the following equation.

$$\nabla f + \lambda \nabla \omega = 0 \qquad (3A.3)$$

That is, the two gradients can be added together in such a way that they cancel each other as long as one of them is scaled. The scaling variable, λ, is called a *Lagrange multiplier*, and instead of using the gradients as shown in Eq. 3A.3, we will restate them as

$$\mathscr{L}(x_1, x_2, \lambda) = f(x_1, x_2) + \lambda \omega(x_1, x_2) \qquad (3A.4)$$

This equation is called the *Lagrange equation* and consists of three variables, x_1, x_2, and λ. When we solve for the optimum values for x_1 and x_2, we will automatically calculate the correct value for λ. To meet the conditions set down in Eq. 3A.3, we simply require that the partial derivative of $\mathscr{L}$ with respect to each of the unknown variables, x_1, x_2, and λ, be equal to zero. That is,

At the optimum:

$$\frac{\partial \mathscr{L}}{\partial x_1} = 0$$

$$\frac{\partial \mathscr{L}}{\partial x_2} = 0 \qquad\qquad (3A.5)$$

$$\frac{\partial \mathscr{L}}{\partial \lambda} = 0$$

To show how this works, solve for the optimum point for the sample problem using Lagrange's method.

$$\mathscr{L}(x_1, x_2, \lambda) = 0.25x_1^2 + x_2^2 + \lambda(5 - x_1 - x_2)$$

$$\frac{\partial \mathscr{L}}{\partial x_1} = 0.5x_1 - \lambda = 0$$

$$\frac{\partial \mathscr{L}}{\partial x_2} = 2x_2 - \lambda = 0 \qquad\qquad (3A.6)$$

$$\frac{\partial \mathscr{L}}{\partial \lambda} = 5 - x_1 - x_2 = 0$$

Note that the last equation in (3A.6) is simply the original constraint equation. The solution to Eq. 3A.6 is

$$x_1 = 4$$

$$x_2 = 1 \qquad\qquad (3A.7)$$

$$\lambda = 2$$

When there is more than one constraint present in the problem, the optimum point can be found in a similar manner to that just used. Suppose there were three constraints to be met, then our problem would be as follows.

Minimize: $\qquad\qquad f(x_1, x_2)$

Subject to: $\qquad\qquad \omega_1(x_1, x_2) = 0$

$$\omega_2(x_1, x_2) = 0 \qquad\qquad (3A.8)$$

$$\omega_3(x_1, x_2) = 0$$

The optimum point would possess the property that the gradient of f and the gradients of ω_1, ω_2, and ω_3 are linearly dependent. That is,

$$\nabla f + \lambda_1 \nabla \omega_1 + \lambda_2 \nabla \omega_2 + \lambda_3 \nabla \omega_3 = 0 \tag{3A.9}$$

Again, we can set up a Lagrangian equation as before.

$$\mathscr{L} = f(x_1, x_2) + \lambda_1 \omega_1(x_1, x_2) + \lambda_2 \omega_2(x_1, x_2) + \lambda_3 \omega_3(x_1, x_2) \tag{3A.10}$$

whose optimum occurs at

$$\frac{\partial \mathscr{L}}{\partial x_1} = 0 \qquad \frac{\partial \mathscr{L}}{\partial x_2} = 0$$

$$\frac{\partial \mathscr{L}}{\partial \lambda_1} = 0 \qquad \frac{\partial \mathscr{L}}{\partial \lambda_2} = 0 \qquad \frac{\partial \mathscr{L}}{\partial \lambda_3} = 0 \tag{3A.11}$$

Up until now, we have assumed that all the constraints in the problem were equality constraints; that is, $\omega(x_1, x_2, \ldots) = 0$. In general, however, optimization problems involve inequality constraints; that is, $g(x_1, x_2, \ldots) \leq 0$, as well as equality constraints. The optimal solution to such problems will not necessarily require all the inequality constraints to be binding. Those that are binding will result in $g(x_1, x_2, \ldots) = 0$ at the optimum.

The fundamental rule that tells when the optimum has been reached is presented in a famous paper by Kuhn and Tucker (reference 3). *The Kuhn–Tucker conditions*, as they are called, are presented here.

Minimize: $f(\mathbf{x})$

Subject to: $\omega_i(\mathbf{x}) = 0 \qquad i = 1, 2, \ldots, N\omega$

$g_i(\mathbf{x}) \leq 0 \qquad i = 1, 2, \ldots, Ng$

$\mathbf{x} = $ vector of real numbers, dimension $= N$

Then, forming the Lagrange function,

$$\mathscr{L}(\mathbf{x}, \lambda, \mu) = f(\mathbf{x}) + \sum_{i=1}^{N\omega} \lambda_i \omega_i(\mathbf{x}) + \sum_{i=1}^{Ng} \mu_i g_i(\mathbf{x})$$

The conditions for an optimum for the point $\mathbf{x}^0$, λ^0, μ^0 are

1. $\dfrac{\partial \mathscr{L}}{\partial x_i}(\mathbf{x}^0, \lambda^0, \mu^0) = 0$ for $i = 1 \ldots N$

2. $\omega_i(\mathbf{x}^0) = 0$ for $i = 1 \ldots N\omega$

3. $g_i(\mathbf{x}^0) \leq 0$ for $i = 1 \ldots Ng$

4. $\left.\begin{array}{l} \mu_i^0 g_i(\mathbf{x}^0) = 0 \\ \mu_i^0 \geq 0 \end{array}\right\}$ for $i = 1 \ldots Ng$

The first condition is simply the familiar set of partial derivatives of the Lagrange function that must equal zero at the optimum. The second and third conditions are simply a restatement of the constraint conditions on the problem. The fourth condition, often referred to as *the complimentary slackness condition,* provides a concise mathematical way to handle the problem of binding and nonbinding constraints. Since the product $\mu_i^0 g_i(\mathbf{x}^0)$ equals zero, either μ_i^0 is equal to zero or $g_i(\mathbf{x}^0)$ is equal to zero, or both are equal to zero. If μ_i^0 is equal to zero, $g_i(\mathbf{x}^0)$ is free to be nonbinding; if μ_i^0 is positive, then $g_i(\mathbf{x}^0)$ must be zero. Thus, we have a clear indication of whether the constraint is binding or not by looking at μ_i^0.

To illustrate how the Kuhn–Tucker equations are used, we will add an inequality constraint to the sample problem used earlier in this appendix. The problem we will solve is as follows.

Minimize: $f(x_1, x_2) = 0.25x_1^2 + x_2^2$

Subject to: $\omega(x_1, x_2) = 5 - x_1 - x_2 = 0$

                        $g(x_1, x_2) = x_1 + 0.2x_2 - 3 \leq 0$

which can be illustrated as in Figure 3.11.

First, set up the Lagrange equation for the problem.

$$\mathcal{L} = f(x_1, x_2) + \lambda[\omega(x_1, x_2)] + \mu[g(x_1, x_2)]$$

$$= 0.25x_1^2 + x_2^2 + \lambda(5 - x_1 - x_2) + \mu(x_1 + 0.2x_2 - 3)$$

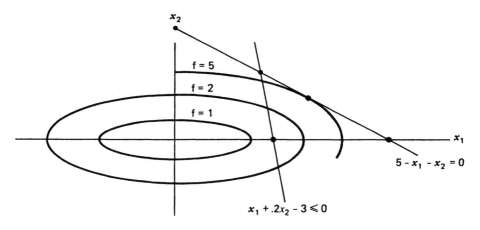

FIG. 3.11 Elliptical objective function with equality and inequality constraints.

The first condition gives

$$\frac{\partial \mathscr{L}}{\partial x_1} = 0.5x_1 - \lambda + \mu = 0$$

$$\frac{\partial \mathscr{L}}{\partial x_2} = 2x_2 - \lambda + 0.2\mu = 0$$

The second condition gives

$$5 - x_1 - x_2 = 0$$

The third condition gives

$$x_1 + 0.2x_2 - 3 \leq 0$$

The fourth condition gives

$$\mu(x_1 + 0.2x_2 - 3) = 0$$

$$\mu \geq 0$$

At this point, we are confronted with the fact that the Kuhn–Tucker conditions only give necessary conditions for a minimum, not a precise, procedure as to how that minimum is to be found. To solve the problem just presented, we must literally experiment with various solutions until we can verify that one of the solutions meets all four conditions. First, let $\mu = 0$, which implies that $g(x_1, x_2)$ can be less than or equal to zero. However, if $\mu = 0$, we can see that the first and second conditions give the same solution as we had previously, without the inequality constraint. But the previous solution violates our inequality constraint; and therefore the four Kuhn–Tucker conditions do not hold with $\mu = 0$. In summary,

---

If $\mu = 0$, then by conditions 1 and 2

$$x_1 = 4$$

$$x_2 = 1$$

$$\lambda = 2$$

but

$$g(x_1, x_2)\Big|_{\substack{x_1 = 4 \\ x_2 = 1}} = 4 + 0.2(1) - 3 = 1.2 \not\leq 0$$

Now we will try a solution in which $\mu > 0$. In this case, $g(x_1, x_2)$ must be exactly zero and our solution can be found by solving for the intersection of $g(x_1, x_2)$ and $\omega(x_1, x_2)$, which occurs at $x_1 = 2.5$, $x_2 = 2.5$. Further, condition 1 gives $\lambda = 5.9375$ and $\mu = 4.6875$, and all four of the Kuhn–Tucker conditions are met. In summary

---

If $\mu > 0$, then by conditions 2 and 3

$$x_1 = 2.5$$
$$x_2 = 2.5$$

by condition 1

$$\lambda = 5.9375$$
$$\mu = 4.6875$$

and

$$g(x_1, x_2)|_{x_1 = x_2 = 2.5} = 2.5 + 0.2(2.5) - 3 = 0$$

All conditions are met.

Considerable insight can be gained into the characteristics of optimal solutions through use of the Kuhn–Tucker conditions. One important insight comes from formulating the optimization problem so that it reflects our standard power system economic dispatch problems. Specifically, we will assume that the objective function consists of a sum of individual cost functions, each of which is a function of only one variable. For example,

$$f(x_1, x_2) = C_1(x_1) + C_2(x_2)$$

Further, we will restrict this problem to have one equality constraint of the form

$$\omega(x_1, x_2) = L - x_1 - x_2 = 0$$

and a set of inequality constraints that act to restrict the problem variables within an upper and lower limit. That is,

$$x_1^- \le x_1 \le x_1^+ \rightarrow \begin{cases} g_1(x_1) = x_1 - x_1^+ \le 0 \\ g_2(x_1) = x_1^- - x_1 \le 0 \end{cases}$$

$$x_2^- \le x_2 \le x_2^+ \rightarrow \begin{cases} g_3(x_2) = x_2 - x_2^+ \le 0 \\ g_4(x_2) = x_2^- - x_2 \le 0 \end{cases}$$

Then the Lagrange function becomes

$$= f(x_1, x_2) + \lambda\omega(x_1, x_2) + \mu_1 g_1(x_1) + \mu_2 g_2(x_1) + \mu_3 g_3(x_2) + \mu_4 g_4(x_2)$$
$$= C_1(x_1) + C_2(x_2) + \lambda(L - x_1 - x_2) + \mu_1(x_1 - x_1^+) + \mu_2(x_1^- - x_1)$$
$$+ \mu_3(x_2 - x_2^+) + \mu_4(x_2^- - x_2)$$

Condition 1 gives

$$C_1'(x_1) - \lambda + \mu_1 - \mu_2 = 0$$
$$C_2'(x_2) - \lambda + \mu_3 - \mu_4 = 0$$

Condition 2 gives

$$L - x_1 - x_2 = 0$$

Condition 3 gives

$$x_1 - x_1^+ \le 0$$
$$x_1^- - x_1 \le 0$$
$$x_2 - x_2^+ \le 0$$
$$x_2^- - x_2 \le 0$$

Condition 4 gives

$$\mu_1(x_1 - x_1^+) = 0 \qquad \mu_1 \ge 0$$
$$\mu_2(x_1^- - x_1) = 0 \qquad \mu_2 \ge 0$$
$$\mu_3(x_2 - x_2^+) = 0 \qquad \mu_3 \ge 0$$
$$\mu_4(x_2^- - x_2) = 0 \qquad \mu_4 \ge 0$$

Case 1

If the optimum solution occurs at values for x_1 and x_2 that are not at either an upper or a lower limit, then all μ values are equal to zero and

$$C_1'(x_1) = C_2'(x_2) = \lambda$$

That is, the incremental costs associated with each variable are equal and this value is exactly the λ we are interested in.

Case 2

Now suppose that the optimum solution requires that x_1 be at its upper limit (i.e., $x_1 - x_1^+ = 0$) and that x_2 is not at its upper or lower limit. Then,

$$\mu_1 \ge 0$$

and μ_2, μ_3, and μ_4 will each equal zero. Then, from condition 1,

$$C'_1(x_1) = \lambda - \mu_1 \rightarrow C'_1(x_1) \leq \lambda$$
$$C'_2(x_2) = \lambda$$

Therefore, the incremental cost associated with the variable that is at its upper limit will always be less than or equal to λ, whereas the incremental cost associated with the variable that is not at limit will exactly equal λ.

Case 3

Now suppose the opposite of Case 2 obtains; that is, let the optimum solution require x_1 to be at its lower limit (i.e., $x_1^- - x_1 = 0$) and again assume that x_2 is not at its upper or lower limit. Then

$$\mu_2 \geq 0$$

and μ_1, μ_3, and μ_4 will each equal zero. Then from condition 1

$$C'_1(x_1) = \lambda + \mu_2 \Rightarrow C'_1(x_1) \geq \lambda$$
$$C'_2(x_2) = \lambda$$

Therefore, the incremental cost associated with a variable at its lower limit will be greater than or equal to λ whereas, again, the incremental cost associated with the variable that is not at limit will equal λ.

Case 4

If the optimum solution requires that both x_1, x_2 are at limit and the equality constraint can be met, then λ and the nonzero μ values are indeterminate. For example, suppose the optimum required that

$$x_1 - x_1^+ = 0$$

and

$$x_2 - x_2^+ = 0$$

Then

$$\mu_1 \geq 0 \qquad \mu_3 \geq 0 \qquad \mu_2 = \mu_4 = 0$$

Condition 1 would give

$$C'_1(x_1) = \lambda - \mu_1$$
$$C'_2(x_2) = \lambda - \mu_3$$

and the specific values for λ, μ_1, and μ_3 would be undetermined. In summary, for the general problem of N variables:

Minimize: $C_1(x_1) + C_2(x_2) + \cdots + C_N(x_N)$

Subject to: $L - x_1 - x_2 - \cdots - x_N = 0$

And:
$$\left. \begin{array}{c} x_i - x_i^+ \leq 0 \\ x_i^- - x_i \leq 0 \end{array} \right\} \quad \text{for } i = 1 \ldots N$$

Let the optimum lie at $x_i = x_i^{opt}$ $i = 1 \ldots N$ and assume that at least one x_i is not at limit. Then,

$$\text{If } x_i^{opt} < x_i^+ \quad \text{and} \quad x_i^{opt} > x_i^-, \quad \text{then} \quad C_i(x_i^{opt}) = \lambda$$
$$\text{If } x_i^{opt} = x_i^+ \qquad\qquad\qquad\qquad\qquad C_i'(x_i^{opt}) \leq \lambda$$
$$\text{If } x_i^{opt} = x_i^- \qquad\qquad\qquad\qquad\qquad C_i'(x_i^{opt}) \geq \lambda$$

Slack Variable Formulation

An alternate approach to the optimization problem with inequality constraints requires that all inequality constraints be made into equality constraints. This is done by adding slack variables in the following way.

If:
$$g(x_1) = x_1 - x_1^+ \leq 0$$
Then:
$$g(x_1, S_1) = x_1 - x_1^+ + S_1^2 = 0$$

We add S_1^2 rather than S_1 so that S_1 need not be limited in sign.

Making all inequality constraints into equality constraints eliminates the need for conditions 3 and 4 of the Kuhn–Tucker conditions. However, as we will see shortly, the result is essentially the same. Let us use our two-variable problem again.

Minimize: $f(x_1, x_2) = C_1(x_1) + C_2(x_2)$

Subject to: $\omega(x_1, x_2) = L - x_1 - x_2 = 0$

And:
$$g_1(x_1) = x_1 - x_1^+ \leq 0 \quad \text{or} \quad g_1(x_1, S_1) = x_1 - x_1^+ + S_1^2 = 0$$
$$g_2(x_1) = x_1^- - x_1 \leq 0 \qquad\qquad g_2(x_1, S_2) = x_1^- - x_1 + S_2^2 = 0$$
$$g_3(x_2) = x_2 - x_2^+ \leq 0 \qquad\qquad g_3(x_2, S_3) = x_2 - x_2^+ + S_3^2 = 0$$
$$g_4(x_2) = x_2^- - x_2 \leq 0 \qquad\qquad g_4(x_2, S_4) = x_2^- - x_2 + S_4^2 = 0$$

The resulting Lagrange function is

$$\mathcal{L} = f(x_1, x_2) + \lambda_0 \omega(x_1, x_2) + \lambda_1 g_1(x_1, S_1)$$
$$+ \lambda_2 g_2(x_1, S_2) + \lambda_3 g_3(x_2, S_3) + \lambda_4 g_4(x_2, S_4)$$

Note that all constraints are now equality constraints, so we have used only λ values as Lagrange multipliers.

Condition 1 gives:
$$\frac{\partial \mathscr{L}}{\partial x_1} = C'_1(x_1) - \lambda_0 + \lambda_1 - \lambda_2 = 0$$

$$\frac{\partial \mathscr{L}}{\partial x_2} = C'_2(x_2) - \lambda_0 + \lambda_3 - \lambda_4 = 0$$

$$\frac{\partial \mathscr{L}}{\partial S_1} = 2\lambda_1 S_1 = 0$$

$$\frac{\partial \mathscr{L}}{\partial S_2} = 2\lambda_2 S_2 = 0$$

$$\frac{\partial \mathscr{L}}{\partial S_3} = 2\lambda_3 S_3 = 0$$

$$\frac{\partial \mathscr{L}}{\partial S_4} = 2\lambda_4 S_4 = 0$$

Condition 2 gives:
$$L - x_1 - x_2 = 0$$

$$(x_1 - x_1^+ + S_1^2) = 0$$

$$(x_1^- - x_1 + S_2^2) = 0$$

$$(x_2 - x_2^+ + S_3^2) = 0$$

$$(x_2^- - x_2 + S_4^2) = 0$$

We can see that the derivatives of the Lagrange function with respect to the slack variables provide us once again with a complimentary slackness rule. For example, if $2\lambda_1 S_1 = 0$, then either $\lambda_1 = 0$ and S_1 is free to be any value or $S_1 = 0$ and λ_1 is free (or λ_1 and S_1 can both be zero). Since there are as many problem variables whether one uses the slack variable form or the inequality constraint form, there is little advantage to either, other than perhaps a conceptual advantage to the student.

Dual Variables

Another way to solve an optimization problem is to use a technique that solves for the Lagrange variables directly and then solves for the problem variables themselves. This formulation is known as a "dual solution" and in it the Lagrange multipliers are called "dual variables." We shall use the example just solved to demonstrate this technique.

The presentation up to now has been concerned with the solution of what is formally called the "primal problem," which was stated in Eq. 3A.2 as:

Minimize: $\qquad\qquad f(x_1, x_2) = 0.25x_1^2 + x_2^2$

Subject to: $\qquad\qquad \omega(x_1, x_2) = 5 - x_1 - x_2$

and its Lagrangian function is:

$$\mathcal{L}(x_1, x_2, \lambda) = 0.25x_1^2 + x_2^2 + \lambda(5 - x_1 - x_2)$$

If we define a dual function, $q(\lambda)$, as:

$$q(\lambda) = \min_{x_1 x_2} \mathcal{L}(x_1, x_2, \lambda) \qquad (3A.12)$$

Then the "dual problem" is to find

$$q^*(\lambda) = \max_{\lambda \geq 0} q(\lambda) \qquad (3A.13)$$

The solution, in the case of the dual problem involves two separate optimization problems. The first requires us to take an initial set of values for x_1 and x_2 and then find the value of λ which maximizes $q(\lambda)$. We then take this value of λ and, holding it constant, we find values of x_1 and x_2 which minimize $\mathcal{L}(x_1, x_2, \lambda)$. This process is repeated or iterated until the solution is found.

In the case of convex objective functions, such as the example used in this appendix, this procedure is guaranteed to solve to the same optimum as the primal problem solution presented earlier.

The reader will note that in the case of the functions presented in Eq. 3A.2, we can simplify the procedure above by eliminating x_1 and x_2 from the problem altogether, in which case we can find the maximum of $q(\lambda)$ directly. If we express the problem variables in terms of the Lagrange multiplier (or dual variable), we obtain:

$$x_1 = 2\lambda$$

$$x_2 = \frac{\lambda}{2}$$

We now eliminate the original problem variables from the Lagrangian function:

$$q(\lambda) = -\left(\frac{5}{4}\right)\lambda^2 + 5\lambda$$

We can use the dual variable to solve our problem as follows:

$$\frac{\partial}{\partial \lambda} q(\lambda) = 0 = \left(\frac{5}{2}\right)\lambda - 5$$

or

$$\lambda = 2$$

Therefore, the value of the dual variable is $q^*(\lambda) = 5$. The values of the primal variables are $x_1 = 4$ and $x_2 = 1$.

In the economic dispatch problem dealt with in this chapter, one cannot eliminate the problem variables since the generating unit cost functions may be piecewise linear or other complex functions. In this case, we must use the dual optimization algorithm described earlier; namely, we first optimize on λ and then on the problem variables, and then go back and update λ, etc. Since the dual problem requires that we find

$$q^*(\lambda) = \max_{\lambda \geq 0} q(\lambda)$$

and we do not have an explicit function in λ (as we did above), we must adopt a slightly different strategy. In the case of economic dispatch or other problems where we cannot eliminate the problem variables, we find a way to adjust λ so as to move $q(\lambda)$ from its initial value to one which is larger. The simplest way to do this is to use a gradient adjustment so that

$$\lambda^1 = \lambda^0 + \left[\frac{d}{d\lambda} q(\lambda)\right]\alpha$$

where α merely causes the gradient to behave well. A more useful way to apply the gradient technique is to let λ be adjusted upwards at one rate and downward at a much slower rate; for example:

$$\alpha = 0.5 \quad \text{when} \quad \frac{d}{d\lambda} q(\lambda) \text{ is positive}$$

and

$$\alpha = 0.1 \quad \text{when} \quad \frac{d}{d\lambda} q(\lambda) \text{ is negative}$$

The closeness to the final solution in the dual optimization method is measured by noting the relative size of the "gap" between the primal function and the dual function. The primal optimization problem can be solved directly in the case of the problem stated in Eq. 3A.2 and the optimal value will be called J^* and it is defined as:

$$J^* = \min \mathscr{L} \tag{3A.14}$$

This value will be compared to the optimum value of the dual function, q^*. The difference between them is called the "duality gap." A good measure of the closeness to the optimal solution is the "relative duality gap," defined as:

$$\frac{J^* - q^*}{q^*} \tag{3A.15}$$

TABLE 3.4   Dual Optimization

Iteration	λ	x_1	x_2	ω	J^*	q^*	$\dfrac{J^* - q^*}{q^*}$
1	0	0	0	5.0	5.0	0	—
2	2.5	5.0	1.25	-1.25	5.0	4.6875	0.0666
3	2.375	4.75	1.1875	-0.9375	5.0	4.8242	0.0364
4	2.2813	4.5625	1.1406	-0.7031	5.0	4.9011	0.0202
5	2.2109	4.4219	1.1055	-0.5273	5.0	4.9444	0.01124
⋮							
20	2.0028	4.0056	1.0014	-0.007	5.0	5.0	0

For a convex problem with continuous variables, the duality gap will become zero at the final solution. When we again take up the dual optimization method in Chapter 5, we will be dealing with nonconvex problems with noncontinuous variables and the duality gap will never actually go to zero.

Using the dual optimization approach on the problem given in Eq. 3A.2 and starting at $\lambda = 0$, we obtain the results shown in Table 3.4. As can be seen, this procedure converges to the correct answer.

A special note about lambda search. The reader should note that the dual technique, when applied to economic dispatch, is the same as the lambda search technique we introduced earlier in this chapter to solve the economic dispatch problem.

APPENDIX 3B
Dynamic-Programming Applications

The application of digital methods to solve a wide variety of control and dynamics optimization problems in the late 1950s led Dr. Richard Bellman and his associates to the development of dynamic programming. These techniques are useful in solving a variety of problems and can greatly reduce the computational effort in finding optimal trajectories or control policies.

The theoretical mathematical background, based on the calculus of variations, is somewhat difficult. The applications are not, however, since they depend on a willingness to express the particular optimization problem in terms appropriate for a dynamic-programming (DP) formulation.

In the scheduling of power generation systems, DP techniques have been developed for the following.

- The economic dispatch of thermal systems.
- The solution of hydrothermal economic-scheduling problems.
- The practical solution of the unit commitment problem.

This text will touch on all three areas.

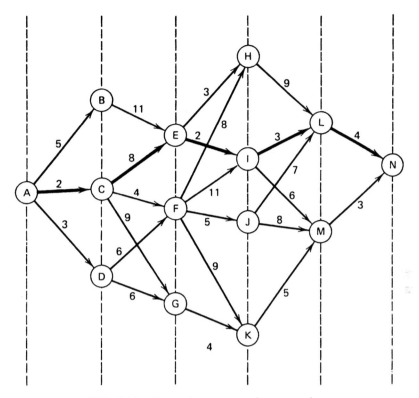

FIG. 3.12   Dynamic-programming example.

First, however, it will be as well to introduce some of the notions of DP by means of some one-dimensional examples. Figure 3.12 represents the cost of transporting a unit shipment from node A to node N. The values on the arcs are the costs, or values, of shipping the unit from the originating node to the terminating node of the arc. The problem is to find the minimum cost route from A to N. The method to be illustrated is that of dynamic programming. The first two examples are from reference 18 and are used by permission.

Starting at A, the minimum cost path to N is ACEILN.
Starting at C, the least cost path to N is   CEILN.
Starting at E, the least cost path to N is   EILN
Starting at I, the least cost path to N is   ILN.
Starting at L, the least cost path to N is   LN.

The same type of statements could be made about the maximum cost path from A to N (ABEHLN). That is, the maximum cost to N, starting from any node on the original maximal path, is contained in that original path.

The choice of route is made in sequence. There are various stages traversed. The optimum sequence is called the *optimal policy*; any subsequence is a *subpolicy*. From this it may be seen that the optimal policy (i.e., the minimum cost route) contains only optimal subpolicies. This is the *Theorem of optimality*.

> An optimal policy must contain only optimal subpolicies.

In reference 20, Bellman and Dreyfus call it the "Principle of optimality" and state it as

> A policy is optimal if, at a stated stage, whatever the preceding decisions may have been, the decisions still to be taken constitute an optimal policy when the result of the previous decisions is included.

We continue with the same example, only now let us find the minimum cost path. Figure 3.13 identifies the stages (I, II, III, IV, V). At the terminus of each stage, there is a set of choices of nodes $\{X_i\}$ to be chosen $[\{X_3\} = \{H, I, J, K\}]$. The symbol $V_a(X_i, X_i + 1)$ represents the "cost" of traversing stage $a(= I, \ldots, V)$ and depends on the variables selected from the sets $\{X_i\}$ and $\{X_i + 1\}$. That is, the cost, V_a, depends on the starting and terminating nodes. Finally, $f_a(X_i)$ is the minimum cost for stages I through a to arrive at some particular node X_i at the end of that stage, starting from A. The numbers in the node circles in Figure 3.13 represent this minimum cost.

$$\{X_0\}: A \qquad \{X_2\}: E, F, G \qquad \{X_4\}: L, M$$
$$\{X_1\}: B, C, D \qquad \{X_3\}: H, I, J, K \qquad \{X_5\}: N$$

$f_I(X_1)$: Minimum cost for the first stage is obvious:

$$f_I(B) = V_1(A, B) = 5$$
$$f_I(C) = V_1(A, C) = 2$$
$$f_I(D) = V_1(A, D) = 3$$

$f_{II}(X_2)$: Minimum cost for stages I and II as a function of X_2:

$$f_{II}(E) = \min_{\{X_1\}} [f_1(X_1) + V_{II}(X_1, E)]$$
$$= \min[5 + 11, \quad 2 + 8, \quad 3 + \infty] = 10$$
$$X_1 = B \quad = C \quad = D \qquad X_1 = C$$

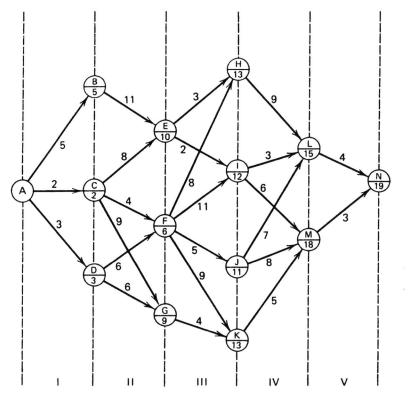

FIG. 3.13 Dynamic-programming example showing minimum cost at each node.

The cost is infinite for node D since there is no path from D to E:

$$f_{II}(F) = \min_{\{X_1\}} [f_I(X_1) + V_{II}(X_1, F)] = \min[\infty, 6, 9] = 6, X_1 = C$$

$$f_{II}(G) = \min_{\{X_1\}} [f_I(X_1) + V_{II}(X_1, G)] = \min[\infty, 11, 9] = 9, X_1 = D$$

Thus, at each stage we should record the minimum cost and the termination starting the stage in order to achieve the minimum cost path for each of the nodes terminating the current stage.

(X_2)	E	F	G
$f_{II}(X_2)$	10	6	9
Path X_0X_1	AC	AC	AD

$f_{III}(X_3)$: Minimum cost of stages I, II, and III as a function of X_3:

$$f_{III}(H) = \min_{\{X_2\}} [f_{II}(X_2) + V_{III}(X_2, H)] = \min[13, 14, \infty] = 13 \quad \text{with} \quad X_2 = E$$

In general,

$$f_{III}(X_3) = \min_{\{X_2\}} [f_{II}(X_2) + V_{III}(X_2, X_3)]$$

Giving,

X_3	H	I	J	K
$f_{III}(X_3)$	13	12	11	13
Path $X_0 X_1 X_2$	ACE	ACE	ACF	ADG

f_{IV}: Minimum cost of stages I through IV as a function of X_4:

$$f_{IV}(X_4) = \min_{\{X_3\}} [f_{III}(X_3) + V_{IV}(X_3, V_4)]$$

$$f_{IV}(L) = \min[13 + 9, 12 + 3, 11 + 7, 13 + \infty] = 15, \qquad X_3 = I$$
$$X_3 = H \quad = I \quad = J \quad = K$$

$$f_{IV}(M) = [13 + \infty, 12 + 6, 11 + 8, 13 + 5] = 18 \qquad X_3 = I \quad \text{or} \quad K$$
$$X_3 = H \quad = I \quad = J \quad = K$$

f_V: Minimum cost of I through V as a function of X_5:

$$f_V(N) = \min_{\{X_4\}} [f_{IV}(X_4) + V_V(X_4, X_5)]$$

$$= \min[15 + 4, 18 + 3] = 19 \qquad X_4 = L$$
$$X_4 = L \quad = M$$

Tracing back, the path of minimum cost is found as follows:

Stage 1	$\{X_i\}$	f_i
1	B, Ⓒ, D	5, ②, 3
2	Ⓔ, F, G	⑩, 6, 9
3	H, Ⓘ, J, K	13, ⑫, 11, 13
4	Ⓛ, M	⑮, 18
5	Ⓝ	⑲

It would be possible to carry out this procedure in the opposite direction just as easily.

An Allocation Problem

Table 3.5 lists the profits to be made in each of four ventures as a function of the investment in the particular venture. Given a limited amount of money to

TABLE 3.5 Profit Versus Investment

Investment Amount	Profit from Venture			
	I	II	III	IV
0	0	0	0	0
1	0.28	0.25	0.15	0.20
2	0.45	0.41	0.25	0.33
3	0.65	0.55	0.40	0.42
4	0.78	0.65	0.50	0.48
5	0.90	0.75	0.65	0.53
6	1.02	0.80	0.73	0.56
7	1.13	0.85	0.82	0.58
8	1.23	0.88	0.90	0.60
9	1.32	0.90	0.96	0.60
10	1.38	0.90	1.00	0.60

allocate, the problem is to find the optimal investment allocation. The only restriction is that investments must be made in integer amounts. For instance, if one had 10 units to invest and the policy were to put 3 in I, 1 in II, 5 in III, and 1 in IV, then

$$\text{Profit} = 0.65 + 0.25 + 0.65 + 0.20 = 1.75$$

The problem is to find an allocation policy that yields the maximum profit. Let

X_1, X_2, X_3, X_4 be investments in I through IV

$V(X_1), V(X_2), V(X_3), V(X_4)$ be profits

$X_1 + X_2 + X_3 + X_4 = 10$ is the constraint; that is,
10 units must be invested

To transform this into a multistage problem, let the stages be

$$X_1, U_1, U_2, A$$

where

$U_1 = X_1 + X_2 \qquad U_1 \le A \qquad U_2 \le A$

$U_2 = U_1 + X_3 \qquad \{A\} = 0, 1, 2, 3, \ldots, 10$

$A = U_2 + X_4$

The total profit is

$$f(X_1, X_2, X_3, X_4) = V_1(X_1) + V_2(X_2) + V_3(X_3) + V_4(X_4)$$

which can be written

$$f(X_1, U_1, U_2, A) = V_1(X_1) + V_2(U_1 - X_1) + V_3(U_2 - U_1) + V_4(A - U_2)$$

At the second stage, we can compute

$$f_2(U_1) = \max_{X_1 = 0, 1, \ldots, U_1} [V_1(X_1) + V_2(U_1 - X_1)]$$

$X_1, X_2,$ or U_1	$V_1(X_1)$	$V_2(X_2)$	$f_2(U_1)$	Optimal Subpolicies for I & II
0	0	0	0	0, 0
1	0.28	0.25	0.28	1, 0
2	0.45	0.41	0.53	1, 1
3	0.65	0.55	0.70	2, 1
4	0.78	0.65	0.90	3, 1
5	0.90	0.75	1.06	3, 2
6	1.02	0.80	1.20	3, 3
7	1.13	0.85	1.33	4, 3
8	1.23	0.88	1.45	5, 3
9	1.32	0.90	1.57	6, 3
10	1.38	0.90	1.68	7, 3

Next, at the third stage,

$$f_3(U_2) = \max_{U_1 = 0, 1, 2, \ldots, U_2} [f_2(U_1) + V_3(U_2 - U_1)]$$

$U_1, U_2,$ or X_3	$f_2(U_1)$	$V_3(X_3)$	$f_3(U_2)$	Optimal Subpolicies For I & II	For I, II, & III
0	0	0	0	0, 0	0, 0, 0
1	0.28	0.15	0.28	1, 0	1, 0, 0
2	0.53	0.25	0.53	1, 1	1, 1, 0
3	0.70	0.40	0.70	2, 1	2, 1, 0
4	0.90	0.50	0.90	3, 1	3, 1, 0
5	1.06	0.62	1.06	3, 2	3, 2, 0
6	1.20	0.73	1.21	3, 3	3, 2, 1
7	1.33	0.82	1.35	4, 3	3, 3, 1
8	1.45	0.90	1.48	5, 3	4, 3, 1
9	1.57	0.96	1.60	6, 3	5, 3, 1 or 3, 3, 3
10	1.68	1.00	1.73	7, 3	4, 3, 3

Finally, the last stage is

$$f_4(A) = \max_{\{U_2\}} [f_3(U_2) + V_4(A - U_2)]$$

U_2, A or X_4	$f_3(U_2)$	$V_4(X_4)$	$f_4(A)$	Optimal Subpolicy for I, II, & III	Optimal Policy
0	0	0	0	0, 0, 0	0, 0, 0, 0
1	0.28	0.20	0.28	1, 0, 0	1, 0, 0, 0
2	0.53	0.33	0.53	1, 1, 0	1, 1, 0, 0
3	0.70	0.42	0.73	2, 1, 0	1, 1, 0, 1
4	0.90	0.48	0.90	3, 1, 0	3, 1, 0, 0 or 2, 1, 0, 1
5	1.06	0.53	1.10	3, 2, 0	3, 1, 0, 1
6	1.21	0.56	1.26	3, 2, 1	3, 2, 0, 1
7	1.35	0.58	1.41	3, 3, 1	3, 2, 1, 1
8	1.48	0.60	1.55	4, 3, 1	3, 3, 1, 1
9	1.60	0.60	1.68	5, 3, 1 or 3, 3, 3	4, 3, 1, 1 or 3, 3, 1, 2
10	1.73	0.60	1.81	4, 3, 3	4, 3, 1, 2

Consider the procedure and solutions:

1. It was not necessary to enumerate all possible solutions. Instead, we used an orderly, stagewise search, the form of which was the same at each stage.
2. The solution was obtained not only for $A = 10$, but for the complete set of A values $\{A\} = 0, 1, 2, \ldots, 10$.
3. The optimal policy contains only optimal subpolicies. For instance, $A = 10$, (4, 3, 1, 2) is the optimal policy. For stages I, II, III, and $U_2 = 8$, (4, 3, 1) is the optimal subpolicy. For stages I and II, and $U_1 = 7$, (4, 3) is the optimal subpolicy. For stage I only, $X_1 = 4$ fixes the policy.
4. Notice also, that by storing the intermediate results, we could work a number of different variations of the same problem with the data already computed.

PROBLEMS

3.1 Assume that the fuel inputs in MBtu per hour for units 1 and 2, which are both on-line, are given by

$$H_1 = 8P_1 + 0.024P_1^2 + 80$$
$$H_2 = 6P_2 + 0.04P_2^2 + 120$$

where

H_n = fuel input to unit n in MBtu per hour (millions of Btu per hour)

P_n = unit output in megawatts

a. Plot the input–output characteristics for each unit expressing input in MBtu per hour and output in megawatts. Assume that the minimum loading of each unit is 20 MW and that the maximum loading is 100 MW.

b. Calculate the net heat rate in Btu per kilowatt-hour, and plot against output in megawatts.

c. Assume that the cost of fuel is 1.5 ₹/MBtu. Calculate the incremental production cost in ₹/MWh of each unit, and plot against output in megawatts.

3.2   Dispatch with Three-Segment Piecewise Linear Incremental Heat Rate Function

Given: Two generating units with incremental heat rate curves (IHR) specified as three connected line segments (four points as shown in Figure 3.14).

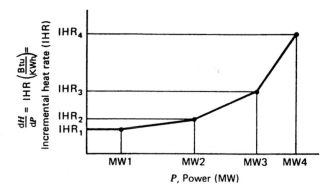

FIG. 3.14   Piecewise linear incremental heat rate curve for Problem 3.2.

Unit 1:

Point	MW	IHR (Btu/kWh)
1	100	7000
2	200	8200
3	300	8900
4	400	11000

Fuel cost for unit 1 = 1.60 ₹/MBtu

Unit 2:

Point	MW	IHR (Btu/kWh)
1	150	7500
2	275	7700
3	390	8100
4	450	8500

Fuel cost for unit 2 = 2.10 ₹/MBtu

Both units are running. Calculate the optimum schedule (i.e., the unit megawatt output for each unit) for various total megawatt values to be supplied by the units. Find the schedule for these total megawatt values:

$$300 \text{ MW}, \quad 500 \text{ MW}, \quad 700 \text{ MW}, \quad 840 \text{ MW}$$

Notes: Piecewise linear increment cost curves are quite common in digital computer executions of economic dispatch. The problem is best solved by using a "search" technique. In such a technique, the incremental cost is given a value and the units are scheduled to meet this incremental cost. The megawatt outputs for the units are added together and compared to the desired total. Depending on the difference, and whether the resulting total is above or below the desired total, a new value of incremental cost is "tried." This is repeated until the incremental cost is found that gives the correct desired value. The trick is to search in an efficient manner so that the number of iterations is minimized.

3.3 Assume the system load served by the two units of Problem 3.1 varies from 50 to 200 MW. For the data of Problem 3.1, plot the outputs of units 1 and 2 as a function of total system load when scheduling generation by equal incremental production costs. Assume that both units are operating.

3.4 As an exercise, obtain the optimum loading of the two generating units in Problem 3.1 using the following technique. The two units are to deliver 100 MW. Assume both units are on-line and delivering power. Plot the total fuel cost for 100 MW of delivered power as generation is shifted from one unit to the other. Find the minimum cost. The optimum schedule should check with the schedule obtained by equal incremental production costs.

3.5 This problem demonstrates the complexity involved when we must commit (turn on) generating units, as well as dispatch them economically. This problem is known as the *unit commitment problem* and is the subject of Chapter 5.

Given the two generating units in Problem 3.1, assume that they are both off-line at the start. Also, assume that load starts at 50 MW and increases to 200 MW. The most economic schedule to supply this varying load will require committing one unit first, followed by commitment of the second unit when the load reaches a higher level.

Determine which unit to commit first and at what load the remaining unit should be committed. Assume no "start-up" costs for either unit.

3.6 The system to be studied consists of two units as described in Problem 3.1. Assume a daily load cycle as follows.

Time Band	Load (MW)
0000–0600	50
0600–1800	150
1800–0000	50

Also, assume that a cost of 180 ℞ is incurred in taking either unit off-line and returning it to service after 12 h. Consider the 24-h period from 0600 one morning to 0600 the next morning.

a. Would it be more economical to keep both units in service for this 24-h period or to remove one of the units from service for the 12-h period from 1800 one evening to 0600 the next morning?

b. What is the economic schedule for the period of time from 0600 to 1800 (load = 150 MW)?

c. What is the economic schedule for the period of time from 1800 to 0600 (load = 50 MW)?

3.7 Assume that all three of the thermal units described below are running. Find the economic dispatch schedules as requested in each part. Use the method and starting conditions given.

Unit Data	Minimum (MW)	Maximum (MW)	Fuel Cost (℞/MBtu)
$H_1 = 225 + 8.4P_1 + 0.0025P_1^2$	45	350	0.80
$H_2 = 729 + 6.3P_2 + 0.0081P_2^2$	45	350	1.02
$H_3 = 400 + 7.5P_3 + 0.0025P_3^2$	47.5	450	0.90

a. Use the lambda-iteration method to find the economic dispatch for a total demand of 450 MW.

b. Use the base-point and participation factor method to find the economic schedule for a demand of 495 MW. Start from the solution to part a.

c. Use a gradient method to find the economic schedule for a total demand of 500 MW, assuming the initial conditions (i.e., loadings) on the three units are

$$P_1 = P_3 = 100 \text{ MW} \quad \text{and} \quad P_2 = 300 \text{ MW}$$

Give the individual unit loadings and cost per hour, as well as the total cost per hour to supply each load level. (MBtu = millions of Btu; H_j = heat input in MBtu/h; P_i = electric power output in MW; $i = 1, 2, 3$.)

3.8 Thermal Scheduling with Straight-Line Segments for Input–Output Curves

The following data apply to three thermal units. Compute and sketch the input–output characteristics and the incremental heat rate characteristics. Assume the unit input–output curves consist of straight-line segments between the given power points.

Unit No.	Power Output (MW)	Net Heat Rate (Btu/kWh)
1	45	13512.5
	300	9900.0
	350	9918.0
2	45	22764.5
	200	11465.0
	300	11060.0
	350	11117.9
3	47.5	16039.8
	200	10000.0
	300	9583.3
	450	9513.9

Fuel costs are:

Unit No.	Fuel Cost (R/MBtu)
1	0.61
2	0.75
3	0.75

Compute the economic schedule for system demands of 300, 460, 500, and 650 MW, assuming all three units are on-line. Give unit loadings and costs per hour as well as total costs in R per hour.

3.9 Environmental Dispatch

Recently, there has been concern that optimum *economic* dispatch was not the best environmentally. The principles of economic dispatch can fairly easily be extended to handle this problem. The following is a problem based on a real situation that occurred in the midwestern United States in 1973. Other cases have arisen with "NO_x" emission in Los Angeles.

Two steam units have input–output curves as follows.

$$H_1 = 400 + 5P_1 + 0.01P_1^2, \qquad \text{MBtu/h, } 20 \le P_1 \le 200 \text{ MW}$$

$$H_2 = 600 + 4P_2 + 0.015P_2^2, \qquad \text{MBtu/h, } 20 \le P_2 \le 200 \text{ MW}$$

The units each burn coal with a heat content of 11,500 Btu/lb that costs 13.50 R per ton (2000 lb). The combustion process in each unit results in 11.75% of the coal by weight going up the stack as fly ash.

a. Calculate the net heat rates of both units at 200 MW.

b. Calculate the incremental heat rates; schedule each unit for optimum *economy* to serve a total load of 250 MW with both units on-line.

c. Calculate the cost of supplying that load.

d. Calculate the rate of emission of fly ash for that case in pounds (lb) per hour, assuming no fly ash removal devices are in service.

e. Unit 1 has a precipitator installed that removes 85% of the fly ash; unit 2's precipitator is 89% efficient. Reschedule the two units for the minimum *total fly ash emission rate* with both on-line to serve a 250 MW load.

f. Calculate the rate of emission of ash and the cost for this schedule to serve the 250 MW load. What is the cost penalty?

g. Where does all that fly ash go?

3.10   Take the generation data shown in Example 3A. Ignore the generation limits and solve for the economic dispatch using the gradient method and Newton's method. Solve for a total generation of 900 MW in each case.

3.11   You have been assigned the job of building an oil pipeline from the West Coast of the United States to the East Coast. You are told that any one of the three West Coast sites is satisfactory and any of the three East Coast sites is satisfactory. The numbers in Figure 3.15 represent relative cost in hundreds of millions $R(R \cdot 10^8)$. Find the cheapest West Coast to East Coast pipeline.

3.12   **The Stagecoach Problem**

A mythical salesman who had to travel west by stagecoach, through unfriendly country, wished to take the safest route. His starting point and destination were fixed, but he had considerable choice as to which states he would travel through en route. The possible stagecoach routes

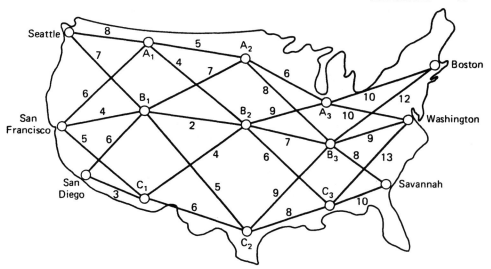

FIG. 3.15 Possible oil pipeline routes for Problem 3.11.

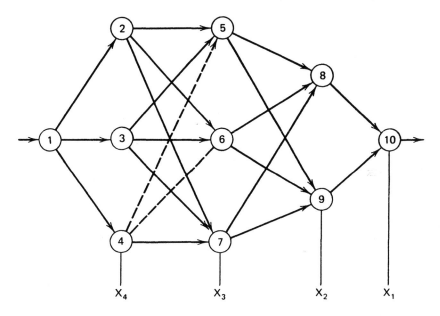

FIG. 3.16 Possible stagecoach routes for Problem 3.12.

are shown in Figure 3.16. After some thought, the salesman deduced a clever way of determining his safest route. Life insurance policies were offered to passengers, and since the cost of each policy was based on a careful evaluation of the safety of that run, the safest route should be the one with the cheapest policy. The cost of the standard policy on the

$i \backslash j$	1	2	3	4	5	6	7	8	9	10
1		2	4	3						
2					7	4	6			
3					3	2	4			
4					4	1	5			
5								1	4	
6								6	3	
7								3	2	
8										3
9										4
10										

FIG. 3.17 Cost to go from state i to state j in Problem 3.12. Costs not shown are infinite.

stagecoach run from state i to state j, denoted as C_{ij}, is given in Figure 3.17. Find the safest path(s) for the salesman to take.

3.13 Economic Dispatch Problem

Consider three generating units that do not have convex input–output functions. (This is the type of problem one encounters when considering valve points in the dispatch problem.)

Unit 1:

$$H_1(P_1) = \begin{cases} 80 + 8P_1 + 0.024P_1^2 & 20\text{ MW} \le P_1 \le 60\text{ MW} \\ 196.4 + 3P_1 + 0.075P_1^2 & 60\text{ MW} \le P_1 \le 100\text{ MW} \end{cases}$$

Generation limits are 20 MW $\le P_1 \le$ 100 MW.

Unit 2:

$$H_2(P_2) = \begin{cases} 120 + 6P_2 + 0.04P_2^2 & 20\text{ MW} \le P_2 \le 40\text{ MW} \\ 157.335 + 3.3333P_2 + 0.08333P_2^2 & 40\text{ MW} \le P_2 \le 100\text{ MW} \end{cases}$$

Generation limits are 20 MW $\le P_2 \le$ 100 MW.

Unit 3:

$$H_3(P_3) = \begin{cases} 100 + 4.6666P_3 + 0.13333P_3^2 & 20\text{ MW} \le P_3 \le 50\text{ MW} \\ 316.66 + 2P_3 + 0.1P_3^2 & 50\text{ MW} \le P_3 \le 100\text{ MW} \end{cases}$$

Generation limits are 20 MW $\leq P_3 \leq$ 100 MW. Fuel costs = 1.5 ₹/MBtu for all units.

a. Plot the cost function for each unit (see Problem 3.1).

b. Plot the incremental cost function for each unit.

c. Find the most economical dispatch for the following total demands assuming all units are on-line:

$$P_D = 100 \text{ MW}$$

$$P_D = 140 \text{ MW}$$

$$P_D = 180 \text{ MW}$$

$$P_D = 220 \text{ MW}$$

$$P_D = 260 \text{ MW}$$

where

$$P_D = P_1 + P_2 + P_3$$

Solve using dynamic programming and discrete load steps of 20 MW, starting at 20 MW through 100 MW for each unit.

d. Can you solve these dispatch problems without dynamic programming? If you think you know how, try solving for $P_D = 100$ MW.

3.14 Given: the two generating units below with piecewise linear cost functions $F(P)$ as shown.

Unit 1: $P_1^{\min} = 25$ MW   and $P_1^{\max} = 200$ MW

P_1(MW)	$F_1(P_1)$(₹h)
25	289.0
100	971.5
150	1436.5
200	1906.5

Unit 2: $P_2^{\min} = 50$ MW   and $P_2^{\max} = 400$ MW

P_2(MW)	$F_2(P_2)$(₹h)
50	3800
100	4230
200	5120
400	6960

Find the optimum generation schedule for a total power delivery of 350 MW (assume both generators are on-line).

3.15   Given: two generator units with piecewise linear incremental cost functions as shown.

Unit 1: $P_1^{\min} = 100$ MW   and $P_1^{\max} = 400$ MW

P_1(MW)	$\dfrac{d}{dP_1} F_1(P_1)$(R/MWh)
100	6.5
200	7.0
300	8.0
400	11.0

Unit 2: $P_2^{\min} = 120$ MW   and $P_2^{\max} = 300$ MW

P_2(MW)	$\dfrac{d}{dP_2} F_2(P_2)$(R/MWh)
120	8.0
150	8.3
200	9.0
300	12.5

a. Find the optimum schedule for a total power delivery of 500 MW.
b. Now assume that there are transmission losses in the system and the incremental losses for the generators are:

$$\frac{dP_{\text{loss}}}{dP_1} = -0.05263$$

and

$$\frac{dP_{\text{loss}}}{dP_2} = 0.04762$$

Find the optimum schedule for a total power delivery of 650 MW; that is, 650 equals the load plus the losses.

FURTHER READING

Since this chapter introduces several optimization concepts, it would be useful to refer to some of the general works on optimization such as references 1 and 2. The importance of the Kuhn–Tucker theorem is given in their paper (reference 3). A very thorough discussion of the Kuhn–Tucker theorem is found in Chapter 1 of reference 4.

For an overview of recent power system optimization practices see references 5 and 6. Several other applications of optimization have been presented. Reference 7 discusses the allocation of regulating margin while dispatching generator units. References 8–11 discuss how to formulate the dispatch problem as one that minimizes air pollution from power plants.

Reference 12 explains how dynamic economic dispatch is developed. Reference 13 is a good review of recent work in economic dispatch. References 14 and 15 show how special problems can be incorporated into economic dispatch, while references 16 and 17 show how altogether different, nonconventional algorithms can be applied to economic dispatch. References 18–21 are an overview of dynamic programming, which is introduced in one of the appendices of this chapter.

1. *Application of Optimization Methods in Power System Engineering*, IEEE Tutorial Course Text 76CH1107-2-PWR, IEEE, New York, 1976.

2. Wilde, P. J., Beightler, C. S., *Foundations of Organization*, Prentice-Hall, Englewood Cliffs, NJ, 1967.

3. Kuhn, H. W., Tucker, A. W., "Nonlinear Programming," in *Second Berkeley Symposium on Mathematical Programming Statistics and Probability*, 1950, University of California Press, Berkeley, 1951.

4. Wismer, D. A., *Optimization Methods for Large-Scale Systems with Applications*, McGraw-Hill, New York, 1971.

5. IEEE Committee Report, "Present Practices in the Economic Operation of Power Systems," *IEEE Transactions on Power Apparatus and Systems*, Vol. PAS-90, July/August 1971, pp. 1768–1775.

6. Najaf-Zadeh, K., Nikolas, J. T., Anderson, S. W., "Optimal Power System Operation Analysis Techniques," *Proceedings of the American Power Conference*, 1977.

7. Stadlin, W. O., "Economic Allocation of Regulating Margin," *IEEE Transactions on Power Apparatus and Systems*, Vol. PAS-90, July/August 1971, pp. 1777–1781.

8. Gent, M. R., Lamont, J. W., "Minimum Emission Dispatch," *IEEE Transactions on Power Apparatus and Systems*, Vol. PAS-90, November/December 1971, pp. 2650–2660.

9. Sullivan, R. L., "Minimum Pollution Dispatching," IEEE Summer Power Meeting, Paper C-72-468, 1972.

10. Friedman, P. G., "Power Dispatch Strategies for Emission and Environmental Control," *Proceedings of the Instrument Society of America*, Vol. 16, 1973.

11. Gjengedal, T., Johansen, S., Hansen, O., "A Qualitative Approach to Economic–Environment Dispatch—Treatment of Multiple Pollutants," *IEEE Transactions on Energy Conversion*, Vol. 7, No. 3, September 1992, pp. 367–373.

12. Ross, D. W., "Dynamic Economic Dispatch," *IEEE Transactions on Power Apparatus and Systems*, November/December 1980, pp. 2060–2068.

13. Chowdhury, B. H., Rahman, S., "A Review of Recent Advances in Economic Dispatch," *IEEE Transactions on Power Systems*, Vol. 5, No. 4, November 1990, pp. 1248–1259.

14. Bobo, D. R., Mauzy, D. M., "Economic Generation Dispatch with Responsive Spinning Reserve Constraints," 1993 *IEEE Power Industry Computer Applications Conference*, 1993, pp. 299–303.

15. Lee, F. N., Breipohl, A. M., "Reserve Constrained Economic Dispatch with

Prohibited Operating Zones," *IEEE Transactions on Power Systems*, Vol. 8, No. 1, February 1993, pp. 246–254.

16. Walters, D. C., Sheble, G. B., "Genetic Algorithm Solution of Economic Dispatch with Valve Point Loading," *IEEE Transactions on Power Systems*, Vol. 8, No. 3, August 1993, pp. 1325–1332.

17. Wong, K. P., Fung, C. C., "Simulated Annealing Bbased Economic Dispatch Algorithm," *IEE Proceedings, Part C: Generation, Transmission and Distribution*, Vol. 140, No. 6, November 1993, pp. 509–515.

18. Kaufmann, A., *Graphs, Dynamic Programming and Finite Games*, Academic Press, New York, 1967.

19. Howard, R. A., *Dynamic Programming and Markov Processes*, Wiley and Technology Press, New York, 1960.

20. Bellman, R. E., Dreyfus, S. E., *Applied Dynamic Programming*, Princeton University Press, Princeton, NJ, 1962.

21. Larson, R. E., Costi, J. L., *Principles of Dynamic Programming*, M. Decker, New York, 1978–1982.

4 Transmission System Effects

As we saw in the previous chapter, the transmission network's incremental power losses may cause a bias in the optimal economic scheduling of the generators. The *coordination equations* include the effects of the incremental transmission losses and complicate the development of the proper schedule. The network elements lead to two other, important effects:

1. The total real power loss in the network increases the total generation demand, and
2. The generation schedule may have to be adjusted by shifting generation to reduce flows on transmission circuits because they would otherwise become overloaded.

It is the last effect that is the most difficult to include in optimum dispatching. In order to include constraints on flows through the network elements, the flows must be evaluated as an integral part of the scheduling effort. This means we must solve the power flow equations along with the generation scheduling equations. (Note that earlier texts, papers, and even the first edition of this book referred to these equations as the "load flow" equations.)

If the constraints on flows in the networks are ignored, then it is feasible to use what are known as *loss formulae* that relate the total and incremental, real power losses in the network to the power generation magnitudes. Development of loss formulae is an art that requires knowledge of the power flows in the network under numerous "typical" conditions. Thus, there is no escaping the need to understand the methods involved in formulating and solving the power flow equations for an AC transmission system.

When the complete transmission system model is included in the development of generation schedules, the process is usually imbedded in a set of computer algorithms known as the *optimal power flow* (or OPF). The complete OPF is capable of establishing schedules for many controllable quantities in the bulk power system (i.e., the generation and transmission systems), such as transformer tap positions, VAR generation schedules, etc. We shall defer a detailed examination of the OPF until Chapter 13.

Another useful set of data that are obtainable when the transmission network is incorporated in the scheduling process is the incremental cost of power at various points in the network. With no transmission effects considered (that is, ignoring all incremental losses and any constraints on power flows), the network is assumed to be a single node and the incremental cost of power is equal to λ everywhere. That is,

$$\frac{dF_i}{dP_i} = \lambda$$

Including the effect of incremental losses will cause the incremental cost of real power to vary throughout the network. Consider the arrangement in Figure 3.2 and assume that the coordination equations have been solved so the values of dF_i/dP_i and λ are known. Let the "penalty factor" of bus i be defined as

$$Pf_i = \frac{1}{\left(1 - \dfrac{\partial P_{\text{loss}}}{\partial P_i}\right)}$$

so that the relationship between the incremental costs at any two buses, i and j, is

$$Pf_i F_i' = Pf_j F_j'$$

where $F_k' = dF_k/dP_k$ is the bus incremental cost. There is no requirement that bus i is a generator bus. If the network effects are included using a network model or a loss formula, bus i might be a load bus or a point where power is delivered to an interconnected system. The incremental cost (or "value") of power at bus i is then,

$$\text{Incremental cost at } i = F_i' = (Pf_j/Pf_i)F_j'$$

where j is any real generator bus where the incremental cost of production is known. So if we can develop a network model to be used in optimum generation scheduling that includes all of the buses, or at least those that are of importance, and if the incremental losses $(\partial P_L/\partial P_k)$ can be evaluated, the coordination equations can be used to compute the incremental cost of power at any point of delivery.

When the schedule is determined using a complete power flow model by using an OPF, the flow constraints can be included and they may affect the value of the incremental cost of power in parts of the network. Rather than attempt a mathematical demonstration, consider a system in which most of the low cost generation is in the north, most of the load is in the south along with some higher cost generating units, and the northern and southern areas are interconnected by a relatively low capacity transmission network. The network north-to-south transfer capability limits the power that can be delivered from the northern area to satisfy the higher load demands. Under a schedule that is constrained by this transmission flow limitation, the southern area's generation would need to be increased above an unconstrained, optimal level in order to satisfy some of the load in that region. The constrained economic schedule would split the system into two regions with a higher incremental cost in the southern area. In most actual cases where transmission does constrain the economic schedule, the effect of the constraints is much more significant than the effects of incremental transmission losses.

This chapter develops the power flow equations and outlines methods of solution. Operations control centers frequently use a version of the power flow equations known as the "decoupled power flow." The power flow equations form the basis for the development of loss formulae. Scheduling methods frequently use penalty factors to incorporate the effect of incremental real power losses in dispatch. These can be developed from the loss formulae or directly from the power flow relationships.

Power flow is the name given to a network solution that shows currents, voltages, and real and reactive power flows at every bus in the system. It is normally assumed that the system is balanced and the common use of the term power flow implies a positive sequence solution only. Full three-phase power-flow solution techniques are available for special-purpose calculations. As used here, we are only interested in balanced solutions. Power flow is not a single calculation such as $E = IR$ or $\mathbf{E} = [Z]\mathbf{I}$ involving linear circuit analysis. Such circuit analysis problems start with a given set of currents or voltages, and one must solve for the linearly dependent unknowns. In the power-flow problem we are given a nonlinear relationship between voltage and current at each bus and we must solve for all voltages and currents such that these nonlinear relationships are met. The nonlinear relationships involve, for example, the real and reactive power consumption at a bus, or the generated real power and scheduled voltage magnitude at a generator bus. As such, the power flow gives us the electrical response of the transmission system to a particular set of loads and generator unit outputs. Power flows are an important part of power system design procedures (system planning). Modern digital computer power-flow programs are routinely run for systems with up to 5000 or more buses and also are used widely in power system control centers to study unique operating problems and to provide accurate calculations of bus penalty factors. Present, state-or-the-art system control centers use the power flow as a key, central element in the scheduling of generation, monitoring of the system, and development of interchange transactions. OPF programs are used to develop optimal economic schedules and control settings that will result in flows that are within the capabilities of the elements of the system, including the transmission network, and bus voltage magnitudes that are within acceptable tolerances.

4.1   THE POWER FLOW PROBLEM AND ITS SOLUTION

The power flow problem consists of a given transmission network where all lines are represented by a Pi-equivalent circuit and transformers by an ideal voltage transformer in series with an impedance. Generators and loads represent the boundary conditions of the solution. Generator or load real and reactive power involves products of voltage and current. Mathematically, the power flow requires a solution of a system of simultaneous nonlinear equations.

4.1.1   The Power Flow Problem on a Direct Current Network

The problems involved in solving a power flow can be illustrated by the use of direct current (DC) circuit examples. The circuit shown in Figure 4.1 has a resistance of $0.25\ \Omega$ tied to a constant voltage of 1.0 V (called the *reference voltage*). We wish to find the voltage at bus 2 that results in a net inflow of 1.2 W. *Buses* are electrical nodes. Power is said to be "injected" into a network; therefore, loads are simply negative injections.

The current from bus 2 to bus 1 is

$$I_{21} = (E_2 - 1.0) \times 4 \tag{4.1}$$

Power P_2 is

$$P_2 = 1.2 = E_2 I_{21} = E_2(E_2 - 1) \times 4 \tag{4.2}$$

or

$$4E_2^2 - 4E_2 - 1.2 = 0 \tag{4.3}$$

The solutions to this quadratic equation are $E_2 = 1.24162$ V and $E_2 = -0.24162$ V. Note that 1.2 W enter bus 2, producing a current of 0.96648 A ($E_2 = 1.24162$), which means that 0.96648 W enter the reference bus and 0.23352 W are consumed in the 0.25-Ω resistor.

Let us complicate the problem by adding a third bus and two more lines (see Figure 4.2). The problem is more complicated because we cannot simply write out the solutions using a quadratic formula. The admittance equations are

$$\begin{bmatrix} I_1 \\ I_2 \\ I_3 \end{bmatrix} = \begin{bmatrix} 14 & -4 & -10 \\ -4 & 9 & -5 \\ -10 & -5 & 15 \end{bmatrix} \begin{bmatrix} E_1 \\ E_2 \\ E_3 \end{bmatrix} \tag{4.4}$$

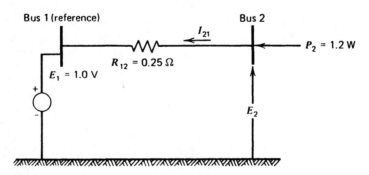

FIG. 4.1   Two-bus DC network.

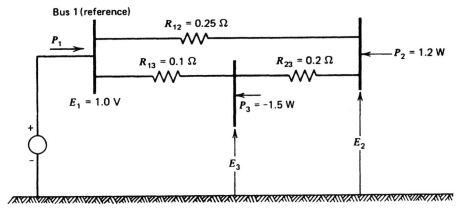

FIG. 4.2   Three-bus DC network.

In this case, we know the power injected at buses 2 and 3 and we know the voltage at bus 1. To solve for the unknowns $(E_2, E_3$ and $P_1)$, we write Eqs. 4.5, 4.6, and 4.7. The solution procedure is known as the *Gauss–Seidel procedure*, wherein a calculation for a new voltage at each bus is made, based on the most recently calculated voltages at all neighbouring buses.

Bus 2:
$$I_2 = \frac{P_2}{E_2} = -4(1.0) + 9E_2 - 5E_3$$

$$E_2^{new} = \frac{1}{9}\left(\frac{1.2}{E_2^{old}} + 4 + 5E_3^{old}\right) \tag{4.5}$$

where E_2^{old} and E_3^{old} are the initial values for E_2 and E_3, respectively.

Bus 3:
$$I_3 = \frac{P_3}{E_3} = -10(1.0) - 5E_2^{new} + 15E_3$$

$$E_3^{new} = \frac{1}{15}\left[\frac{-1.5}{E_3^{old}} + 10 + 5E_2^{new}\right] \tag{4.6}$$

where E_2^{new} is the voltage found in solving Eq. 4.5, and E_3^{old} is the initial value of E_3.

Bus 1:
$$P_1 = E_1 I_1^{new} = 1.0 I_1^{new} = 14 - 4E_2^{new} - 10E_3^{new} \tag{4.7}$$

The Gauss–Seidel method first assumes a set of voltages at buses 2 and 3 and then uses Eqs. 4.5 and 4.6 to solve for new voltages. The new voltages are compared to the voltage's most recent values, and the process continues until

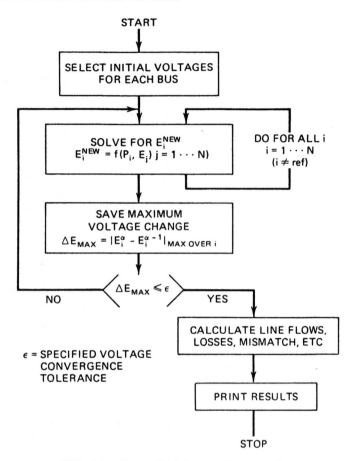

FIG. 4.3 Gauss–Seidel power-flow solution.

the change in voltage is very small. This is illustrated in the flowchart in Figure 4.3 and in Eqs. 4.8 and 4.9.

First iteration:

$$E_2^{(0)} = E_3^{(0)} = 1.0$$

$$E_2^{(1)} = \frac{1}{9}\left(\frac{1.2}{1.0} + 4 + 5\right) = 1.133$$

$$E_3^{(1)} = \frac{1}{15}\left[\frac{-1.5}{1.0} + 10 + 5(1.133)\right] = 0.944$$

(4.8)

$$\Delta E_{\max} = 0.133 \text{ too large}$$

Note: In calculating $E_3^{(1)}$ we used the new value of E_2 found in the first correction.

Second iteration: $E_2^{(2)} = \dfrac{1}{9}\left[\dfrac{1.2}{1.133} + 4 + 5(0.944)\right] = 1.087$

$$E_3^{(2)} = \dfrac{1}{15}\left[\dfrac{-1.5}{0.944} + 10 + 5(1.087)\right] = 0.923 \qquad (4.9)$$

$$\Delta E_{max} = 0.046$$

And so forth until $\Delta E_{max} < \varepsilon$.

4.1.2   The Formulation of the AC Power Flow

AC power flows involve several types of bus specifications, as shown in Figure 4.4. Note that $[P, \theta]$, $[Q, |E|]$, and $[Q, \theta]$ combinations are generally not used.

The transmission network consists of complex impedances between buses and from the buses to ground. An example is given in Figure 4.5. The equations are written in matrix form as

$$
\begin{bmatrix} I_1 \\ I_2 \\ I_3 \\ I_4 \end{bmatrix} =
\begin{bmatrix}
y_{12} & -y_{12} & 0 & 0 \\
-y_{12} & (y_{12} + y_{2g} + y_{23}) & -y_{23} & 0 \\
0 & -y_{23} & (y_{23} + y_{3g} + y_{34}) & -y_{34} \\
0 & 0 & -y_{34} & (y_{34} + y_{4g})
\end{bmatrix}
\begin{bmatrix} E_1 \\ E_2 \\ E_3 \\ E_4 \end{bmatrix}
$$

$$(4.10)$$

(All I's, E's, y's complex)

This matrix is called the *network Y matrix*, which is written as

$$
\begin{bmatrix} I_1 \\ I_2 \\ I_3 \\ I_4 \end{bmatrix} =
\begin{bmatrix}
Y_{11} & Y_{12} & Y_{13} & Y_{14} \\
Y_{21} & Y_{22} & Y_{23} & Y_{24} \\
Y_{31} & Y_{32} & Y_{33} & Y_{34} \\
Y_{41} & Y_{42} & Y_{43} & Y_{44}
\end{bmatrix}
\begin{bmatrix} E_1 \\ E_2 \\ E_3 \\ E_4 \end{bmatrix}
\qquad (4.11)
$$

The rules for forming a Y matrix are

> If a line exists from i to j
>
> $$Y_{ij} = -y_{ij}$$
>
> and
>
> $$Y_{ii} = \sum_j y_{ij} + y_{ig}$$
>
> j over all lines connected to i.

Bus Type	P	Q	$\lvert E \rvert$	θ	Comments
Load	✓	✓			Usual load representation
Voltage Controlled	✓		✓		Assume $\lvert E \rvert$ is held constant no matter what Q is
Generator or Synchronous Condenser	✓		✓ when $Q^- < Q_g < Q^+$		Generator or synchronous condenser ($P = 0$) has VAR limits
	✓	✓ when $Q_g < Q^-$ $Q_g > Q^+$			Q^- minimum VAR limit Q^+ maximum VAR limit $\lvert E \rvert$ is held as long as Q_g is within limit
Fixed Z to Ground					Only Z is given
Reference			✓	✓	"Swing bus" must adjust net power to hold voltage constant (essential for solution)

FIG. 4.4 Power-flow bus specifications (quantities checked are the bus boundary conditions).

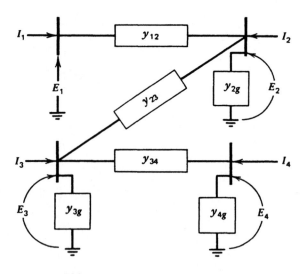

FIG. 4.5 Four-bus AC network.

The equation of net power injection at a bus is usually written as

$$\frac{P_k - jQ_k}{E_k^*} = \sum_{\substack{j=1 \\ j \neq k}}^{n} Y_{kj} E_j + Y_{kk} E_k \tag{4.12}$$

4.1.2.1 The Gauss–Seidel Method

The voltages at each bus can be solved for by using the Gauss–Seidel method. The equation in this case is

$$E_k^{(\alpha)} = \frac{1}{Y_{kk}} \frac{(P_k - jQ_k)}{E_k^{(\alpha-1)*}} - \frac{1}{Y_{kk}} \left[\sum_{j<k} Y_{kj} E_j^{(\alpha)} + \sum_{j>k} Y_{kj} E_j^{(\alpha-1)} \right] \tag{4.13}$$

Voltage at
iteration α

The Gauss–Seidel method was the first AC power-flow method to be developed for solution on digital computers. This method is characteristically long in solving due to its slow convergence and often difficulty is experienced with unusual network conditions such as negative reactance branches. The solution procedure is the same as shown in Figure 4.3.

4.1.2.2 The Newton–Raphson Method

One of the disadvantages of the Gauss–Seidel method lies in the fact that each bus is treated independently. Each correction to one bus requires subsequent correction to all the buses to which it is connected. The Newton–Raphson method is based on the idea of calculating the corrections while taking account of all the interactions.

Newton's method involves the idea of an error in a function $f(x)$ being driven to zero by making adjustments Δx to the independent variable associated with the function. Suppose we wish to solve

$$f(x) = K \tag{4.14}$$

In Newton's method, we pick a starting value of x and call it x^0. The error is the difference between K and $f(x^0)$. Call the error ε. This is shown in Figure 4.6 and given in Eq. 4.15.

$$f(x^0) + \varepsilon = K \tag{4.15}$$

To drive the error to zero, we use a Taylor expansion of the function about x^0.

$$f(x^0) + \frac{df(x^0)}{dx} \Delta x + \varepsilon = K \tag{4.16}$$

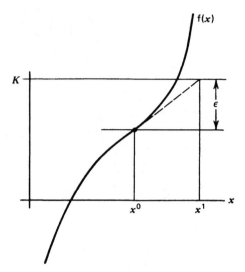

FIG. 4.6 Newton's method.

Setting the error to zero, we calculate

$$\Delta x = \left(\frac{\mathrm{df}(x^0)}{\mathrm{d}x}\right)^{-1} [K - \mathrm{f}(x^0)] \qquad (4.17)$$

When we wish to solve a load flow, we extend Newton's method to the multivariable case (the multivariable case is called the *Newton–Raphson method*). An equation is written for each bus "*i*."

$$P_i + jQ_i = E_i I_i^* \qquad (4.18)$$

where

$$I_i = \sum_{k=1}^{N} Y_{ik} E_k$$

then

$$P_i + jQ_i = E_i \left(\sum_{k=1}^{N} Y_{ik} E_k\right)^*$$

$$= |E_i|^2 Y_{ii}^* + \sum_{\substack{k=1 \\ k \neq i}}^{N} Y_{ik}^* E_i E_k^*$$

As in the Gauss–Seidel method, a set of starting voltages is used to get things going. The $P + jQ$ calculated is subtracted from the scheduled $P + jQ$ at the bus, and the resulting errors are stored in a vector. As shown in the following, we will assume that the voltages are in polar coordinates and that we are going to adjust each voltage's magnitude and phase angle as separate independent variables. Note that at this point, two equations are written for each bus: one for real power and one for reactive power. For each bus,

$$\Delta P_i = \sum_{k=1}^{N} \frac{\partial P_i}{\partial \theta_k} \Delta \theta_k + \sum_{k=1}^{N} \frac{\partial P_i}{\partial |E_k|} \Delta |E_k|$$

$$\Delta Q_i = \sum_{k=1}^{N} \frac{\partial Q_i}{\partial \theta_k} \Delta \theta_k + \sum_{k=1}^{N} \frac{\partial Q_i}{\partial |E_k|} \Delta |E_k|$$

(4.19)

All the terms are arranged in a matrix (the Jacobian matrix) as follows.

$$\begin{bmatrix} \Delta P_1 \\ \\ \Delta Q_1 \\ \\ \Delta P_2 \\ \\ \Delta Q_2 \\ \vdots \end{bmatrix} = \underbrace{\begin{bmatrix} \dfrac{\partial P_1}{\partial \theta_1} & \dfrac{\partial P_1}{\partial |E_1|} & \cdots \\ \\ \dfrac{\partial Q_1}{\partial \theta_1} & \dfrac{\partial Q_1}{\partial |E_1|} & \cdots \\ \\ \vdots & \vdots & \vdots \end{bmatrix}}_{\text{Jacobian matrix}} \begin{bmatrix} \Delta \theta_1 \\ \\ \Delta |E_1| \\ \\ \vdots \end{bmatrix}$$

(4.20)

The Jacobian matrix in Eq. 4.20 starts with the equation for the real and reactive power at each bus. This equation, Eq. 4.18, is repeated below:

$$P_i + jQ_i = E_i \sum_{k=1}^{N} Y_{ik}^* E_k^*$$

This can be expanded as:

$$P_i + jQ_i = \sum_{k=1}^{N} |E_i||E_k|(G_{ik} - jB_{ik})\varepsilon^{j(\theta_i - \theta_k)}$$

$$= \sum_{k=1}^{N} \{|E_i||E_k|[G_{ik}\cos(\theta_i - \theta_k) + B_{ik}\sin(\theta_i - \theta_k)]$$

$$+ j[|E_i||E_k|[G_{ik}\sin(\theta_i - \theta_k) - B_{ik}\cos(\theta_i - \theta_k)]]\}$$

(4.21)

where

θ_i, θ_k = the phase angles at buses i and k, respectively;

$|E_i|, |E_k|$ = the bus voltage magnitudes, respectively

$G_{ik} + jB_{ik} = Y_{ik}$ is the ik term in the Y matrix of the power system.

The general practice in solving power flows by Newton's method has been to use

$$\frac{\Delta|E_i|}{|E_i|}$$

instead of simply $\Delta|E_i|$; this simplifies the equations. The derivatives are:

$$\frac{\partial P_i}{\partial \theta_k} = |E_i||E_k|[G_{ik} \sin (\theta_i - \theta_k) - B_{ik} \cos (\theta_i - \theta_k)]$$

$$\frac{\partial P_i}{\left(\dfrac{\partial|E_k|}{|E_k|}\right)} = |E_i||E_k|[G_{ik} \cos (\theta_i - \theta_k) + B_{ik} \sin (\theta_i - \theta_k)]$$

$$\frac{\partial Q_i}{\partial \theta_k} = -|E_i||E_k|[G_{ik} \cos (\theta_i - \theta_k) + B_{ik} \sin (\theta_i - \theta_k)]$$

$$\frac{\partial Q_i}{\left(\dfrac{\partial|E_i|}{|E_i|}\right)} = |E_i||E_k|[G_{ik} \sin (\theta_i - \theta_k) - B_{ik} \cos (\theta_i - \theta_k)]$$

(4.22)

For $i = k$:

$$\frac{\partial P_i}{\partial \theta_i} = -Q_i - B_{ii}E_i^2$$

$$\frac{\partial P_i}{\left(\dfrac{\partial|E_i|}{|E_i|}\right)} = P_i + G_{ii}E_i^2$$

$$\frac{\partial Q_i}{\partial \theta_i} = P_i - G_{ii}E_i^2$$

$$\frac{\partial Q_i}{\left(\dfrac{\partial|E_i|}{|E_i|}\right)} = Q_i - B_{ii}E_i^2$$

Equation 4.20 now becomes

$$\begin{bmatrix} \Delta P_1 \\ \Delta Q_1 \\ \Delta P_2 \\ \Delta Q_2 \\ \vdots \end{bmatrix} = [J] \begin{bmatrix} \Delta\theta_1 \\ \dfrac{\Delta|E_1|}{|E_1|} \\ \Delta\theta_2 \\ \dfrac{\Delta|E_2|}{|E_2|} \\ \vdots \end{bmatrix}$$

(4.23)

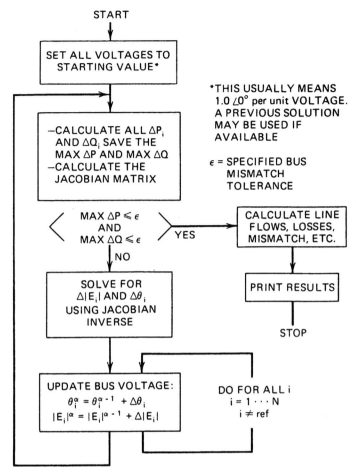

FIG. 4.7 Newton–Raphson power-flow solution.

The solution to the Newton–Raphson power flow runs according to the flowchart in Figure 4.7. Note that solving for $\Delta\theta$ and $\Delta|E|$ requires the solution of a set of linear equations whose coefficients make up the Jacobian matrix. The Jacobian matrix generally has only a few percent of its entries that are nonzero. Programs that solve an AC power flow using the Newton–Raphson method are successful because they take advantage of the Jacobian's "sparsity." The solution procedure uses Gaussian elimination on the Jacobian matrix and does not calculate J^{-1} explicitly. (See reference 3 for introduction to "sparsity" techniques.)

EXAMPLE 4A

The six-bus network shown in Figure 4.8 will be used to demonstrate several aspects of load flows and transmission loss factors. The voltages and flows

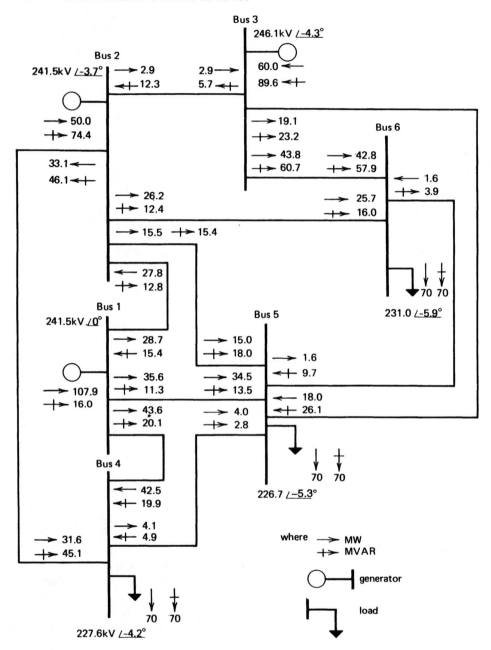

FIG. 4.8 Six-bus network base case AC power flow.

shown are for the "base case" of 210 MW total load. The impedance values and other data for this system may be found in the appendix of this chapter.

4.1.3 The Decoupled Power Flow

The Newton power flow is the most robust power flow algorithm used in practice. However, one drawback to its use is the fact that the terms in the Jacobian matrix must be recalculated each iteration, and then the entire set of linear equations in Eq. 4.23 must also be resolved each iteration.

Since thousands of complete power flows are often run for a planning or operations study, ways to speed up this process were sought. Reference 11 shows the development of a technique known as the "fast decoupled power flow" (it is often referred to as the "Stott decoupled power flow," in reference to its first author).

Starting with the terms in the Jacobian matrix (see Eq. 4.22), the following simplications are made:

- Neglect and interaction between P_i and any $|E_k|$ (it was observed by power system engineers that real power was little influenced by changes in voltage magnitude—so this effect was incorporated in the algorithm). Then, all the derivatives

$$\frac{\partial P_i}{\left(\dfrac{\partial |E_k|}{|E_k|}\right)}$$

will be considered to be zero.

- Neglect any interaction between Q_i and θ_k (see the note above—a similar observation was made on the insensitivity of reactive power to changes in phase angle). Then, all the derivatives

$$\frac{\partial Q_i}{\partial \theta_k}$$

are also considered to be zero.

- Let $\cos(\theta_i - \theta_j) \cong 1$ which is a good approximation since $(\theta_i - \theta_j)$ is usually small.
- Assume that

$$G_{ik} \sin(\theta_i - \theta_k) \ll B_{ik}$$

- Assume that

$$Q_i \ll B_{ii}|E_i|^2$$

This leaves the derivatives as:

$$\frac{\partial P_i}{\partial \theta_k} = -|E_i||E_k|B_{ik} \tag{4.24}$$

$$\frac{\partial Q_i}{\left(\frac{\partial |E_k|}{|E_k|}\right)} = -|E_i||E_k|B_{ik} \tag{4.25}$$

If we now write the power flow adjustment equations as:

$$\Delta P_i = \left(\frac{\partial P_i}{\partial \theta_k}\right)\Delta\theta_k \tag{4.26}$$

$$\Delta Q_i = \left[\frac{\partial Q_i}{\left(\frac{\partial |E_k|}{|E_k|}\right)}\right]\frac{\Delta |E_k|}{|E_k|} \tag{4.27}$$

then, substituting Eq. 4.24 into Eq. 4.26, and Eq. 4.25 into Eq. 4.27, we obtain:

$$\Delta P_i = -|E_i||E_k|B_{ik}\,\Delta\theta_k \tag{4.28}$$

$$\Delta Q_i = -|E_i||E_k|B_{ik}\,\frac{\Delta |E_k|}{|E_k|} \tag{4.29}$$

Further simplification can then be made:

- Divide Eqs. 4.28 and 4.29 by $|E_i|$.
- Assume $|E_k| \cong 1$ in Eq. 4.28.

which results in:

$$\frac{\Delta P_i}{|E_i|} = -B_{ik}\Delta\theta_k \tag{4.30}$$

$$\frac{\Delta Q_i}{|E_i|} = -B_{ik}\Delta |E_k| \tag{4.31}$$

We now build Eqs. 4.30 and 4.31 into two matrix equations:

$$\begin{bmatrix} \dfrac{\Delta P_1}{|E_1|} \\[2ex] \dfrac{\Delta P_2}{|E_2|} \\[2ex] \vdots \end{bmatrix} = \begin{bmatrix} -B_{11} & -B_{12} & \cdots \\ -B_{21} & -B_{22} & \cdots \\ \vdots & & \end{bmatrix}\begin{bmatrix} \Delta\theta_1 \\ \Delta\theta_2 \\ \vdots \end{bmatrix} \tag{4.32}$$

$$
\begin{bmatrix}
\dfrac{\Delta Q_1}{|E_1|} \\[2ex]
\dfrac{\Delta Q_2}{|E_2|} \\[2ex]
\vdots
\end{bmatrix}
=
\begin{bmatrix}
-B_{11} & -B_{12} & \cdots \\
-B_{21} & -B_{22} & \cdots \\
& \vdots &
\end{bmatrix}
\begin{bmatrix}
\Delta|E_1| \\[1ex]
\Delta|E_2| \\[1ex]
\vdots
\end{bmatrix}
\tag{4.33}
$$

Note that both Eqs. 4.32 and 4.33 use the same matrix. Further simplification, however, will make them different.

Simplifying the $\Delta P - \Delta\theta$ relationship of Eq. 4.32:

● Assume $r_{ik} \ll x_{ik}$; this changes $-B_{ik}$ to $-1/x_{ik}$.
● Eliminate all shunt reactances to ground.
● Eliminate all shunts to ground which arise from autotransformers.

Simplifying the $\Delta Q - \Delta|E|$ relationship of Eq. 4.33:

● Omit all effects from phase shift transformers.

The resulting equations are:

$$
\begin{bmatrix}
\dfrac{\Delta P_1}{|E_1|} \\[2ex]
\dfrac{\Delta P_2}{|E_2|} \\[2ex]
\vdots
\end{bmatrix}
= [B']
\begin{bmatrix}
\Delta\theta_1 \\[1ex]
\Delta\theta_2 \\[1ex]
\vdots
\end{bmatrix}
\tag{4.34}
$$

$$
\begin{bmatrix}
\dfrac{\Delta Q_1}{|E_1|} \\[2ex]
\dfrac{\Delta Q_2}{|E_2|} \\[2ex]
\vdots
\end{bmatrix}
= [B'']
\begin{bmatrix}
\Delta|E_1| \\[1ex]
\Delta|E_2| \\[1ex]
\vdots
\end{bmatrix}
\tag{4.35}
$$

where the terms in the matrices are:

$$
B'_{ik} = -\frac{1}{x_{ik}}, \text{ assuming a branch from } i \text{ to } k \text{ (zero otherwise)}
$$

$$
B'_{ii} = \sum_{k=1}^{N} \frac{1}{x_{ik}}
$$

$$B''_{ik} = -B_{ik} = -\frac{x_{ik}}{r_{ik}^2 + x_{ik}^2}$$

$$B''_{ii} = \sum_{k=1}^{N} -B_{ik}$$

The decoupled power flow has several advantages and disadvantages over the Newton power flow. (Note: Since the introduction and widespread use of the decoupled power flow, the Newton power flow is often referred to as the "full Newton" power flow.)

Advantages:

- B' and B'' are constant; therefore, they can be calculated once and, except for changes to B'' resulting from generation VAR limiting, they are not updated.
- Since B' and B'' are each about one-quarter of the number of terms in $[J]$ (the full Newton power flow Jacobian matrix), there is much less arithmetic to solve Eqs. 4.34 and 4.35.

Disadvantages:

- The decoupled power flow algorithm may fail to converge when some of the underlying assumptions (such as $r_{ik} \ll x_{ik}$) do not hold. In such cases, one must switch to using the full Newton power flow.

Note that Eq. 4.34 is often referred to as the $P-\theta$ Eq. and Eq. 4.35 as the $Q-E$ (or $Q-V$) equation.

A flowchart of the algorithm is shown in Figure 4.9. A comparison of the convergence of the Gauss–Seidel, the full Newton and the decoupled power flow algorithms is shown in Figure 4.10.

4.1.4   The "DC" Power Flow

A further simplification of the power flow algorithm involves simply dropping the $Q-V$ equation (Eq. 4.35) altogether. This results in a completely linear, noniterative, power flow algorithm. To carry this out, we simply assume that all $|E_i| = 1.0$ per unit. Then Eq. 4.34 becomes:

$$\begin{bmatrix} \Delta P_1 \\ \Delta P_2 \\ \vdots \end{bmatrix} = [B'] \begin{bmatrix} \Delta\theta_1 \\ \Delta\theta_2 \\ \vdots \end{bmatrix} \tag{4.36}$$

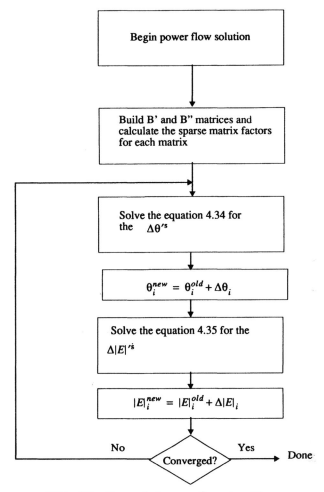

FIG. 4.9 Decoupled power flow algorithm.

where the terms in B' are as described previously. The DC power flow is only good for calculating MW flows on transmission lines and transformers. It gives no indication of what happens to voltage magnitudes, or MVAR or MVA flows. The power flowing on each line using the DC power flow is then:

$$P_{ik} = \frac{1}{x_{ik}} (\theta_i - \theta_k) \tag{4.37}$$

and

$$P_i = \sum_{\substack{k = \text{buses} \\ \text{connected to } i}}^{N} P_k$$

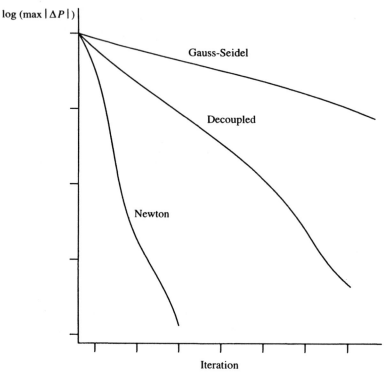

FIG. 4.10   Comparison of three power flow algorithm convergence characteristics.

EXAMPLE 4B

The megawatt flows on the network in Figure 4.11 will be solved using the DC power flow. The B' matrix equation is:

$$\begin{bmatrix} 7.5 & -5.0 \\ -5.0 & 9.0 \end{bmatrix}\begin{bmatrix} \theta_1 \\ \theta_2 \end{bmatrix} = \begin{bmatrix} P_1 \\ P_2 \end{bmatrix}$$

$$\theta_3 = 0$$

Note that all megawatt quantities and network quantities are expressed in pu (per unit on 100 MVA base). All phase angles will then be in radians.

The solution to the preceding matrix equation is:

$$\begin{bmatrix} \theta_1 \\ \theta_2 \end{bmatrix} = \begin{bmatrix} 0.2118 & 0.1177 \\ 0.1177 & 0.1765 \end{bmatrix}\begin{bmatrix} 0.65 \\ -1.00 \end{bmatrix} = \begin{bmatrix} 0.02 \\ -0.1 \end{bmatrix}$$

The resulting flows are shown in Figure 4.12 and calculated using Eq. 4.37. Note that all flows in Figure 4.12 were converted to actual megawatt values.

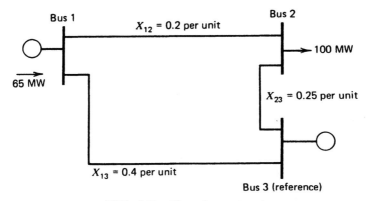

FIG. 4.11   Three-bus network.

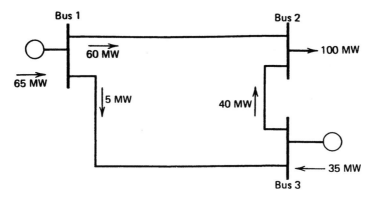

FIG. 4.12   Three-bus network showing flows calculated by DC power flow.

EXAMPLE 4C

The network of Example 4A was solved using the DC power flow with resulting power flows as shown in Figure 4.13. The DC power flow is useful for rapid calculations of real power flows, and, as will be shown later, it is very useful in security analysis studies.

4.2   TRANSMISSION LOSSES

4.2.1   A Two-Generator System

We are given the power system in Figure 4.14. The losses on the transmission line are proportional to the square of the power flow. The generating units are identical, and the production cost is modeled using a quadratic equation. If both units were loaded to 250 MW, we would fall short of the 500 MW load value by 12.5 MW lost on the transmission line, as shown in Figure 4.15.

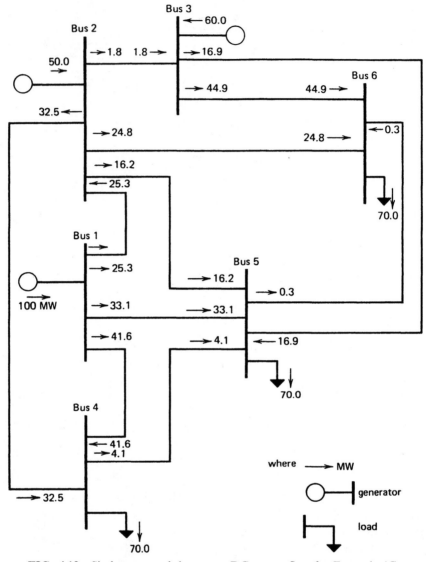

FIG. 4.13 Six-bus network base case DC power flow for Example 4C.

Where should the extra 12.5 MW be generated? Solve the Lagrange equation that was given in Chapter 3.

$$\mathcal{L} = F_1(P_1) + F_2(P_2) + \lambda(500 + P_{loss} - P_1 - P_2) \qquad (4.38)$$

where

$$P_{loss} = 0.0002P_1^2$$

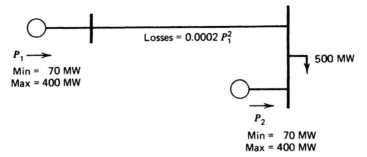

FIG. 4.14 Two-generator system.

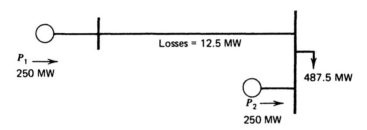

FIG. 4.15 Two-generator system with both generators at 250 MW output.

then

$$\frac{\partial \mathscr{L}}{\partial P_1} = \frac{dF_1(P_1)}{dP_1} - \lambda \left(1 - \frac{\partial P_{\text{loss}}}{\partial P_1} \right) = 0$$

$$\frac{\partial \mathscr{L}}{\partial P_2} = \frac{dF_2(P_2)}{dP_2} - \lambda \left(1 - \frac{\partial P_{\text{loss}}}{\partial P_2} \right) = 0 \qquad (4.39)$$

$$P_1 + P_2 - 500 - P_{\text{loss}} = 0$$

Substituting into Eq. 4.39,

$$7.0 + 0.004 P_1 - \lambda(1 - 0.0004 P_1) = 0$$

$$7.0 + 0.004 P_2 - \lambda = 0$$

$$P_1 + P_2 - 500 - 0.0002 P_1^2 = 0$$

Solution:
$$P_1 = 178.882$$

$$P_2 = 327.496$$

Production cost:
$$F_1(P_1) + F_2(P_2) = 4623.15 \text{R/h}$$

Losses:
$$6.378 \text{ MW}$$

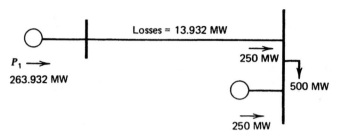

FIG. 4.16   Two-generator system with generator 1 supplying all losses.

Suppose we had decided simply to ignore the economic influence of losses and ran unit 1 up until it supplied all the losses. It would need to be run at 263.932 MW, as shown in Figure 4.16. In this case, the total production cost would be

$$F_1(263.932) + F_2(250) = 4661.84 \, ₹/h$$

Note that the optimum dispatch tends toward supplying the losses from the unit close to the load, and it also resulted in a lower value of losses. Also note that best economics are not necessarily attained at minimum losses. The minimum loss solution for this case would simply run unit 1 down and unit 2 up as far as possible. The result is unit 2 on high limit.

$$P_1 = 102.084 \, \text{MW}$$
$$P_2 = 400.00 \, \text{MW (high limit)}$$

The minimum loss production cost would be

$$F_1(102.084) + F_2(400) = 4655.43 \, ₹/h$$
$$\text{Min losses} = 2.084 \, \text{MW}$$

4.2.2   Coordination Equations, Incremental Losses, and Penalty Factors

The classic Lagrange multiplier solution to the economic dispatch problem was given in Chapter 3. This is repeated here and expanded.

Minimize: $\mathscr{L} = F_T + \lambda\phi$

Where: $$F_T = \sum_{i=1}^{N} F_i(P_i)$$

$$\phi = P_{\text{load}} + P_{\text{loss}}(P_1, P_2 \ldots P_N) - \sum_{i=1}^{N} P_i$$

Solution: $\dfrac{\partial\mathscr{L}}{\partial P_i} = 0$ for all $P_{i\min} \leq P_i \leq P_{i\max}$

Then

$$\frac{\partial \mathscr{L}}{\partial P_1} = \frac{dF_i}{dP_i} - \lambda\left(1 - \frac{\partial P_{\text{loss}}}{\partial P_i}\right) = 0$$

The equations are rearranged

$$\left(\frac{1}{1 - \dfrac{\partial P_{\text{loss}}}{\partial P_i}}\right)\frac{dF_i(P_i)}{dP_i} = \lambda \qquad (4.40)$$

where

$$\frac{\partial P_{\text{loss}}}{\partial P_i}$$

is called the *incremental loss* for bus i, and

$$Pf_i = \left(\frac{1}{1 - \dfrac{\partial P_{\text{loss}}}{\partial P_i}}\right)$$

is called the *penalty factor* for bus i. Note that if the losses increase for an increase in power from bus i, the incremental loss is positive and the penalty factor is greater than unity.

When we did not take account of transmission losses, the economic dispatch problem was solved by making the incremental cost at each unit the same. We can still use this concept by observing that the penalty factor, Pf_i, will have the following effect. For $Pf_i > 1$ (positive increase in P_i results in increase in losses)

$$Pf_i \frac{dF_i(P_i)}{dP_i}$$

acts as if

$$\frac{dF_i(P_i)}{dP_i}$$

had been slightly increased (moved up). For $Pf_i < 1$ (positive increase in P_i results in decrease in losses)

$$Pf_i \frac{dF_i(P_i)}{dP_i}$$

acts as if

$$\frac{dF_i(P_i)}{dP_i}$$

had been slightly decreased (moved down). The resulting set of equations look like

$$Pf_i \frac{dF_i(P_i)}{dP_i} = \lambda \qquad \text{for all } P_{i\min} \leq P_i \leq P_{i\max} \qquad (4.41)$$

and are called *coordination equations*. The P_i values that result when penalty factors are used will be somewhat different from the dispatch which ignores the losses (depending on the Pf_i and $dF_i(P_i)/dP_i$ values). This is illustrated in Figure 4.17.

4.2.3   The B Matrix Loss Formula

The B matrix loss formula was originally introduced in the early 1950s as a practical method for loss and incremental loss calculations. At the time, automatic dispatching was performed by analog computers and the loss formula was "stored" in the analog computers by setting precision potentiometers. The equation for the B matrix loss formula is as follows.

$$P_{\text{loss}} = \mathbf{P}^T[B]\mathbf{P} + B_0^T P + B_{00} \qquad (4.42)$$

where

$$\mathbf{P} = \text{vector of all generator bus net MW}$$

$$[B] = \text{square matrix of the same dimension as } \mathbf{P}$$

$$B_0 = \text{vector of the same length as } \mathbf{P}$$

$$B_{00} = \text{constant}$$

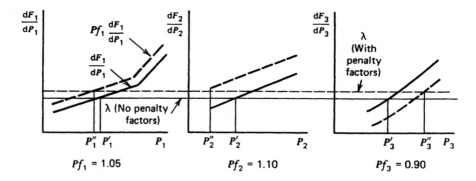

P_i' = Dispatch ignoring losses
P_i'' = Dispatch with penalty factors

FIG. 4.17   Economic dispatch, with and without penalty factors.

This can be written:

$$P_{loss} = \sum_i \sum_j P_i B_{ij} P_j + \sum_i B_{i0} P_i + B_{00} \tag{4.43}$$

Before we discuss the calculation of the B coefficients, we will discuss how the coefficients are used in an economic dispatch calculation. Substitute Eq. 4.43 into Eqs. 3.7, 3.8, and 3.9.

$$\phi = - \sum_{i=1}^{N} P_i + P_{load} + \left(\sum_i \sum_j P_i B_{ij} P_j + \sum_i B_{i0} P_i + B_{00} \right) \tag{4.44}$$

Then

$$\frac{\partial \mathscr{L}}{\partial P_i} = \frac{dF_i(P_i)}{dP_i} - \lambda \left(1 - 2 \sum_j B_{ij} P_j - B_{i0} \right) \tag{4.45}$$

Note that the presence of the incremental losses has coupled the coordination equations; this makes solution somewhat more difficult. A method of solution that is often used is shown in Figure 4.18.

EXAMPLE 4D

The B matrix loss formula for the network in Example 4A is given here. (Note that all P_i values must be per unit on 100 MVA base, which results in P_{loss} in per unit on 100 MVA base.)

$$P_{loss} = [P_1 \quad P_2 \quad P_3] \begin{bmatrix} 0.0676 & 0.00953 & -0.00507 \\ 0.00953 & 0.0521 & 0.00901 \\ -0.00507 & 0.00901 & 0.0294 \end{bmatrix} \begin{bmatrix} P_1 \\ P_2 \\ P_3 \end{bmatrix}$$

$$+ [-0.0766 \quad -0.00342 \quad 0.0189] \begin{bmatrix} P_1 \\ P_2 \\ P_3 \end{bmatrix} + 0.040357$$

From the base case power flow we have

$$P_1 = 107.9 \text{ MW}$$

$$P_2 = 50.0 \text{ MW}$$

$$P_2 = 60.0 \text{ MW}$$

$$P_{loss} = 7.9 \text{ MW (as calculated by the power flow)}$$

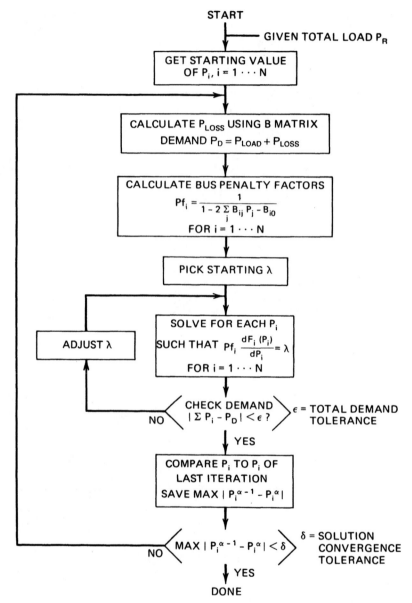

FIG. 4.18 Economic dispatch with updated penalty factors.

With these generation values placed in the B matrix, we see a very close agreement with the power flow calculation.

$$P_{\text{loss}} = \begin{bmatrix} 1.079 & 0.50 & 0.60 \end{bmatrix} \begin{bmatrix} 0.0676 & 0.00953 & -0.00507 \\ 0.00953 & 0.0521 & 0.00901 \\ -0.00507 & 0.00901 & 0.0294 \end{bmatrix} \begin{bmatrix} 1.079 \\ 0.50 \\ 0.60 \end{bmatrix}$$

$$+ \begin{bmatrix} -0.0766 & -0.00342 & 0.0189 \end{bmatrix} \begin{bmatrix} 1.079 \\ 0.50 \\ 0.60 \end{bmatrix} + 0.040357$$

$$= 0.07877 \text{ pu (or 7.877 MW) loss}$$

EXAMPLE 4E

Let the fuel cost curves for the three units in the six-bus network of Example 4A be given as

$$F_1(P_1) = 213.1 + 11.669P_1 + 0.00533P_1^2 \; \text{R/h}$$

$$F_2(P_2) = 200.0 + 10.333P_2 + 0.00889P_2^2 \; \text{R/h}$$

$$F_3(P_3) = 240.0 + 10.833P_3 + 0.00741P_3^2 \; \text{R/h}$$

with unit dispatch limits

$$50.0 \text{ MW} \le P_1 \le 200 \text{ MW}$$

$$37.5 \text{ MW} \le P_2 \le 150 \text{ MW}$$

$$45.0 \text{ MW} \le P_3 \le 180 \text{ MW}$$

A computer program using the method of Figure 4.17 was run using:

$$P_{\text{load}} \text{ (total load to be supplied)} = 210 \text{ MW}$$

The resulting iterations (Table 4.1) show how the program must redispatch again and again to account for the changes in losses and penalty factors.

Note that the flowchart of Figure 4.18 shows a "two-loop" procedure. The "inner" loop adjusts λ until total demand is met; then the outer loop recalculates the penalty factors. (Under some circumstances the penalty factors are quite sensitive to changes in dispatch. If the incremental costs are relatively "flat," this procedure may be unstable and special precautions may need to be employed to insure convergence.)

TABLE 4.1 Iterations for Example 4E

Iteration	λ	P_{loss}	P_D	P_1	P_2	P_3
1	12.8019	17.8	227.8	50.00	85.34	92.49
2	12.7929	11.4	221.4	74.59	71.15	75.69
3	12.8098	9.0	219.0	73.47	70.14	75.39
4	12.8156	8.8	218.8	73.67	69.98	75.18
5	12.8189	8.8	218.8	73.65	69.98	75.18
6	12.8206	8.8	218.8	73.65	69.98	75.18

4.2.4 Exact Methods of Calculating Penalty Factors

4.2.4.1 A Discussion of Reference Bus Versus Load Center Penalty Factors

The B matrix assumes that all load currents conform to an equivalent total load current and that the equivalent load current is the negative of the sum of all generator currents. When incremental losses are calculated, something is implied.

$$\text{Total loss} = \mathbf{P}^T[B]\mathbf{P} + \mathbf{B}_0^T\mathbf{P} + B_{00}$$

$$\text{Incremental loss at generator bus } i = \frac{\partial P_{loss}}{\partial P_i}$$

The incremental loss is the change in losses when an increment is made in generation output. As just derived, the incremental loss for bus i assumed that all the other generators remained fixed. By the original assumption, however, the load currents all conform to each other and always balance with the generation; then the implication in using a B matrix is that an *incremental increase in generator output is matched by an equivalent increment in load.*

An alternative approach to economic dispatch is to use a reference bus that always moves when an increment in generation is made. Figure 4.19 shows a

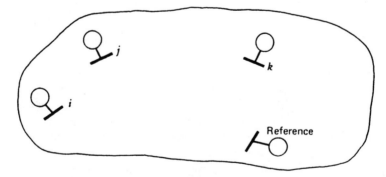

FIG. 4.19 Power system with reference generator.

power system with several generator buses and a reference-generator bus. Suppose we change the generation on bus i by ΔP_i,

$$P_i^{new} = P_i^{old} + \Delta P_i \qquad (4.46)$$

Furthermore, we will assume that *load stays constant* and that to compensate for the increase in ΔP_i, the reference bus just drops off by ΔP_{ref}.

$$P_{ref}^{new} = P_{ref}^{old} + \Delta P_{ref} \qquad (4.47)$$

If nothing else changed, ΔP_{ref} would be the negative of ΔP_i; however, the flows on the system can change as a result of the two generation adjustments. The change in flow is apt to cause a change in losses so that ΔP_{ref} is not necessarily equal to ΔP_i. That is,

$$\Delta P_{ref} = -\Delta P_i + \Delta P_{loss} \qquad (4.48)$$

Next, we can define β_i as the ratio of the negative change in the reference-bus power to the change ΔP_i.

$$\beta_i = \frac{-\Delta P_{ref}}{\Delta P_i} = \frac{(\Delta P_i - \Delta P_{loss})}{\Delta P_i} \qquad (4.49)$$

or

$$\beta_i = 1 - \frac{\partial P_{loss}}{\partial P_i} \qquad (4.50)$$

We can define economic dispatch as follows.

> All generators are in economic dispatch when a shift of ΔP MW from any generator to the reference bus results in no change in net production cost; where ΔP is arbitrarily small.

That is, if

$$\text{Total production cost} = \sum F_i(P_i)$$

then the change in production cost with a shift ΔP_i from plant i is

$$\Delta \text{Production cost} = \frac{dF_i(P_i)}{dP_i} \Delta P_i + \frac{dF_{ref}(P_{ref})}{dP_{ref}} \Delta P_{ref} \qquad (4.51)$$

but

$$\Delta P_{ref} = -\beta_i \Delta P_i$$

then

$$\Delta \text{Production cost} = \frac{dF_i(P_i)}{dP_i} \Delta P_i - \beta_i \frac{dF_{\text{ref}}(P_{\text{ref}})}{dP_{\text{ref}}} \Delta P_i \qquad (4.52)$$

To satisfy the economic conditions,

$$\Delta \text{Production cost} = 0$$

or

$$\frac{dF_i(P_i)}{dP_i} = \beta_i \frac{dF_{\text{ref}}(P_{\text{ref}})}{dP_{\text{ref}}} \qquad (4.53)$$

which could be written as

$$\frac{1}{\beta_i} \frac{dF_i(P_i)}{dP_i} = \frac{dF_{\text{ref}}(P_{\text{ref}})}{dP_{\text{ref}}} \qquad (4.54)$$

This is very similar to Eq. 4.40. To obtain an economic dispatch solution, pick a value of generation on the reference bus and then set all other generators according to Eq. 4.54, and check for total demand and readjust reference generation as needed until a solution is reached.

Note further that this method is exactly the first-order gradient method with losses.

$$\Delta F_T = \sum_{i \neq \text{ref}} \left[\frac{dF_i}{dP_i} - \beta_i \frac{dF_{\text{ref}}}{dP_{\text{ref}}} \right] \Delta P_i \qquad (4.55)$$

4.2.4.2 Reference-Bus Penalty Factors Direct from the AC Power Flow

The reference-bus penalty factors may be derived using the Newton–Raphson power flow. What we wish to know is the ratio of change in power on the reference bus when a change ΔP_i is made.

Where P_{ref} is a function of the voltage magnitude and phase angle on the network, when a change in ΔP_i is made, all phase angles and voltages in the network will change. Then

$$\Delta P_{\text{ref}} = \sum_i \frac{\partial P_{\text{ref}}}{\partial \theta_i} \Delta \theta_i + \sum_i \frac{\partial P_{\text{ref}}}{\partial |E_i|} \Delta |E_i|$$

$$= \sum_i \frac{\partial P_{\text{ref}}}{\partial \theta_i} \frac{\partial \theta_i}{\partial P_i} \Delta P_i + \sum_i \frac{\partial P_{\text{ref}}}{\partial |E_i|} \frac{\partial |E_i|}{\partial P_i} \Delta P_i \qquad (4.56)$$

To carry out the matrix manipulations, we will also need the following.

$$\Delta P_{\text{ref}} = \sum_i \frac{\partial P_{\text{ref}}}{\partial \theta_i} \Delta \theta_i + \sum_i \frac{\partial P_{\text{ref}}}{\partial |E_i|} \Delta |E_i|$$

$$= \sum_i \frac{\partial P_{\text{ref}}}{\partial \theta_i} \frac{\partial \theta_i}{\partial Q_i} \Delta Q_i + \sum_i \frac{\partial P_{\text{ref}}}{\partial |E_i|} \frac{\partial |E_i|}{\partial Q_i} \Delta Q_i \qquad (4.57)$$

The terms $\partial P_{ref}/\partial\theta_i$ and $\partial P_{ref}/|E_i|$ *are derived by differentiating Eq.* 4.18 *for the reference bus. The terms* $\partial\theta_i/\partial P_i$ and $\partial|E_i|/\partial P_i$ are from the inverse Jacobian matrix (see Eq. 4.20). We can write Eqs. 4.56 and 4.57 for every bus i in the network. The resulting equation is

$$
\begin{bmatrix} \dfrac{\partial P_{ref}}{\partial P_1} & \dfrac{\partial P_{ref}}{\partial Q_1} & \dfrac{\partial P_{ref}}{\partial P_2} & \dfrac{\partial P_{ref}}{\partial Q_2} & \cdots & \dfrac{\partial P_{ref}}{\partial P_N} & \dfrac{\partial P_{ref}}{\partial Q_N} \end{bmatrix}
$$
$$
= \begin{bmatrix} \dfrac{\partial P_{ref}}{\partial \theta_1} & \dfrac{\partial P_{ref}}{\partial |E_1|} & \dfrac{\partial P_{ref}}{\partial \theta_2} & \dfrac{\partial P_{ref}}{\partial |E_2|} & \cdots & \dfrac{\partial P_{ref}}{\partial \theta_N} & \dfrac{\partial P_{ref}}{\partial |E_N|} \end{bmatrix} [J^{-1}] \qquad (4.58)
$$

By transposing we get

$$
\begin{bmatrix} \dfrac{\partial P_{ref}}{\partial P_1} \\[2mm] \dfrac{\partial P_{ref}}{\partial Q_1} \\[2mm] \dfrac{\partial P_{ref}}{\partial P_2} \\[2mm] \dfrac{\partial P_{ref}}{\partial Q_2} \\[2mm] \vdots \\[2mm] \dfrac{\partial P_{ref}}{\partial P_N} \\[2mm] \dfrac{\partial P_{ref}}{\partial Q_N} \end{bmatrix} = [J^{T-1}] \begin{bmatrix} \dfrac{\partial P_{ref}}{\partial \theta_1} \\[2mm] \dfrac{\partial P_{ref}}{\partial |E_1|} \\[2mm] \dfrac{\partial P_{ref}}{\partial \theta_2} \\[2mm] \dfrac{\partial P_{ref}}{\partial |E_2|} \\[2mm] \vdots \\[2mm] \dfrac{\partial P_{ref}}{\partial \theta_N} \\[2mm] \dfrac{\partial P_{ref}}{\partial |E_N|} \end{bmatrix} \qquad (4.59)
$$

In practice, instead of calculating J^{T-1} explicitly, we use Gaussian elimination on J^T in the same way we operate on J in the Newton power flow solution.

APPENDIX
Power Flow Input Data for Six-Bus System

Figure 4.20 lists the input data for the six-bus sample system used in the examples in Chapter 4. The impedances are per unit on a base of 100 MVA. The generation cost functions are contained in Example 4E.

Line Data

From bus	To bus	R(pu)	X(pu)	BCAP[a] (pu)
1	2	0.10	0.20	0.02
1	4	0.05	0.20	0.02
1	5	0.08	0.30	0.03
2	3	0.05	0.25	0.03
2	4	0.05	0.10	0.01
2	5	0.10	0.30	0.02
2	6	0.07	0.20	0.025
3	5	0.12	0.26	0.025
3	6	0.02	0.10	0.01
4	5	0.20	0.40	0.04
5	6	0.10	0.30	0.03

[a] BCAP = half total line charging suseptance.

Bus Data

Bus number	Bus type	Voltage schedule (pu V)	P_{gen} (pu MW)	P_{load} (pu MW)	Q_{load} (pu MVAR)
1	Swing	1.05			
2	Gen.	1.05	0.50	0.0	0.0
3	Gen.	1.07	0.60	0.0	0.0
4	Load		0.0	0.7	0.7
5	Load		0.0	0.7	0.7
6	Load		0.0	0.7	0.7

FIG. 4.20   Input data for six-bus sample power system.

PROBLEMS

4.1   The circuit elements in the 138 kV circuit in Figure 4.21 are in per unit on a 100 MVA base with the nominal 138 kV voltage as base. The $P + jQ$ load is scheduled to be 170 MW and 50 MVAR.

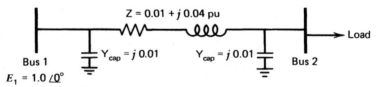

FIG. 4.21   Two-bus AC system for Problem 4.1.

a. Write the Y matrix for this two-bus system.

b. Assume bus 1 as the reference bus and set up the Gauss–Seidel correction equation for bus 2. (Use $1.0 \angle 0°$ as the initial voltage on bus 2.) Carry out two or three iterations and show that you are converging.

c. Apply the "DC" load flow conventions to this circuit and solve for the phase angle at bus 2 for the same load real power of 1.7 per unit.

4.2 Given the network in Figure 4.22 (base = 100 MVA):

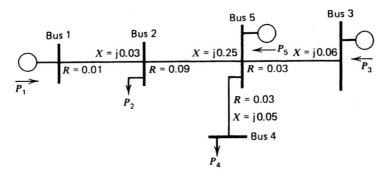

FIG. 4.22 Five-bus network for Problem 4.2.

a. Develop the $[B']$ matrix for this system.

$$
\begin{bmatrix} P_1 \\ P_2 \\ P_3 \\ P_4 \\ P_5 \end{bmatrix} = [B'] \begin{bmatrix} \theta_1 \\ \theta_2 \\ \theta_3 \\ \theta_4 \\ \theta_5 \end{bmatrix}
\qquad
\begin{array}{l} P \text{ in per unit MW} \\ \theta \text{ in radians (rad)} \end{array}
$$

b. Assume bus 5 as the reference bus. To carry out a "DC" load flow, we will set $\theta_5 = 0$ rad. Row 5 and column 5 will be zeroed.

$$
\begin{bmatrix} P_1 \\ P_2 \\ P_3 \\ P_4 \\ P_5 \end{bmatrix} =
\begin{bmatrix}
 & & & & 0 \\
 & & & & 0 \\
 & \text{Remainder} & & & 0 \\
 & \text{of } B' & & & 0 \\
 & & & & 0 \\
0 & 0 & 0 & 0 & 0
\end{bmatrix}
\begin{bmatrix} \theta_1 \\ \theta_2 \\ \theta_3 \\ \theta_4 \\ \theta_5 \end{bmatrix}
$$

Solve for the $[B']^{-1}$ matrix.

$$
\begin{bmatrix} \theta_1 \\ \theta_2 \\ \theta_3 \\ \theta_4 \\ \theta_5 \end{bmatrix} = [B']^{-1} \begin{bmatrix} P_1 \\ P_2 \\ P_3 \\ P_4 \\ P_5 \end{bmatrix}
$$

c. Calculate the phase angles for the set of power injections.

$$P_1 = 100 \text{ MW generation}$$

$$P_2 = 120 \text{ MW load}$$

$$P_3 = 150 \text{ MW generation}$$

$$P_4 = 200 \text{ MW load}$$

d. Calculate P_5 according to the "DC" load flow.

e. Calculate all power flows on the system using the phase angles found in part c.

f. (Optional) Calculate the reference-bus penalty factors for buses 1, 2, 3, and 4. Assume all bus voltage magnitudes are 1.0 per unit.

4.3 Given the following loss formula (use P values in MW):

$$
B_{ij} = \begin{matrix} & 1 & 2 & 3 \\ 1 & \\ 2 & \\ 3 & \end{matrix}
\begin{bmatrix}
1.36255 \times 10^{-4} & 1.753 \times 10^{-5} & 1.8394 \times 10^{-4} \\
1.754 \times 10^{-5} & 1.5448 \times 10^{-4} & 2.82765 \times 10^{-4} \\
1.8394 \times 10^{-4} & 2.82765 \times 10^{-4} & 1.6147 \times 10^{-3}
\end{bmatrix}
$$

B_{i0} and B_{00} are neglected. Assume three units are on-line and have the following characteristics.

Unit 1: $H_1 = 312.5 + 8.25P_1 + 0.005P_1^2$, MBu/h

$$50 \le P_1 \le 250 \text{ MW}$$

Fuel cost $= 1.05 \text{ R/MBtu}$

Unit 2: $H_2 = 112.5 + 8.25P_2 + 0.005P_2^2$, MBtu/h

$$5 \le P_2 \le 150 \text{ MW}$$

Fuel cost $= 1.217 \text{ R/MBtu}$

Unit 3:
$$H_3 = 50 + 8.25P_3 + 0.005P_3^2, \text{ MBtu/h}$$
$$15 \le P_3 \le 100 \text{ MW}$$
$$\text{Fuel cost} = 1.1831 \text{ R/MBtu}$$

a. No Losses Used in Scheduling

  i. Calculate the optimum dispatch and total cost neglecting losses for $P_D = 190$ MW.*

  ii. Using this dispatch and the loss formula, calculate the system losses.

b. Losses Included in Scheduling

  i. Find the optimum dispatch for a total generation of $P_D = 190$ MW* using the coordination equations and the loss formula.

  ii. Calculate the cost rate.

  iii. Calculate the total losses using the loss formula.

  iv. Calculate the resulting load supplied.

4.4 All parts refer to the three-bus system shown in Figure 4.23.

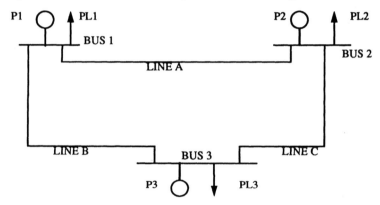

FIG. 4.23 Network for Problem 4.4.

Data for this problem is as follows:

Unit 1: $P_1 = 570$ MW

Unit 2: $P_2 = 330$ MW

Unit 3: $P_3 = 200$ MW

Loads:

$$P_{L1} = 200 \text{ MW}$$
$$P_{L2} = 400 \text{ MW}$$
$$P_{L3} = 500 \text{ MW}$$

* $P_{demand} = P_1 + P_2 + P_3 = P_D$
  $P_{loss} = $ power loss
  $P_{load} = P_D - P_{loss} = $ net load

Transmission line data:

P_{loss} in line A $= 0.02P_A^2$ (where $P_A = P$ flow from bus 1 to bus 2)

P_{loss} in line B $= 0.02P_B^2$ (where $P_B = P$ flow from bus 1 to bus 3)

P_{loss} in line C $= 0.02P_C^2$ (where $P_C = P$ flow from bus 2 to bus 3)

Note: the above data are for P_{loss} in per unit when power flows P_A or P_B or P_C are in per unit.

Line reactances:

$$X_A = 0.2 \text{ per unit}$$

$$X_B = 0.3333 \text{ per unit}$$

$$X_C = 0.05 \text{ per unit}$$

(assume 100-MVA base when converting to per unit).

a. Find how the power flows distribute using the DC power flow approximation. Use bus 3 as the reference.

b. Calculate the total losses.

c. Calculate the incremental losses for bus 1 and bus 2 as follows: assume that ΔP_1 is balanced by an equal change on the reference bus. Use the DC power flow data from part a and calculate the change in power flow on all three lines ΔP_A, ΔP_B, and ΔP_C. Now calculate the line incremental loss as:

$$\Delta P_{\text{loss}_A} = \left(\frac{\partial}{\partial P_A}\right)\Delta P_A = (0.04P_A)\Delta P_A$$

Similarly, calculate for lines B and C.

d. Find the bus penalty factors calculated from the line incremental losses found in part c.

4.5 The three-bus, two-generator power system shown in Figure 4.24 is to be dispatched to supply the 500-MW load. Each transmission line has losses

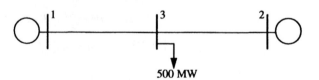

500 MW

FIG. 4.24 Circuit for Problem 4.5.

that are given by the equations below.

$$P_{\text{loss}_{13}} = 0.0001P_1^2$$

$$P_{\text{loss}_{23}} = 0.0002P_2^2$$

$$F_1(P_1) = 500 + 8P_1 + 0.002P_1^2$$

$$50 \text{ MW} < P_1 < 500 \text{ MW}$$

$$F_2(P_2) = 400 + 7.9P_2 + 0.0025P_2^2$$

$$50 \text{ MW} < P_2 < 500 \text{ MW}$$

You are to attempt to solve for both the economic dispatch of this system and the "power flow." The power flow should show what power enters and leaves each bus of the network. If you use an iterative solution, show at least two complete iterations. You may use the following initial conditions: $P_1 = 250$ MW and $P_2 = 250$ MW.

FURTHER READING

The basic papers on solution of the power flow can be found in references 1–5. The development of the loss-matrix equations is based on the work of Kron (reference 6), who developed the reference-frame transformation theory. Other developments of the transmission-loss formula are seen in references 7 and 8. Meyer's paper (9) is representative of recent adaptation of sparsity programming methods to calculation of the loss matrix.

The development of the reference-bus penalty factor method can be seen in references 10 and 11. Reference 12 gives an excellent derivation of the reference-bus penalty factors derived from the Newton power-flow equations. Reference 12 provides an excellent summary of recent developments in power system dispatch.

1. Ward, J. B., Hale, H. W., "Digital Computer Solution of Power-Flow Problems," *AIEE Transactions, Part III Power Apparatus and Systems*, Vol. 75, June 1956, pp. 398–404.

2. VanNess, J. E., "Iteration Methods for Digital Load Flow Studies," *AIEE Transactions on Power Apparatus and Systems*, Vol. 78A, August 1959, pp. 583–588.

3. Tinney, W. F., Hart, C. E., "Power Flow Solution by Newton's Method," *IEEE Transactions on Power Apparatus and Systems*, Vol. PAS-86, November 1967, pp. 1449–1460.

4. Stott, B., Alsac, O., "Fast Decoupled Load Flow," *IEEE Transactions on Power Apparatus and Systems*, Vol. PAS-93, May/June 1974, pp. 859–869.

5. Stott, B., "Review of Load-Flow Calculation Methods," *Proceedings of the IEEE*, Vol. 62, No. 2, July 1974, pp. 916–929.

6. Kron, G., "Tensorial Analysis of Integrated Transmission Systems—Part I: The Six Basic Reference Frames," *AIEE Transactions*, Vol. 70, Part I, 1951, pp. 1239–1248.

7. Kirchmayer, L. K., Stagg, G. W., "Analysis of Total and Incremental Losses in Transmission Systems," *AIEE Transactions*, Vol. 70, Part I, 1951, pp. 1179–1205.

8. Early, E. D., Watson, R. E., "A New Method of Determining Constants for the General Transmission Loss Equation," *AIEE Transactions on Power Apparatus and Systems*, Vol. PAS-74, February 1956, pp. 1417–1423.

9. Meyer, W. S., "Efficient Computer Solution for Kron and Kron Early-Loss Formulas," *Proceedings of the 1973 PICA Conference*, IEEE 73 CHO 740-1, PWR, pp. 428–432.

10. Shipley, R. B., Hochdorf, M., "Exact Economic Dispatch—Digital Computer Solution," *AIEE Transactions on Power Apparatus and Systems*, Vol. PAS-75, November 1956, pp. 1147–1152.

11. Dommel, H. W., Tinney, W. F., "Optimal Power Flow Solutions," *IEEE Transactions on Power Apparatus and Systems*, Vol. PAS-87, October 1968, pp. 1866–1876.

12. Happ, H. H., "Optimal Power Dispatch," *IEEE Transactions on Power Apparatus and Systems*, Vol. PAS-93, May/June 1974, pp. 820–830.

5 Unit Commitment

5.1 INTRODUCTION

Because human activity follows cycles, most systems supplying services to a large population will experience cycles. This includes transportation systems, communication systems, as well as electric power systems. In the case of an electric power system, the total load on the system will generally be higher during the daytime and early evening when industrial loads are high, lights are on, and so forth, and lower during the late evening and early morning when most of the population is asleep. In addition, the use of electric power has a weekly cycle, the load being lower over weekend days than weekdays. But why is this a problem in the operation of an electric power system? Why not just simply commit enough units to cover the maximum system load and leave them running? Note that to "commit" a generating unit is to "turn it on;" that is, to bring the unit up to speed, synchronize it to the system, and connect it so it can deliver power to the network. The problem with "commit enough units and leave them on line" is one of economics. As will be shown in Example 5A, it is quite expensive to run too many generating units. A great deal of money can be saved by turning units off (decommitting them) when they are not needed.

EXAMPLE 5A

Suppose one had the three units given here:

Unit 1:	Min $= 150$ MW
	Max $= 600$ MW
	$H_1 = 510.0 + 7.2P_1 + 0.00142P_1^2$ MBtu/h
Unit 2:	Min $= 100$ MW
	Max $= 400$ MW
	$H_2 = 310.0 + 7.85P_2 + 0.00194P_2^2$ MBtu/h
Unit 3:	Min $= 50$ MW
	Max $= 200$ MW
	$H_3 = 78.0 + 7.97P_3 + 0.00482P_3^2$ MBtu/h

with fuel costs:

$$\text{Fuel cost}_1 = 1.1 \, \text{R/MBtu}$$

$$\text{Fuel cost}_2 = 1.0 \, \text{R/MBtu}$$

$$\text{Fuel cost}_3 = 1.2 \, \text{R/MBtu}$$

If we are to supply a load of 550 MW, what unit or combination of units should be used to supply this load most economically? To solve this problem, simply try all combinations of the three units. Some combinations will be infeasible if the sum of all maximum MW for the units committed is less than the load or if the sum of all minimum MW for the units committed is greater than the load. For each feasible combination, the units will be dispatched using the techniques of Chapter 3. The results are presented in Table 5.1.

Note that the least expensive way to supply the generation is not with all three units running, or even any combination involving two units. Rather, the optimum commitment is to only run unit 1, the most economic unit. By only running the most economic unit, the load can be supplied by that unit operating closer to its best efficiency. If another unit is committed, both unit 1 and the other unit will be loaded further from their best efficiency points such that the net cost is greater than unit 1 alone.

Suppose the load follows a simple "peak-valley" pattern as shown in Figure 5.1a. If the operation of the system is to be optimized, units must be shut down as the load goes down and then recommitted as it goes back up. We would like to know which units to drop and when. As we will show later, this problem is far from trivial when real generating units are considered. One approach to this solution is demonstrated in Example 5B, where a simple priority list scheme is developed.

TABLE 5.1 Unit Combinations and Dispatch for 550-MW Load of Example 5A

Unit 1	Unit 2	Unit 3	Max Generation	Min Generation	P_1	P_2	P_3	F_1	F_2	F_3	Total Generation Cost $F_1 + F_2 + F_3$
Off	Off	Off	0	0				Infeasible			
Off	Off	On	200	50				Infeasible			
Off	On	Off	400	100				Infeasible			
Off	On	On	600	150	0	400	150	0	3760	1658	5418
On	Off	Off	600	150	550	0	0	5389	0	0	5389
On	Off	On	800	200	500	0	50	4911	0	586	5497
On	On	Off	1000	250	295	255	0	3030	2440	0	5471
On	On	On	1200	300	267	233	50	2787	2244	586	5617

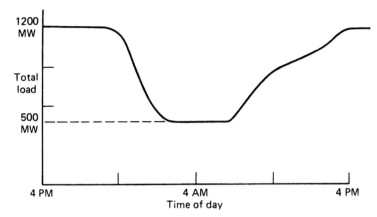

FIG. 5.1a   Simple "peak-valley" load pattern.

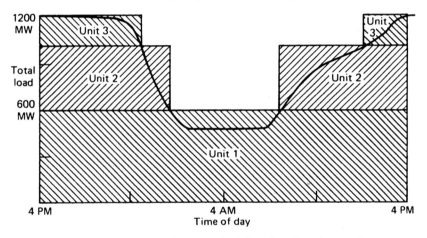

FIG. 5.1b   Unit commitment schedule using shut-down rule.

EXAMPLE 5B

Suppose we wish to know which units to drop as a function of system load. Let the units and fuel costs be the same as in Example 5A, with the load varying from a peak of 1200 MW to a valley of 500 MW. To obtain a "shut-down rule," simply use a brute-force technique wherein all combinations of units will be tried (as in Example 5A) for each load value taken in steps of 50 MW from 1200 to 500. The results of applying this brute-force technique are given in Table 5.2. Our shut-down rule is quite simple.

When load is above 1000 MW, run all three units; between 1000 MW and 600 MW, run units 1 and 2; below 600 MW, run only unit 1.

TABLE 5.2   "Shut-down Rule" Derivation for Example 5B

	Optimum Combination		
Load	Unit 1	Unit 2	Unit 3
1200	On	On	On
1150	On	On	On
1100	On	On	On
1050	On	On	On
1000	On	On	Off
950	On	On	Off
900	On	On	Off
850	On	On	Off
800	On	On	Off
750	On	On	Off
700	On	On	Off
650	On	On	Off
600	On	Off	Off
550	On	Off	Off
500	On	Off	Off

Figure 5.1b shows the unit commitment schedule derived from this shut-down rule as applied to the load curve of Figure 5.1a.

So far, we have only obeyed one simple constraint: *Enough units will be committed to supply the load.* If this were all that was involved in the unit commitment problem—that is, just meeting the load—we could stop here and state that the problem was "solved." Unfortunately, other constraints and other phenomena must be taken into account in order to claim an optimum solution. These constraints will be discussed in the next section, followed by a description of some of the presently used methods of solution.

5.1.1   Constraints in Unit Commitment

Many constraints can be placed on the unit commitment problem. The list presented here is by no means exhaustive. Each individual power system, power pool, reliability council, and so forth, may impose different rules on the scheduling of units, depending on the generation makeup, load-curve characteristics, and such.

5.1.2   Spinning Reserve

Spinning reserve is the term used to describe the total amount of generation available from all units synchronized (i.e., spinning) on the system, minus the present load and losses being supplied. Spinning reserve must be carried so that the loss of one or more units does not cause too far a drop in system frequency (see Chapter 9). Quite simply, if one unit is lost, there must be ample reserve on the other units to make up for the loss in a specified time period.

Spinning reserve must be allocated to obey certain rules, usually set by regional reliability councils (in the United States) that specify how the reserve is to be allocated to various units. Typical rules specify that reserve must be a given percentage of forecasted peak demand, or that reserve must be capable of making up the loss of the most heavily loaded unit in a given period of time. Others calculate reserve requirements as a function of the probability of not having sufficient generation to meet the load.

Not only must the reserve be sufficient to make up for a generation-unit failure, but the reserves must be allocated among fast-responding units and slow-responding units. This allows the automatic generation control system (see Chapter 9) to restore frequency and interchange quickly in the event of a generating-unit outage.

Beyond spinning reserve, the unit commitment problem may involve various classes of "scheduled reserves" or "off-line" reserves. These include quick-start diesel or gas-turbine units as well as most hydro-units and pumped-storage hydro-units that can be brought on-line, synchronized, and brought up to full capacity quickly. As such, these units can be "counted" in the overall reserve assessment, as long as their time to come up to full capacity is taken into account.

Reserves, finally, must be spread around the power system to avoid transmission system limitations (often called "bottling" of reserves) and to allow various parts of the system to run as "islands," should they become electrically disconnected.

EXAMPLE 5C

Suppose a power system consisted of two isolated regions: a western region and an eastern region. Five units, as shown in Figure 5.2, have been committed to supply 3090 MW. The two regions are separated by transmission tie lines that can together transfer a maximum of 550 MW in either direction. This is also shown in Figure 5.2. What can we say about the allocation of spinning reserve in this system?

The data for the system in Figure 5.2 are given in Table 5.3. With the exception of unit 4, the loss of any unit on this system can be covered by the spinning reserve on the remaining units. Unit 4 presents a problem, however. If unit 4 were to be lost and unit 5 were to be run to its maximum of 600 MW, the eastern region would still need 590 MW to cover the load in that region. The 590 MW would have to be transmitted over the tie lines from the western region, which can easily supply 590 MW from its reserves. However, the tie capacity of only 550 MW limits the transfer. Therefore, the loss of unit 4 cannot

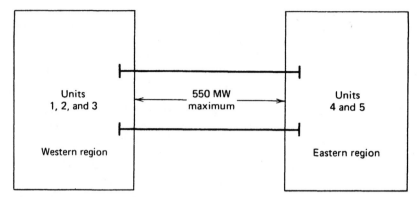

FIG. 5.2 Two-region system.

TABLE 5.3 **Data for the System in Figure 5.2**

Region	Unit	Unit Capacity (MW)	Unit Output (MW)	Regional Generation (MW)	Spinning Reserve	Regional Load (MW)	Inter-change (MW)
Western	1	1000	900 ⎫		100		
	2	800	420 ⎬	1740	380	1900	160 in
	3	800	420 ⎭		380		
Eastern	4	1200	1040 ⎫	1350	160	1190	160 out
	5	600	310 ⎭		290		
Total	1–5	4400	3090	3090	1310	3090	

be covered even though the entire system has ample reserves. The only solution to this problem is to commit more units to operate in the eastern region.

5.1.3 Thermal Unit Constraints

Thermal units usually require a crew to operate them, especially when turned on and turned off. A thermal unit can undergo only gradual temperature changes, and this translates into a time period of some hours required to bring the unit on-line. As a result of such restrictions in the operation of a thermal plant, various constraints arise, such as:

- **Minimum up time:** once the unit is running, it should not be turned off immediately.
- **Minimum down time:** once the unit is decommitted, there is a minimum time before it can be recommitted.

- **Crew constraints:** if a plant consists of two or more units, they cannot both be turned on at the same time since there are not enough crew members to attend both units while starting up.

In addition, because the temperature and pressure of the thermal unit must be moved slowly, a certain amount of energy must be expended to bring the unit on-line. This energy does not result in any MW generation from the unit and is brought into the unit commitment problem as a *start-up cost*.

The start-up cost can vary from a maximum "cold-start" value to a much smaller value if the unit was only turned off recently and is still relatively close to operating temperature. There are two approaches to treating a thermal unit during its down period. The first allows the unit's boiler to cool down and then heat back up to operating temperature in time for a scheduled turn on. The second (called *banking*) requires that sufficient energy be input to the boiler to just maintain operating temperature. The costs for the two can be compared so that, if possible, the best approach (cooling or banking) can be chosen.

$$\text{Start-up cost when cooling} = C_c(1 - \varepsilon^{-t/\alpha}) \times F + C_f$$

where

C_c = cold-start cost (MBtu)

F = fuel cost

C_f = fixed cost (includes crew expense, maintenance expenses) (in ℝ)

α = thermal time constant for the unit

t = time (h) the unit was cooled

$$\text{Start-up cost when banking} = C_t \times t \times F + C_f$$

where

C_t = cost (MBtu/h) of maintaining unit at operating temperature

Up to a certain number of hours, the cost of banking will be less than the cost of cooling, as is illustrated in Figure 5.3.

Finally, the capacity limits of thermal units may change frequently, due to maintenance or unscheduled outages of various equipment in the plant; this must also be taken into account in unit commitment.

5.1.4   Other Constraints

5.1.4.1   Hydro-Constraints

Unit commitment cannot be completely separated from the scheduling of hydro-units. In this text, we will assume that the hydrothermal scheduling (or "coordination") problem can be separated from the unit commitment problem. We, of course, cannot assert flatly that our treatment in this fashion will always result in an optimal solution.

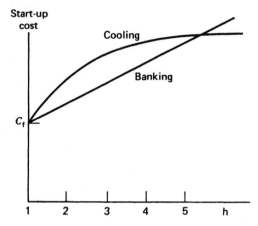

FIG. 5.3   Time-dependent start-up costs.

5.1.4.2   Must Run

Some units are given a must-run status during certain times of the year for reason of voltage support on the transmission network or for such purposes as supply of steam for uses outside the steam plant itself.

5.1.4.3 Fuel Constraints

We will treat the "fuel scheduling" problem briefly in Chapter 6. A system in which some units have limited fuel, or else have constraints that require them to burn a specified amount of fuel in a given time, presents a most challenging unit commitment problem.

5.2   UNIT COMMITMENT SOLUTION METHODS

The commitment problem can be very difficult. As a theoretical exercise, let us postulate the following situation.

- We must establish a loading pattern for M periods.
- We have N units to commit and dispatch.
- The M load levels and operating limits on the N units are such that any one unit can supply the individual loads and that any combination of units can also supply the loads.

Next, assume we are going to establish the commitment by enumeration (brute force). The total number of combinations we need to try each hour is,

$$C(N, 1) + C(N, 2) + \ldots + C(N, N - 1) + C(N, N) = 2^N - 1$$

where $C(N, j)$ is the combination of N items taken j at a time. That is,

$$C(N, j) = \left[\frac{N!}{(N-j)!\,j!} \right]$$

$$j! = 1 \times 2 \times 3 \times \ldots \times j$$

For the total period of M intervals, the maximum number of possible combinations is $(2^N - 1)^M$, which can become a horrid number to think about.

For example, take a 24-h period (e.g., 24 one-hour intervals) and consider systems with 5, 10, 20, and 40 units. The value of $(2^N - 1)^{24}$ becomes the following.

N	$(2^N - 1)^{24}$
5	6.2×10^{35}
10	1.73×10^{72}
20	3.12×10^{144}
40	(Too big)

These very large numbers are the upper bounds for the number of enumerations required. Fortunately, the constraints on the units and the load-capacity relationships of typical utility systems are such that we do not approach these large numbers. Nevertheless, the real practical barrier in the optimized unit commitment problem is the high dimensionality of the possible solution space.

The most talked-about techniques for the solution of the unit commitment problem are:

- Priority-list schemes,
- Dynamic programming (DP),
- Lagrange relation (LR).

5.2.1   Priority-List Methods

The simplest unit commitment solution method consists of creating a priority list of units. As we saw in Example 5B, a simple shut-down rule or priority-list scheme could be obtained after an exhaustive enumeration of all unit combinations at each load level. The priority list of Example 5B could be obtained in a much simpler manner by noting the full-load average production cost of each unit, where the full-load average production cost is simply the net heat rate at full load multiplied by the fuel cost.

EXAMPLE 5D

Construct a priority list for the units of Example 5A. (Use the same fuel costs as in Example 5A.) First, the full-load average production cost will be calculated:

Unit	Full Load Average Production Cost (R/MWh)
1	9.79
2	9.48
3	11.188

A strict priority order for these units, based on the average production cost, would order them as follows:

Unit	R/MWh	Min MW	Max MW
2	9.48	100	400
1	9.79	150	600
3	11.188	50	200

and the commitment scheme would (ignoring min up/down time, start-up costs, etc.) simply use only the following combinations.

Combination	Min MW from Combination	Max MW from Combination
2 + 1 + 3	300	1200
2 + 1	250	1000
2	100	400

Note that such a scheme would not completely parallel the shut-down sequence described in Example 5B, where unit 2 was shut down at 600 MW leaving unit 1. With the priority-list scheme, both units would be held on until load reached 400 MW, then unit 1 would be dropped.

Most priority-list schemes are built around a simple shut-down algorithm that might operate as follows.

- At each hour when load is dropping, determine whether dropping the next unit on the priority list will leave sufficient generation to supply the load plus spinning-reserve requirements. If not, continue operating as is; if yes, go on to the next step.

- Determine the number of hours, H, before the unit will be needed again. That is, assuming that the load is dropping and will then go back up some hours later.

- If H is less than the minimum shut-down time for the unit, keep commitment as is and go to last step; if not, go to next step.

- Calculate two costs. The first is the sum of the hourly production costs for the next H hours with the unit up. Then recalculate the same sum for the unit down and add in the start-up cost for either cooling the unit or banking it, whichever is less expensive. If there is sufficient savings from shutting down the unit, it should be shut down, otherwise keep it on.

- Repeat this entire procedure for the next unit on the priority list. If it is also dropped, go to the next and so forth.

Various enhancements to the priority-list scheme can be made by grouping of units to ensure that various constraints are met. We will note later that dynamic-programming methods usually create the same type of priority list for use in the DP search.

5.2.2 Dynamic-Programming Solution

5.2.2.1 Introduction

Dynamic programming has many advantages over the enumeration scheme, the chief advantage being a reduction in the dimensionality of the problem. Suppose we have found units in a system and any combination of them could serve the (single) load. There would be a maximum of $2^4 - 1 = 15$ combinations to test. However, if a strict priority order is imposed, there are only four combinations to try:

Priority 1 unit
Priority 1 unit + Priority 2 unit
Priority 1 unit + Priority 2 unit + Priority 3 unit
Priority 1 unit + Priority 2 unit + Priority 3 unit + Priority 4 unit

The imposition of a priority list arranged in order of the full-load average-cost rate would result in a theoretically correct dispatch and commitment only if:

1. No load costs are zero.
2. Unit input–output characteristics are linear between zero output and full load.
3. There are no other restrictions.
4. Start-up costs are a fixed amount.

In the dynamic-programming approach that follows, we assume that:

1. A *state* consists of an array of units with specified units operating and the rest off-line.
2. The start-up cost of a unit is independent of the time it has been off-line (i.e., it is a fixed amount).
3. There are no costs for shutting down a unit.
4. There is a strict priority order, and in each interval a specified minimum amount of capacity must be operating.

A feasible state is one in which the committed units can supply the required load and that meets the minimum amount of capacity each period.

5.2.2.2   Forward DP Approach

One could set up a dynamic-programming algorithm to run backward in time starting from the final hour to be studied, back to the initial hour. Conversely, one could set up the algorithm to run forward in time from the initial hour to the final hour. The forward approach has distinct advantages in solving generator unit commitment. For example, if the start-up cost of a unit is a function of the time it has been off-line (i.e., its temperature), then a forward dynamic-program approach is more suitable since the previous history of the unit can be computed at each stage. There are other practical reasons for going forward. The initial conditions are easily specified and the computations can go forward in time as long as required. A forward dynamic-programming algorithm is shown by the flowchart in Figure 5.4.

The recursive algorithm to compute the minimum cost in hour K with combination I is,

$$F_{\text{cost}}(K, I) = \min_{\{L\}} [P_{\text{cost}}(K, I) + S_{\text{cost}}(K - 1, L: K, I) + F_{\text{cost}}(K - 1, L)] \quad (5.1)$$

where

$$F_{\text{cost}}(K, I) = \text{least total cost to arrive at state } (K, I)$$

$$P_{\text{cost}}(K, I) = \text{production cost for state } (K, I)$$

$$S_{\text{cost}}(K - 1, L: K, I) = \text{transition cost from state } (K - 1, L) \text{ to state } (K, I)$$

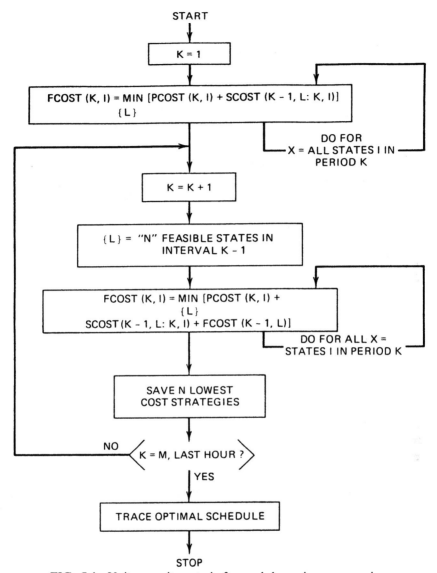

FIG. 5.4   Unit commitment via forward dynamic programming.

State (K, I) is the Ith combination in hour K. For the forward dynamic-programming approach, we define a *strategy* as the transition, or path, from one state at a given hour to a state at the next hour.

Note that two new variables, X and N, have been introduced in Figure 5.4.

X = number of states to search each period

N = number of strategies, or paths, to save at each step

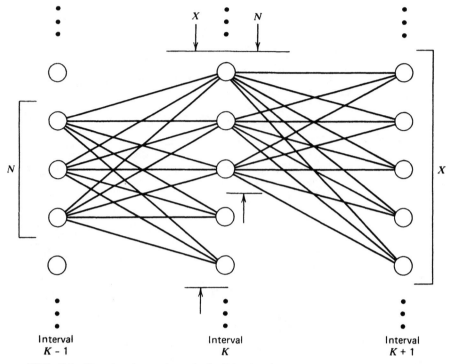

FIG. 5.5   Restricted search paths in DP algorithm with $N = 3$ and $X = 5$.

These variables allow control of the computational effort (see Figure 5.5). For complete enumeration, the maximum number of the value of X or N is $2^n - 1$.

For example, with a simple priority-list ordering, the upper bound on X is n, the number of units. Reducing the number N means that we are discarding the highest cost schedules at each time interval and saving only the lowest N paths or strategies. There is no assurance that the theoretical optimal schedule will be found using a reduced number of strategies and search range (the X value); only experimentation with a particular program will indicate the potential error associated with limiting the values of X and N below their upper bounds.

EXAMPLE 5E

For this example, the complete search range will be used and three cases will be studied. The first is a priority-list schedule, the second is the same example with complete enumeration. Both of the first two cases ignore hot-start costs and minimum up and down times. The third case includes the hot-start costs, as well as the minimum up and down times. Four units are to be committed to serve an 8-h load pattern. Data on the units and the load pattern are contained in Table 5.4.

TABLE 5.4   Unit Characteristics, Load Pattern, and Initial Status for the Cases in Example 5E

Unit	Max (MW)	Min (MW)	Incremental Heat Rate (Btu/kWh)	No-Load Cost (R/h)	Full-Load Ave. Cost (R/mWh)	Minimum Times (h) Up	Down
1	80	25	10440	213.00	23.54	4	2
2	250	60	9000	585.62	20.34	5	3
3	300	75	8730	684.74	19.74	5	4
4	60	20	11900	252.00	28.00	1	1

Unit	Initial Conditions Hours Off-Line ($-$) or On-Line ($+$)	Start-Up Costs Hot (R)	Cold (R)	Cold Start (h)
1	-5	150	350	4
2	8	170	400	5
3	8	500	1100	5
4	-6	0	0.02	0

Load Pattern Hour	Load (MW)
1	450
2	530
3	600
4	540
5	400
6	280
7	290
8	500

In order to make the required computations more efficiently, a simplified model of the unit characteristics is used. In practical applications, two- or three-section stepped incremental curves might be used, as shown in Figure 5.6. For our example, only a single step between minimum and the maximum power points is used. The units in this example have linear $F(P)$ functions:

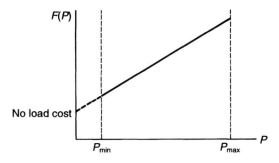

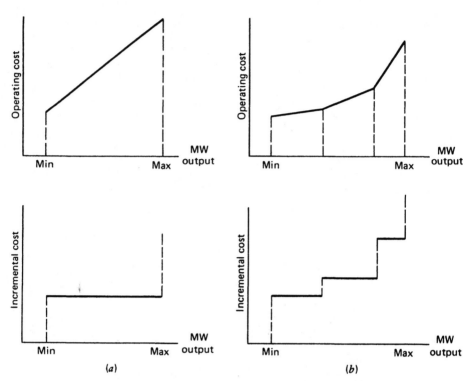

FIG. 5.6   (a) Single-step incremental cost curve and (b) multiple-step incremental cost curve.

The $F(P)$ function is:

$$F(P) = \text{No-load cost} + \text{Inc cost} \times P$$

Note, however, that the unit must operate within its limits. Start-up costs for the first two cases are taken as the cold-start costs. The priority order for the four units in the example is: unit 3, unit 2, unit 1, unit 4. For the first two cases, the minimum up and down times are taken as 1 h for all units.

In all three cases we will refer to the capacity ordering of the units. This is shown in Table 5.5, where the unit combinations or states are ordered by maximum net capacity for each combination.

Case 1

In Case 1, the units are scheduled according to a strict priority order. That is, units are committed in order until the load is satisfied. The total cost for the interval is the sum of the eight dispatch costs plus the transitional costs for starting any units. In this first case, a maximum of 24 dispatches must be considered.

TABLE 5.5 Capacity Ordering of the Units

State	Unit Combination[a]	Maximum Net Capacity for Combination
15	1 1 1 1	690
-14	1 1 1 0	630
13	0 1 1 1	610
12	0 1 1 0	550
11	1 0 1 1	440
10	1 1 0 1	390
9	1 0 1 0	380
8	0 0 1 1	360
7	1 1 0 0	330
6	0 1 0 1	310
5	0 0 1 0	300
4	0 1 0 0	250
3	1 0 0 1	140
2	1 0 0 0	80
1	0 0 0 1	60
0	0 0 0 0	0
	Unit 1 2 3 4	

[a] 1 = Committed (unit operating).
0 = Uncommitted (unit shut down).

For Case 1, the only states examined each hour consist of:

State No.	Unit Status	Capacity (MW)
5	0 0 1 0	300
12	0 1 1 0	550
14	1 1 1 0	630
15	1 1 1 1	690

Note that this is the priority order; that is, state $5 =$ unit 3, state $12 =$ units $3 + 2$, state $14 =$ unit $3 + 2 + 1$, and state $15 =$ units $3 + 2 + 1 + 4$. For the first 4 h, only the last three states are of interest. The sample calculations illustrate the technique. All possible commitments start from state 12 since this was given as the initial condition. For hour 1, the minimum cost is state 12, and so on. The results for the priority-ordered case are as follows.

Hour	State with Min Total Cost	Pointer for Previous Hour
1	12 (9208)	12
2	12 (19857)	12
3	14 (32472)	12
4	12 (43300)	14
⋮	⋮	⋮

Note that state 13 is not reachable in this strict priority ordering.

Sample Calculations for Case 1

$$F_{cost}(J, K) = \min_{\{L\}} [P_{cost}(J, K) + S_{cost}(J - 1, L: J, K) + F_{cost}(J - 1, L)]$$

Allowable states are

$$\{\ \} = \{0010, 0110, 1110, 1111\} = \{5, 12, 14, 15\}$$

In hour $0\{L\} = \{12\}$, initial condition.

$J = 1$:   1st hour

$$\frac{K}{15} \quad F_{cost}(1, 15) = P_{cost}(1,15) + S_{cost}(0, 12: 1, 15)$$

$$= 9861 + 350 = 10211$$

$$14 \quad F_{cost}(1, 14) = 9493 + 350 = 9843$$

$$12 \quad F_{cost}(1, 12) = 9208 + 0 = 9208$$

$J = 2$:   2nd hour

Feasible states are $\{12, 14, 15\} = \{K\}$, so $X = 3$. Suppose two strategies are saved at each stage, so $N = 2$, and $\{L\} = \{12, 14\}$,

$$\frac{K}{15} \quad F_{cost}(2, 15) = \min_{\{12,14\}} [P_{cost}(2, 15) + S_{cost}(1, L: 2, 15) + F_{cost}(1, L)]$$

$$= 11301 + \min \begin{bmatrix} (350 + 9208) \\ (0 + 9843) \end{bmatrix} = 20859$$

and so on.

Case 2

In Case 2, complete enumeration is tried with a limit of $(2^4 - 1) = 15$ dispatches each of the eight hours, so that there is a theoretical maximum of $15^8 = 2.56 \cdot 10^9$ possibilities. Fortunately, most of these are not feasible because they do not supply sufficient capacity, and can be discarded with little analysis required.

Figure 5.7 illustrates the computational process for the first 4 h for Case 2. On the figure itself, the circles denote states each hour. The numbers within the circles are the "pointers." That is, they denote the state number in the previous hour that provides the path to that particular state in the current hour. For example, in hour 2, the minimum costs for states 12, 13, 14, and 15, all result from transitions from state 12 in hour 1. Costs shown on the connections are the start-up costs. At each state, the figures shown are the hourly cost/total cost.

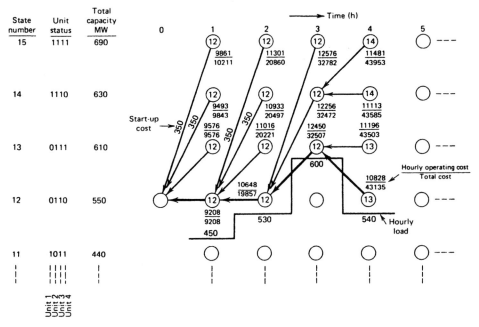

FIG. 5.7 Example 5E, Cases 1 and 2 (first 4 h).

In Case 2, the true optimal commitment is found. That is, it is less expensive to turn on the less efficient peaking unit, number 4, for hour 3, than to start up the more efficient unit 1 for that period. By hour 3, the difference in total cost is ₨165, or ₨0.104/MWh. This is not an insignificant amount when compared with the fuel cost per MWh for an average thermal unit with a net heat rate of 10,000 Btu/kWh and a fuel cost of ₨2.00 MBtu. A savings of ₨165 every 3 h is equivalent to ₨481,800/yr.

The total 8-h trajectories for Cases 1 and 2 are shown in Figure 5.8. The neglecting of start-up and shut-down restrictions in these two cases permits the shutting down of all but unit 3 in hours 6 and 7. The only difference in the two trajectories occurs in hour 3, as discussed in the previous paragraph.

Case 3

In case 3, the original unit data are used so that the minimum shut-down and operating times are observed. The forward dynamic-programming algorithm was repeated for the same 8-h period. Complete enumeration was used. That is, the upper bound on X shown in the flowchart was 15. Three different values for N, the number of strategies saved at each stage, were taken as 4, 8, and 10. The same trajectory was found for values of 8 and 10. This trajectory is shown in Figure 5.9. However, when only four strategies were saved, the procedure flounders (i.e., fails to find a feasible path) in

150

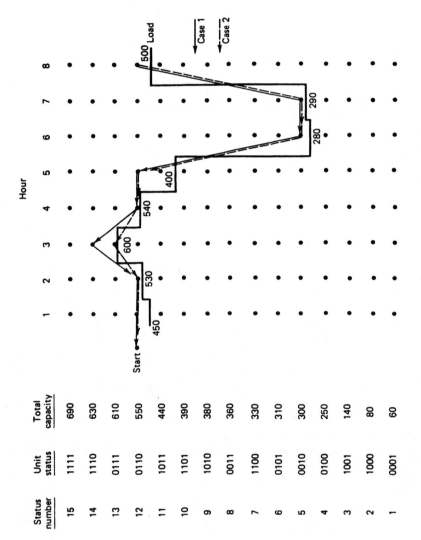

Status number	Unit status	Total capacity
15	1111	690
14	1110	630
13	0111	610
12	0110	550
11	1011	440
10	1101	390
9	1010	380
8	0011	360
7	1100	330
6	0101	310
5	0010	300
4	0100	250
3	1001	140
2	1000	80
1	0001	60

FIG. 5.8 Example 5E, Cases 1 and 2 (complete solutions).

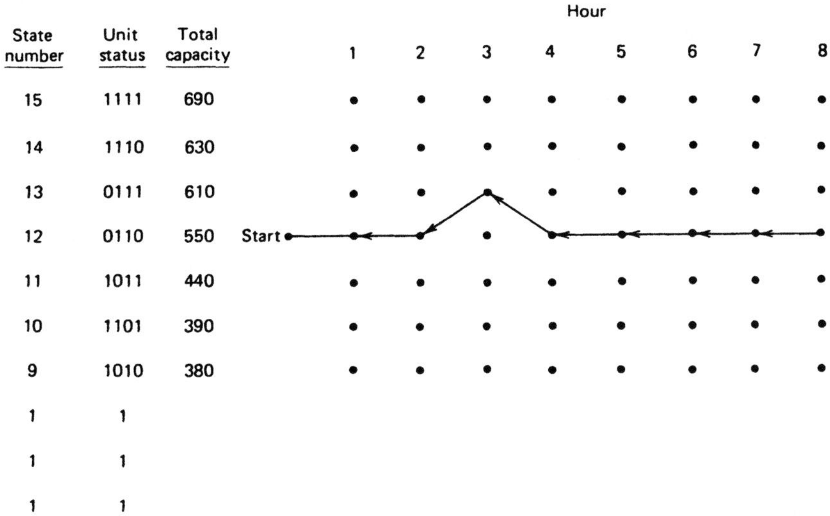

State number	Unit status	Total capacity	Hour							
			1	2	3	4	5	6	7	8
15	1111	690	•	•	•	•	•	•	•	•
14	1110	630	•	•	•	•	•	•	•	•
13	0111	610	•	•	•	•	•	•	•	•
12	0110	550	Start •	•	•	•	•	•	•	•
11	1011	440	•	•	•	•	•	•	•	•
10	1101	390	•	•	•	•	•	•	•	•
9	1010	380	•	•	•	•	•	•	•	•
1	1									
1	1									
1	1									

FIG. 5.9   Example 5E, Case 3.

hour 8, because the lowest cost strategies in hour 7 have shut down units that cannot be restarted in hour 8 because of minimum unit downtime rules.

The practical remedy for this deficiency in the method shown in Figure 5.4 is to return to a period prior to the low-load hours and temporarily keep more (i.e., higher cost) strategies. This will permit keeping a nominal number of strategies at each stage. The other alternative is, of course, the method used here: run the entire period with more strategies saved.

These cases can be summarized in terms of the total costs found for the 8-h period, as shown in Table 5.6. These cases illustrate the forward dynamic-programming method and also point out the problems involved in the practical application of the method.

TABLE 5.6   Summary of Cases 1–3

Case	Conditions	Total Cost (R)
1	Priority order. Up and down times neglected	73439
2	Enumeration ($X \leq 15$) with 4 strategies (N) saved. Up and down times neglected	73274
3	$X \leq 15$. Up and down times observed	
	$N = 4$ strategies	No solution
	$B = 8$ strategies	74110
	$N = 10$ strategies	74110

5.2.3   Lagrange Relaxation Solution

The dynamic-programming method of solution of the unit commitment problem has many disadvantages for large power systems with many generating units. This is because of the necessity of forcing the dynamic-programming solution to search over a small number of commitment states to reduce the number of combinations that must be tested in each time period.

In the Lagrange relaxation technique these disadvantages disappear (although other technical problems arise and must be addressed, as we shall see). This method is based on a dual optimization approach as introduced in Appendix 3A and further expanded in the appendix to this chapter. (The reader should be familiar with both of these appendices before proceeding further.)

We start by defining the variable U_i^t as:

$$U_i^t = 0 \text{ if unit } i \text{ is off-line during period } t$$

$$U_i^t = 1 \text{ if unit } i \text{ is on-line during period } t$$

We shall now define several constraints and the objective function of the unit commitment problem:

1. Loading constraints:

$$P_{\text{load}}^t - \sum_{i=1}^{N} P_i^t U_i^t = 0 \quad \text{for } t = 1 \dots T \tag{5.2}$$

2. Unit limits:

$$U_i^t P_i^{\text{min}} \le P_i^t \le U_i^t P_i^{\text{max}} \quad \text{for } i = 1 \dots N \quad \text{and} \quad t = 1 \dots T \tag{5.3}$$

3. Unit minimum up- and down-time constraints. Note that other constraints can easily be formulated and added to the unit commitment problem. These include transmission security constraints (see Chapter 11), generator fuel limit constraints, and system air quality constraints in the form of limits on emissions from fossil-fired plants, spinning reserve constraints, etc.

4. The objective function is:

$$\sum_{t=1}^{T} \sum_{i=1}^{N} [F_i(P_i^t) + \text{Start up cost}_{i,t}] U_i^t = F(P_i^t, U_i^t) \tag{5.4}$$

We can then form the Lagrange function similar to the way we did in the economic dispatch problem:

$$\mathcal{L}(P, U, \lambda) = F(P_i^t, U_i^t) + \sum_{t=1}^{T} \lambda^t \left(P_{\text{load}}^t - \sum_{i=1}^{N} P_i^t U_i^t \right) \tag{5.5}$$

The unit commitment problem requires that we minimize the Lagrange function above, subject to the local unit constraints 2 and 3, which can be applied to each unit separately. Note:

1. The cost function, $F(P_i^t, U_i^t)$, together with constraints 2 and 3 are each *separable over units*. That is, what is done with one unit does not affect the cost of running another unit, as far as the cost function and the unit limits (constraint 2) and the unit up- and down-time (constraint 3) are concerned.
2. Constraints 1 are *coupling constraints* across the units so that what we do to one unit affects what will happen on other units if the coupling constraints are to be met.

The Lagrange relaxation procedure solves the unit commitment problem by "relaxing" or temporarily ignoring the coupling constraints and solving the problem as if they did not exist. This is done through the dual optimization procedure as explained in the appendix of this chapter. The dual procedure attempts to reach the constrained optimum by maximizing the Lagrangian with respect to the Lagrange multipliers, while minimizing with respect to the other variables in the problem; that is:

$$q^*(\lambda) = \max_{\lambda^t} q(\lambda) \tag{5.6}$$

where

$$q(\lambda) = \min_{P_i^t, U_i^t} \mathscr{L}(P, U, \lambda) \tag{5.7}$$

This is done in two basic steps:

Step 1   Find a value for each λ^t which moves $q(\lambda)$ toward a larger value.
Step 2   Assuming that the λ^t found in step 1 are now fixed, find the minimum of $\mathscr{L}$ by adjusting the values of P^t and U^t.

The adjustment of the λ^t values will be dealt with at a later time in this section; assume then that a value has been chosen for all the λ^t and that they are now to be treated as fixed numbers. We shall minimize the Lagrangian as follows. First, we rewrite the Lagrangian as:

$$\mathscr{L} = \sum_{t=1}^{T} \sum_{i=1}^{N} [F_i(P_i^t) + \text{Start up cost}_{i,t}]U_i^t + \sum_{t=1}^{T} \lambda^t \left(P_{\text{load}}^t - \sum_{i=1}^{N} P_i^t U_i^t \right) \tag{5.8}$$

This is now rewritten as:

$$\mathscr{L} = \sum_{t=1}^{T} \sum_{i=1}^{N} [F_i(P_i^t) + \text{Start up cost}_{i,t}]U_i^t + \sum_{t=1}^{T} \lambda^t P_{\text{load}}^t - \sum_{t=1}^{T} \sum_{i=1}^{N} \lambda^t P_i^t U_i^t \tag{5.9}$$

The second term above is constant and can be dropped (since the λ^t are fixed). Finally, we write the Lagrange function as:

$$\mathscr{L} = \sum_{i=1}^{N} \left(\sum_{t=1}^{T} \{ [F_i(P_i^t) + \text{Start up cost}_{i,t}] U_i^t - \lambda^t P_i^t U_i^t \} \right) \qquad (5.10)$$

Here, we have achieved our goal of separating the units from one another. The term inside the outer brackets; that is:

$$\sum_{t=1}^{T} \{ [F_i(P_i^t) + \text{Start up cost}_{i,t}] U_i^t - \lambda^t P_i^t U_i^t \}$$

can be solved separately for each generating unit, without regard for what is happening on the other generating units. The minimum of the Lagrangian is found by solving for the minimum for each generating unit over all time periods; that is:

$$\min q(\lambda) = \sum_{i=1}^{N} \min \sum_{t=1}^{T} \{ [F_i(P_i^t) + \text{Start up cost}_{i,t}] U_i^t - \lambda^t P_i^t U_i^t \} \qquad (5.11)$$

Subject to

$$U_i^t P_i^{\min} \le P_i^t \le U_i^t P_i^{\max} \quad \text{for } t = 1 \ldots T$$

and the up- and down-time constraints. This is easily solved as a dynamic-programming problem in one variable. This can be visualized in the figure below, which shows the only two possible states for unit i (i.e., $U_i^t = 0$ or 1):

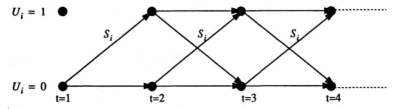

where S_i is the start-up cost for unit i.

At the $U_i^t = 0$ state, the value of the function to minimized is trivial (i.e., it equals zero); at the state where $U_i^t = 1$, the function to be minimized is (the start-up cost is dropped here since the minimization is with respect to P_i^t):

$$\min [F_i(P_i) - \lambda^t P_i^t] \qquad (5.12)$$

The minimum of this function is found by taking the first derivative:

$$\frac{d}{dP_i^t} [F_i(P_i) - \lambda^t P_i^t] = \frac{d}{dP_i^t} F_i(P_i^t) - \lambda^t = 0 \qquad (5.13)$$

The solution to this equation is

$$\frac{d}{dP_i^t} F_i(P_i^{opt}) = \lambda^t \qquad (5.14)$$

There are three cases to be concerned with depending on the relation of P_i^{opt} and the unit limits:

1. If $P_i^{opt} \leq P_i^{min}$. then:

$$\min [F_i(P_i) - \lambda^t P_i^t] = F_i(P_i^{min}) - \lambda^t P_i^{min} \qquad (5.15a)$$

2. If $P_i^{min} \leq P_i^{opt} \leq P_i^{max}$, then:

$$\min [F_i(P_i) - \lambda^t P_i^t] = F_i(P_i^{opt}) - \lambda^t P_i^{opt} \qquad (5.15b)$$

3. If $P_i^{opt} \geq P_i^{max}$, then:

$$\min [F_i(P_i) - \lambda^t P_i^t] = F_i(P_i^{max}) - \lambda^t P_i^{max} \qquad (5.15c)$$

The solution of the two-state dynamic program for each unit proceeds in the normal manner as was done for the forward dynamic-programming solution of the unit commitment problem itself. Note that since we seek to minimize $[F_i(P_i) - \lambda^t P_i^t]$ at each stage and that when $U_i^t = 0$ this value goes to zero, then the only way to get a value lower is to have

$$[F_i(P_i) - \lambda^t P_i^t] < 0$$

The dynamic program should take into account all the start-up costs, S_i, for each unit, as well as the minimum up and down time for the generator. Since we are solving for each generator independently, however, we have avoided the dimensionality problems that affect the dynamic-programming solution.

5.2.3.1   Adjusting λ

So far, we have shown how to schedule generating units with fixed values of λ^t for each time period. As shown in the appendix to this chapter, the adjustment of λ^t must be done carefully so as to maximize $q(\lambda)$. Most references to work on the Lagrange relaxation procedure use a combination of gradient search and various heuristics to achieve a rapid solution. Note that unlike in the appendix, the λ here is a vector of values, each of which must be adjusted. Much research in recent years has been aimed at ways to speed the search for the correct values of λ for each hour. In Example 5D, we shall use the same technique of adjusting λ for each hour that is used in the appendix. For the unit commitment problem solved in Example 5D, however, the λ adjustment factors are different:

$$\lambda^t = \lambda^t + \left[\frac{d}{d\lambda} q(\lambda)\right]\alpha \qquad (5.16)$$

where

$$\alpha = 0.01 \quad \text{when} \quad \frac{d}{d\lambda} q(\lambda) \text{ is positive} \tag{5.17}$$

and

$$\alpha = 0.002 \quad \text{when} \quad \frac{d}{d\lambda} q(\lambda) \text{ is negative} \tag{5.18}$$

Each λ^t is treated separately. The reader should consult the references listed at the end of this chapter for more efficient methods of adjusting the λ values. The overall Lagrange relaxation unit commitment algorithm is shown in Figure 5.10.

Reference 15 introduces the use of what this text called the "relative duality gap" or $(J^* - q^*)/q^*$. The relative duality gap is used in Example 5D as a measure of the closeness to the solution. Reference 15 points out several useful things about dual optimization applied to the unit commitment problem.

1. For large, real-sized, power-system unit commitment calculations, the duality gap does become quite small as the dual optimization proceeds, and its size can be used as a stopping criterion. The larger the problem (larger number of generating units), the smaller the gap.
2. The convergence is unstable at the end, meaning that some units are being switched in and out, and the process never comes to a definite end.
3. There is no guarantee that when the dual solution is stopped, it will be at a feasible solution.

All of the above are demonstrated in Example 5D. The duality gap is large at the beginning and becomes progressively smaller as the iterations progress. The solution reaches a commitment schedule when at least enough generation is committed so that an economic dispatch can be run, and further iterations only result in switching marginal units on and off. Finally, the loading constraints are not met by the dual solution when the iterations are stopped.

Many of the Lagrange relaxation unit commitment programs use a few iterations of a dynamic-programming algorithm to get a good starting point, then run the dual optimization iterations, and finally, at the end, they use heuristic logic or restricted dynamic programming to get to a final solution. The result is a solution that is not limited to search windows, such as had to be done in strict application of dynamic programming.

EXAMPLE 5D

In this example, a three-generator, four-hour unit commitment problem will be solved. The data for this problem are as follows. Given the three generating

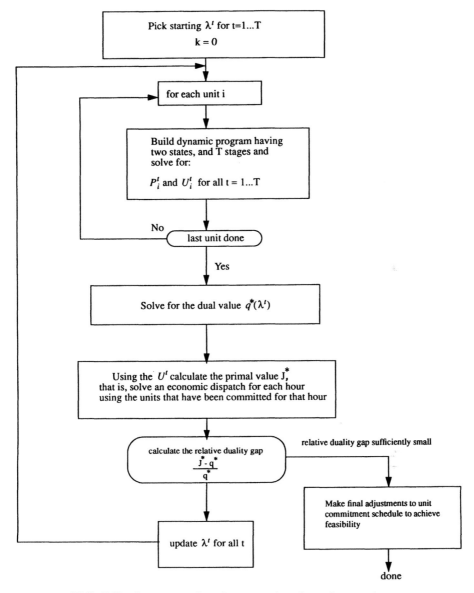

FIG. 5.10 Lagrange relaxation procedure for unit commitment.

units below:

$$F_1(P_1) = 500 + 10P_1 + 0.002P_1^2 \quad \text{and} \quad 100 < P_1 < 600$$

$$F_2(P_2) = 300 + 8P_2 + 0.0025P_2^2 \quad \text{and} \quad 100 < P_2 < 400$$

$$F_3(P_3) = 100 + 6P_3 + 0.005P_3^2 \quad \text{and} \quad 50 < P_3 < 200$$

Load:

t	P_{load}^t(MW)
1	170
2	520
3	1100
4	330

No start-up costs, no minimum up- or down-time constraints.

This example is solved using the Lagrange relaxation technique. Shown below are the results of several iterations, starting from an initial condition where all the λ^t values are set to zero. An economic dispatch is run for each hour, provided there is sufficient generation committed that hour. If there is not enough generation committed, the total cost for that hour is set arbitrarily to 10,000. Once each hour has enough generation committed, the primal value J^* simply represents the total generation cost summed over all hours as calculated by the economic dispatch.

The dynamic program for each unit with a $\lambda^t = 0$ for each hour will always result in all generating units off-line.

Iteration 1

Hour	λ	u_1	u_2	u_3	P_1	P_2	P_3	$P_{load}^t - \sum_{i=1}^{N} P_i^t U_i^t$	P_1^{edc}	P_2^{edc}	P_3^{edc}
1	0	0	0	0	0	0	0	170	0	0	0
2	0	0	0	0	0	0	0	520	0	0	0
3	0	0	0	0	0	0	0	1100	0	0	0
4	0	0	0	0	0	0	0	330	0	0	0

$$q(\lambda) = 0.0, \qquad J^* = 40,000, \quad \text{and} \quad \frac{J^* - q^*}{q^*} = \text{undefined}$$

In the next iteration, the λ^t values have been increased. To illustrate the use of dynamic programming to schedule each generator, we will detail the DP steps for unit 3:

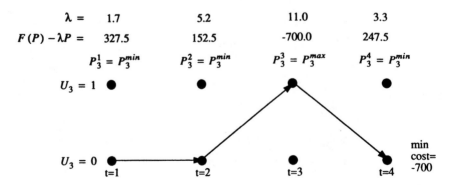

The result is to schedule unit 3 off during hours 1, 2, and 4 and on during hour 3. Further, unit 3 is scheduled to be at its maximum of 200 MW during hour 3. The results, after all the units have been scheduled by DP, are as follows.

Iteration 2

Hour	λ	u_1	u_2	u_3	P_1	P_2	P_3	$P_{load}^t - \sum\limits_{i=1}^{N} P_i^t U_i^t$	P_1^{edc}	P_2^{edc}	P_3^{edc}
1	1.7	0	0	0	0	0	0	170	0	0	0
2	5.2	0	0	0		0	0	520	0	0	0
3	11.0	0	1	1	0	400	200	500	0	0	0
4	3.3	0	0	0	0	0	0	330	0	0	0

$$q(\lambda) = 14{,}982, \qquad J^* = 40{,}000, \quad \text{and} \quad \frac{J^* - q^*}{q^*} = 1.67$$

Iteration 3

Hour	λ	u_1	u_2	u_3	P_1	P_2	P_3	$P_{load}^t - \sum\limits_{i=1}^{N} P_i^t U_i^t$	P_1^{edc}	P_2^{edc}	P_3^{edc}
1	3.4	0	0	0	0	0	0	170	0	0	0
2	10.4	0	1	1	0	400	200	-80	0	320	200
3	16.0	1	1	1	600	400	200	-100	500	400	200
4	6.6	0	0	0	0	0	0	330	0	0	0

$$q(\lambda) = 18{,}344, \qquad J^* = 36{,}024, \quad \text{and} \quad \frac{J^* - q^*}{q^*} = 0.965$$

Iteration 4

Hour	λ	u_1	u_2	u_3	P_1	P_2	P_3	$P_{load}^t - \sum\limits_{i=1}^{N} P_i^t U_i^t$	P_1^{edc}	P_2^{edc}	P_3^{edc}
1	5.1	0	0	0	0	0	0	170	0	0	0
2	10.24	0	1	1	0	400	200	-80	0	320	200
3	15.8	1	1	1	600	400	200	-100	500	400	200
4	9.9	0	1	1	0	380	200	-250	0	130	200

$$q(\lambda) = 19{,}214, \qquad J^* = 28{,}906, \quad \text{and} \quad \frac{J^* - q^*}{q^*} = 0.502$$

Iteration 5

Hour	λ	u_1	u_2	u_3	P_1	P_2	P_3	$P_{load}^t - \sum_{i=1}^{N} P_i^t U_i^t$	P_1^{edc}	P_2^{edc}	P_3^{edc}
1	6.8	0	0	0	0	0	0	170	0	0	0
2	10.08	0	1	1	0	400	200	−80	0	320	200
3	15.6	1	1	1	600	400	200	−100	500	400	200
4	9.4	0	0	1	0	0	200	130	0	0	0

$$q(\lambda) = 19.532, \qquad J^* = 36{,}024, \quad \text{and} \quad \frac{J^* - q^*}{q^*} = 0.844$$

Iteration 6

Hour	λ	u_1	u_2	u_3	P_1	P_2	P_3	$P_{load}^t - \sum_{i=1}^{N} P_i^t U_i^t$	P_1^{edc}	P_2^{edc}	P_3^{edc}
1	8.5	0	0	1	0	0	200	−30	0	0	170
2	9.92	0	1	1	0	384	200	−64	0	320	200
3	15.4	1	1	1	600	400	200	−100	500	400	200
4	10.7	0	1	1	0	400	200	−270	0	130	200

$$q(\lambda) = 19{,}442, \qquad J^* = 20{,}170, \quad \text{and} \quad \frac{J^* - q^*}{q^*} = 0.037$$

The commitment schedule does not change significantly with further iterations, although it is not by any means stable. Further iterations do reduce the duality gap somewhat, but these iterations are unstable in that unit 2 is on the borderline between being committed and not being committed, and is switched in and out with no final convergence. After 10 iterations, $q(\lambda) = 19{,}485$, $J^* = 20{,}017$, and $(J^* - q^*)/q^* = 0.027$. This latter value will not go to zero, nor will the solution settle down to a final value; therefore, the algorithm must stop when $(J^* - q^*)/q^*$ is sufficiently small (e.g., less than 0.05 in this case).

APPENDIX
Dual Optimization on a Nonconvex Problem

We introduced the concept of dual optimization in Appendix 3A and pointed out that when the function to be optimized is convex, and the variables are continuous, then the maximization of the dual function gives the identical result as minimizing the primal function. Dual optimization is also used in solving the unit commitment problem. However, in the unit commitment problem there are variables that must be restricted to two values: 1 or 0. These 1–0 variables cause a great deal of trouble and are the reason for the difficulty in solving the unit commitment problem.

The application of the dual optimization technique to the unit commitment problem has been given the name "Lagrange relaxation" and the formulation of the unit commitment problem using this method is shown in the text in Section 5.2.3. In this appendix, we illustrate this technique with a simple geometric problem. The problem is structured with 1–0 variables which makes it clearly nonconvex. Its form is generally similar to the form of the unit commitment problems, but that is incidental for now.

The sample problem to be solved is given below. It illustrates the ability of the dual optimization technique to solve the unit commitment problem. Given:

$$J(x_1, x_2, u_1, u_2) = (0.25x_1^2 + 15)u_1 + (0.255x_2^2 + 15)u_2 \qquad (5A.1)$$

subject to:

$$\omega = 5 - x_1 u_1 - x_2 u_2 \qquad (5A.2)$$

and

$$0 \le x_1 \le 10 \qquad (5A.3)$$

$$0 \le x_2 \le 10 \qquad (5A.4)$$

where x_1 and x_2 are continuous real numbers, and:

$$u_1 = 1 \quad \text{or} \quad 0$$

$$u_2 = 1 \quad \text{or} \quad 0$$

Note that in this problem we have two functions, one in x_1 and the other in x_2. The functions were chosen to demonstrate certain phenomena in a dual optimization. Note that the functions are numerically close and only differ by a small, constant amount. Each of these functions is multiplied by a 1–0 variable and combined into the overall objective function. There is also a constraint that combines the x_1 and x_2 variables again with the 1–0 variables. There are four possible solutions.

1. If u_1 and u_2 are both zero, the problem cannot have a solution since the equality constraint cannot be satisfied.
2. If $u_1 = 1$ and $u_2 = 0$, we have the trivial solution that $x_1 = 5$ and x_2 does not enter into the problem anymore. The objective function is 21.25.
3. If $u_1 = 0$ and $u_2 = 1$, then we have the trivial result that $x_2 = 5$ and x_1 does not enter into the problem. The objective function is 21.375.
4. If $u_1 = 1$ and $u_2 = 1$, we have a simple Lagrange function of:

$$\mathscr{L}(x_1, x_2, \lambda) = (0.25x_1^2 + 15) + (0.255x_2^2 + 15) + \lambda(5 - x_1 - x_2) \qquad (5A.5)$$

The resulting optimum is at $x_1 = 2.5248$, $x_2 = 2.4752$, and $\lambda = 1.2642$, with an objective function value of 33.1559. Therefore, we know the optimum value for this problem; namely, $u_1 = 1$, $u_2 = 0$, and $x_1 = 5$.

What we have done, of course, is to enumerate all possible combinations of the 1–0 variables and then optimize over the continuous variables. When there are more than a few 1–0 variables, this cannot be done because of the large number of possible combinations. However, there is a systematic way to solve this problem using the dual formulation.

The Lagrange relaxation method solves problems such as the one above, as follows. Define the Lagrange function as:

$$\mathcal{L}(x_1, x_2, u_1, u_2, \lambda) = (0.25x_1^2 + 15)u_1 + (0.255x_2^2 + 15)u_2$$
$$+ \lambda(5 - x_1u_1 - x_2u_2) \qquad (5A.6)$$

As shown in Appendix 3A, we define $q(\lambda)$ as:

$$q(\lambda) = \min_{x_1, x_2, u_1, u_2} \mathcal{L} \qquad (5A.7)$$

where x_1, x_2, u_1, u_2 obey the limits and the 1–0 conditions as before. The dual problem is then to find

$$q^*(\lambda) = \max_{\lambda \geq 0} q(\lambda) \qquad (5A.8)$$

This is different from the dual optimization approach used in the Appendix 3A because of the presence of the 1–0 variables. Because of the presence of the 1–0 variables we cannot eliminate variables; therefore, we keep all the variables in the problem and proceed in alternating steps as shown in the Appendix 3A.

Step 1   Pick a value for λ^k and consider it fixed. Now the Lagrangian function can be minimized. This is much simpler than the situation we had before since we are trying to minimize

$$(0.25x_1^2 + 15)u_1 + (0.255x_2^2 + 15)u_2 + \lambda^k(5 - x_1u_1 - x_2u_2)$$

where the value of λ^k is fixed.

We can then rearrange the equation above as:

$$(0.25x_1^2 + 15 - x_1\lambda^k)u_1 + (0.255x_2^2 + 15 - x_2\lambda^k)u_2 + \lambda^k 5$$

The last term above is fixed and we can ignore it. The other terms are now given in such a way that the minimization of this function is relatively easy. Note that the minimization is now over two terms, each being multiplied by a 1–0 variable. Since these two terms are summed in the Lagrangian, we can minimize the entire function by minimizing each term separately. Since each term is the product of a function in x and λ (which is fixed), and these are all multiplied by the 1–0 variable u, then the minimum will be zero (that is with $u = 0$) or it will be negative, with $u = 1$ and the value of x set so that the term inside the parentheses is negative. Looking at the first term, the optimum value of x_1 is found by (ignore u_1 for a moment):

$$\frac{d}{dx_1}(0.25x_1^2 + 15 - x_1\lambda^k) = 0 \qquad (5A.9)$$

If the value of x_1 which satisfies the above falls outside the limits of 0 and 10 for x_1, we force x_1 to the limit violated. If the term in the first brackets

$$(0.25x_1^2 + 15 - x_1\lambda^k)$$

is positive, then we can minimize the Lagrangian by merely setting $u_1 = 0$; otherwise $u_1 = 1$.

Looking at the second term, the optimum value of x_2 is found by (again, ignore u_2):

$$\frac{d}{dx_2}(0.255x_2^2 + 15 - x_2\lambda^k) = 0 \qquad (5A.10)$$

and if the value of x_2 which satisfies the above value falls outside the 0 to 10 limits on x_2, we set it to the violated limit. Similarly, the term in the second brackets

$$(0.255x_2^2 + 15 - x_2\lambda^k)$$

is evaluated. If it is positive, then we minimize the Lagrangian by making $u_2 = 0$; otherwise $u_2 = 1$. We have now found the minimum value of $\mathcal{L}$ with a specified fixed value of λ^k.

Step 2   Assume that the variables x_1, x_2, u_1, u_2 found in step 1 are fixed and find a value for λ that maximizes the dual function. In this case, we cannot solve for the maximum since $q(\lambda)$ is unbounded with respect to λ. Instead, we form the gradient of $q(\lambda)$ with respect to λ and we adjust λ so as to move in the direction of increasing $q(\lambda)$. That is, given

$$\nabla q = \frac{dq}{d\lambda} \qquad (5A.11)$$

which for our problem is

$$\frac{dq}{d\lambda} = 5 - x_1 u_1 - x_2 u_2 \qquad (5A.12)$$

we adjust λ according to

$$\lambda^{k+1} = \lambda^k + \frac{dq}{d\lambda}\alpha \qquad (5A.13)$$

where α is a multiplier chosen to move λ only a short distance. (This is simply a gradient search method as was introduced in Chapter 3). Note also, that if both u_1 and u_2 are zero, the gradient will be 5, indicating a positive value telling us to increase λ. Eventually, increasing λ will result in a negative value for

$$(0.25x_1^2 + 15 - x_1\lambda^k)$$

or for

$$(0.255x_2^2 + 15 - x_2\lambda^k)$$

or for both, and this will cause u_1 or u_2, or both, to be set to 1. Once the value of λ is increased, we go back to step 1 and find the new values for x_1, x_2, u_1, u_2 again.

The real difficulty here is in not increasing λ by too much. In the example presented above, the following scheme was imposed on the adjustment of λ:

- If $\dfrac{dq}{d\lambda}$ is positive, then use $\alpha = 0.2$.

- If $\dfrac{dq}{d\lambda}$ is negative, then use $\alpha = 0.005$.

This lets λ approach the solution slowly, and if it overshoots, it backs up very slowly. This is a common technique to make a gradient "behave."

We must also note that, given the few variables we have, and given the fact that two of them are 1–0 variables, the value of λ will not converge to the value needed to minimize the Lagrangian. In fact, it is seldom possible to find a λ^k that will make the problem feasible with respect to the equality constraint. However, when we have found the values for u_1 and u_2 at any iteration, we can then calculate the minimum of $J(x_1, x_2, u_1, u_2)$ by solving for the minimum of

$$[(0.25x_1^2 + 15)u_1 + (0.255x_2^2 + 15)u_2 + \lambda(5 - x_1u_1 - x_2u_2)]$$

using the techniques in Appendix 3A (since the u_1 and u_2 variables are now known).

The solution to this minimum will be at $x_1 = \overline{x_1}$, $x_2 = \overline{x_2}$ and $\lambda = \bar{\lambda}$. For the case where u_1 and u_2 are both zero, we shall arbitrarily set this value to a large value (here we set it to 50). We shall call this minimum value

TABLE 5.7 Dual Optimization on a Sample Problem

Iteration	λ	u_1	u_2	x_1	x_2	q^*	ω	$\bar{\lambda}$	$\overline{x}_1$	$\overline{x}_2$	J^*	$\dfrac{J^* - q^*}{q^*}$
1	0	0	0	0	0	0	5.0	—	—	—	50.0	—
2	1.0	0	0	2.0	1.9608	5.0	5.0	—	—	—	50.0	9.0
3	2.0	0	0	4.0	3.9216	10.0	5.0	—	—	—	50.0	4.0
4	3.0	0	0	6.0	5.8824	15.0	5.0	—	—	—	50.0	2.33
5	4.0	1	1	8.0	7.8431	18.3137	-10.8431	1.2624	2.5248	2.4752	33.1559	0.8104
6	3.9458	1	1	7.8916	7.7368	18.8958	-10.6284	1.2624	2.5248	2.4752	33.1559	0.7546
7	3.8926	1	0	7.7853	7.6326	19.3105	-2.7853	2.5	5.0	—	21.25	0.1004
8	3.8787	1	0	7.7574	7.6053	19.3491	-2.7574	2.5	5.0	—	21.25	0.0982

165

$J^*(\overline{x_1}, \overline{x_2}, u_1, u_2)$ and we shall observe that it starts out with a large value, and decreases, while the dual value $q^*(\lambda)$ starts out with a value of zero, and increases. Since there are 1–0 variables in this problem, the primal values and the dual values never become equal. The value $J^* - q^*$ is called the duality gap and we shall call the value

$$\frac{J^* - q^*}{q^*}$$

the relative duality gap.

The presence of the 1–0 variables causes the algorithm to oscillate around a solution with one or more of the 1–0 variables jumping from 1 to 0 to 1, etc. In such cases, the user of the Lagrange relaxation algorithm must stop the algorithm, based on the value of the relative duality gap.

The iterations starting from $\lambda = 0$ are shown in Table 5.7. The table shows eight iterations and illustrates the slow approach of λ toward the threshold when both of the 1–0 variables flip from 0 to 1. Also note that ω became negative and the value of λ must now be decreased. Eventually, the optimal solution is reached and the relative duality gap becomes small. However, as is typical with the dual optimization on a problem with 1–0 variables, the solution is not stable and if iterated further it exhibits further changes in the 1–0 variables as λ is adjusted. Both the q^* and J^* values and the relative duality gap are shown in Table 5.7.

PROBLEMS

5.1 Given the unit data in Tables 5.8 and 5.9, use forward dynamic-programming to find the optimum unit commitment schedules covering the 8-h period. Table 5.9 gives all the combinations you need, as well as the operating cost for each at the loads in the load data. A " × " indicates that a combination cannot supply the load. The starting conditions are: at the beginning of the first period units 1 and 2 are up, units 3 and 4 are down and have been down for 8 h.

5.2 Table 5.10 presents the unit characteristics and load pattern for a five-unit, four-time-period problem. Each time period is 2 h long. The input–output characteristics are approximated by a straight line from min to max generation, so that the incremental heat rate is constant. Unit no-load and start-up costs are given in terms of heat energy requirements.

a. Develop the priority list for these units and solve for the optimum unit commitment. Use a strict priority list with a search range of three ($X = 3$) and save no more than three strategies ($N = 3$). Ignore min up-/min down-times for units.

b. Solve the same commitment problem using the strict priority list with $X = 3$ and $N = 3$ as in part a, but obey the min up/min down time rules.

TABLE 5.8   Unit Commitment Data for Problem 5.1

Unit	Max (MW)	Min (MW)	Incremental Heat Rate (Btu/kWh)	No-Load Energy Input (MBtu/h)	Start-Up Energy (MBtu)
1	500	70	9950	300	800
2	250	40	10200	210	380
3	150	30	11000	120	110
4	150	30	11000	120	110

Load data (all time periods = 2 h);

Time Period	Load (MW)
1	600
2	800
3	700
4	950

Start-up and shut-down rules:

Unit	Minimum Up Time (h)	Minimum Down Time (h)
1	2	2
2	2	2
3	2	4
4	2	4

Fuel cost = 1.00 ₨/MBtu.

TABLE 5.9   Unit Combinations and Operating Cost for Problem 5.1

Combination	Unit 1	Unit 2	Unit 3	Unit 4	Load 600 MW	Load 700 MW	Load 800 MW	Load 950 MW
A	1	1	0	0	6505	7525	×	×
B	1	1	1	0	6649	7669	8705	×
C	1	1	1	1	6793	7813	8833	10475

1 = up; 0 = down.

c. (Optional) Find the optimum unit commitment without use of a strict priority list (i.e., all 32 unit on/off combinations are valid). Restrict the search range to decrease your effort. Obey the min up-/min down-time rules.

When using a dynamic-programming method to solve a unit commitment problem with minimum up- and down-time rules, one must save an additional piece of information at each state, each hour. This information

TABLE 5.10   The Unit Characteristic and Load Pattern for Problem 5.2

Unit	Max (MW)	Net Full-Load Heat Rate (Btu/kWh)	Incremental Heat Rate (Btu/kWh)	Min (MW)	No-Load Cost (MBtu/h)	Start-Up Cost (MBtu)	Min Up/Down Time (h)
1	200	11000	9900	50	220	400	8
2	60	11433	10100	15	80	150	8
3	50	12000	10800	15	60	105	4
4	40	12900	11900	5	40	0	4
5	25	13500	12140	5	34	0	4

	Load Pattern	
Hours	Load (MW)	Conditions
1–2	250	1. Initially (prior to hour 1), only unit 1 is on and has been
3–4	320	on for 4 h.
5–6	110	2. Ignore losses, spinning reserve, etc. The only requirement
7–8	75	is that the generation be able to supply the load.
		3. Fuel costs for all units may be taken as 1.40 ₽/MBtu

simply tells us whether any units are ineligible to be shut down or started up at that state. If such units exist at a particular state, the transition cost, S_{cost}, to a state that violates the start-up/shut-down rules should be given a value of infinity.

5.3   Lagrange Relaxation Problem

Given the three generating units below:

$$F_1(P_1) = 30 + 10P_1 + 0.002P_1^2 \quad \text{and} \quad 100 < P_1 < 600$$
$$F_2(P_2) = 20 + 8P_2 + 0.0025P_2^2 \quad \text{and} \quad 100 < P_2 < 400$$
$$F_3(P_3) = 10 + 6P_3 + 0.005P_3^2 \quad \text{and} \quad 50 < P_3 < 200$$

Load:

t	P_{load}^t (MW)
1	300
2	500
3	1100
4	400

No start-up costs, no minimum up- or down-time constraints.

a. Solve for the unit commitment by conventional dynamic programming.

b. Set up and carry out four iterations of the Lagrange relaxation method. Let the initial values of λ^t be zero for $t = 1 \ldots 4$.

c. Resolve with the added condition that the third generator has a minimum up time of 2 h.

FURTHER READING

Some good introductory references to the unit commitment problem are found in references 1–3. A survey of the state-of-the-art (as of 1975) of unit commitment solutions is found in reference 4. References 5 and 6 provide a good look at two commercial unit commitment programs in present use.

References 7–11 deal with unit commitment as an integer-programming problem. Much of the pioneering work in this area was done by Garver (reference 7), who also sounded a note of pessimism in a discussion of reference 8, written together with Happ in 1968. Further research (references 9–11) has refined the unit commitment solution by integer programming but has never really overcome the Garver–Happ limitations presented in the 1968 discussion, thus leaving dynamic programming and Lagrange relaxation as the only viable solution techniques to large-scale unit commitment problems.

The reader should see references 12 and 13 for a discussion of valve-point loading and for a thorough development of economic dispatch via dynamic programming.

Reference 14 provides the reader with a good overview of unit commitment scheduling. References 15, 16, and 17 are recommended for an understanding of the Lagrange relaxation method, while references 18–21 cover some of the special problems encountered in unit commitment scheduling.

1. Baldwin, C. J., Dale, K. M., Dittrich, R. F., "A Study of Economic Shutdown of Generating Units in Daily Dispatch," *AIEE Transactions on Power Apparatus and Systems*, Vol. PAS-78, December 1959, pp. 1272–1284.

2. Burns, R. M., Gibson, C. A., "Optimization of Priority Lists for a Unit Commitment Program," IEEE Power Engineering Society Summer Meeting, Paper A-75-453-1, 1975.

3. Davidson, P. M., Kohbrman, F. J., Master, G. L., Schafer, G. R., Evans, J. R., Lovewell, K. M., Payne, T. B., "Unit Commitment Start-Stop Scheduling in the Pennsylvania–New Jersey–Maryland Interconnection," 1967 PICA Conference Proceedings, IEEE, 1967, pp. 127–132.

4. Gruhl, J., Schweppe, F., Ruane, M., "Unit Commitment Scheduling of Electric Power Systems," *Systems Engineering for Power: Status and Prospects*, Henniker, NH, U.S. Government Printing Office, Washington, DC, 1975.

5. Pang, C. K., Chen, H. C., "Optimal Short-Term Thermal Unit Commitment," *IEEE Transactions on Power Apparatus and Systems*, Vol. PAS-95, July/August 1976, pp. 1336–1346.

6. Happ, H. H., Johnson, P. C., Wright, W. J., "Large Scale Hydro-Thermal Unit Commitment—Method and Results," *IEEE Transactions on Power Apparatus and Systems*, Vol. PAS-90, May/June 1971, pp. 1373–1384.

7. Garver, L. L., "Power Generation Scheduling by Integer Programming—

Development of Theory," *AIEE Transactions on Power Apparatus and Systems*, Vol. PAS-82, February 1963, pp. 730–735.

8. Muckstadt, J. A., Wilson, R. C., "An Application of Mixed-Integer Programming Duality to Scheduling Thermal Generating Systems," *IEEE Transactions on Power Apparatus and Systems*, Vol. PAS-87, December 1968, pp. 1968–1978.

9. Ohuchi, A., Kaji, I., "A Branch-and-Bound Algorithm for Start-up and Shut-down Problems of Thermal Generating Units," *Electrical Engineering in Japan*, Vol. 95, No. 5, 1975, pp. 54–61.

10. Dillon, T. S., Egan, G. T., "Application of Combinational Methods to the Problems of Maintenance Scheduling and Unit Commitment in Large Power Systems," *Proceedings of IFAC Symposium on Large Scale Systems Theory and Applications*, Udine, Italy, 1976.

11. Dillon, T. S., Edwin, K. W., Kochs, H. D., Taud, R. J., "Integer Programming Approach to the Problem of Optimal Unit Commitment with Probabilistic Reserve Determination," *IEEE Transactions on Power Apparatus and Systems*, Vol. 97, November/December 1978, pp. 2154–2166.

12. Happ, H. H., Ille, W. B., Reisinger, R. M., "Economic System Operation Considering Valve Throttling Losses, I—Method of Computing Valve-Loop Heat Rates on Multivalve Turbines," *IEEE Transactions on Power Apparatus and Systems*, Vol. PAS-82, February 1963, pp. 609–615.

13. Ringlee, R. J., Williams, D. D., "Economic Dispatch Operation Considering Valve Throttling Losses, II—Distribution of System Loads by the Method of Dynamic Programming," *IEEE Transactions on Power Apparatus and Systems*, Vol. PAS-82, February 1963, pp. 615–622.

14. Cohen, A. I., Sherkat, V. R., "Optimization-Based Methods for Operations Scheduling, *Proceedings IEEE*, December 1987, pp. 1574–1591.

15. Bertsekas, D., Lauer, G. S., Sandell, N. R., Posbergh, T. A., "Optimal Short-Term Scheduling of Large-Scale Power Systems," *IEEE Transactions on Automatic Control*, Vol. AC-28, No. 1, January 1983, pp. 1–11.

16. Merlin, A., Sandrin, P., "A New Method for Unit Commitment at Electricité de France," *IEEE Transactions on Power Apparatus and Systems*, Vol. PAS-102, May 1983, pp. 1218–1225.

17. Zhuang, F., Galiana, F. D., "Towards a More Rigorous and Practical Unit Commitment by Lagrangian Relaxation," *IEEE Transactions on Power Systems*, Vol. 3, No. 2, May 1988, pp. 763–773.

18. Lee, F. N., Chen, Q., Breipohl, A., "Unit Commitment Risk with Sequential Rescheduling," *IEEE Transactions on Power Systems*, Vol. 6, No. 3, August 1991, pp. 1017–1023.

19. Vemuri, S., Lemonidis, L., "Fuel Constrained Unit Commitment," *IEEE Transactions on Power Systems*, Vol. 7, No. 1, February 1992, pp. 410–415.

20. Wang, C., Shahidepour, S. M., "Optimal Generation Scheduling with Ramping Costs," *1993 IEEE Power Industry Computer Applications Conference*, pp. 11–17.

21. Shaw, J. J., "A Direct Method for Security-Constrained Unit Commitment," IEEE Transactions Paper 94 SM 591-8 PWRS, presented at the IEEE Power Engineering Society Summer Meeting, San Francisco, CA, July 1994.

6 Generation with Limited Energy Supply

6.1 INTRODUCTION

The economic operation of a power system requires that expenditures for fuel be minimized over a period of time. When there is no limitation on the fuel supply to any of the plants in the system, the economic dispatch can be carried out with only the present conditions as data in the economic dispatch algorithm. In such a case, the fuel costs are simply the incoming price of fuel with, perhaps, adjustments for fuel handling and maintenance of the plant.

When the energy resource available to a particular plant (be it coal, oil, gas, water, or nuclear fuel) is a limiting factor in the operation of the plant, the entire economic dispatch calculation must be done differently. Each economic dispatch calculation must account for what happened before and what will happen in the future.

This chapter begins the development of solutions to the dispatching problem "over time." The techniques used are an extension of the familiar Lagrange formulation. Concepts involving slack variables and penalty functions are introduced to allow solution under certain conditions.

The example chosen to start with is a fixed fuel supply that must be paid for, whether or not it is consumed. We might have started with a limited fuel supply of natural gas that must be used as boiler fuel because it has been declared as "surplus." The take-or-pay fuel supply contract is probably the simplest of these possibilities.

Alternatively, we might have started directly with the problem of economic scheduling of hydroelectric plants with their stored supply of water or with light-water-moderated nuclear reactors supplying steam to drive turbine generators. Hydroelectric plant scheduling involves the scheduling of water flows, impoundments (storage), and releases into what usually prove to be a rather complicated hydraulic network (namely, the watershed). The treatment of nuclear unit scheduling requires some understanding of the physics involved in the reactor core and is really beyond the scope of this current text (the methods useful for optimizing the unit outputs are, however, quite similar to those used in scheduling other limited energy systems).

6.2   TAKE-OR-PAY FUEL SUPPLY CONTRACT

Assume there are N normally fueled thermal plants plus one turbine generator, fueled under a "*take-or-pay*" *agreement*. We will interpret this type of agreement as being one in which the utility agrees to use a minimum amount of fuel during a period (the "take") or, failing to use this amount, it agrees to pay the minimum charge. This last clause is the "pay" part of the "take-or-pay" contract.

While this unit's cumulative fuel consumption is below the minimum, the system excluding this unit should be scheduled to minimize the total fuel cost, subject to the constraint that the total fuel consumption for the period for this particular unit is equal to the specified amount. Once the specified amount of fuel has been used, the unit should be scheduled normally. Let us consider a special case where the minimum amount of fuel consumption is also the maximum. The system is shown in Figure 6.1. We will consider the operation of the system over j_{max} time intervals j where $j = 1, \ldots, j_{max}$, so that

$$P_{1j}, P_{2j}, \ldots, P_{Tj} \quad \text{(power outputs)}$$

$$F_{1j}, F_{2j}, \ldots, F_{Nj} \quad \text{(fuel cost rate)}$$

and

$$q_{T1}, q_{T2}, \ldots, q_{Tj} \quad \text{(take-or-pay fuel input)}$$

are the power outputs, fuel costs, and take-or-pay fuel inputs, where

$$P_{ij} \triangleq \text{power from } i^{th} \text{ unit in the } j^{th} \text{ time interval}$$

$$F_{ij} \triangleq \mathbb{R}/h \text{ cost for } i^{th} \text{ unit during the } j^{th} \text{ time interval}$$

$$q_{Tj} \triangleq \text{fuel input for unit } T \text{ in } j^{th} \text{ time interval}$$

$$F_{Tj} \triangleq \mathbb{R}/h \text{ cost for unit } T \text{ in } j^{th} \text{ time interval}$$

$$P_{\text{load } j} \triangleq \text{total load in the } j^{th} \text{ time interval}$$

$$n_j \triangleq \text{Number of hours in the } j^{th} \text{ time interval}$$

Mathematically, the problem is as follows:

$$\min \sum_{j=1}^{j_{max}} \left(n_j \sum_{i=1}^{N} F_{ij} \right) + \sum_{j=1}^{j_{max}} n_j F_{Tj} \tag{6.1}$$

subject to

$$\phi = \sum_{j=1}^{j_{max}} n_j q_{Tj} - q_{\text{TOT}} = 0 \tag{6.2}$$

and

$$\psi_j = P_{\text{load } j} - \sum_{i=1}^{N} P_{ij} - P_{Tj} = 0 \quad \text{for } j = 1 \ldots j_{max} \tag{6.3}$$

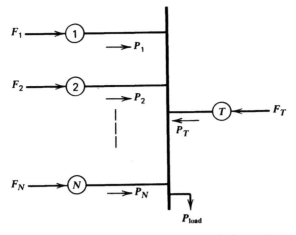

FIG. 6.1 $N + 1$ unit system with take-or-pay fuel supply at unit T.

or, in words,

> We wish to determine the minimum production cost for units 1 to N subject to constraints that ensure that fuel consumption is correct and also subject to the set of constraints to ensure that power supplied is correct each interval.

Note that (for the present) we are ignoring high and low limits on the units themselves. It should also be noted that the term

$$\sum_{j=1}^{j_{\max}} n_j F_{Tj}$$

is constant because the total fuel to be used in the "T" plant is fixed. Therefore, the total cost of that fuel will be constant and we can drop this term from the objective function.

The Lagrange function is

$$\mathscr{L} = \sum_{j=1}^{j_{\max}} n_j \sum_{i=1}^{N} F_{ij} + \sum_{j=1}^{j_{\max}} \lambda_j \left(P_{\text{load } j} - \sum_{i=1}^{N} P_{ij} - P_{Tj} \right) + \gamma \left(\sum_{j=1}^{j_{\max}} n_j q_{Tj} - q_{\text{TOT}} \right) \quad (6.4)$$

The independent variables are the powers P_{ij} and P_{Tj}, since $F_{ij} = F_i(P_{ij})$ and

$q_{Tj} = q_T(P_{Tj})$. For any given time period, $j = k$,

$$\frac{\partial \mathscr{L}}{\partial P_{ik}} = 0 = n_k \frac{dF_{ik}}{dP_{ik}} - \lambda_k \qquad \text{for } i = 1 \ldots N \tag{6.5}$$

and

$$\frac{\partial \mathscr{L}}{\partial P_{Tk}} = -\lambda_k + \gamma n_k \frac{dq_{Tk}}{dP_{Tk}} = 0 \tag{6.6}$$

Note that if one analyzes the dimensions of γ, it would be ℞ per unit of q (e.g., ℞/ft^3, ℞/bbl, ℞/ton). As such, γ has the units of a "fuel price" expressed in volume units rather than MBtu as we have used up to now. Because of this, γ is often referred to as a "pseudo-price" or "shadow price." In fact, once it is realized what is happening in this analysis, it becomes obvious that we could solve fuel-limited dispatch problems by simply adjusting the price of the limited fuel(s); thus, the terms "pseudo-price" and "shadow price" are quite meaningful.

Since γ appears unsubscripted in Eq. 6.6, γ would be expected to be a constant value over all the time periods. This is true unless the fuel-limited machine is constrained by fuel-storage limitations. We will encounter such limitations in hydroplant scheduling in Chapter 7. The appendix to Chapter 7 shows when to expect a constant γ and when to expect a discontinuity in γ.

Figure 6.2a shows how the load pattern may look. The solution to a fuel-limited dispatching problem will require dividing the load pattern into time intervals, as in Figure 6.2b, and assuming load to be constant during each interval. Assuming all units are on-line for the period, the optimum dispatch could be done using a simple search procedure for γ, as is shown in Figure 6.3. Note that the procedure shown in Figure 6.3 will only work if the fuel-limited unit does not hit either its high or its low limit in any time interval.

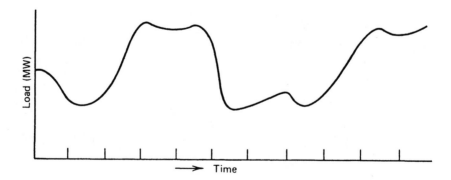

FIG. 6.2a   Load pattern.

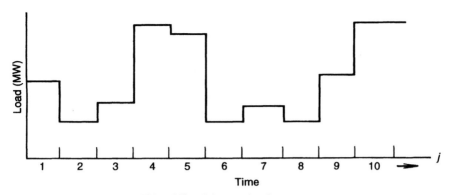

FIG. 6.2b   Discrete load pattern.

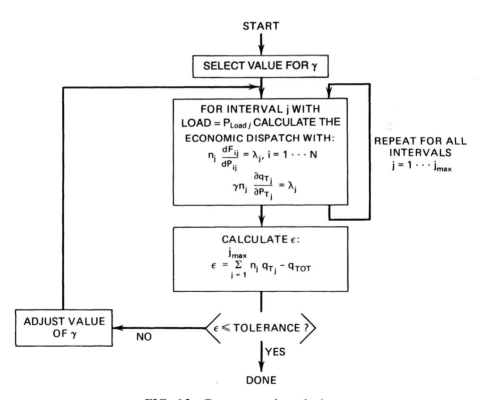

FIG. 6.3   Gamma search method.

6.3   COMPOSITE GENERATION PRODUCTION COST FUNCTION

A useful technique to facilitate the take-or-pay fuel supply contract procedure is to develop a composite generation production cost curve for all the non-fuel-constrained units. For example, suppose there were N non-fuel constrained units to be scheduled with the fuel-constrained unit as shown in Figure 6.4. Then a composite cost curve for units $1, 2, \ldots, N$ can be developed.

$$F_s(P_s) = F_1(P_1) + \ldots + F_N(P_N) \tag{6.7}$$

where

$$P_s = P_1 + \ldots + P_N$$

and

$$\frac{dF_1}{dP_1} = \frac{dF_2}{dP_2} = \ldots = \frac{dF_N}{dP_N} = \lambda$$

If one of the units hits a limit, its output is held constant, as in Chapter 3, Eq. 3.6.

A simple procedure to allow one to generate $F_s(P_s)$ consists of adjusting λ from λ^{min} to λ^{max} in specified increments, where

$$\lambda^{min} = \min\left(\frac{dF_i}{dP_i}, i = 1 \ldots N\right)$$

$$\lambda^{max} = \max\left(\frac{dF_i}{dP_i}, i = 1 \ldots N\right)$$

At each increment, calculate the total fuel consumption and the total power output for all the units. These points represent points on the $F_s(P_s)$ curve. The points may be used directly by assuming $F_s(P_s)$ consists of straight-line segments between the points, or a smooth curve may be fit to the points using a least-squares fitting program. Be aware, however, that such smooth curves may have undesirable properties such as nonconvexity (e.g., the first derivative is not monotonically increasing). The procedure to generate the points on $F_s(P_s)$ is shown in Figure 6.5.

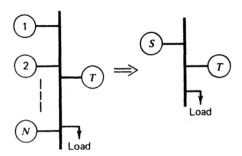

FIG. 6.4   Composite generator unit.

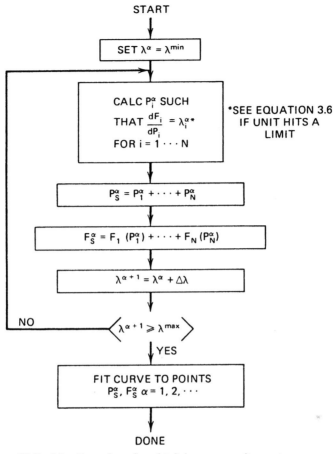

FIG. 6.5 Procedure for obtaining composite cost curve.

EXAMPLE 6A

The three generating units from Example 3A are to be combined into a composite generating unit. The fuel costs assigned to these units will be

$$\text{Fuel cost for unit } 1 = 1.1 \text{ ₹/MBtu}$$

$$\text{Fuel cost for unit } 2 = 1.4 \text{ ₹/MBtu}$$

$$\text{Fuel cost for unit } 3 = 1.5 \text{ ₹/MBtu}$$

Figure 6.6a shows the individual unit incremental costs, which range from 8.3886 to 14.847 ₹/MWh. A program was written based on Figure 6.5, and λ was stepped from 8.3886 to 14.847.

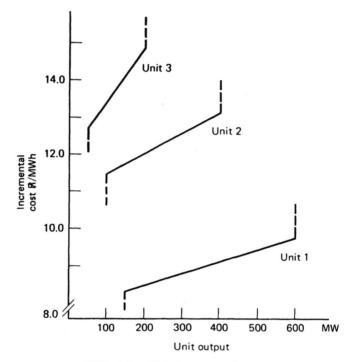

FIG. 6.6a Unit incremental costs.

TABLE 6.1 **Lambda Steps Used in Constructing a Composite Cost Curve for Example 6A**

Step	λ	P_s	F_s	F_s Approx
1	8.3886	300.0	4077.12	4137.69
2	8.7115	403.4	4960.92	4924.39
3	9.0344	506.7	5878.10	5799.07
4	9.3574	610.1	6828.66	6761.72
5	9.6803	713.5	7812.59	7812.35
6	10.0032	750.0	8168.30	8204.68
7	11.6178	765.6	8348.58	8375.29
8	11.9407	825.0	9048.83	9044.86
9	12.2636	884.5	9768.28	9743.54
10	12.5866	943.9	10506.92	10471.31
11	12.9095	1019.4	11469.56	11436.96
12	13.2324	1088.4	12369.40	12360.58
13	13.5553	1110.67	12668.51	12668.05
14	13.8782	1133.00	12974.84	12979.63
15	14.2012	1155.34	13288.37	13295.30
16	14.5241	1177.67	13609.12	13615.09
17	14.8470	1200.00	13937.00	13938.98

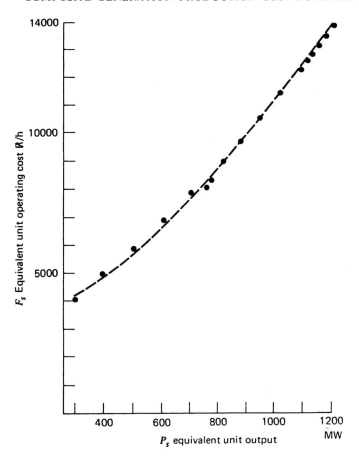

FIG. 6.6b Equivalent unit input/output curve.

At each increment, the three units are dispatched to the same λ and then outputs and generating costs are added as shown in Figure 6.5. The results are given in Table 6.1. The result, called F_s approx in Table 6.1 and shown in Figure 6.6b, was calculated by fitting a second-order polynomial to the P_s and F_s points using a least-squares fitting program. The equivalent unit function is

$$F_s \text{ approx}(P_s) = 2352.65 + 4.7151 P_s + 0.0041168 P_s^2$$

$$(\text{R/h}) \qquad 300 \text{ MW} \le P_s \le 1200 \text{ MW}$$

The reader should be aware that when fitting a polynomial to a set of points, many choices can be made. The preceding function is a good fit to the total operating cost of the three units, but it is not that good at approximating the incremental cost. More-advanced fitting methods should be used if one desires to match total operating cost as well as incremental cost. See Problem 6.2 for an alternative procedure.

EXAMPLE 6B

Find the optimal dispatch for a gas-fired steam plant given the following.

Gas-fired plant:

$$H_T(P_T) = 300 + 6.0P_T + 0.0025P_T^2 \text{ MBtu/h}$$

$$\text{Fuel cost for gas} = 2.0 \text{ R/ccf (where 1 ccf} = 10^3 \text{ ft}^3)$$

$$\text{The gas is rated at 1100 Btu/ft}^3$$

$$50 \leq P_T \leq 400$$

Composite of remaining units:

$$H_s(P_s) = 200 + 8.5P_s + 0.002P_s^2 \text{ MBtu/h}$$

$$\text{Equivalent fuel cost} = 0.6 \text{ R/MBtu}$$

$$50 \leq P_s \leq 500$$

The gas-fired plant must burn $40 \cdot 10^6$ ft^3 of gas. The load pattern is shown in Table 6.2. If the gas constraints are ignored, the optimum economic schedule for these two plants appears as is shown in Table 6.3. Operating cost of the composite unit over the entire 24-h period is 52,128.03 R. The total gas consumption is $21.8 \cdot 10^6$ ft^3. Since the gas-fired plant must burn $40 \cdot 10^6$ ft^3 of gas, the cost will be 2.0 R/1000 ft^3 × $40 \cdot 10^6$ ft^3, which is 80,000 R for the gas. Therefore, the total cost will be 132,128.03 R. The solution method shown in Figure 6.3 was used with γ values ranging from 0.500 to 0.875. The final value for γ is 0.8742 R/ccf with an optimal schedule as shown in Table 6.4. This schedule has a fuel cost for the composite unit of 34,937.47 R. Note that the gas unit is run much harder and that it does not hit either limit in the optimal

TABLE 6.2 Load Pattern

Time Period	Load
1. 0000–0400	400 MW
2. 0400–0800	650 MW
3. 0800–1200	800 MW
4. 1200–1600	500 MW
5. 1600–2000	200 MW
6. 2000–2400	300 MW

Where: $n_j = 4$, $j = 1 \ldots 6$.

**TABLE 6.3   Optimum Economic Schedule
(Gas Constraints Ignored)**

Time Period	P_s	P_T
1	350	50
2	500	150
3	500	300
4	450	50
5	150	50
6	250	50

TABLE 6.4   Optimal Schedule (Gas Constraints Met)

Time Period	P_s	P_T
1	197.3	202.6
2	353.2	296.8
3	446.7	353.3
4	259.7	240.3
5	72.6	127.4
6	135.0	165.0

schedule. Further, note that the total cost is now

$$34{,}937.47 \text{ R} + 80{,}000 \text{ R} = 114{,}937.4 \text{ R}$$

so we have lowered the total fuel expense by properly scheduling the gas plant.

6.4   SOLUTION BY GRADIENT SEARCH TECHNIQUES

An alternative solution procedure to the one shown in Figure 6.3 makes use of Eqs. 6.5 and 6.6.

$$n_k \frac{dF_{ik}}{dP_{ik}} = \lambda_k$$

and

$$\lambda_k = \gamma n_k \frac{dq_{Tk}}{dP_{Tk}}$$

then

$$\gamma = \left(\frac{\dfrac{dF_{ik}}{dP_{ik}}}{\dfrac{dq_{Tk}}{dP_{Tk}}} \right) \tag{6.8}$$

For an optimum dispatch, γ will be constant for all hours j, $j = 1 \ldots j_{\max}$.

We can make use of this fact to obtain an optimal schedule using the procedures shown in Figure 6.7a or Figure 6.7b. Both these procedures attempt to adjust fuel-limited generation so that γ will be constant over time. The algorithm shown in Figure 6.7a differs from the algorithm shown in Figure 6.7b in the way the problem is started and in the way various time intervals are

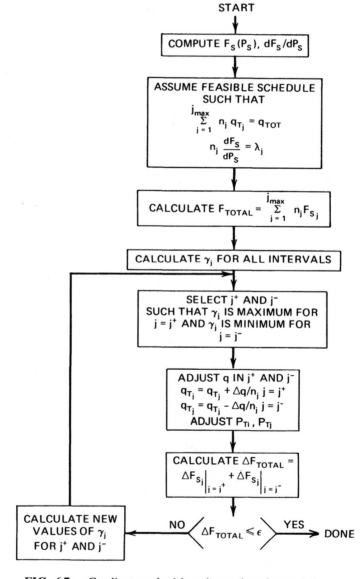

FIG. 6.7a Gradient method based on relaxation technique.

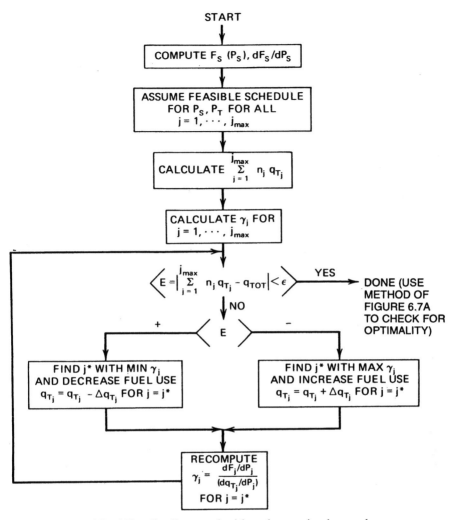

FIG. 6.7b Gradient method based on a simple search.

selected for adjustment. The algorithm in Figure 6.7a requires an initial feasible but not optimal schedule and then finds an optimal schedule by "pairwise" trade-offs of fuel consumption while maintaining problem feasibility. The algorithm in Figure 6.7b does not require an initial feasible fuel usage schedule but achieves this while optimizing. These two methods may be called gradient methods because q_{Tj} is treated as a vector and the γ_j values indicate the gradient of the objective function with respect to q_{Tj}. The method of Figure 6.7b should be followed by that of Figure 6.7a to insure optimality.

EXAMPLE 6C

Use the method of Figure 6.7b to obtain an optimal schedule for the problem given in Example 6B. Assume that the starting schedule is the economic dispatch schedule shown in Example 6B.

Initial Dispatch

	Time Period					
	1	2	3	4	5	6
P_s	350	500	500	450	150	250
P_T	50	150	300	50	50	50
γ	1.0454	1.0266	0.9240	1.0876	0.9610	1.0032

$\sum q_T = 21.84 \cdot 10^6$ ft^3.

Since we wish to burn $40.0 \cdot 10^6$ ft^3 of gas, the error is negative; therefore, we must increase fuel usage in the time period having maximum γ, that is, period 4. As a start, increase P_T to 150 MW and drop P_s to 350 MW in period 4.

Result of Step 1

	Time Period					
	1	2	3	4	5	6
P_s	350	500	500	350	150	250
P_T	50	150	300	150	50	50
γ	1.0454	1.0266	0.9240	0.9680	0.9610	1.0032

$\sum q_T = 24.2 \cdot 10^6$ ft^3.

The error is still negative, so we must increase fuel usage in the period with maximum γ, which is now period 1. Increase P_T to 200 MW and drop P_s to 200 MW in period 1.

Result of Step 2

	Time Period					
	1	2	3	4	5	6
P_s	200	500	500	350	150	250
P_T	200	150	300	150	50	50
γ	0.8769	1.0266	0.9240	0.9680	0.9610	1.0032

$\sum q_T = 27.8 \cdot 10^6$ ft^3.

and so on. After 11 steps, the schedule looks like this:

	Time Period					
	1	2	3	4	5	6
P_s	200	350	450	250	75	140
P_T	200	300	350	250	125	160
γ	0.8769	0.8712	0.8772	0.8648	0.8767	0.8794

$\sum q_T = 40.002 \cdot 10^6 \text{ ft}^3$.

which is beginning to look similar to the optimal schedule generated in Example 6A.

6.5 HARD LIMITS AND SLACK VARIABLES

This section takes account of hard limits on the take-or-pay generating unit. The limits are

$$P_T \geq P_{T\min} \tag{6.9}$$

and

$$P_T \leq P_{T\max} \tag{6.10}$$

These may be added to the Lagrangian by the use of two constraint functions and two new variables called *slack variables* (see Appendix 3A). The constraint functions are

$$\psi_{1j} = P_{Tj} - P_{T\max} + S_{1j}^2 \tag{6.11}$$

and

$$\psi_{2j} = P_{T\min} - P_{Tj} + S_{2j}^2 \tag{6.12}$$

where S_{1j} and S_{2j} are slack variables that may take on any real value including zero.

The new Lagrangian then becomes

$$\mathscr{L} = \sum_{j=1}^{j_{\max}} n_j \sum_{i=1}^{N} F_{ij} + \sum_{j=1}^{j_{\max}} \lambda_j \left(P_{\text{load } j} - \sum_{i=1}^{N} P_{ij} - P_{Tj} \right) + \gamma \left(\sum_{j=1}^{j_{\max}} n_j q_{Tj} - Q_{\text{TOT}} \right)$$
$$+ \sum_{j=1}^{j_{\max}} \alpha_{1j}(P_{Tj} - P_{T\max} + S_{1j}^2) + \sum_{j=1}^{j_{\max}} \alpha_{2j}(P_{T\min} - P_{Tj} + S_{2j}^2) \tag{6.13}$$

where α_{1j}, α_{2j} are Lagrange multipliers. Now, the first partial derivatives for the kth period are

$$\frac{\partial \mathscr{L}}{\partial P_{ik}} = 0 = n_k \frac{dF_{ik}}{dP_{ik}} - \lambda_k$$

$$\frac{\partial \mathscr{L}}{\partial P_{Tk}} = 0 = -\lambda_k + \alpha_{1k} - \alpha_{2k} + \gamma n_k \frac{dq_{Tk}}{dP_{Tk}}$$

$$\frac{\partial \mathscr{L}}{\partial S_{1k}} = 0 = 2\alpha_{1k}S_{1k} \tag{6.14}$$

$$\frac{\partial \mathscr{L}}{\partial S_{2k}} = 0 = 2\alpha_{2k}S_{2k}$$

As we noted in Appendix 3A, when the constrained variable (P_{T_k} in this case) is within bounds, the new Lagrange multipliers $\alpha_{1k} = \alpha_{2k} = 0$ and S_{1k} and S_{2k} are nonzero. When the variable is limited, one of the slack variables, S_{1k} or S_{2k}, becomes zero and the associated Lagrange multiplier will take on a nonzero value.

Suppose in some interval k, $P_{Tk} = P_{max}$, then $S_{1k} = 0$ and $\alpha_{1k} \neq 0$. Thus,

$$-\lambda_k + \alpha_{1k} + \gamma n_k \frac{dq_{Tk}}{dP_{Tk}} = 0 \tag{6.15}$$

and if

$$\lambda_k > \gamma n_k \frac{dq_{Tk}}{dP_{Tk}}$$

the value of α_{1k} will take on the value just sufficient to make the equality true.

EXAMPLE 6D

Repeat Example 6B with the maximum generation on P_T reduced to 300 MW. Note that the optimum schedule in Example 6A gave a $P_T = 353.3$ MW in the third time period. When the limit is reduced to 300 MW, the gas-fired unit will have to burn more fuel in other time periods to meet the $40 \cdot 10^3$ ft^3 gas consumption constraint.

TABLE 6.5   Resulting Optimal Schedule with $P_{Tmax} = 300$ MW

Time Period j	P_{sj}	P_{Tj}	λ_j	$\gamma_{nj} \dfrac{\partial q_T}{\partial P_{Tj}}$	α_{1j}
1	183.4	216.6	5.54	5.54	0
2	350.0	300.0	5.94	5.86	0.08
3	500.0	300.0	6.3	5.86	0.44
4	245.4	254.6	5.69	5.69	0
5	59.5	140.5	5.24	5.24	0
6	121.4	178.6	5.39	5.39	0

Table 6.5 shows the resulting optimal schedule where $\gamma = 0.8603$ and total cost $= 122{,}984.83$ ₽.

6.6 FUEL SCHEDULING BY LINEAR PROGRAMMING

Figure 6.8 shows the major elements in the chain making up the delivery system that starts with raw-fuel suppliers and ends up in delivery of electric power to individual customers. The basic elements of the chain are as follows.

The suppliers: These are the coal, oil, and gas companies with which the utility must negotiate contracts to acquire fuel. The contracts are usually written for a long term (10 to 20 yr) and may have stipulations, such as the minimum and maximum limits on the quantity of fuel delivered over a specified time period. The time period may be as long as a year, a month, a week, a day, or even for a period of only a few minutes. Prices may change, subject to the renegotiation provisions of the contracts.

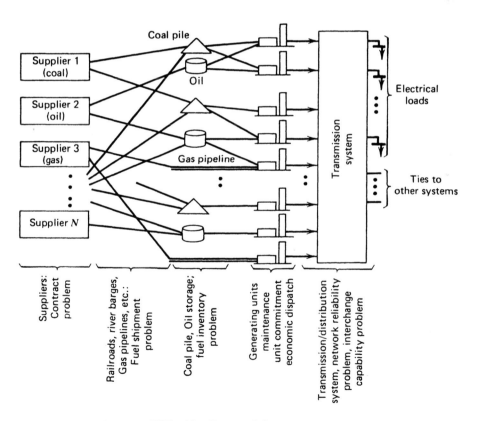

FIG. 6.8 Energy delivery system.

Transportation: Railroads, unit trains, river barges, gas-pipeline companies, and such, all present problems in scheduling of deliveries of fuel.

Inventory: Coal piles, oil storage tanks, underground gas storage facilities. Inventories must be kept at proper levels to forestall fuel shortages when load levels exceed forecast or suppliers or shippers are unable to deliver. Price fluctuations also complicate the decisions on when and how much to add or subtract from inventories.

The remainder of the system—generators, transmission, and loads—are covered in other chapters.

One of the most useful tools for solving large fuel-scheduling problems is linear programming (LP). If the reader is not familiar with LP, an easily understood algorithm is provided in the appendix of this chapter.

Linear programming is an optimization procedure that minimizes a linear objective function with variables that are also subject to linear constraints. Because of this limitation, any nonlinear functions either in the objective or in the constraint equations will have to be approximated by linear or piecewise linear functions.

To solve a fuel-scheduling problem with linear programming, we must break the total time period involved into discrete time increments, as was done in Example 6B. The LP solution will then consist of an objective function that is made up of a sum of linear or piecewise linear functions, each of which is a function of one or more variables from only one time step. The constraints will be linear functions of variables from each time step. Some constraints will be made up of variables drawn from one time step whereas others will span two or more time steps. The best way to illustrate how to set up an LP to solve a fuel-scheduling problem will be to use an example.

EXAMPLE 6E

We are given two coal-burning generating units that must both remain on-line for a 3-wk period. The combined output from the two units is to supply the following loads (loads are assumed constant for 1 wk).

Week	Load (MW)
1	1200
2	1500
3	800

The two units are to be supplied by one coal supplier who is under contract to supply 40,000 tons of coal per week to the two plants. The plants have existing coal inventories at the start of the 3-wk period. We must solve for the following.

1. How should each plant be operated each week?
2. How should the coal deliveries be made up each week?

The data for the problem are as follows.

Coal: Heat value = 11,500 Btu/lb = 23 MBtu/ton (1 ton = 2000 lb)

Coal can all be delivered to one plant or the other or it can be split, some going to one plant, some to the other, as long as the total delivery in each week is equal to 40,000 tons. The coal costs 30 ₹/ton or 1.3 ₹/MBtu.

Inventories: Plant 1   has an initial inventory of 70,000 tons; its final inventory is not restricted

Plant 2   has an initial inventory of 70,000 tons; its final inventory is not restricted

Both plants have a maximum coal storage capacity of 200,000 tons of coal.

Generating units:

Unit	Min (MW)	Max (MW)	Heat Input at Min (MBtu/h)	Heat Input at Max (MBtu/h)
1	150	600	1620	5340
2	400	1000	3850	8750

The input versus output function will be approximated by a linear function for each unit:

$$H_1(P_1) = 380.0 + 8.267P_1$$

$$H_2(P_2) = 583.3 + 8.167P_2$$

The unit cost curves are

$$F_1(P_1) = 1.3 \text{ ₹/MBtu} \times H_1(P_1) = 495.65 + 10.78P_1 \text{ (₹/h)}$$

$$F_2(P_2) = 1.3 \text{ ₹/MBtu} \times H_2(P_2) = 760.8 + 10.65P_2 \text{ (₹/h)}$$

The coal consumption q(tons/h) for each unit is

$$q_1(P_1) = \frac{1}{23}\left(\frac{\text{tons}}{\text{MBtu}}\right) \times H_1(P_1) = 16.52 + 0.3594P_1 \text{ tons/h}$$

$$q_2(P_2) = \frac{1}{23}\left(\frac{\text{tons}}{\text{MBtu}}\right) \times H_2(P_2) = 25.36 + 0.3551P_2 \text{ tons/h}$$

To solve this problem with linear programming, assume that the units are to be operated at a constant rate during each week and that the coal deliveries will each take place at the beginning of each week. Therefore, we will set up the problem with 1-wk time periods and the generating unit cost functions and coal consumption functions will be multiplied by 168 h to put them on a "per week" basis; then,

$$
\begin{aligned}
F_1(P_1) &= 83{,}269.2 + 1811P_1 \text{ R/wk} \\
F_2(P_2) &= 127{,}814.4 + 1789P_2 \text{ R/wk} \\
q_1(P_1) &= 2775.4 + 60.4P_1 \text{ tons/wk} \\
q_2(P_2) &= 4260.5 + 59.7P_2 \text{ tons/wk}
\end{aligned}
\tag{6.16}
$$

We are now ready to set up the objective function and the constraints for our linear programming solution.

Objective function: To minimize the operating cost over the 3-wk period. The objective function is

$$
\begin{aligned}
\text{Minimize } Z = {}& F_1[P_1(1)] + F_2[P_2(1)] + F_1[(P_1(2)] + F_2[P_2(2)] \\
& + F_1[P_1(3)] + F_2[P_2(2)]
\end{aligned}
\tag{6.17}
$$

where $P_i(j)$ is the power output of the ith unit during the jth week, $j = 1 \ldots 3$.

Constraints: During each time period, the total power delivered from the units must equal the scheduled load to be supplied; then

$$
\begin{aligned}
P_1(1) + P_2(1) &= 1200 \\
P_1(2) + P_2(2) &= 1500 \\
P_1(3) + P_2(3) &= 800
\end{aligned}
\tag{6.18}
$$

Similarly, the coal deliveries, D_1 and D_2, made to plant 1 and plant 2, respectively, during each week must sum to 40,000 tons; then

$$D_1(1) + D_2(1) = 40,000$$
$$D_1(2) + D_2(2) = 40,000 \qquad (6.19)$$
$$D_1(3) + D_2(3) = 40,000$$

The volume of coal at each plant at the beginning of each week plus the delivery of coal to that plant minus the coal burned at the plant will give the coal remaining at the beginning of the next week. Letting V_1 and V_2 be the volume of coal in each coal pile at the beginning of the week, respectively, we have the following set of equations governing the two coal piles.

$$V_1(1) + D_1(1) - q_1(1) = V_1(2)$$
$$V_2(1) + D_2(1) - q_2(1) = V_2(2)$$
$$V_1(2) + D_1(2) - q_1(2) = V_1(3)$$
$$V_2(2) + D_2(2) - q_2(2) = V_2(3) \qquad (6.20)$$
$$V_1(3) + D_1(3) - q_1(3) = V_1(4)$$
$$V_2(3) + D_2(3) - q_2(3) = V_2(4)$$

where $V_i(j)$ is the volume of coal in the ith coal pile at the beginning of the jth week.

To set these equations up for the linear-programming solutions, substitute the $q_1(P_1)$ and $q_2(P_2)$ equations from 6.16 into the equations of 6.20. In addition, all constant terms are placed on the right of the equal sign and all variable terms on the left; this leaves the constraints in the standard form for inclusion in the LP. The result is

$$D_1(1) - 60.4P_1(1) - V_1(2) = 2775.4 - V_1(1)$$
$$D_2(1) - 59.7P_2(1) - V_2(2) = 4260.5 - V_2(1)$$
$$V_1(2) + D_1(2) - 60.4P_1(2) - V_1(3) = 2775.4$$
$$V_2(2) + D_2(2) - 59.7P_2(2) - V_2(3) = 4260.5 \qquad (6.21)$$
$$V_1(3) + D_1(3) - 60.4P_1(3) - V_1(4) = 2775.4$$
$$V_2(3) + D_2(3) - 59.7P_2(3) - V_2(4) = 4260.5$$

Note: $V_1(1)$ and $V_2(1)$ are constants that will be set when we start the problem.

The constraints from Eqs. 6.18, 6.19, and 6.21 are arranged in a matrix, as shown in Figure 6.9. Each variable is given an upper and lower bound in keeping with the "upper bound" solution shown in the appendix of this chapter. The $P_1(t)$ and $P_2(t)$ variables are given the upper and lower bounds corresponding

FIG. 6.9 Linear-programming constraint matrix for Example 6E.

Problem Variable	D1(1)	P1(1)	D2(1)	P2(1)	V1(2)	D1(2)	P1(2)	V2(2)	D2(2)	P2(2)	V1(3)	D1(3)	P1(3)	V2(3)	D2(3)	P2(3)	V1(4)	V2(4)	Constraint Units
LP Variable	X_1	X_2	X_3	X_4	X_5	X_6	X_7	X_8	X_9	X_{10}	X_{11}	X_{12}	X_{13}	X_{14}	X_{15}	X_{16}	X_{17}	X_{18}	
Constraint																			
1	1	1																	1200
2			1	1															40000
3	1	−60.4			−1														2775.4 − $V_1(1)$
4			1	−59.7				−1											4260.5 − $V_2(1)$
5					1	1													1500
6								1	1										40000
7					1	1	−60.4				−1								2775.4
8								1	1	−59.7				−1					4260.5
9											1	1							800
10														1	1				40000
11											1	1	−60.4				−1		2775.4
12														1	1	−59.7		−1	4260.5
Variable min.	0	150	0	400	0	0	150	0	0	400	0	0	150	0	0	400	0	0	
Variable max.	40000	600	40000	1000	200000	40000	600	200000	40000	1000	200000	40000	600	200000	40000	1000	200000	200000	

Week 1: columns X_1–X_4. Week 2: columns X_5–X_{10}. Week 3: columns X_{11}–X_{16}. Final Conditions: columns X_{17}–X_{18}.

to the upper and lower limits on the generating units. $D_1(t)$ and $D_2(t)$ are given upper and lower bounds of 40,000 and zero. $V_1(t)$ and $V_2(t)$ are given upper and lower bounds of 200,000 and zero.

Solution: The solution to this problem was carried out with a computer program written to solve the upper bound LP problem using the algorithm shown in the Appendix. The first problem solved had coal storage at the beginning of the first week of

$$V_1(1) = 70,000 \text{ tons}$$

$$V_2(1) = 70,000 \text{ tons}$$

The solution is:

Time Period	V_1	D_1	P_1	V_2	D_2	P_2
1	70000.0	0	200	70000.0	40000.0	1000
2	55144.6	0	500	46039.5	40000.0	1000
3	22169.2	19013.5	150	22079.0	20986.5	650
4	29347.3					

Optimum cost = 6,913,450.8 ₽.

In this case, there are no constraints on the coal deliveries to either plant and the system can run in the most economic manner. Since unit 2 has a lower incremental cost, it is run at its maximum when possible. Furthermore, since no restrictions were placed on the coal pile levels at the end of the third week, the coal deliveries could have been shifted a little from unit 2 to unit 1 with no effect on the generation dispatch.

The next case solved was purposely structured to create a fuel shortage at unit 2. The beginning inventory at plant 2 was set to 50,000 tons, and a requirement was imposed that at the end of the third week the coal pile at unit 2 be no less than 8000 tons. The solution was made by changing the right-hand side of the fourth constraint from $-65,739.5$ (i.e., $4260.5 - 70,000$) to -45739.5 (i.e., $4260.5 - 50,000$) and placing a lower bound on $V_2(4)$ (i.e., variable X_{18}) of 8000. The solution is:

Time Period	V_1	D_1	P_1	V_2	D_2	P_2
1	70000.0	0	200	50000.0	40000.0	1000
2	55144.6	0	500	26039.5	40000.0	1000
3	22169.2	0	300.5276	2079.0	40000.0	499.4724
4	1241.9307			8000.0		

Optimum cost = 6,916,762.4 ₽.

Note that this solution requires unit 2 to drop off its generation in order to meet the end-point constraint on its coal pile. In this case, all the coal must be delivered to plant 2 to minimize the overall cost.

The final case was constructed to show the interaction of the fuel deliveries and the economic dispatch of the generating units. In this case, the initial coal piles were set to 10,000 tons and 150,000 tons, respectively. Furthermore, a restriction of 30,000 tons minimum in the coal pile at unit 1 at the end of the third week was imposed.

To obtain the most economic operation of the two units over the 3-wk period, the coal deliveries will have to be adjusted to insure both plants have sufficient coal. The solution was obtained by setting the right-hand side of the third and fourth constraint equations to -7224.6 and -145739.5, respectively, as well as imposing a lower bound of 30,000 on $V_1(4)$ (i.e., variable X_{17}). The solution is:

Time Period	V_1	D_1	P_1	V_2	D_2	P_2
1	10000.0	4855.4	200	150000.0	35144.6	1000
2	0.0	40000.0	500	121184.1	0	1000
3	7024.6	40000.0	150	57223.6	0	650
4	35189.2			14158.1		

Optimum cost = 6,913,450.8 R.

The LP was able to find a solution that allowed the most economic operation of the units while still directing enough coal to unit 1 to allow it to meet its end-point coal pile constraint. Note that, in practice, we would probably not wish to let the coal pile at unit 1 go to zero. This could be prevented by placing an appropriate lower bound on all the volume variables (i.e., X_5, X_8, X_{11}, X_{14}, X_{17}, and X_{18}).

This example has shown how a fuel-management problem can be solved with linear programming. The important factor in being able to solve very large fuel-scheduling problems is to have a linear-programming code capable of solving large problems having perhaps tens of thousands of constraints and as many, or more, problem variables. Using such codes, elaborate fuel-scheduling problems can be optimized out over several years and play a critical role in utility fuel-management decisions.

APPENDIX
Linear Programming

Linear programming is perhaps the most widely applied mathematical programming technique. Simply stated, linear programming seeks to find the optimum value of a linear objective function while meeting a set of linear constraints. That is, we wish to find the optimum set of x values that minimize the following objective function:

$$Z = c_1 x_1 + c_2 x_2 + \ldots + c_N x_N$$

subject to a set of linear constraints:

$$a_{11} x_1 + a_{12} x_2 + \ldots + a_{1N} x_N \leq b_1$$
$$a_{21} x_1 + a_{22} x_2 + \ldots + a_{2N} x_N \leq b_2$$
$$\vdots$$

In addition, the variables themselves may have specified upper and lower limits.

$$x_i^{\min} \leq x_i \leq x_i^{\max} \qquad i = 1 \ldots N$$

There are a variety of solutions to the LP problem. Many of these solutions are tailored to a particular type of problem. This appendix will not try to develop the theory of alternate LP solution methods. Rather, it will present a simple LP algorithm that can be used (or programmed on a computer) to solve the applicable power-system sample problems given in this text.

The algorithm is presented in its simplest form. There are alternative formulations, and these will be indicated when appropriate. If the student has access to a standard LP program, such a standard program may be used to solve any of the problems in this book.

The LP technique presented here is properly called an *upper-bounding dual linear programming algorithm*. The "upper-bounding" part of its name refers to the fact that variable limits are handled implicitly in the algorithm. "Dual" refers to the theory behind the way in which the algorithm operates. For a complete explanation of the primal and dual algorithms, refer to the references cited at the end of this chapter.

In order to proceed in an orderly fashion to solve a dual upper-bound linear programming problem, we must first add what is called a *slack variable* to each constraint. The slack variable is so named because it equals the difference or slack between a constraint and its limit. By placing a slack variable into an inequality constraint, we can transform it into an equality constraint. For example, suppose we are given the following constraint.

$$2x_1 + 3x_2 \leq 15 \qquad (6A.1)$$

We can transform this constraint to an equality constraint by adding a slack variable, x_3.

$$2x_1 + 3x_2 + x_3 = 15 \qquad (6A.2)$$

If x_1 and x_2 were to be given values such that the sum of the first two terms in Eq. 6A.2 added up to less than 15, we could still satisfy Eq. 6A.2 by setting x_3 to the difference. For example, if $x_1 = 1$ and $x_2 = 3$, then $x_3 = 4$ would satisfy Eq. 6A.2. We can go even further, however, and restrict the values of x_3 so that Eq. 6A.2 still acts as an inequality constraint such as Eq. 6A.1. Note that when the first two terms of Eq. 6A.2 add to exactly 15, x_3 must be set to zero. By restricting x_3 to always be a positive number, we can force Eq. 6A.2 to yield the same effect as Eq. 6A.1. Thus,

$$\left.\begin{array}{r} 2x_1 + 3x_2 + x_3 = 15 \\ 0 \le x_3 \le \infty \end{array}\right\} \quad \text{is equivalent to: } 2x_1 + 3x_2 \le 15$$

For a "greater than or equal to" constraint, we merely change the bounds on the slack variable:

$$\left.\begin{array}{r} 2x_1 + 3x_2 + x_3 = 15 \\ -\infty \le x_3 \le 0 \end{array}\right\} \quad \text{is equivalent to: } 2x_1 + 3x_2 \ge 15$$

Because of the way the dual upper-bounding algorithm is initialized, we will always require slack variables in every constraint. In the case of an equality constraint, we will add a slack variable and then require its upper and lower bounds to both equal zero.

To solve our linear programming algorithm, we must arrange the objective function and constraints in a tabular form as follows.

$$
\begin{array}{llll}
a_{11}x_1 + a_{12}x_2 + \ldots + x_{\text{slack}_1} & & = b_1 \\
a_{21}x_1 + a_{22}x_2 + \ldots & + x_{\text{slack}_2} & = b_2 & \qquad (6A.3)\\
c_1 x_1 + c_2 x_2 + \ldots & -Z = 0
\end{array}
$$

$$\underbrace{\qquad\qquad\qquad\qquad}$$

Basis variables

Because we have added slack variables to each constraint, we automatically have arranged the set of equations into what is called *canonical* form. In canonical form, there is at least one variable in each constraint whose coefficient is zero in all the other constraints. These variables are called the *basis variables*. The entire solution procedure for the linear programming algorithm centers on performing "pivot" operations that can exchange a nonbasis variable for a basis variable. A pivot operation may be shown by using our tableau in Eq. 6A.3. Suppose we wished to exchange variable x_1, a nonbasis variable, for x_{slack_2}, a slack variable. This could be accomplished by "pivoting" on column 1, row 2. To carry out the pivoting operation we execute the following steps.

Pivoting on Column 1, Row 2

Step 1   Multiply row 2 by $1/a_{21}$. That is, each $a_{2j}, j = 1 \ldots N$ in row 2 becomes

$$a'_{2j} = \frac{a_{2j}}{a_{21}} \qquad j = 1 \ldots N$$

and

$$b_2 \text{ becomes } b'_2 = \frac{b_2}{a_{21}}$$

Step 2   For each row i ($i \neq 2$), multiply row 2 by a_{i1} and subtract from row i. That is, each coefficient a_{ij} in row i ($i \neq 2$) becomes

$$a'_{ij} = a_{ij} - a_{i1}a'_{2j} \qquad j = 1 \ldots N$$

and

$$b_i \text{ becomes } b'_i = b_i - a_{i1}b'_2$$

Step 3   Last of all, we also perform the same operations in step 2 on the cost row. That is, each coefficient c_j becomes

$$c'_j = c_j - c_1 a'_{2j} \qquad j = 1 \ldots N$$

The result of carrying out the pivot operation will look like this:

$$a'_{12}x_2 + \ldots x_{\text{slack}_1} + a'_{1s_2}x_{\text{slack}_2} \qquad = b'_1$$
$$x_1 + a'_{22}x_2 + \ldots \qquad + a'_{2s_2}x_{\text{slack}_2} \qquad = b'_2$$
$$c'_2x_2 + \ldots \qquad + c'_{s2}x_{\text{slack}_2} - Z = Z'$$

Notice that the new basis for our tableau is formed by variable x_1 and x_{slack_1}, x_{slack_2} no longer has zero coefficients in row 1 or the cost row.

The dual upper-bounding algorithm proceeds in simple steps wherein variables that are in the basis are exchanged for variables out of the basis. When an exchange is made, a pivot operation is carried out at the appropriate row and column. The nonbasis variables are held equal to either their upper or their lower value, while the basis variables are allowed to take any value without respect to their upper or lower bounds. The solution terminates when all the basis variables are within their respective limits.

In order to use the dual upper-bound LP algorithm, follow these rules.

Start:

1. Each variable that has a nonzero coefficient in the cost row (i.e., the objective function) must be set according to the following rule.

$$\text{If } C_j > 0, \qquad \text{set } x_j = x_j^{\min}$$

$$\text{If } C_j < 0, \qquad \text{set } x_j = x_j^{\max}$$

2. If $C_j = 0$, x_j may be set to any value, but for convenience set it to its minimum also.
3. Add a slack variable to each constraint. Using the x_j values from steps 1 and 2, set the slack variables to make each constraint equal to its limit.

Variable Exchange:

1. Find the basis variable with the greatest violation; this determines the row to be pivoted on. Call this row R. If there are no limit violations among the basis variables, we are done. The most-violated variable leaves the basis and is set equal to the limit that was violated.
2. Select the variable to enter the basis using one of the following column selection procedures.

Column Selection Procedure P1 (Most-violated variable below its minimum)

Given constraint row R, whose basis variable is below its minimum and is the worst violation. Pick column S, so that, $c_S/(-a_{R,S})$ is minimum for all S that meet the following rules:

a. S is not in the current basis.
b. $a_{R,S}$ is not equal to zero.
c. If x_S is at its minimum, then $a_{R,S}$ must be negative and c_S must be positive or zero.
d. If x_S is at its maximum, then $a_{R,S}$ must be positive and c_S must be negative or zero.

Column Selection Procedure P2 (Most-violated variable above its maximum)

Given constraint row R, whose basis variable is above its maximum and is the worst violation. Pick column S, so that, $c_S/a_{R,S}$ is the minimum for all S that meet the following rules:

a. S is not in the current basis.
b. $a_{R,S}$ is not already zero.

c. If x_S is at its minimum, then $a_{R,S}$ must be positive and c_S must be positive or zero.

d. If x_S is at its maximum, then $a_{R,S}$ must be negative and c_S must be negative or zero.

3. When a column has been selected, pivot at the selected row R (from step 1) and column S (from step 2). The pivot column's variable, S, goes into the basis.

If no column fits the column selection criteria, we have an infeasible solution. That is, there are no values for $x_1 \ldots x_N$ that will satisfy all constraints

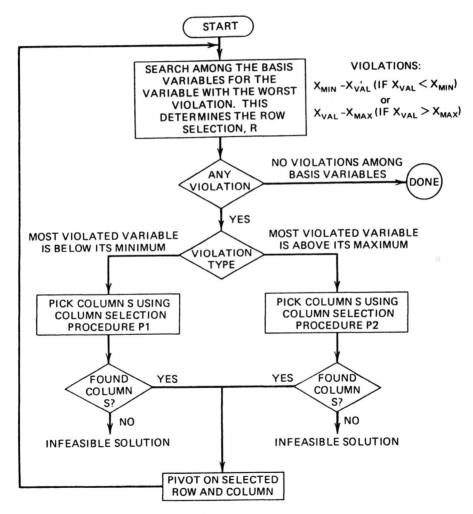

FIG. 6.10   Dual upper-bound linear programming algorithm.

simultaneously. In some problems, the cost coefficient c_S associated with column S will be zero for several different values of S. In such a case, $c_S/a_{R,S}$ will be zero for each such S and none of them will be the minimum. The fact that c_S is zero means that there will be no increase in cost if any of the S values are pivoted into the basis; therefore, the algorithm is indifferent to which one is chosen.

Setting the Variables after Pivoting

1. All nonbasis variables, except x_S, remain as they were before pivoting.
2. The most violated variable is set to the limit that was violated.
3. Since all nonbasis variables are determined, we can proceed to set each basis variable to whatever value is required to make the constraints balance. Note that this last step may move all the basis variables to new values, and some may now end up violating their respective limits (including the x_S variable).

Go back to step 1 of the variable exchange procedure.

These steps are shown in flowchart form in Figure 6.10. To help you understand the procedures involved, a sample problem is solved using the dual upper-bounding algorithm. The sample problem, shown in Figure 6.11, consists of a two-variable objective with one equality constraint and one inequality constraint.

First, we must put the equations into canonical form by adding slack variables x_3 and x_4. These variables are given limits corresponding to the type of constraint into which they are placed, x_3 is the slack variable in the equality

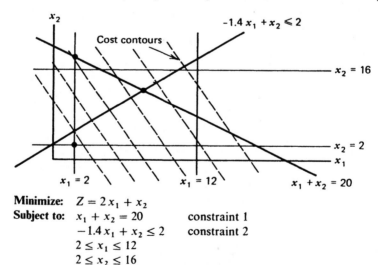

Minimize: $Z = 2x_1 + x_2$
Subject to: $x_1 + x_2 = 20$ constraint 1
$-1.4x_1 + x_2 \le 2$ constraint 2
$2 \le x_1 \le 12$
$2 \le x_2 \le 16$

FIG. 6.11   Sample linear programming problem.

constraint, so its limits are both zero; x_4 is in an inequality constraint, so it is restricted to be a positive number. To start the problem, the objective function must be set to the minimum value it can attain, and the algorithm will then seek the minimum constrained solution by increasing the objective just enough to reach the constrained solution. Thus, we set both x_1 and x_2 at their minimum values since the cost coefficients are both positive. These conditions are shown here:

Constraint 1:	$x_1 + x_2$		$+ x_3$		$= 20 \leftarrow$ R	
Constraint 2:	$-1.4x_1 + x_2$			$+ x_4$	$= 2$	
Cost:	$2x_1 + x_2$			$-Z = 0$		
					$0 \leq x_3 \leq 0$	
					$0 \leq x_4 \leq \infty$	
Minimum:	2	2	0	0		
Present value:	2	2	16	2.8	6	
Maximum:	12	16	0	∞		
			Basis variable 1	Basis variable 2		
			$\uparrow$			
			Worst-violated variable			

We can see from these conditions that variable x_3 is the worst-violated variable and that it presently exceeds its maximum limit of zero. Thus, we must use column procedure P2 on constraint number 1. This is summarized as follows:

Using selection procedure P2 on constraint 1:

$$i = 1 \quad a_1 > 0 \quad x_1 = x_1^{min} \quad c_1 > 0 \quad \text{then} \quad \frac{c_1}{a_1} = \frac{2}{1} = 2$$

$$i = 2 \quad a_2 > 0 \quad x_2 = x_2^{min} \quad c_1 > 0 \quad \text{then} \quad \frac{c_2}{a_2} = \frac{1}{1} = 1$$

$$\min c_i/a_i \text{ is } 1 \text{ at } i = 2$$

Pivot at column 2, row 1

To carry out the required pivot operations on column 2, row 1, we need merely subtract the first constraint from the second constraint and from the objective function. This results in:

Constraint 1:	x_1	$+ x_2 + x_3$		$=$	20
Constraint 2:	$-2.4x_1$	$- x_3$	$+ x_4$	$= -18 \leftarrow$ R
Cost:	x_1	$- x_3$	$-Z =$	-20
Minimum:	2	2 0	0	
Present value:	2	18 0	-13.2 22	
Maximum:	12	16 0	∞	

	Basis variable 1	Basis variable 2
		↖
		Worst-violated variable

We can see now that the variable with the worst violation is x_4 and that x_4 is below its minimum. Thus, we must use selection procedure P1 as follows:

Using selection procedure P1 on constraint 2:

$i = 1 \quad a_1 < 0 \quad x_1 = x_1^{min} \qquad c_1 > 0 \quad \text{then} \quad \dfrac{c_1}{-a_1} = \dfrac{1}{-(-2.4)} = 0.4166$

$i = 3 \quad a_3 < 0 \quad x_3 = x_3^{min} = x_3^{max} \quad c_3 < 0 \quad \text{then } x_3 \text{ is not eligible}$

Pivot at column 1, row 2

After pivoting, this results in:

Constraint 1:		$x_2 + 0.5833x_3 + 0.4166x_4$		$=$	12.5
Constraint 2:	x_1	$+\ 0.4166x_3 - 0.4166x_4$		$=$	7.5
Cost:		$-\ 1.4166x_3 + 0.4166x_4$	$-\ Z =$		-27.5
Minimum:	2	2	0	0	
Present value:	7.5	12.5	0	0	-27.5
Maximum:	12	16	0	∞	
	Basis variable 1	Basis variable 2			

At this point, we have no violations among the basis variables, so the algorithm can stop at the optimum.

$$\left.\begin{array}{l} x_1 = 7.5 \\ x_2 = 12.5 \end{array}\right\} \text{cost} = 27.5$$

See Figure 6.11 to verify that this is the optimum. The dots in Figure 6.11 show the solution points beginning at the starting point $x_1 = 2$, $x_2 = 2$, cost = 6.0, then going to $x_1 = 2$, $x_2 = 18$, cost = 22.0, and finally to the optimum $x_1 = 7.5$, $x_2 = 12.5$, cost = 27.5.

How does this algorithm work? At each step, two decisions are made.

1. Select the most-violated variable.
2. Select a variable to enter the basis.

The first decision will allow the procedure to eliminate, one after the other, those constraint violations that exist at the start, as well as those that happen during the variable-exchange steps. The second decision (using the column selection procedures) guarantees that the rate of increase in cost, to move the violated variable to its limit, is minimized. Thus, the algorithm starts from a minimum cost, infeasible solution (constraints violated), toward a minimum cost, feasible solution, by minimizing the rate of cost increase at each step.

PROBLEMS

6.1 Three units are on-line all 720 h of a 30-day month. Their characteristics are as follows:

$$H_1 = 225 + 8.47P_1 + 0.0025P_1^2, \quad 50 \leq P_1 \leq 350$$

$$H_2 = 729 + 6.20P_2 + 0.0081P_2^2, \quad 50 \leq P_2 \leq 350$$

$$H_3 = 400 + 7.20P_3 + 0.0025P_3^2, \quad 50 \leq P_3 \leq 450$$

In these equations, the H_i are in MBtu/h and the P_i are in MW.

Fuel costs for units 2 and 3 are 0.60 Ʀ/MBtu. Unit 1, however, is operated under a take-or-pay fuel contract where 60,000 tons of coal are to be burned and/or paid for in each 30-day period. This coal costs 12 Ʀ/ton delivered and has an average heat content of 12,500 Btu/lb (1 ton = 2000 lb).

The system monthly load-duration curve may be approximated by three steps as follows.

Load (MW)	Duration (h)	Energy (MWh)
800	50	40000
500	550	275000
300	120	36000
Total	720	351000

a. Compute the economic schedule for the month assuming all three units are on-line all the time and that the coal must be consumed. Show the MW loading for each load period, the MWh of each unit, and the value of gamma (the pseudo-fuel cost).

b. What would be the schedule if unit 1 was burning the coal at 12 Ʀ/ton with no constraint to use 60,000 tons? Assume the coal may be purchased on the spot market for that price and compute all the data asked for in part a. In addition, calculate the amount of coal required for the unit.

6.2 Refer to Example 6A, where three generating units are combined into a single composite generating unit. Repeat the example, except develop an equivalent incremental cost characteristic using only the incremental characteristics of the three units. Using this composite incremental characteristic plus the zero-load intercept costs of the three units, develop the total cost characteristic of the composite. (Suggestion: Fit the composite incremental cost data points using a linear approximation and a least-squares fitting algorithm.)

6.3 Refer to Problem 3.8, where three generator units have input–output curves specified as a series of straight-line segments. Can you develop a composite input–output curve for the three units? Assume all three units are on-line and that the composite input–output curve has as many linear segments as needed.

6.4 Refer to Example 6E. The first problem solved in Example 6E left the end-point restrictions at zero to 200,000 tons for both coal piles at the end of the 3-wk period. Resolve the first problem [$V_1(1) = 70,000$ and $V_2(1) = 70,000$] with the added restriction that the final volume of coal at plant 2 at the end of the third week be at least 20,000 tons.

6.5 Refer to Example 6E. In the second case solved with the LP algorithm (starting volumes equal to 70,000 and 50,000 for plant 1 and plant 2, respectively), we restricted the final volume of the coal pile at plant 2 to be 8000 tons. What is the optimum schedule if this final volume restriction is relaxed (i.e., the final coal pile at plant 2 could go to zero)?

6.6 Using the linear programming problem in the text shown in Example 6E, run a linear program to find the following:

   1. The coal unloading machinery at plant 2 is going to be taken out for maintenance for one week. During the maintenance work, no coal can be delivered to plant 2. The plant management would like to know if this should be done in week 2 or week 3. The decision will be based on the overall three-week total cost for running both plants.
   2. Could the maintenance be done in week 1? If not, why not?

Use as initial conditions those found in the beginning of the sample LP executions found in the text; i.e., $V_1(1) = 70,000$ and $V_2(2) = 70,000$.

6.7 The "Cut and Shred Paper Company" of northern Minnesota has two power plants. One burns coal and the other burns natural gas supplied by the Texas Gas Company from a pipeline. The paper company has ample supplies of coal from a mine in North Dakota and it purchases gas as take-or-pay contracts for fixed periods of time. For the 8-h time period shown below, the paper company must burn $15 \cdot 10^6$ ft^3 of gas.
   The fuel costs to the paper company are

   Coal: 0.60 $/MBtu

   Gas: 2.0 $/ccf (where 1 ccf = 1000 ft^3)
                the gas is rated at 1100 Btu/ft^3

Input–output characteristics of generators:

Unit 1 (coal unit): $H_1(P_1) = 200 + 8.5P_1 + 0.002P_1^2$ MBtu/h

$50 < P_1 < 500$

Unit 2 (gas unit): $H_2(P_2) = 300 + 6.0P_2 + 0.0025P_2^2$ MBtu/h

$50 < P_2 < 400$

Load (both load periods are 4 h long):

Period	Load (MW)
1	400
2	650

Assume both units are on-line for the entire 8 h.

Find the most economic operation of the paper company power plants, over the 8 h, which meets the gas consumption requirements.

6.8 Repeat the example in the Appendix, replacing the $x_1 + x_2 = 20$ constraint with:

$$x_1 + x_2 < 20$$

Redraw Figure 6.11 and show the admissible, convex region.

6.9 An oil-fired power plant (Figure 6.12) has the following fuel consumption curve.

$$q(\text{bbl/h}) = \begin{cases} 50 + P + 0.005P^2 & \text{for } 100 \le P \le 500 \text{ MW} \\ 0 & \text{for } P = 0 \end{cases}$$

The plant is connected to an oil storage tank with a maximum capacity of 4000 bbl. The tank has an initial volume of oil of 3000 bbl. In addition, there is a pipeline supplying oil to the plant. The pipeline terminates in the same storage tank and must be operated by contract at 500 bbl/h. The oil-fired power plant supplies energy into a system, along with other units. The other units have an equivalent cost curve of

$$F_{eq} = 300 + 6P_{eq} + 0.0025P_{eq}^2$$

$$50 \le P_{eq} \le 700 \text{ MW}$$

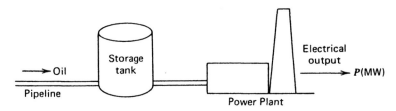

FIG. 6.12   Oil-fired power plant with storage tank for Problem 6.9.

The load to be supplied is given as follows:

Period	Load (MW)
1	400
2	900
3	700

Each time period is 2 h in length. Find the oil-fired plant's schedule using dynamic programming, such that the operating cost on the equivalent plant is minimized and the final volume in the storage tank is 2000 bbl at the end of the third period. When solving, you may use 2000, 3000, and 4000 bbl as the storage volume states for the tank. The q versus P function values you will need are included in the following table.

q(bbl/h)	P(MW)
0	0
200	100.0
250	123.6
500	216.2
750	287.3
1000	347.2
1250	400.0
1500	447.7
1800	500.0

The plant may be shut down for any of the 2-h periods with no start-up or shut-down costs.

FURTHER READING

There has not been a great deal of research work on fuel scheduling as specifically applied to power systems. However, the fuel-scheduling problem for power systems is not really that much different from other "scheduling" problems, and, for this type of problem, a great deal of literature exists.

References 1–4 are representative of efforts in applying scheduling techniques to the power system fuel-scheduling problem. References 5–8 are textbooks on linear programming that the authors have used. There are many more texts that cover LP and its variations. The reader is encouraged to study LP independently of this text if a great deal of use is to be made of LP. Many computing equipment and independent software companies have excellent LP codes that can be used, rather than writing one's own code. Reference 8 is the basis for the algorithm in the appendix to this chapter. References 9–11 give recent techniques used.

1. Trefny, F. J., Lee, K. Y., "Economic Fuel Dispatch," *IEEE Transactions on Power Apparatus and Systems*, Vol. 100, July 1981, 3468–3477.

2. Seymore, G. F., "Fuel Scheduling for Electric Power Systems," in A. M. Erisman, K. W. Noves, M. H. Dwarakanath (eds.), *Electric Power Problems: The Mathematical Challenge*, SIAM, Philadelphia, 1980, pp. 378–392.

3. Lamont, J. W., Lesso, W. G. "An Approach to Daily Fossil Fuel Management," in A. M. Erisman, K. W. Noves, M. H. Dwarakanath (eds), *Electric Power Problems: The Mathematical Challenge*, SIAM, Philadelphia, 1980, pp. 414–425.

4. Lamont, J. W., Lesso, W. G., Rantz, M., "Daily Fossil Fuel Management," 1979 *PICA Conference Proceedings*, IEEE Publication, 79CH1381-3-PWR, pp. 228–235.

5. Lasdon, L. S., *Optimization Theory for Large Systems*, Macmillan, New York, 1970.

6. Hadley, G., *Linear Programming*, Addison-Wesley, Reading, MA, 1962.

7. Wagner, H. M., *Principle of Operations Research with Application to Managerial Decisions*, Prentice-Hall, Englewood Cliffs, NJ, 1975.

8. Wagner, H. M., "The Dual Simplex Algorithm for Bounded Variables," *Naval Research Logistics Quarterly*, Vol. 5, 1958, pp. 257–261.

9. Rosenberg, L. D., Williams, D. A., Campbell, J. D., "Fuel Scheduling and Accounting," *IEEE Transactions on Power Systems*, Vol. 5, No. 2, May 1990, pp. 682–688.

10. Lee, F. N., "Adaptive Fuel Allocation Approach to Generation Dispatch using Pseudo Fuel Prices," *IEEE Transactions on Power Systems*, Vol. 7, No. 2, May 1992, pp. 487–496.

11. Sherkat, V. R., Ikura, Y., "Experience with Interior Point Optimization Software for a Fuel Planning Application," 1993 *IEEE Power Industry Computer Applications Conference*, pp. 89–96.

7 Hydrothermal Coordination

7.1 INTRODUCTION

The systematic coordination of the operation of a system of hydroelectric generation plants is usually more complex than the scheduling of an all-thermal generation system. The reason is both simple and important. That is, the hydroelectric plants may very well be coupled both electrically (i.e., they all serve the same load) and hydraulically (i.e., the water outflow from one plant may be a very significant portion of the inflow to one or more other, downstream plants).

No two hydroelectric systems in the world are alike. They are all different. The reasons for the differences are the natural differences in the watersheds, the differences in the manmade storage and release elements used to control the water flows, and the very many different types of natural and manmade constraints imposed on the operation of hydroelectric systems. River systems may be simple with relatively few tributaries (e.g., the Connecticut River), with dams in series (hydraulically) along the river. River systems may encompass thousands of acres, extend over vast multinational areas, and include many tributaries and complex arrangements of storage reservoirs (e.g., the Columbia River basin in the Pacific Northwest).

Reservoirs may be developed with very large storage capacity with a few high-head plants along the river. Alternatively, the river may have been developed with a larger number of dams and reservoirs, each with smaller storage capacity. Water may be intentionally diverted through long raceways that tunnel through an entire mountain range (e.g., the Snowy Mountain scheme in Australia). In European developments, auxiliary reservoirs, control dams, locks, and even separate systems for pumping water back upstream have been added to rivers.

However, the one single aspect of hydroelectric plants that differentiates the coordination of their operation more than any other is the existence of the many, and highly varied, constraints. In many hydrosystems, the generation of power is an adjunct to the control of flood waters or the regular, scheduled release of water for irrigation. Recreation centers may have developed along the shores of a large reservoir so that only small surface water elevation changes are possible. Water release in a river may well have to be controlled so that the river is navigable at all times. Sudden changes, with high-volume releases of water, may be prohibited because the release could result in a large wave traveling downstream with potentially damaging effects. Fish ladders may be needed. Water releases may be dictated by international treaty.

To repeat: all hydrosystems are different.

7.1.1 Long-Range Hydro-Scheduling

The coordination of the operation of hydroelectric plants involves, of course, the scheduling of water releases. The *long-range hydro-scheduling problem* involves the long-range forecasting of water availability and the scheduling of reservoir water releases (i.e., "drawdown") for an interval of time that depends on the reservoir capacities.

Typical long-range scheduling goes anywhere from 1 wk to 1 yr or several years. For hydro schemes with a capacity of impounding water over several seasons, the long-range problem involves meteorological and statistical analyses.

Nearer-term water inflow forecasts might be based on snow melt expectations and near-term weather forecasts. For the long-term drawdown schedule, a basic policy selection must be made. Should the water be used under the assumption that it will be replaced at a rate based on the statistically expected (i.e., mean value) rate, or should the water be released using a "worst-case" prediction. In the first instance, it may well be possible to save a great deal of electric energy production expense by displacing thermal generation with hydro-generation. If, on the other hand, a worst-case policy was selected, the hydroplants would be run so as to minimize the risk of violating any of the hydrological constraints (e.g., running reservoirs too low, not having enough water to navigate a river). Conceivably, such a schedule would hold back water until it became quite likely that even worst-case rainfall (runoff, etc.) would still give ample water to meet the constraints.

Long-range scheduling involves optimizing a policy in the context of unknowns such as load, hydraulic inflows, and unit availabilities (steam and hydro). These unknowns are treated statistically, and long-range scheduling involves optimization of statistical variables. Useful techniques include:

1. Dynamic programming, where the entire long-range operation time period is simulated (e.g., 1 yr) for a given set of conditions.
2. Composite hydraulic simulation models, which can represent several reservoirs.
3. Statistical production cost models.

The problems and techniques of long-range hydro-scheduling are outside the scope of this text, so we will end the discussion at this point and continue with short-range hydro-scheduling.

7.1.2 Short-Range Hydro-Scheduling

Short-range hydro-scheduling (1 day to 1 wk) involves the hour-by-hour scheduling of all generation on a system to achieve minimum production cost for the given time period. In such a scheduling problem, the load, hydraulic inflows, and unit availabilities are assumed known. A set of starting conditions (e.g., reservoir levels) is given, and the optimal hourly schedule that minimizes a desired objective, while meeting hydraulic steam, and electric system constraints, is sought. Part of the hydraulic constraints may involve meeting "end-point" conditions at the end of the scheduling interval in order to conform to a long-range, water-release schedule previously established.

7.2 HYDROELECTRIC PLANT MODELS

To understand the requirements for the operation of hydroelectric plants, one must appreciate the limitations imposed on operation of hydro-resources by flood control, navigation, fisheries, recreation, water supply, and other demands on the water bodies and streams, as well as the characteristics of energy conversion from the potential energy of stored water to electric energy. The amount of energy available in a unit of stored water, say a cubic foot, is equal to the product of the weight of the water stored (in this case, 62.4 lb) times the height (in feet) that the water would fall. One thousand cubic feet of water falling a distance of 42.5 ft has the energy equivalent to 1 kWh. Correspondingly, 42.5 ft^3 of water falling 1000 ft also has the energy equivalent to 1 kWh.

Consider the sketch of a reservoir and hydroelectric plant shown in Figure 7.1. Let us consider some overall aspects of the falling water as it travels from the reservoir through the penstock to the inlet gates, through the hydraulic turbine down the draft tube and out the tailrace at the plant exit. The power that the water can produce is equal to the rate of water flow in cubic feet per

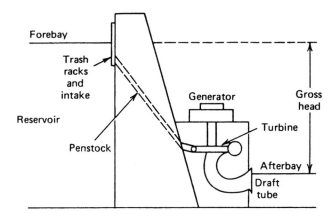

FIG. 7.1   Hydroplant components.

second times a conversion coefficient that takes into account the net head (the distance through which the water falls, less the losses in head caused by the flow) times the conversion efficiency of the turbine generator. A flow of 1 ft^3/sec falling 100 ft has the power equivalent of approximately 8.5 kW. If the flow-caused loss in head was 5%, or 5 ft, then the power equivalent for a flow of 1 ft^3 of water per second with the net drop of $100 - 5$, or 95 ft, would have the power equivalent of slightly more than 8 kW (8.5 × 95%). Conversion efficiencies of turbine generators are typically in the range of 85 to 90% at the best efficiency operating point for the turbine generator, so 1 ft^3/sec falling 100 ft would typically develop about 7 kW at most.

Let us return to our description of the hydroelectric plant as illustrated in Figure 7.1. The hydroelectric project consists of a body of water impounded by a dam, the hydroplant, and the exit channel or lower water body. The energy available for conversion to electrical energy of the water impounded by the dam is a function of the gross head; that is, the elevation of the surface of the reservoir less the elevation of the afterbay, or downstream water level below the hydroelectric plant. The head available to the turbine itself is slightly less than the gross head, due to the friction losses in the intake, penstock, and draft tube. This is usually expressed as the *net head* and is equal to the gross head less the flow losses (measured in feet of head). The flow losses can be very significant for low head (10 to 60 ft) plants and for plants with long penstocks (several thousand feet). The water level at the afterbay is influenced by the flow out of the reservoir, including plant release and any spilling of water over the top of the dam or through bypass raceways. During flooding conditions such as spring runoff, the rise in afterbay level can have a significant and adverse effect on the energy and capacity or power capacity of the hydroplant.

The type of turbine used in a hydroelectric plant depends primarily on the design head for the plant. By far the largest number of hydroelectric projects use reaction-type turbines. Only two types of reaction turbines are now in common use. For medium heads (that is, in the range from 60 to 1000 ft), the Francis turbine is used exclusively. For the low-head plants (that is, for design heads in the range of 10 to 60 ft), the propeller turbine is used. The more modern propeller turbines have adjustable pitch blading (called *Kaplan turbines*) to improve the operating efficiency over a wide range of plant net head. Typical turbine performance results in an efficiency at full gate loading of between 85 to 90%. The Francis turbine and the adjustable propeller turbine may operate at 65 to 125% of rated net head as compared to 90 to 110% for the fixed propeller.

Another factor affecting operating efficiency of hydro-units is the MW loading. At light unit loadings, efficiency may drop below 70% (these ranges are often restricted by vibration and cavitation limits) and at full gate may rise to about 87%. If the best use of the hydro-resource is to be obtained, operation of the hydro-unit near its best efficiency gate position and near the designed head is necessary. This means that unit loading and control of reservoir forebay are necessary to make efficient use of hydro-resources. Unit loading should be

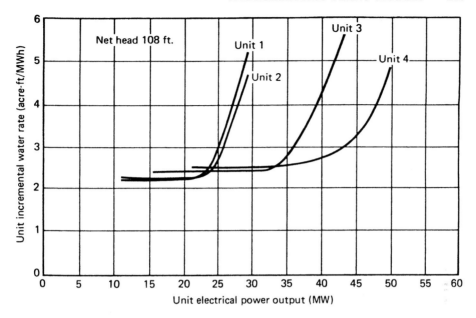

FIG. 7.2   Incremental water rate versus power output.

near best efficiency gate position, and water-release schedules must be co-
ordinated with reservoir inflows to maintain as high a head on the turbines as
the limitations on forebay operations will permit.

Typical plant performance for a medium head, four-unit plant in South
America is illustrated in Figure 7.2. The incremental "water rate" is expressed
in acre-feet per megawatt hour.* The rise in incremental water rate with
increasing unit output results primarily from the increased hydraulic losses with
the increased flow. A composite curve for multiple unit operation at the plant
would reflect the mutual effects of hydraulic losses and rise in afterbay with
plant discharge. Very careful attention must be given to the number of units
run for a given required output. One unit run at best efficiency will usually use
less water than two units run at half that load.

High-head plants (typically over 1000 ft) use impulse or Pelton turbines. In
such turbines, the water is directed into spoon-shaped buckets on the wheel by
means of one or more water jets located around the outside of the wheel.

In the text that follows, we will assume a characteristic giving the relationship
between water flow through the turbine, q, and power output, $P(MW)$, where
q is expressed in ft^3/sec or acre-ft/h. Furthermore, we will not be concerned
with what type of turbine is being used or the characteristics of the reservoir,
other than such limits as the reservoir head or volume and various flows.

---

* An *acre-foot* is a common unit of water volume. It is the amount of water that will cover 1 acre
to a depth of 1 ft ($43,560 \, ft^3$). It also happens to be nearly equal to half a cubic foot per second
flow for a day ($43,200 \, ft^3$). An acre-foot is equal to $1.2335 \cdot 10^3 \, m^3$.

7.3 SCHEDULING PROBLEMS

7.3.1 Types of Scheduling Problems

In the operation of a hydroelectric power system, three general categories of problems arise. These depend on the balance between the hydroelectric generation, the thermal generation, and the load.

Systems without any thermal generation are fairly rare. The economic scheduling of these systems is really a problem in scheduling water releases to satisfy all the hydraulic constraints and meet the demand for electrical energy. Techniques developed for scheduling hydrothermal systems may be used in some systems by assigning a pseudo-fuel cost to some hydroelectric plant. Then the schedule is developed by minimizing the production "cost" as in a conventional hydrothermal system. In all hydroelectric systems, the scheduling could be done by simulating the water system and developing a schedule that leaves the reservoir levels with a maximum amount of stored energy. In geographically extensive hydroelectric systems, these simulations must recognize water travel times between plants.

Hydrothermal systems where the hydroelectric system is by far the largest component may be scheduled by economically scheduling the system to produce the minimum cost for the thermal system. These are basically problems in scheduling energy. A simple example is illustrated in the next section where the hydroelectric system cannot produce sufficient energy to meet the expected load.

The largest category of hydrothermal systems include those where there is a closer balance between the hydroelectric and thermal generation resources and those where the hydroelectric system is a small fraction of the total capacity. In these systems, the schedules are usually developed to minimize thermal-generation production costs, recognizing all the diverse hydraulic constraints that may exist. The main portion of this chapter is concerned with systems of this type.

7.3.2 Scheduling Energy

Suppose, as in Figure 7.3, we have two sources of electrical energy to supply a load, one hydro and another steam. The hydroplant can supply the load

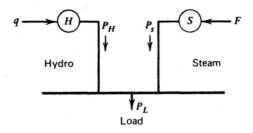

FIG. 7.3 Two-unit hydrothermal system.

by itself for a limited time. That is, for any time period j,

$$P_{Hj}^{\max} \geq P_{\text{load } j} \qquad j = 1 \ldots j_{\max} \tag{7.1}$$

However, the energy available from the hydroplant is insufficient to meet the load.

$$\sum_{j=1}^{j_{\max}} P_{Hj} n_j \leq \sum_{j=1}^{j_{\max}} P_{\text{load } j} n_j \qquad n_j = \text{number of hours in period } j$$

$$\sum_{j=1}^{j_{\max}} n_j = T_{\max} = \text{total interval} \tag{7.2}$$

We would like to use up the entire amount of energy from the hydroplant in such a way that the cost of running the steam plant is minimized. The steam-plant energy required is

$$\sum_{j=1}^{j_{\max}} P_{\text{load } j} n_j - \sum_{j=1}^{j_{\max}} P_{Hj} n_j = E \tag{7.3}$$

| Load | Hydro- | Steam |
| energy | energy | energy |

We will not require the steam unit to run for the entire interval of $T_{\max}$ hours. Therefore,

$$\sum_{j=1}^{N_s} P_{Sj} n_j = E \qquad N_s = \text{number of periods the steam plant is run} \tag{7.4}$$

Then

$$\sum_{j=1}^{N_s} n_j \leq T_{\max}$$

the scheduling problem becomes

$$\text{Min } F_T = \sum_{j=1}^{N_s} F(P_{Sj}) n_j \tag{7.5}$$

subject to

$$\sum_{j=1}^{N_s} P_{Sj} n_j - E = 0 \tag{7.6}$$

and the Lagrange function is

$$\mathscr{L} = \sum_{j=1}^{N_s} F(P_{Sj}) n_j + \alpha \left(E - \sum_{j=1}^{N_s} P_{sj} n_j \right) \tag{7.7}$$

Then

$$\frac{\partial \mathscr{L}}{\partial P_{sj}} = \frac{dF(P_{sj})}{dP_{sj}} - \alpha = 0 \qquad \text{for } j = 1 \ldots N_s$$

or   (7.8)

$$\frac{dF(P_{sj})}{dP_{sj}} = \alpha \qquad \text{for } j = 1 \ldots N_s$$

This means that the steam plant should be run at constant incremental cost for the entire period it is on. Let this optimum value of steam-generated power be P_s^*, which is the same for all time intervals the steam unit is on. This type of schedule is shown in Figure 7.4.

The total cost over the interval is

$$F_T = \sum_{j=1}^{N_s} F(P_s^*)n_j = F(P_s^*) \sum_{j=1}^{N_s} n_j = F(P_s^*)T_s \qquad (7.9)$$

where

$$T_s = \sum_{j=1}^{N_s} n_j = \text{the total run time for the steam plant}$$

Let the steam-plant cost be expressed as

$$F(P_s) = A + BP_s + CP_s^2 \qquad (7.10)$$

then

$$F_T = (A + BP_s^* + CP_s^{*2})T_s \qquad (7.11)$$

also note that

$$\sum_{j=1}^{N_s} P_{sj}n_j = \sum_{j=1}^{N_s} P_s^*n_j = P_s^*T_s = E \qquad (7.12)$$

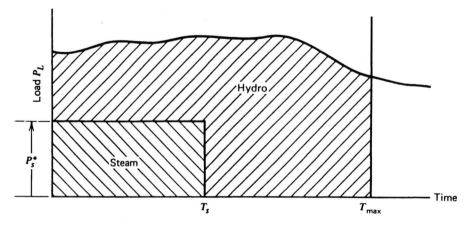

FIG. 7.4   Resulting optimal hydrothermal schedule.

Then

$$T_s = \frac{E}{P_s^*} \tag{7.13}$$

and

$$F_T = (A + BP_s^* + CP_s^{*2})\left(\frac{E}{P_s^*}\right) \tag{7.14}$$

Now we can establish the value of P_s^* by minimizing F_T:

$$\frac{dF_T}{dP_s^*} = \frac{-AE}{P_s^{*2}} + CE = 0 \tag{7.15}$$

or

$$P_s^* = \sqrt{A/C} \tag{7.16}$$

which means the unit should be operated at its maximum efficiency point long enough to supply the energy needed, E. Note, if

$$F(P_s) = A + BP_s + CP_s^2 = f_c \times H(P_s) \tag{7.17}$$

where f_c is the fuel cost, then the heat rate is

$$\frac{H(P_s)}{P_s} = \frac{1}{f_c}\left(\frac{A}{P_s} + B + CP_s\right) \tag{7.18}$$

and the heat rate has a minimum when

$$\frac{d}{dP_s}\left[\frac{H(P_s)}{P_s}\right] = 0 = \frac{-A}{P_s^2} + C \tag{7.19}$$

giving best efficiency at

$$P_s = \sqrt{A/C} = P_s^* \tag{7.20}$$

EXAMPLE 7A

A hydroplant and a steam plant are to supply a constant load of 90 MW for 1 wk (168 h). The unit characteristics are

Hydroplant: $q = 300 + 15P_H$ acre-ft/h

$0 \le P_H \le 100$ MW

Steam plant: $H_s = 53.25 + 11.27P_s + 0.0213P_s^2$

$12.5 \le P_s \le 50$ MW

Part 1

Let the hydroplant be limited to 10,000 MWh of energy. Solve for T_s^*, the run time of the steam unit. The load is $90 \times 168 = 15,120$ MWh, requiring 5120 MWh to be generated by the steam plant.

The steam plant's maximum efficiency is at $\sqrt{53.25/0.0213} = 50$ MW. Therefore, the steam plant will need to run for 5120/50 or 102.4 h. The resulting schedule will require the steam plant to run at 50 MW and the hydroplant at 40 MW for the first 102.4 h of the week and the hydroplant at 90 MW for the remainder.

Part 2

Instead of specifying the energy limit on the steam plant, let the limit be on the volume of water that can be drawn from the hydroplants' reservoir in 1 wk. Suppose the maximum drawdown is 250,000 acre-ft, how long should the steam unit run?

To solve this we must account for the plant's q versus P characteristic. A different flow will take place when the hydroplant is operated at 40 MW than when it is operated at 90 MW. In this case,

$$q_1 = [300 + 15(40)] \times T_s \text{ acre-ft}$$
$$q_2 = [300 + 15(90)] \times (168 - T_s) \text{ acre-ft}$$

and

$$q_1 + q_2 = 250,000 \text{ acre-ft}$$

Solving for T_s we get 36.27 h.

7.4 THE SHORT-TERM HYDROTHERMAL SCHEDULING PROBLEM

A more general and basic short-term hydrothermal scheduling problem requires that a given amount of water be used in such a way as to minimize the cost of running the thermal units. We will use Figure 7.5 in setting up this problem.

The problem we wish to set up is the general, short-term hydrothermal scheduling problem where the thermal system is represented by an equivalent unit, P_s, as was done in Chapter 6. In this case, there is a single hydroelectric plant, P_H. We assume that the hydroplant is not sufficient to supply all the load demands during the period and that there is a maximum total volume of water that may be discharged throughout the period of $T_{\max}$ hours.

In setting up this problem and the examples that follow, we assume all spillages, s_j, are zero. The only other hydraulic constraint we will impose initially is that the total volume of water discharged must be exactly as

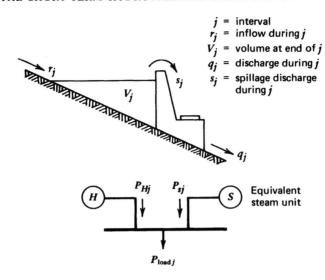

j = interval
r_j = inflow during j
V_j = volume at end of j
q_j = discharge during j
s_j = spillage discharge during j

Equivalent steam unit

$P_{load\,j}$

FIG. 7.5 Hydrothermal system with hydraulic constraints.

defined. Therefore, the mathematical scheduling problem may be set up as follows:

Problem:
$$\text{Min } F_T = \sum_{j=1}^{j_{max}} n_j F_j \tag{7.21}$$

Subject to:
$$\sum_{j=1}^{j_{max}} n_j q_j = q_{TOT} \qquad \text{total water discharge}$$

$$P_{load\,j} - P_{Hj} - P_{sj} = 0 \qquad \text{load balance for } j = 1 \ldots j_{max}$$

where
$$n_j = \text{length of } j^{\text{th}} \text{ interval}$$

$$\sum_{j=1}^{j_{max}} n_j = T_{max}$$

and the loads are constant in each interval. Other constraints could be imposed, such as:

$$V_j|_{j=0} = V_s \qquad \text{starting volume}$$

$$V_j|_{j=j_{max}} = V_E \qquad \text{ending volume}$$

$$q_{min} \le q_j \le q_{max} \qquad \text{flow limits for } j = 1 \ldots j_{max}$$

$$q_j = Q_j \qquad \text{fixed discharge for a particular hour}$$

Assume constant head operation and assume a q versus P characteristic is available, as shown in Figure 7.6, so that

$$q = q(P_H) \tag{7.22}$$

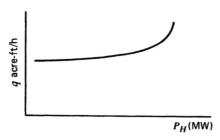

FIG. 7.6 Hydroelectric unit input–output characteristic for constant head.

We now have a similar problem to the take-or-pay fuel problem. The Lagrange function is

$$\mathscr{L} = \sum_{j=1}^{j_{\max}} [n_j F(P_{sj}) + \lambda_j(P_{\text{load } j} - P_{Hj} - P_{sj})] + \gamma\left[\sum_{j=1}^{j_{\max}} n_j q_j(P_{Hj}) - q_{\text{TOT}}\right] \quad (7.23)$$

and for a specific interval $j = k$,

$$\frac{\partial \mathscr{L}}{\partial P_{sk}} = 0$$

gives

$$n_k \frac{dF_{sk}}{dP_{sk}} = \lambda_k \quad (7.24)$$

and

$$\frac{\partial \mathscr{L}}{\partial P_{Hk}} = 0$$

gives

$$\gamma n_k \frac{dq_k}{dP_{Hk}} = \lambda_k \quad (7.25)$$

This is solved using the same techniques shown in Chapter 6.

Suppose we add the network losses to the problem. Then at each hour,

$$P_{\text{load } j} + P_{\text{loss } j} - P_{Hj} - P_{sj} = 0 \quad (7.26)$$

and the Lagrange function becomes

$$\mathscr{L} = \sum_{j=1}^{j_{\max}} [n_j F(P_{sj}) + \lambda_j(P_{\text{load } j} + P_{\text{loss } j} - P_{Hj} - P_{sj})]$$

$$+ \gamma\left[\sum_{j=1}^{j_{\max}} n_j q_j(P_{Hj}) - q_{\text{TOT}}\right] \quad (7.27)$$

with resulting coordination equations (hour k):

$$n_k \frac{dF(P_{sk})}{dP_{sk}} + \lambda_k \frac{\partial P_{\text{loss }k}}{\partial P_{sk}} = \lambda_k \qquad (7.28)$$

$$\gamma n_k \frac{dq(P_{Hk})}{dP_{Hk}} + \lambda_k \frac{\partial P_{\text{loss }k}}{\partial P_{Hk}} = \lambda_k \qquad (7.29)$$

This gives rise to a more complex scheduling solution requiring three loops, as shown in Figure 7.7. In this solution procedure, ε_1 and ε_2 are the respective tolerances on the load balance and water balance relationships.

Note that this problem ignores volume and hourly discharge rate constraints.

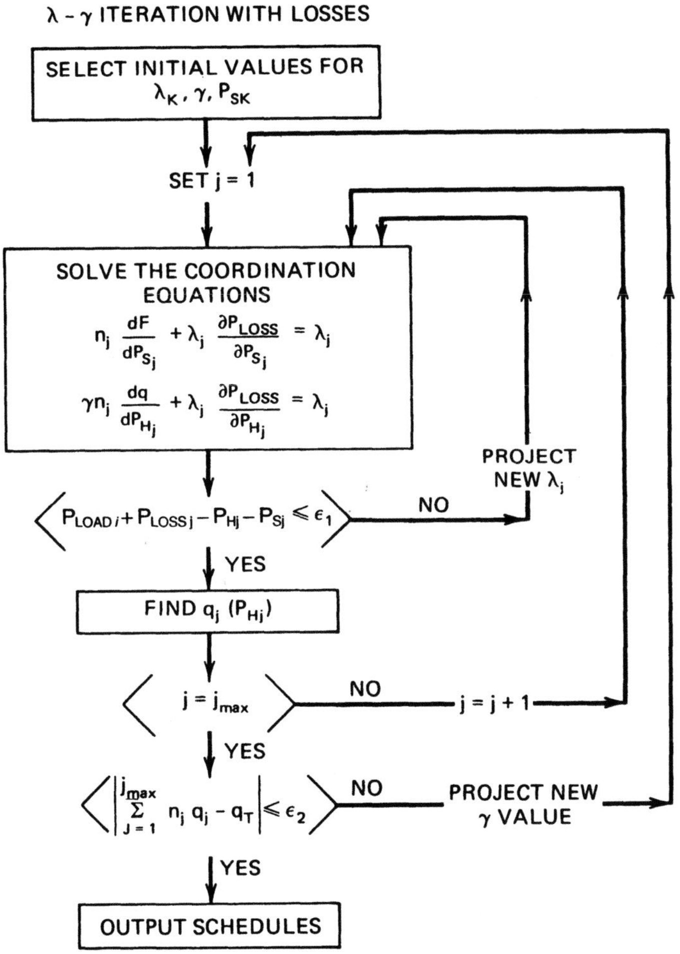

FIG. 7.7 A λ–γ iteration scheme for hydrothermal scheduling.

As a result, the value of γ will be constant over the entire scheduling period as long as the units remain within their respective scheduling ranges. The value of γ would change if a constraint (i.e., $V_j = V_{max}$, etc.) were encountered. This would require that the scheduling logic recognize such constraints and take appropriate steps to adjust γ so that the constrained variable does not go beyond its limit. The appendix to this chapter gives a proof that γ is constant when no storage constraints are encountered. As usual, in any gradient method, care must be exercised to allow constrained variables to move off their constraints if the solution so dictates.

EXAMPLE 7B

A load is to be supplied from a hydroplant and a steam system whose characteristics are given here.

Equivalent steam system: $\quad H = 500 + 8.0P_s + 0.0016P_s^2 \quad$ (MBtu/h)

$$\text{Fuel cost} = 1.15 \,\text{R/MBtu}$$

$$150 \text{ MW} \leq P_s \leq 1500 \text{ MW}$$

Hydroplant: $\quad q = 330 + 4.97P_H$ acre-ft/h

$$0 \leq P_H \leq 1000 \text{ MW}$$

$$q = 5300 + 12(P_H - 1000) + 0.05(P_H - 1000)^2 \text{ acre-ft/h}$$

$$1000 < P_H < 1100 \text{ MW}$$

The hydroplant is located a good distance from the load. The electrical losses are

$$P_{loss} = 0.00008P_h^2 \text{ MW}$$

The load to be supplied is connected at the steam plant and has the following schedule:

$$2400{-}1200 = 1200 \text{ MW}$$

$$1200{-}2400 = 1500 \text{ MW}$$

The hydro-unit's reservoir is limited to a drawdown of 100,000 acre-ft over the entire 24-h period. Inflow to the reservoir is to be neglected. The optimal schedule for this problem was found using a program written using Figure 7.7. The results are:

Time Period	P steam	P hydro	Hydro-Discharge (acre-ft/h)
2400–1200	567.4	668.3	3651.5
1200–2400	685.7	875.6	4681.7

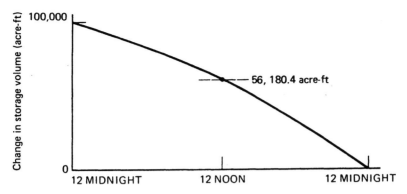

FIG. 7.8   Change in storage volume (=cumulative discharge) versus time for Example 7B.

The optimal value for γ is 2.028378 ₽/acre-ft. The storage in the hydroplant's reservoir goes down in time as shown in Figure 7.8. No natural inflows or spillage are assumed to occur.

7.5   SHORT-TERM HYDRO-SCHEDULING: A GRADIENT APPROACH

The following is an outline of a first-order gradient approach, as shown in Figure 6.7a, to the problem of finding the optimum schedule for a hydrothermal power system. We assume a single equivalent thermal unit with a convex input–output curve and a single hydroplant. Let:

j = the interval = 1, 2, 3, ..., j_{max}

V_j = storage volume at the end of interval j

q_j = discharge rate during interval j

r_j = inflow rate into the storage reservoir during interval j

P_{sj} = steam generation during j^{th} interval.

s_j = spillage discharge rate during interval j

$P_{loss\ j}$ = losses, assumed here to be zero

$P_{load\ j}$ = received power during the jth interval (load)

P_{Hj} = hydro-generation during the jth hour

Next, we let the discharge from the hydroplant be a function of the hydro-power output only. That is, a constant head characteristic is assumed.

Then,

$$q_j(P_{Hj}) = q_j$$

so that to a first order,*

$$\Delta q_j = \frac{dq_j}{dP_H} \Delta P_{Hj}$$

The total cost for fuel over the $j = 1, 2, 3, \ldots, j_{max}$ intervals is

$$F_T = \sum_{j=1}^{j_{max}} n_j F_j(P_{sj})$$

This may be expanded in a Taylor series to give the change in fuel cost for a change in steam-plant schedule.

$$\Delta F_T = \sum_{j=1}^{j_{max}} n_j [F'_j \Delta P_{sj} + \tfrac{1}{2} F''_j (\Delta P_{sj})^2 + \ldots]$$

To the first order this is

$$\Delta F_T = \sum_{j=1}^{j_{max}} n_j F'_j \Delta P_{sj}$$

In any given interval, the electrical powers must balance:

$$P_{\text{load } j} - P_{sj} - P_{Hj} = 0$$

so that,

$$\Delta P_{sj} = -\Delta P_{Hj}$$

or

$$\Delta P_{sj} = -\frac{\Delta q_j}{\left(\dfrac{dq_j}{dP_{Hj}}\right)}$$

Therefore,

$$\Delta F_T = -\sum_{j=1}^{j_{max}} n_j \left[\frac{\left(\dfrac{dF_j}{dP_{sj}}\right)}{\left(\dfrac{dq_j}{dP_{Hj}}\right)}\right] \Delta q_j = -\sum_{j=1}^{j_{max}} n_j \gamma_j \Delta q_j$$

where

$$\gamma_j = \frac{\left(\dfrac{dF_j}{dP_{sj}}\right)}{\left(\dfrac{dq_j}{dP_{Hj}}\right)}$$

* ΔP_s and ΔF designate changes in the quantities P_s and F.

The variables γ_j are the incremental water values in the various intervals and give an indication of how to make the "moves" in the application of the first-order technique. That is, the "steepest descent" to reach minimum fuel cost (or the best period to release a unit of water) is the period with the maximum value of γ. The values of water release, Δq_j, must be chosen to stay within the hydraulic constraints. These may be determined by use of the *hydraulic continuity equation*:

$$V_j = V_{j-1} + (r_j - q_j - s_j)n_j$$

to compute the reservoir storage each interval. We must also observe the storage limits,

$$V_{min} \le V_j \le V_{max}$$

We will assume spillage is prohibited so that all $s_j = 0$, even though there may well be circumstances where allowing $s_j > 0$ for some j might reduce the thermal system cost.

The discharge flow may be constrained both in rate and in total. That is,

$$q_{min} \le q_j \le q_{max}$$

and

$$\sum_{j=1}^{j_{max}} n_j q_j = q_{TOT}$$

The flowchart in Figure 6.7a illustrates the application of this method. Figure 7.9 illustrates a typical trajectory of storage volume versus time and illustrates the special rules that must be followed when constraints are taken. Whenever a constraint is reached (that is, storage V_j is equal to V_{min} or V_{max}), one must choose intervals in a more restricted manner than as shown in Figure 6.7a. This is summarized here.

1. **No Constraints Reached**
   Select the pair of intervals j^- and j^+ anywhere from $j = 1 \ldots j_{max}$.

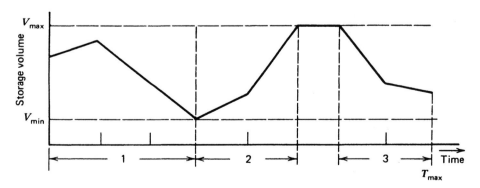

FIG. 7.9 Storage volume trajectory.

2. A Constraint Is Reached

Option A: Choose the j^- and j^+ within one of the subintervals. That is, choose both j^- and j^+ from periods 1, 2, or 3 in Figure 7.9. This will guarantee that the constraint is not violated. For example, choosing a time j^+ within period 1 to increase release, and choosing j^- also in period 1 to decrease release, will mean no net release change at the end of subinterval 1, so the V_{min} constraint will not be violated.

Option B: Choose j^- and j^+ from different subintervals so that the constraint is no longer reached. For example, choosing j^+ within period 2 and j^- within period 1 will mean the V_{min} and V_{max} limits are no longer reached at all.

Other than these special rules, one can apply the flowchart of Figure 6.7a exactly as shown (while understanding that q is water rather than fuel as in Figure 6.7a).

EXAMPLE 7C

Find an optimal hydro-schedule using the gradient technique of section 7.5. The hydroplant and equivalent steam plant are the same as Example 7B, with the following additions.

Load pattern:

First day	2400–1200 =	1200 MW
	1200–2400 =	1500 MW
Second day	2400–1200 =	1100 MW
	1200–2400 =	1800 MW
Third day	2400–1200 =	950 MW
	1200–2400 =	1300 MW

Hydro-reservoir:

1. 100,000 acre-ft at the start.

2. Must have 60,000 acre-ft at the end of schedule.

3. Reservoir volume is limited as follows:

$$60{,}000 \text{ acre-ft} \leq V \leq 120{,}000 \text{ acre-ft}$$

4. There is a constant inflow into the reservoir of 2000 acre-ft/h over the entire 3-day period

The initial schedule has constant discharge; thereafter, each update or "step" in the gradient calculations was carried out by entering the j^+, j^- and Δq into a computer terminal that then recalculated all period γ values, flows, and so forth. The results of running this program are shown in Figure 7.10.

```
        INITIAL SCHEDULE ( CONSTANT DISCHARGE )
 J        Ps         PH        GAMMA          VOLUME        DISCHARGE
 1      752.20     447.80     2.40807       93333.3        2555.555
 2     1052.20     447.80     2.63020       86666.7        2555.555
 3      652.20     447.80     2.33402       90000.0        2555.555
 4     1352.20     447.80     2.85233       73333.4        2555.555
 5      502.20     447.80     2.22296       66666.7        2555.555
 6      852.20     447.80     2.48211       60000.1        2555.555

    TOTAL OPERATING COST FOR ABOVE SCHEDULE =      719725.50  R
ENTER JMAX,JMIN,DELQ
4,5,1000

 J        Ps         PH        GAMMA          VOLUME        DISCHARGE
 1      752.20     447.80     2.40807       93333.3        2555.555
 2     1052.20     447.80     2.63020       86666.7        2555.555
 3      652.20     447.80     2.33402       90000.0        2555.555
 4     1150.99     649.01     2.70335       61333.4        3555.555
 5      703.41     246.59     2.37194       66666.7        1555.555
 6      852.20     447.80     2.48211       50000.1        2555.555

    TOTAL OPERATING COST FOR ABOVE SCHEDULE =      713960.75  R
ENTER JMAX,JMIN,DELQ
4,3,400-

 J        Ps         PH        GAMMA          VOLUME        DISCHARGE
 1      752.20     447.80     2.40807       93333.3        2555.555
 2     1052.20     447.80     2.63020       86666.7        2555.555
 3      732.69     367.31     2.39362       84800.0        2155.555
 4     1070.51     729.49     2.64376       61333.4        3955.555
 5      703.41     246.59     2.37194       56666.7        1555.555
 6      852.20     447.80     2.48211       60000.1        2555.555

    TOTAL OPERATING COST FOR ABOVE SCHEDULE =      712474.00  R
ENTER JMAX,JMIN,DELQ
4,5,100

 J        Ps         PH        GAMMA          VOLUME        DISCHARGE
 1      752.20     447.80     2.40807       93333.3        2555.555
 2     1052.20     447.80     2.63020       86666.7        2555.555
 3      732.69     367.31     2.39362       84800.0        2155.555
 4     1050.39     749.61     2.62886       60133.4        4055.555
 5      723.53     226.47     2.38684       66666.7        1455.555
 6      852.20     447.80     2.48211       50000.1        2555.555

    TOTAL OPERATING COST FOR ABOVE SCHEDULE =      712165.75  R
ENTER JMAX,JMIN,DELQ
2,5,10

 J        Ps         PH        GAMMA          VOLUME        DISCHARGE
 1      752.20     447.80     2.40807       93333.3        2555.555
 2     1050.19     449.81     2.62871       86546.7        2565.555
 3      732.69     367.31     2.39362       84680.0        2155.555
 4     1050.39     749.61     2.62886       60013.4        4055.555
 5      725.54     224.46     2.38833       66666.7        1445.555
 6      852.20     447.80     2.48211       60000.1        2555.555

    TOTAL OPERATING COST FOR ABOVE SCHEDULE =      712136.75  R
ENTER JMAX,JMIN,DELQ
4,5,1.111

 J        Ps         PH        GAMMA          VOLUME        DISCHARGE
 1      752.20     447.80     2.40807       93333.3        2555.555
 2     1050.19     449.81     2.62871       86546.7        2565.555
 3      732.69     367.31     2.39362       84680.0        2155.555
 4     1050.17     749.83     2.62870       60000.0        4056.666
 5      725.77     224.23     2.38849       66666.7        1444.444
 6      852.20     447.80     2.48211       60000.0        2555.555

    TOTAL OPERATING COST FOR ABOVE SCHEDULE =      712133.50  R
```

FIG. 7.10 Computer printout for Example 7C. *(Continued on next page)*

```
ENTER JMAX,JMIN,DELQ
2,3,800
```

J	Ps	PH	GAMMA	VOLUME	DISCHARGE
1	752.20	447.80	2.40807	93333.3	2555.555
2	889.22	610.78	2.50953	76946.7	3365.555
3	893.65	206.35	2.51280	84680.0	1355.555
4	1050.17	749.83	2.62870	60000.0	4056.666
5	725.77	224.23	2.38849	65665.7	1444.444
6	852.20	447.80	2.48211	50000.0	2555.555

```
   TOTAL OPERATING COST FOR ABOVE SCHEDULE =      711020.75 R
ENTER JMAX,JMIN,DELQ
4,1,750
```

J	Ps	PH	GAMMA	VOLUME	DISCHARGE
1	903.11	296.89	2.51981	102333.3	1805.555
2	889.22	610.78	2.50953	85946.7	3365.555
3	893.65	206.35	2.51280	93680.0	1355.555
4	899.26	900.74	2.51696	50000.0	4806.665
5	725.77	224.23	2.38849	66666.7	1444.444
6	852.20	447.90	2.48211	60000.1	2555.555

```
   TOTAL OPERATING COST FOR ABOVE SCHEDULE =      710040.75 R
ENTER JMAX,JMIN,DELQ
6,5,400
```

J	Ps	PH	GAMMA	VOLUME	DISCHARGE
1	903.11	296.89	2.51981	102333.3	1805.555
2	889.22	610.78	2.50953	85946.7	3365.555
3	893.65	206.35	2.51280	93680.0	1355.555
4	899.26	900.74	2.51696	60000.0	4806.665
5	806.25	143.75	2.44809	71466.7	1044.444
6	771.72	528.29	2.42252	60000.1	2955.555

```
   TOTAL OPERATING COST FOR ABOVE SCHEDULE =      709877.38 R
```

FIG. 7.10   (*Continued*)

Note that the column labeled VOLUME gives the reservoir volume at the *end* of each 12-h period. Note that after the fifth step, the volume schedule reaches its bottom limit at the end of period 4. The subsequent steps require a choice of j^+ and j^- from either {1, 2, 3, and 4} or from {5, 6}. (P_s, P_H are MW, gamma is R/acre-ft, volume is in acre-ft, discharge is in acre-ft/h.)

Note that the "optimum" schedule is undoubtedly located between the last two iterations. If we were to release less water in any of the first four intervals and more during 5 or 6, the thermal system cost would increase. We can theoretically reduce our operating costs a few fractions of an R by leveling the γ values in each of the two subintervals, {1, 2, 3, 4} and {5, 6}, but the effort is probably not worthwhile.

7.6 HYDRO-UNITS IN SERIES (HYDRAULICALLY COUPLED)

Consider now, a hydraulically coupled system consisting of three reservoirs in series (see Figure 7.11). The discharge from any upstream reservoir is assumed

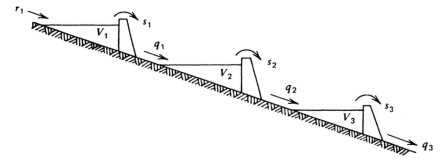

FIG. 7.11 Hydraulically coupled hydroelectric plants.

to flow directly into the succeeding downstream plant with no time lag. The
hydraulic continuity equations are

$$V_{1j} = V_{1j-1} + (r_{1j} - s_{1j} - q_{1j})n_j$$
$$V_{2j} = V_{2j-1} + (q_{1j} + s_{1j} - s_{2j} - q_{2j})n_j$$
$$V_{3j} = V_{3j-1} + (q_{2j} + s_{2j} - s_{3j} - q_{3j})n_j$$

where

$r_j = $ inflow

$V_j = $ reservoir volume

$s_j = $ spill rate over the dam's spillway

$q_j = $ hydroplant discharge

$n_j = $ numbers of hours in each scheduling period

The object is to minimize

$$\sum_{j=1}^{j_{max}} n_j F(P_{sj}) = \text{total cost} \qquad (7.30)$$

subject to the following constraints

$$P_{\text{load } j} - P_{sj} - P_{H1j} - P_{H2j} - P_{H3j} = 0$$

and

$$V_{1j} - V_{1j-1} - (r_{1j} - s_{1j} - q_{1j})n_j = 0$$
$$V_{2j} - V_{2j-1} - (q_{1j} + s_{1j} - s_{2j} - q_{2j})n_j = 0 \qquad (7.31)$$
$$V_{3j} - V_{3j-1} - (q_{2j} + s_{2j} - s_{3j} - q_{3j})n_j = 0$$

All equations in set 7.31 must apply for $j = 1 \ldots j_{max}$.

The Lagrange function would then appear as

$$
\begin{aligned}
\mathscr{L} = \sum_{j=1}^{j_{\max}} \{ &[n_j F(P_{sj}) - \lambda_j (P_{\text{load }j} - P_{sj} - P_{H1j} - P_{H2j} - P_{H3j})] \\
&+ \gamma_{1j}[V_{1j} - V_{1j-1} - (r_{1j} - s_{1j} - q_{1j})n_j] \\
&+ \gamma_{2j}[V_{2j} - V_{2j-1} - (q_{1j} + s_{1j} - s_{2j} - q_{2j})n_j] \\
&+ \gamma_{3j}[V_{3j} - V_{3j-1} - (q_{2j} + s_{2j} - s_{3j} - q_{3j})n_j] \}
\end{aligned}
$$

Note that we could have included more constraints to take care of reservoir volume limits, end-point volume limits, and so forth, which would have necessitated using the Kuhn–Tucker conditions when limits were reached.

Hydro-scheduling with multiple-coupled plants is a formidable task. Lambda–gamma iteration techniques or gradient techniques can be used; in either case, convergence to the optimal solution can be slow. For these reasons, hydro-scheduling for such systems is often done with dynamic programming (see Section 7.8) or linear programming (see Section 7.9).

7.7 PUMPED-STORAGE HYDROPLANTS

Pumped-storage hydroplants are designed to save fuel costs by serving the peak load (a high fuel-cost load) with hydro-energy and then pumping the water back up into the reservoir at light load periods (a lower cost load). These plants may involve separate pumps and turbines or, more recently, reversible pump turbines. Their operation is illustrated by the two graphs in Figure 7.12. The

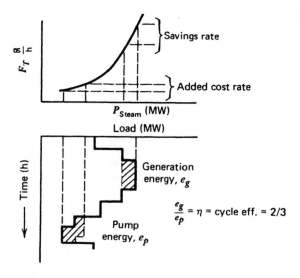

FIG. 7.12 Thermal input–output characteristic and typical daily load cycle.

first is the composite thermal system input–output characteristic and the second is the load cycle.

The pumped-storage plant is operated until the added pumping cost exceeds the savings in thermal costs due to the *peak shaving operations*. Figure 7.12 illustrates the operation on a daily cycle. If

$$\left.\begin{array}{l} e_g = \text{generation, MWh} \\ e_p = \text{pumping load, MWh} \end{array}\right\} \text{ for the same volume of water}$$

then the cycle efficiency is

$$\eta = \frac{e_g}{e_p} \qquad (\eta \text{ is typically about } 0.67)$$

Storage reservoirs have limited storage capability and typically provide 4 to 8 or 10 h of continuous operation as a generator. Pumped-storage plants may be operated on a daily or weekly cycle. When operated on a weekly cycle, pumped-storage plants will start the week (say a Monday morning in the United States) with a full reservoir. The plant will then be scheduled over a weekly period to act as a generator during high load hours and to refill the reservoir partially, or completely, during off-peak periods.

Frequently, special interconnection arrangements may facilitate pumping operations if arrangements are made to purchase low-cost, off-peak energy. In some systems, the system operator will require a complete daily refill of the reservoir when there is any concern over the availability of capacity reserves. In those instances, economy is secondary to reliability.

7.7.1 Pumped-Storage Hydro-Scheduling with a λ–γ Iteration

Assume:

1. Constant head hydro-operation.
2. An equivalent steam unit with convex input–output curve.
3. A 24-h operating schedule, each time intervals equals 1 h.
4. In any one interval, the plant is either pumping or generating or idle (idle will be considered as just a limiting case of pumping or generating).
5. Beginning and ending storage reservoir volumes are specified.
6. Pumping can be done continuously over the range of pump capability.
7. Pump and generating ratings are the same.
8. There is a constant cycle efficiency, η.

The problem is set up ignoring reservoir volume constraints to show that the same type of equations can result as those that arose in the conventional hydro-case. Figure 7.13 shows the water flows and equivalent electrical system.

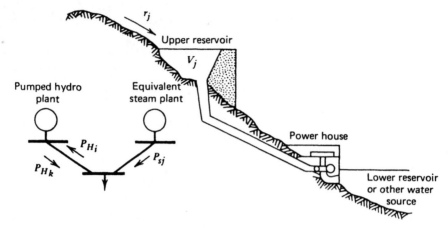

FIG. 7.13 Pumped-storage hydraulic flows and electric system flows.

In some interval, j,

$$r_j = \text{inflow (acre-ft/h)}$$

$$V_j = \text{volume at end of interval (acre-ft)}$$

$$q_j = \text{discharge if generating (acre-ft/h)}$$

or

$$w_j = \text{pumping rate if pumping (acre-ft/h)}$$

Intervals during the day are classified into two sets:

$$\{k\} = \text{intervals of generation}$$

$$\{i\} = \text{intervals of pumping}$$

The reservoir constraints are to be monitored in the computational procedure. The initial and final volumes are

$$V_0 = V_s$$

$$V_{24} = V_e$$

The problem is to minimize the sum of the hourly costs for steam generation over the day while observing the constraints. This total fuel cost for a day is (note that we have dropped n_j here since $n_j = 1$ h):

$$F_T = \sum_{j=1}^{24} F_j(P_{sj})$$

We consider the two sets of time intervals:

1. $\{k\}$: **Generation intervals:**   The electrical and hydraulic constraints are

$$P_{\text{load } k} + P_{\text{loss } k} - P_{sk} - P_{Hk} = 0$$

$$V_k - V_{k-1} - r_k + q_k = 0$$

These give rise to a Lagrange function during a generation hour (interval k) of

$$E_k = F_k + \lambda_k(P_{\text{load } k} + P_{\text{loss } k} - P_{sk} - P_{Hk}) + \gamma_k(V_k - V_{k-1} - r_k + q_k) \quad (7.32)$$

2. $\{i\}$: **Pump intervals:**   Similarly, for a typical pumping interval, i,

$$P_{\text{load } i} + P_{\text{loss } i} - P_{si} + P_{Hi} = 0$$

$$V_i - V_{i-1} - r_i - w_i = 0 \quad (7.33)$$

$$E_i = F_i + \lambda_i(P_{\text{load } i} + P_{\text{loss } i} - P_{si} + P_{Hi}) + \gamma_i(V_i - V_{i-1} - r_i - w_i)$$

Therefore, the total Lagrange function is

$$E = \sum_{\{k\}} E_k + \sum_{\{i\}} E_i + \varepsilon_s(V_0 - V_s) + \varepsilon_e(V_{24} - V_e) \quad (7.34)$$

where the end-point constraints on the storage have been added.

In this formulation, the hours in which no pumped hydro activity takes place may be considered as pump (or generate) intervals with

$$P_{Hi} = P_{Hk} = 0$$

To find the minimum of $F_T = \sum F_j$, we set the first partial derivatives of E to zero.

1. $\{k\}$: **Generation intervals:**

$$\frac{\partial E}{\partial P_{sk}} = 0 = -\lambda_k\left(1 - \frac{\partial P_{\text{loss}}}{\partial P_{sk}}\right) + \frac{dF_k}{dP_{sk}}$$

$$\frac{\partial E}{\partial P_{Hk}} = 0 = -\lambda_k\left(1 - \frac{\partial P_{\text{loss}}}{\partial P_{Hk}}\right) + \gamma_K \frac{dq_k}{dP_{Hk}}$$

$$(7.35)$$

2. $\{i\}$: **Pump intervals:**

$$\frac{\partial E}{\partial P_{si}} = 0 = -\lambda_i\left(1 - \frac{\partial P_{\text{loss}}}{\partial P_{si}}\right) + \frac{dF_i}{dP_{si}}$$

$$\frac{\partial E}{\partial P_{Hi}} = 0 = +\lambda_i\left(1 + \frac{\partial P_{\text{loss}}}{\partial P_{Hk}}\right) - \gamma_i \frac{dw_i}{dP_{Hi}}$$

(7.36)

For the $\partial E/\partial V$, we can consider any interval of the entire day—for instance, the ℓth interval—which is not the first or 24th hour.

$$\frac{\partial E}{\partial V_\ell} = 0 = \gamma_\ell - \gamma_{\ell+1}$$

and for $\ell = 0$ and $= 24$

$$\frac{\partial E}{\partial V_0} = 0 = -\gamma_1 + \varepsilon_s \quad \text{and} \quad \frac{\partial E}{\partial V_{24}} = 0 = \gamma_{24} + \varepsilon_e$$

(7.37)

From Eq. 7.37, it may be seen that γ is a constant. Therefore, it is possible to solve the pumped-storage scheduling problem by means of a λ–γ iteration over the time interval chosen. It is necessary to monitor the calculations to prevent a violation of the reservoir constraints, or else to incorporate them in the formulation.

It is also possible to set up the problem of scheduling the pumped-storage hydroplant in a form that is very similar to the gradient technique used for scheduling conventional hydroplants.

7.7.2 Pumped-Storage Scheduling by a Gradient Method

The interval designations and equivalent electrical system are the same as those shown previously. This time, losses will be neglected. Take a 24-h period and start the schedule with no pumped-storage hydro-activity initially. Assume that the steam system is operated each hour such that

$$\frac{dF_j}{dP_{sj}} = \lambda_j \quad j = 1, 2, 3, \ldots, 24$$

That is, the single, equivalent steam-plant source is realized by generating an economic schedule for the load range covered by the daily load cycle.

Next, assume the pumped-storage plant generates a small amount of power, ΔP_{Hk}, at the peak period k. These changes are shown in Figure 7.14. The change in steam-plant cost is

$$\Delta F_k = \frac{\partial F_k}{\partial P_{sk}} \Delta P_{sk} = -\frac{dF_k}{dP_{sk}} \Delta P_{Hk} \quad \text{or} \quad \Delta F_k = -\lambda_k \Delta P_{Hk}$$

(7.38)

which is the savings due to generating ΔP_{Hk}.

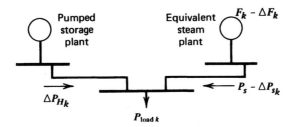

FIG. 7.14 Incremental increase in hydro-generation in hour k.

Next, we assume that the plant will start the day with a given reservoir volume and we wish to end with the same volume. The volume may be measured in terms of the MWh of generation of the plant. The overall operating cycle has an efficiency, η. For instance, if $\eta = 2/3$; 3 MWh of pumping are required to replace 2 MWh of generation water use. Therefore, to replace the water used in generating the ΔP_{Hk} power, we need to pump an amount $(\Delta P_{Hk}/\eta)$.

To do this, search for the lowest cost ($=$ lowest load) interval, i, of the day during which to do the pumping: This changes the steam system cost by an amount

$$\Delta F_i = \frac{\partial F_i}{\partial P_{si}} \Delta P_{si} = \frac{dF_i}{dP_{si}}\left(\frac{\Delta P_{Hk}}{\eta}\right) = \frac{\lambda_i}{\eta} \Delta P_{Hk} \qquad (7.39)$$

The total cost change over the day is then

$$\Delta F_T = \Delta F_k + \Delta F_i$$

$$= \Delta P_{Hk}\left(\frac{\lambda_i}{\eta} - \lambda_k\right) \qquad (7.40)$$

Therefore, the decision to generate in k and replace the water in i is economic if ΔF_T is negative (a decrease in cost); this is true if

$$\lambda_k > \frac{\lambda_i}{\eta} \qquad (7.41)$$

There are practical considerations to be observed, such as making certain that the generation and pump powers required are less than or equal to the pump or generation capacity in any interval. The whole cycle may be repeated until:

1. It is no longer possible to find periods k and i such that $\lambda_k = \lambda_i/\eta$.
2. The maximum or minimum storage constraints have been reached.

When implementing this method, it may be necessary also to do pumping in more than one interval to avoid power requirements greater than the unit rating. This can be done; then the criterion would be

$$\lambda_k > (\lambda_i + \lambda_{i'})/\eta$$

Figure 7.15 shows the way in which a single pump-generate step could be made. In this figure, the maximum capacity is taken as 1500 MW, where the pumped-storage unit is generating or pumping.

These procedures assume that commitment of units does not change as a result of the operation of the pumped-storage hydroplant. It does not presume that the equivalent steam-plant characteristics are identical in the 2 h because the same techniques can be used when different thermal characteristics are present in different hours.

Longer cycles may also be considered. For instance, you could start a schedule for a week and perhaps find that you were using the water on the weekday peaks and filling the reservoir on weekends. In the case where a reservoir constraint was reached, you would split the week into two parts

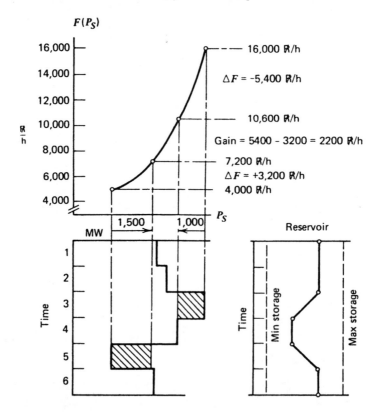

FIG. 7.15   Single step in gradient iteration for a pumped-storage plant. Cycle efficiency is two-thirds. Storage is expressed in equivalent MWh of generation.

and see if you could increase the overall savings by increasing the plant use. Another possibility may be to schedule each day of a week on a daily cycle. Multiple, uncoupled pumped-storage plants could also be scheduled in this fashion. The most reasonable-looking schedules would be developed by running the plants through the scheduling routines in parallel. (Schedule a little on plant 1, then shift to plant 2, etc.) In this way, the plants will all share in the peak shaving. Hydraulically coupled pumped-storage plants and/or pump-back plants combined with conventional hydroplants may be handled similarly.

EXAMPLE 7D

A pumped-storage plant is to operate so as to minimize the operating cost of the steam units to which it is connected. The pumped-storage plant has the following characteristics.

Generating: q positive when generating, P_H is positive and $0 \le P_H \le +300$ MW

$$q(P_H) = 200 + 2P_H \text{ acre-ft } (P_H \text{ in MW})$$

Pumping: q negative when pumping, P_P is negative and -300 MW $\le P_P \le 0$

$$q(P_p) = -600 \text{ acre-ft/h} \quad \text{with} \quad P_p = -300 \text{ MW}$$

Operating restriction: The pumped hydroplant will be allowed to operate only at -300 MW when pumping. Cycle efficiency $\eta = 0.6667$ [the efficiency has already been built into the $q(P_H)$ equations].

The equivalent steam system has the cost curve

$$F(P_s) = 3877.5 + 3.9795P_s + 0.00204P_s^2 \text{ R/h } (200 \text{ MW} \le P_s \le 2500 \text{ MW})$$

Find the optimum pump-generate schedule using the gradient method for the following load schedule and reservoir constraint.

Load Schedule (Each Period is 4 h Long)

Period	Load (MW)
1	1600
2	1800
3	1600
4	500
5	500
6	500

The reservoir starts at 8000 acre-ft and must be at 8000 acre-ft at the end of the sixth period.

Initial Schedule

Period	Load (MW)	P_s	λ	Hydropump/Gen. (+ = gen., − = pump)	Reservoir Volume at End of Period
1	1600	1600	10.5	0	8000
2	1800	1800	11.3	0	8000
3	1600	1600	10.5	0	8000
4	500	500	6.02	0	8000
5	500	500	6.02	0	8000
6	500	500	6.02	0	8000

Starting with $k = 2$ and $i = 4$: $\lambda_2 = 11.3$; $\lambda_4 = 6.02$; $\lambda_4/\eta = 9.03$.

Therefore, it will pay to generate as much as possible during the second period as long as the pump can restore the equivalent acre-ft of water during the fourth period. Therefore, the first schedule adjustment will look like the following.

Period	Load (MW)	P_s	λ	Hydropump/ Gen.	Reservoir Volume at End of Period
1	1600	1600	10.5	0	8000
2	1800	1600	10.5	+ 200	5600
3	1600	1600	10.5	0	5600
4	500	800	7.24	− 300	8000
5	500	500	6.02	0	8000
6	500	500	6.02	0	8000

Next, we can choose to generate another 200 MW from the hydroplant during the first period and restore the reservoir during the fifth period.

Period	Load (MW)	P_s	λ	Hydropump/ Gen.	Reservoir Volume at End of Period
1	1600	1400	9.69	+ 200	5600
2	1800	1600	10.5	+ 200	3200
3	1600	1600	10.5	0	3200
4	500	800	7.24	− 300	5600
5	500	800	7.24	− 300	8000
6	500	500	6.02	0	8000

Finally, we can also generate in the third period and replace the water in the sixth period.

Period	Load (MW)	P_s	λ	Hydropump/ Gen.	Reservoir Volume at End of Period
1	1600	1400	9.69	+200	5600
2	1800	1600	10.50	+200	3200
3	1600	1400	9.69	+200	800
4	500	800	7.24	−300	3200
5	500	800	7.24	−300	5600
6	500	800	7.24	−300	8000

A further savings can be realized by "flattening" the steam generation for the first three periods. Note that the costs for the first three periods as shown in the preceding table would be:

Period	P_s	Cost (₽)	λ	Hydropump/Gen.
1	1400	53788.80	9.69	+200
2	1600	61868.40	10.50	+200
3	1400	53788.80	9.69	+200
4, 5, 6	800	100400.40	7.24	−300
		269846.40		

If we run the hydroplant at full output during the peak (period 2) and then reduce the amount generated during periods 1 and 3, we will achieve a savings.

Period	P_s	Cost (₽)	λ	Hydropump/Gen.
1	1450	55747.50	9.90	+150
2	1500	57747.00	10.10	+300
3	1450	55747.50	9.90	+150
4, 5, 6		100400.40	7.24	−300
		269642.40		

The final reservoir schedule would be:

Period	Reservoir Volume
1	6000
2	2800
3	800
4	3200
5	5600
6	8000

7.8 DYNAMIC-PROGRAMMING SOLUTION TO THE HYDROTHERMAL SCHEDULING PROBLEM

Dynamic programming may be applied to the solution of the hydrothermal scheduling problem. The multiplant, hydraulically coupled systems offer computational difficulties that make it difficult to use that type of system to illustrate the benefits of applying DP to this problem. Instead we will illustrate the application with a single hydroplant operated in conjunction with a thermal system. Figure 7.16 shows a single, equivalent steam plant, P_s, and a hydroplant with storage, P_H, serving a single series of loads, P_L. Time intervals are denoted by j, where j runs between 1 and j_{max}.

Let:

$$r_j = \text{net inflow rate during period } j$$

$$V_j = \text{storage volume at the end of period } j$$

$$q_j = \text{flow rate through the turbine during period } j$$

$$P_{Hj} = \text{power output during period } j$$

$$s_j = \text{spillage rate during period } j$$

$$P_{sj} = \text{steam-plant output}$$

$$P_{\text{load } j} = \text{load level}$$

$$F_j = \text{fuel cost rate for period } j$$

Both starting and ending storage volumes, V_0 and $V_{j_{max}}$, are given, as are the period loads. The steam plant is assumed to be on for the entire period. Its input–output characteristic is

$$F_j = a + bP_{sj} + cP_{sj}^2 \ \text{R/h} \tag{7.42}$$

The water use rate characteristic of the hydroelectric plant is

$$q_j = d + gP_{Hj} + hP_{Hj}^2, \ \text{acre-ft/h} \qquad \text{for } P_{Hj} > 0 \tag{7.43}$$

and

$$= 0 \qquad \text{for } P_{Hj} = 0$$

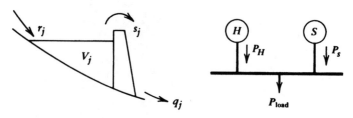

FIG. 7.16 Hydrothermal system model used in dynamic-programming illustration.

The coefficients a through h are constants. We will take the units of water flow rate as acre-ft/h. If each interval, j, is n_j hours long, the volume in storage changes as

$$V_j = V_{j-1} + n_j(r_j - q_j - s_j) \qquad (7.44)$$

Spilling water will not be permitted (i.e., all $s_j = 0$).

If V_i and V_k denote two different volume states, and

$$V_{j-1} = V_i$$
$$V_j = V_k$$

then the rate of flow through the hydro-unit during interval j is

$$q_j = \frac{(V_i - V_k)}{n_j} + r_j$$

where q_j must be nonnegative and is limited to some maximum flow rate, q_{max}, which corresponds to the maximum power output of the hydro-unit. The scheduling problem involves finding the minimum cost trajectory (i.e., the volume at each stage). As indicated in Figure 7.17, numerous feasible trajectories may exist.

The DP algorithm is quite simple. Let:

$\{i\}$ = the volume states at the start of the period j

$\{k\}$ = the states at the end of j

$TC_k(j)$ = the total cost from the start of the scheduling period to the end of period j for the reservoir storage state V_k

$PC(i, j-1: k, j)$ = production cost of the thermal system in period j to go from an initial volume of V_i to an end of period volume V_k.

The forward DP algorithm is then,

$$TC_k(0) = 0$$

and

$$TC_k(j) = \min_{\{i\}} [TC_i(j-1) + PC(i, j-1: k, j)] \qquad (7.45)$$

We must be given the loads and natural inflows. The discharge rate through the hydro-unit is, of course, fixed by the initial and ending storage levels and this, in turn, establishes the values of P_H and P_s. The computation of the thermal production cost follows directly.

There may well be volume states in the set V_k that are unreachable from

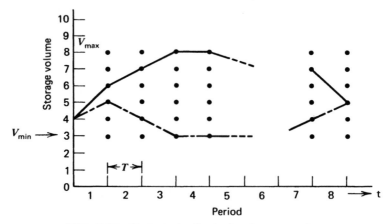

FIG. 7.17 Trajectories for hydroplant operation.

some of the initial volume states V_i because of the operating limits on the hydroplants. There are many variations on the hydraulic constraints that may be incorporated in the DP computation. For example, the discharge rates may be fixed during certain intervals to allow fish ladders to operate or to provide water for irrigation.

Using the volume levels as state variables restricts the number of hydro-power output levels that are considered at each stage, since the discharge rate fixes the value of power. If a variable-head plant is considered, it complicates the calculation of the power level as an average head must be used to establish the value of P_H. This is relatively easy to handle.

EXAMPLE 7E

It is, perhaps, better to use a simple numerical example than to attempt to discuss the DP application generally. Let us consider the two-plant case just described with the steam-plant characteristics as shown in Figure 7.18 with $F = 700 + 4.8P_s + P_s^2/2000$, R/h, and $dF/dP_s = 4.8 + P_s/1000$, R/MWh, for P_s in MW and $200 \le P_s \le 1200$. MW. The hydro-unit is a constant-head plant, shown in Figure 7.19, with

$$q = 260 + 10P_H \text{ for } P_H > 0, \qquad q = 0 \text{ for } P_H = 0$$

where P_H is in MW, and

$$0 \le P_H \le 200 \text{ MW}$$

The discharge rate is in acre-ft/h. There is no spillage, and both initial and final

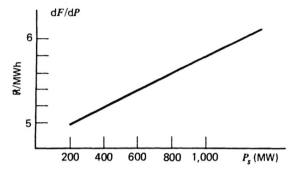

FIG. 7.18 Steam plant incremental cost function.

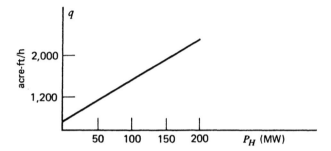

FIG. 7.19 Hydroplant q versus P_H function.

volumes are 10,000 acre-ft. The storage volume limits are 6000 and 18,000 acre-ft. The natural inflow is 1000 acre-ft/h.

The scheduling problem to be examined is for a 24-h day with individual periods taken as 4 h each ($n_j = 4.0$ h). The loads and natural inflows into the storage pond are:

Period j	$P_{\text{load }j}$ (MW)	Inflow Rate $r(j)$ (acre-ft/h)
1	600	1000
2	1000	1000
3	900	1000
4	500	1000
5	400	1000
6	300	1000

Procedure

If this were an actual scheduling problem, we might start the search using a coarse grid on both the time interval and the volume states. This would permit the future refinement of the search for the optimal trajectory after a crude search had established the general neighborhood. Finer grid steps bracketing the range of the coarse steps around the initial optimal trajectory could then be used to establish a better path. The method will work well for problems with convex (concave) functions. For this example, we will limit our efforts to 4-h time steps and storage volume steps that are 2000 acre-ft apart.

During any period, the discharge rate through the hydro-unit is

$$q_j = \frac{(V_{j-1} - V_j)}{4} + 1000 \tag{7.46}$$

The discharge rate must be nonnegative and not greater than 2260 acre-ft/h. For this problem, we may use the equation that relates P_H, the plant output, to the discharge rate, q. In a more general case, we may have to deal with tables that relate P_H, q, and the net hydraulic head.

The DP procedure may be illustrated for the first two intervals as follows. We take the storage volume steps at 6000, 8000, 10,000, ..., 18,000 acre-ft. The initial set of volume states is limited to 10,000 acre-ft. (In this example, volumes will be expressed in 1000 acre-ft to save space.) The table here summarizes the calculations for $j = 1$; the graph in Figure 7.20 shows the trajectories. We need

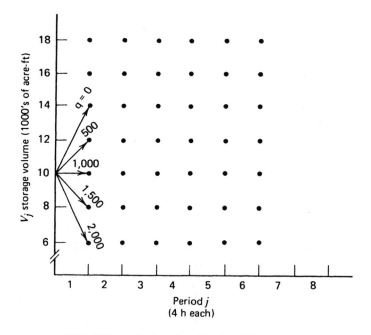

FIG. 7.20   Initial trajectories for DP example.

not compute the data for greater volume states since it is possible to do no more than shut the unit down and allow the natural inflow to increase the amount of water stored.

V_k	$j = 1$ q	$P_L(1) = 600$ MW P_H	P_s	$\{i\} = 10$ $TC_k(j)(\mathbb{R})$
14	0	0	600	15040
12	500	24	576	14523
10	1000	74	526	13453
8	1500	124	476	12392
6	2000	174	426	11342

The tabulation for the second and succeeding intervals is more complex since there are a number of initial volume states to consider. A few are shown in the following table and illustrated in Figure 7.21.

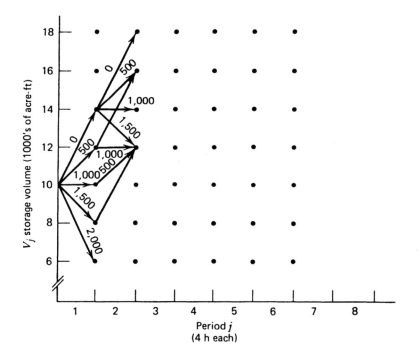

FIG. 7.21 Second-stage trajectories for DP example.

$j = 2$		$P_L = 1000$ MW			$\{i\} = [6, 8, 10, 12, 14]$
V_k	V_i	q	P_H	P_s	$TC_k(j)(\text{R})$
18	14	0	0	1000	39040[a]
16	14	500	24	976	38484[a]
16	12	0	0	1000	38523
14	14	1000	74	926	37334[a]
14	12	500	24	976	37967
14	10	0	0	1000	37453
12	14	1500	124	876	39194[a]
12	12	1000	74	926	39818
12	10	500	24	976	36897
12	8	0	0	1000	36392
$\vdots$	$\vdots$		$\ddots$		$\vdots$
6	10	2000	174	826	33477[a]
6	8	1500	124	876	33546
6	6	1000	74	926	33636

[a] Denotes the minimum cost path.

Finally, in the last period, the following combinations:

$j = 6$		$P_L = 300$ MW			$\{i\} = [6, 8, 10, 12, 14]$
V_k	V_i	q	P_H	P_s	$TC_k(j)(\text{R})$
10	10	1000	74	226	82240.61
10	8	500	24	276	82260.21
10	6	0	0	300	81738.46

are the only feasible combinations since the end volume is set at 10 and the minimum loading for the thermal plant is 200 MW.

The final, minimum cost trajectory for the storage volume is plotted in Figure 7.22. This path is determined to a rather coarse grid of 2000 acre-ft by 4-h steps in time and could be easily recomputed with finer increments.

7.8.1 Extension to Other Cases

The DP method is amenable to application in more complex situations. Longer time steps make it useful to compute seasonal *rule curves*, the long-term storage plan for a system of reservoirs. Variable-head cases may be treated. A sketch of the type of characteristics encountered in variable-head plants is shown in Figure 7.23. In this case, the variation in maximum plant output may be as important as the variation in water use rate as the net head varies.

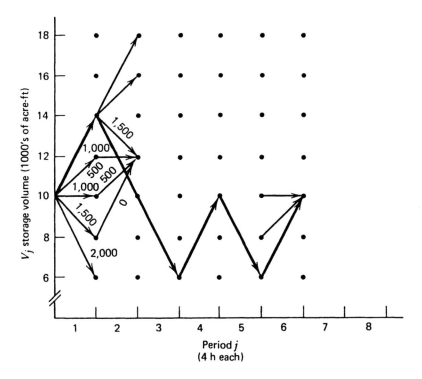

FIG. 7.22   Final trajectory for hydrothermal-scheduling example.

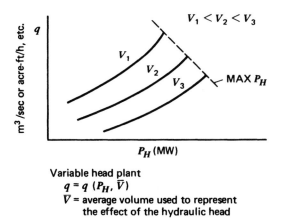

FIG. 7.23   Input–output characteristic for variable-head hydroelectric plant.

7.8.2 Dynamic-Programming Solution to Multiple Hydroplant Problem

Suppose we are given the hydrothermal system shown in Figure 7.24. We have the following hydraulic equations when spilling is constrained to zero

$$V_{1j} = V_{1j-1} + r_{1j} - q_{1j}$$
$$V_{2j} = V_{2j-1} + q_{1j} - q_{2j}$$

and the electrical equation

$$P_{H1}(q_{1j}) + P_{H2}(q_{2j}) + P_{sj} - P_{\text{load }j} = 0$$

There are a variety of ways to set up the DP solution to this program. Perhaps the most obvious would be to again let the reservoir volumes, V_1 and V_2, be the state variables and then run over all feasible combinations. That is,

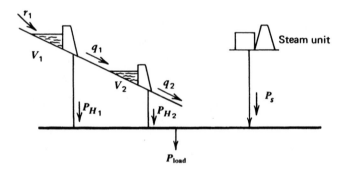

FIG. 7.24 Hydrothermal system with hydraulically coupled hydroelectric plants.

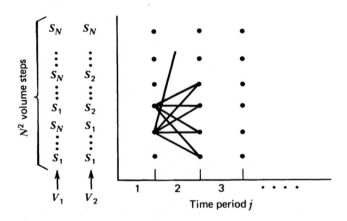

FIG. 7.25 Trajectory combinations for coupled plants.

let V_1 and V_2 both be divided into N volume steps $S_1 \dots S_2$. Then the DP must consider N^2 steps at each time interval, as shown in Figure 7.25.

This procedure might be a reasonable way to solve the multiple hydroplant scheduling problem if the number of volume steps were kept quite small. However, this is not practical when a realistic schedule is desired. Consider, for example, a reservoir volume that is divided into 10 steps ($N = 10$). If there were only one hydroplant, there would be 10 states at each time period, resulting in a possible 100 paths to be investigated at each stage. If there were two reservoirs with 10 volume steps, there would be 100 states at each time interval with a possibility of 10,000 paths to investigate at each stage.

This dimensionality problem can be overcome through the use of a procedure known as *successive approximation*. In this procedure, one reservoir is scheduled while keeping the other's schedule fixed, alternating from one reservoir to the other until the schedules converge. The steps taken in a successive approximation method appear in Figure 7.26.

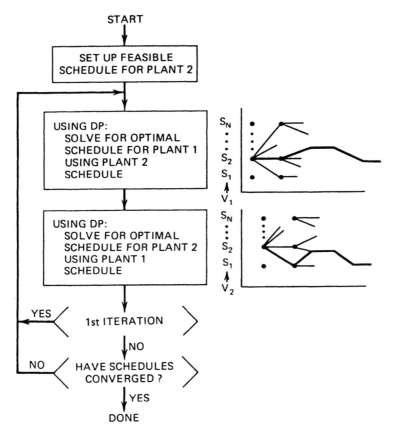

FIG. 7.26 Successive approximation solution.

7.9 HYDRO-SCHEDULING USING LINEAR PROGRAMMING

One of the more useful ways to solve large hydro-scheduling problems is through the use of linear programming. Modern LP codes and computers make this an increasingly useful option. In this section, a simple, single reservoir hydroplant operating in conjunction with a single steam plant, as shown in Figure 7.5, will be modeled using linear programming (see reference 16).

First, we shall show how each of the models needed are expressed as linear models which can be incorporated in an LP. The notation is as follows:

P_{sj} = the steam plant net output at time period j

P_{hj} = the hydroplant net output at time period j

q_j = the turbine discharge at time period j

s_j = the reservoir spill at time period j

V_j = the reservoir volume at time period j

r_j = the net inflow to the reservoir during time period j

sf_k = the slopes of the piecewise linear steam-plant cost function

sh_k = the slopes of the piecewise linear hydroturbine electrical output versus discharge function

sd_k = the slopes of the piecewise linear spill function

$P_{\text{load }j}$ = the net electrical load at time period j

The steam plant will be modeled with a piecewise linear cost function, $F(P_j)$, as shown in Figure 7.27. The three segments shown will be represented as P_{sj1}, P_{sj2}, P_{sj3} where each segment power, P_{sjk}, is measured from the start of the kth segment. Each segment has a slope designated sf_1, sf_2, sf_3; then, the cost function itself is

$$F(P_{sj}) = F(P_s^{\min}) + sf_1 P_{sj1} + sf_2 P_{sj2} + sf_3 P_{sj3} \qquad (7.47)$$

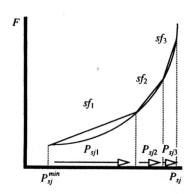

FIG. 7.27   Steam plant piecewise linear cost function.

and

$$0 < P_{sjk} < P_{sjk}^{\max} \qquad \text{for } k = 1, 2, 3 \tag{7.48}$$

and finally

$$P_{sj} = P_s^{\min} + P_{sj1} + P_{sj2} + P_{sj3} \tag{7.49}$$

The hydroturbine discharge versus the net electrical output function is designated $P_h(q_j)$ and is also modeled as a piecewise linear curve. The actual characteristic is usually quite nonlinear, as shown by the dotted line in Figure 7.28. As explained in reference 16, hydroplants are rarely operated close to the low end of this curve, rather they are operated close to their maximum efficiency or full gate flow points. Using the piecewise linear characteristic shown in Figure 7.28, the plant will tend to go to one of these two points.

In this model, the net electrical output is given as a linear sum:

$$P_{hj} = sh_1 q_{j1} + sh_2 q_{j2} \tag{7.50}$$

The spill out of the reservoir is modeled as a function of the reservoir volume and it is assumed that the spill is zero if the volume of water in the reservoir is less than a given limit. This can easily be modeled by the piecewise linear characteristic in Figure 7.29, where the spill is constrained to be zero if the volume of water in the reservoir is less than the first volume segment where

$$s_j = sd_1 V_{j1} + sd_2 V_{j2} + sd_3 V_{j3} \tag{7.53}$$

and

$$0 \le V_{jk} \le V_{jk}^{\max} \qquad \text{for } k = 1, 2, 3 \tag{7.54}$$

then

$$V_j = V_{j1} + V_{j2} + V_{j3} \tag{7.55}$$

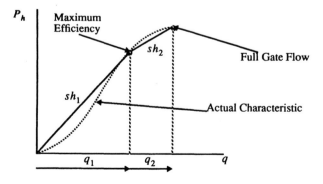

FIG. 7.28   Hydroturbine characteristic.

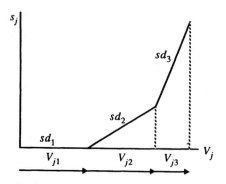

FIG. 7.29   Spill characteristic.

The hydro-scheduling linear program then consists of the following; minimize

$$\sum_{j=1}^{j_{\max}} F(P_{sj})$$

subject to

$$V_j - V_{j-1} - (r_j - s_j - q_j) = 0 \quad \text{for } j = 1 \ldots j_{\max}$$

where

$$P_{hj} = P_h(q_j)$$

$$s_j = s(V_j)$$

and

$$P_{sj} + P_{hj} - P_{\text{load } j} = 0 \qquad \text{for } j = 1 \ldots j_{\max}$$

Note that this simple hydro-scheduling problem will generate eight constraints for each time step:

- Two constraints for the steam-plant characteristic.
- Two constraints for the hydroturbine characteristic.
- Two constraints for the spill characteristic.
- One constraint for the volume continuity equation.
- One constraint for the load balance.

In addition, there are 15 variables for each time step. If the linear program were to be run with 1-h time periods for 1 week, it would have to accommodate a model with 1344 constraints and 2520 variables. This may seem quite large, but is actually well within the capability of modern linear programming codes. Reference 16 reports on a hydro-scheduling model containing about 10,000 constraints and 35,000 variables.

When multiple reservoir/plant models connected by multiple rivers and channels are modeled, there are many more additional constraints and variables needed. Nonetheless, the use of linear programming is common and can be relied upon to give excellent solutions.

APPENDIX
Hydro-Scheduling with Storage Limitations

This appendix expands on the Lagrange equation formulation of the fuel-limited dispatch problem in Chapter 6 and the reservoir-limited hydro-dispatch problem of Chapter 7. The expansion includes generator and reservoir storage limits and provides a proof that the "fuel cost" or "water cost" Lagrange multiplier γ will be constant unless reservoir storage limitations are encountered.

To begin, we will assume that we have a hydro-unit and an equivalent steam unit supplying load as in Figure 7.5. Assume that the scheduling period is broken down into three equal time intervals with load, generation, reservoir inflow, and such, constant within each period. In Chapter 6 (Section 6.2, Eqs. 6.1–6.6) and Chapter 7 (Section 7.4, Eqs. 7.22–7.29) we assumed that the total q was to be fixed at q_{TOT}, that is (see Section 7.4 for definition of variables),

$$q_{TOT} = \sum_{j=1}^{j_{max}} n_j q(P_{Hj}) \tag{7A.1}$$

In the case of a storage reservoir with an initial volume V_0, this constraint is equivalent to fixing the final volume in the reservoir. That is,

$$V_0 + n_1[r_1 - q(P_{H1})] = V_1 \tag{7A.2}$$

$$V_1 + n_2[r_2 - q(P_{H2})] = V_2 \tag{7A.3}$$

$$V_2 + n_3[r_3 - q(P_{H3})] = V_3 \tag{7A.4}$$

Substituting Eq. 7A.2 into Eq. 7A.3 and then substituting the result into Eq. 7A.4, we get

$$V_0 + \sum_{j=1}^{3} n_j r_j - \sum_{j=1}^{3} n_j q(P_{Hj}) = V_3 \tag{7A.5}$$

or

$$V_0 + \sum_{j=1}^{3} n_j r_j - q_{TOT} = V_3 \tag{7A.6}$$

Therefore, fixing q_{TOT} is equivalent to fixing V_3, the final reservoir storage. The optimization problem will be expressed as:

Minimize total steam plant cost:

$$\sum_{j=1}^{3} n_j F_s(P_{sj})$$

Subject to equality constraints:

$$P_{\text{load } j} - P_{sj} - P_{Hj} = 0 \quad \text{for } j = 1, 2, 3$$
$$V_0 + n_1 r_1 - n_1 q(P_{H2}) = V_1$$
$$V_1 + n_2 r_2 - n_2 q(P_{H2}) = V_2$$
$$V_2 + n_3 r_3 - n_3 q(P_{H3}) = V_3$$

And subject to inequality constraints:

$$V_j > V^{\min} \quad V_j < V^{\max}$$
$$P_{sj} > P_s^{\min} \quad P_{sj} < P_s^{\max} \quad \text{for } j = 1, 2, 3$$
$$P_{Hj} = P_H^{\min} \quad P_{Hj} < P_H^{\max}$$

We can now write a Lagrange equation to solve this problem:

$$
\begin{aligned}
\mathscr{L} = &\sum_{j=1}^{3} n_j F_s(P_{sj}) + \sum_{j=1}^{3} \lambda_j(P_{\text{load } j} - P_{sj} - P_{Hj}) \\
&+ \gamma_1[-V_0 - n_1 r_1 + n_1 q(P_{H1}) + V_1] \\
&+ \gamma_2[-V_1 - n_2 r_2 + n_2 q(P_{H2}) + V_2] \\
&+ \gamma_3[-V_2 - n_3 r_3 + n_3 q(P_{H3}) + V_3] \\
&+ \sum_{j=1}^{3} \alpha_j^-(V^{\min} - V_j) + \sum_{j=1}^{3} \alpha_j^+(V_j - V^{\max}) \\
&+ \sum_{j=1}^{3} \mu_{sj}^-(P_s^{\min} - P_{sj}) + \sum_{j=1}^{3} \mu_{sj}^+(P_{sj} - P_2^{\max}) \\
&+ \sum_{j=1}^{3} \mu_{Hj}^-(P_H^{\min} - P_{Hj}) + \sum_{j=1}^{3} \mu_{Hj}^+(P'_{Hj} - P_H^{\max}) \quad \quad (7\text{A}.7)
\end{aligned}
$$

where

n_j, P_{sj}, P_{Hj}, and $q(P_{Hj})$ are as defined in Section 7.4

λ_j, γ_j, α_j^-, α_j^+, μ_{sj}^-, μ_{sj}^+, μ_{Hj}^-, μ_{Hj}^+ are Lagrange multipliers

$V^{\min}$ and $V^{\max}$ are limits on reservoir storage

$P_s^{\min}$, $P_s^{\max}$, and $P_H^{\min}$ are limits on the generator output at the equivalent system and hydroplants, respectively

We can set up the conditions for an optimum using the Kuhn–Tucker equations as shown in Appendix 3A. The first set of conditions are

$$\frac{\partial \mathscr{L}}{\partial P_{sj}} = n_j \frac{dF_s}{dP_{sj}} - \lambda_j - \mu_{sj}^- + \mu_{sj}^+ = 0 \quad \quad (7\text{A}.8)$$

$$\frac{\partial \mathscr{L}}{\partial P_{Hj}} = -\lambda_j + \gamma_j n_j \frac{dq(P_{Hj})}{dP_{Hj}} - \mu_{Hj}^- + \mu_{Hj}^+ = 0 \quad \quad (7\text{A}.9)$$

$$\frac{\partial \mathscr{L}}{\partial V_j} = \gamma_j - \gamma_{j+1} - \alpha_j^- + \alpha_j^+ = 0 \quad \quad (7\text{A}.10)$$

The second and third set of conditions are just the original equality and inequality constraints. The fourth set of conditions are

$$\alpha_j^- (V^{\min} - V_j) = 0 \qquad \alpha_j^- \geq 0 \qquad (7A.11)$$

$$\alpha_j^+ (V_j - V^{\max}) = 0 \qquad \alpha_j^+ \geq 0 \qquad (7A.12)$$

$$\mu_{sj}^- (P_s^{\min} - P_{sj}) = 0 \qquad \mu_{sj}^- \geq 0 \qquad (7A.13)$$

$$\mu_{sj}^+ (P_{sj} - P_s^{\max}) = 0 \qquad \mu_{sj}^+ \geq 0 \qquad (7A.14)$$

$$\mu_{Hj}^- (P_H^{\min} - P_{Hj}) = 0 \qquad \mu_{Hj}^- \geq 0 \qquad (7A.15)$$

$$\mu_{Hj}^+ (P_{Hj} - P_H^{\max}) = 0 \qquad \mu_{Hj}^+ \geq 0 \qquad (7A.16)$$

If we assume that no generation limits are being hit, then μ_{sj}^-, μ_{sj}^+, μ_{Hj}^-, and μ_{Hj}^+ for $j = 1, 2, 3$ are each equal to zero. The solution in Eqs. 7A.8, 7A.9, and 7A.10 is

$$n_j \frac{dF_s}{dP_{sj}} = \lambda_j \qquad (7A.17)$$

$$\gamma_j n_j \frac{dq(P_{Hj})}{dP_{Hj}} = \lambda_j \qquad (7A.18)$$

$$\gamma_j - \gamma_{j+1} = \alpha_j^- - \alpha_j^+ \qquad (7A.19)$$

Now suppose the following volume-limiting solution exists:

$$V_1 > V^{\min} \quad \text{and} \quad V_1 < V^{\max}$$

then by Eq. 7A.11 and Eq. 7A.12

$$\alpha_1^- = 0 \quad \text{and} \quad \alpha_1^+ = 0$$

and

$$V_2 = V^{\min} \quad \text{and} \quad V_2 < V^{\max}$$

then

$$\alpha_2^- > 0 \qquad \alpha_2^+ = 0$$

Then clearly, from Eq. 7A.19,

$$\gamma_1 - \gamma_2 = \alpha_1^- - \alpha_1^+ = 0$$

so

$$\gamma_1 = \gamma_2$$

and

$$\gamma_2 - \gamma_3 = \alpha_2^- - \alpha_2^+ > 0$$

so

$$\gamma_2 > \gamma_3$$

Thus, we see that γ_j will be constant over time unless a storage volume limit is hit. Further, note that this is true regardless of whether or not generator limits are hit.

PROBLEMS

7.1 Given the following steam-plant and hydroplant characteristics:

Steam plant:

Incremental cost $= 2.0 + 0.002P_s$ R/MWh   and $100 \leq P_s \leq 500$ MW

Hydroplant:

Incremental water rate $= 50 + 0.02P_H$ ft^3/sec/MW $0 \leq P_H \leq 500$ MW

Load:

Time Period	P_{load} (MW)
1400–0900	350
0900–1800	700
1800–2400	350

Assume:

- The water input for $P_H = 0$ may also be assumed to be zero, that is

$$q(P_H) = 0 \quad \text{for } P_H = 0$$

- Neglect losses.
- The thermal plant remains on-line for the 24-h period.

Find the optimum schedule of P_s and P_H over the 24-h period that meets the restriction that the total water used is 1250 million ft^3 of water; that is,

$$q_{TOT} = 1.25 \times 10^9 \text{ ft}^3$$

7.2 Assume that the incremental water rate in Problem 7.1 is constant at 60 ft^3/sec/MW and that the steam unit is not necessarily on all the time. Further, assume that the thermal cost is

$$F(P_s) = 250 + 2P_s + P_s^2/1000$$

Repeat Problem 7.1 with the same water constraint.

7.3 Gradient Method for Hydrothermal Scheduling

A thermal-generation system has a composite fuel cost characteristic that may be approximated by

$$F = 700 + 4.8P_s + P_s^2/2000, \text{ R/h}$$

for

$$200 \leq P_s \leq 1200 \text{ MW}$$

The system load may also be supplied by a hydro-unit with the following characteristics:

$$q(P_H) = 0 \quad \text{when } P_H = 0$$

$$q(P_H) = 260 + 10P_H, \text{ acre-ft/h} \quad \text{for } 0 < P_H \leq 200 \text{ MW}$$

$$q(P_H) = 2260 + 10(P_H - 200) + 0.028(P_H - 200)^2 \text{ acre/h}$$
$$\text{for } 200 < P_H \leq 250 \text{ MW}$$

The system load levels in chronological order are as follows:

Period	P_{load} (MW)
1	600
2	1000
3	900
4	500
5	400
6	500

Each period is 4 h long.

7.3.1   Assume the thermal unit is on-line all the time and find the optimum schedule (the values of P_s and P_H for each period) such that the hydroplant uses 23,500 acre-ft of water. There are no other hydraulic constraints or storage limits, and you may turn the hydro-unit off when it will help.

7.3.2   Now, still assuming the thermal unit is on-line each period, use a gradient method to find the optimum schedule given the following conditions on the hydroelectric plant.

    a. There is a constant inflow into the storage reservoir of 1000 acre-ft/h.

    b. The storage reservoir limits are

$$V_{max} = 18{,}000 \text{ acre-ft}$$

and

$$V_{min} = 6000 \text{ acre-ft}$$

    c. The reservoir starts the day with a level of 10,000 acre-ft, and we wish to end the day with 10,500 acre-ft in storage.

7.4   Hydrothermal Scheduling using Dynamic Programming

Repeat Example 7E except the hydroelectric unit's water rate characteristic is now one that reflects a variable head. This characteristic also exhibits a maximum capability that is related to the net head. That is,

$$q = 0 \quad \text{for } P_H = 0$$

$$q = 260 + 10P_H\left(1.1 - \frac{\bar{V}}{100{,}000}\right) \quad \text{acre-ft/h}$$

for

$$0 < P_H \leq 2000\left(0.9 + \frac{\bar{V}}{100{,}000}\right) \text{ MW}$$

where

$$\bar{V} = \text{average reservoir volume}$$

For this problem, assume constant rates during a period so that

$$\bar{V} = \tfrac{1}{2}(V_k + V_i)$$

where

$$V_k = \text{end of period volume}$$

$$V_i = \text{start of period volume}$$

The required data are

Fossil unit: On-line entire time

$$F = 770 + 5.28P_s + 0.55 \times 10^{-3}P_s^2 \text{ R/h}$$

for

$$200 \le P_s \le 1200 \text{ MW}$$

Hydro-storage and inflow:

$$r = 1000 \text{ acre-ft/h inflow}$$

$$6000 \le V \le 18,000 \text{ acre-ft storage limits}$$

$$V = 10,000 \text{ acre-ft initially}$$

and

$$V = 10,000 \text{ acre-ft at end of period}$$

Load for 4-h periods:

J: Period	P_{load} (MW)
1	600
2	1000
3	900
4	500
5	400
6	300

Find the optimal schedule with storage volumes calculated at least to the nearest 500 acre-ft.

7.5   Pumped-Storage Plant Scheduling Problem

A thermal generation system has a composite fuel-cost characteristic as follows:

$$F = 250 + 1.5P_s + P_s^2/200 \text{ R/h}$$

for

$$200 \le P_s \le 1200 \text{ MW}$$

In addition, it has a pumped-storage plant with the following characteristics:

1. Maximum output as a generator = 180 MW (the unit may generate between 0 and 180 MW).
2. Pumping load = 200 MW (the unit may only pump at loads of 100 or 200 MW).

3. The cycle efficiency is 70% (that is, for every 70 MWh generated, 100 MWh of pumping energy are required).
4. The reservoir storage capacity is equivalent to 1600 MWh of generation.

The system load level in chronological order is the same as that in Problem 7.3.

a. Assume the reservoir is full at the start of the day and must be full at the end of the day. Schedule the pumped-storage plant to minimize the thermal system costs.

b. Repeat the solution to part a, assuming that the storage capacity of the reservoir is unknown and that it should be at the same level at the end of the day. How large should it be for minimum thermal production cost?

Note: In solving these problems you may assume that the pumped-storage plant may operate for partial time periods. That is, it does not have to stay at a constant output or pumping load for the entire 4-h load period.

7.6 The "Light Up Your Life Power Company" operates one hydro-unit and four thermal-generating units. The on/off schedule of all units, as well as the MW output of the units, is to be determined for the load schedule given below.

Thermal unit data (fuel cost = 1.0 \$/MBTU):

Unit No.	Max (MW)	Min (MW)	Incremental Heat Rate (Btu/kWh)	No-load Energy Input (MBtu/hr)	Start Up (MBtu)	Min Up Time (h)	Min Down Time (h)
1	500	70	9950	300	800	4	4
2	250	40	10200	210	380	4	4
3	150	30	11000	120	110	4	8
4	150	30	11000	120	110	4	16

Hydroplant data:

$$Q(P_h) = 1000 + 25P_h \text{ acre-ft/h}$$

where

$$0 < P_h < 200 \text{ MW}$$

min up and down time for the hydroplant is 1 h.

Load data (each time period is 4 h):

Time Period	P_{load} (MW)
1	600
2	800
3	700
4	1150

The starting conditions are: units 1 and 2 are running and have been up for 4 h, units 3, 4, and the hydro-unit are down and have been for 16 h.

Find the schedule of the four thermal units and the hydro-unit that minimizes thermal production cost if the hydro-units starts with a full reservoir and must use 24,000 acre-ft of water over the 16-h period.

7.7 The "Lost Valley Paper Company" of northern Maine operates a very large paper plant and adjoining facilities. All of the power supplied to the paper plant must come from its own hydroplant and a group of thermal-generation facilities that we shall lump into one equivalent generating plant. The operation of the hydro-facility is tightly governed by the Maine Department of Natural Resources.

Hydroplant data:

$$Q(P_h) = 250 + 25P_h \text{ acre-ft/h}$$

and

$$0 < P_h < 500 \text{ MW}$$

Equivalent steam-plant data:

$$F(P_s) = 600 + 5P_s + 0.005P_s^2 \text{ \$/h}$$

and

$$100 < P_s < 1000 \text{ MW}$$

Load data (each period is 4 h):

Time Period	P_{load} (MW)
1	800
2	1000
3	500

The Maine Department of Natural Resources had stated that for the 12-h period above, the hydroplant starts at a full reservoir containing 20,000 acre-ft of water and ends with a reservoir that is empty. Assume that there is no inflow to the reservoir and that both units are on-line for the entire 12 h.

Find the optimum schedule for the hydroplant using dynamic programming. Use only three volume states for this schedule: 0, 10,000, and 20,000 acre-ft.

FURTHER READING

The literature relating to hydrothermal scheduling is extensive. For the reader desiring a more complete guide to these references, we suggest starting with reference 1, which is a bibliography covering 1959 through 1972, prepared by a working group of the Power Engineering Society of IEEE.

References 2 and 3 contain examples of simulation methods applied to the scheduling of large hydroelectric systems. The five-part series of papers by Bernholz and Graham (reference 4) presents a fairly comprehensive package of techniques for optimization of short-range hydrothermal schedules applied to the Ontario Hydro system. Reference 5 is an example of optimal scheduling on the Susquehanna River.

A theoretical development of the hydrothermal scheduling equations is contained in reference 6. This 1964 reference should be reviewed by any reader contemplating undertaking a research project in hydrothermal scheduling methods. It points out clearly the impact of the constraints and their effects on the pseudomarginal value of hydroelectric energy.

Reference 7 illustrates an application of gradient-search methods to the coupled plants in the Ontario system. Reference 8 illustrates the application of dynamic-programming techniques to this type of hydrothermal system in a tutorial fashion. References 9 and 10 contain examples of methods for scheduling pumped-storage hydroelectric plants in a predominantly thermal system. References 11–16 show many recent scheduling techniques.

This short reference list is only a sample. The reader should be aware that a literature search in hydrothermal-scheduling methods is a major undertaking. We suggest the serious student of this topic start with reference 1 and its predecessors and successors.

1. "Description and Bibliography of Major Economy-Security Functions, Parts I, II, and III," IEEE Working Group Report, *IEEE Transactions on Power Apparatus and Systems*, Vol. PAS-100, January 1981, pp. 211–235.

2. Bruderell, R. N., Gilbreath, J. H., "Economic Complementary Operation of Hydro Storage and Steam Power in the Integrated TVA System," *AIEE Transactions*, Vol. 78, June 1959, pp. 136–150.

3. Hildebrand, C. E., "The Analysis of Hydroelectric Power-Peaking and Poundage by Computer," *AIEE Transactions*, Vol. 79, Part III, December 1960, pp. 1023–1029.

4. Bernholz, B., Graham, L. J., "Hydrothermal Economic Scheduling," a five-part series:

   a. "Part I. Solution by Incremental Dynamic Programming," *AIEE Transactions*, Vol. 79, Part III, December 1960, pp. 921–932.

   b. "Part II. Extension of Basic Theory," *AIEE Transactions*, Vol. 81, Part III, January 1962, pp. 1089–1096.

c. "Part III. Scheduling the Thermal System using Constrained Steepest Descent," *AIEE Transactions*, Vol. 81, Part III, February 1962, pp. 1096–1105.

d. "Part IV. A Continuous Procedure for Maximizing the Weighted Output of a Hydroelectric Generating Station," *AIEE Transactions*, Vol. 81, Part III, February 1962, pp. 1105–1107.

e. "Part V. Scheduling a Hydrothermal System with Interconnections," *AIEE Transactions*, Vol. 82, Part III, June 1963, pp. 249–255.

5. Anstine, L. T., Ringlee, R. J., "Susquenhanna River Short-Range Hydrothermal Coordination," *AIEE Transactions*, Vol. 82, Part III, April 1963, pp. 185–191.

6. Kirchmayer, L. K., Ringlee, R. J., "Optimal Control of Thermal Hydro-System Operation," IFAC Proceedings, 1964, pp. 430/1–430/6.

7. Bainbridge, E. S., McNamee, J. M., Robinson, D. J., Nevison, R. D., "Hydrothermal Dispatch with Pumped Storage," *IEEE Transactions on Power Apparatus and Systems*, Vol. PAS-85, May 1966, pp. 472–485.

8. Engles, L., Larson, R. E., Peschon, J., Stanton, K. N., "Dynamic Programming Applied to Hydro and Thermal Generation Scheduling," A paper contained in the IEEE Tutorial Course Text, 76CH1107-2-PWR, IEEE, New York, 1976.

9. Bernard, P. J., Dopazo, J. F., Stagg, G. W., "A Method for Economic Scheduling of a Combined Pumped Hydro and Steam-Generating System," *IEEE Transactions on Power Apparatus and Systems*, Vol. PAS-83, January 1964, pp. 23–30.

10. Kennedy, T., Mabuce, E. M., "Dispatch of Pumped Storage on an Interconnected Hydrothermal System," *IEEE Transactions on Power Apparatus and Systems*, Vol. PAS-84, June 1965, pp. 446–457.

11. Duncan, R. A., Seymore, G. E., Streiffert, D. L., Engberg, D. J., "Optimal Hydrothermal Coordination For Multiple Reservoir River Systems," *IEEE Transactions on Power Apparatus and Systems*, Vol. PAS-104, No. 5, May 1985, pp. 1154–1159.

12. Johannesen, A., Gjelsvik, A., Fosso, O. B., Flatabo, N., "Optimal Short Term Hydro Scheduling including Security Constraints," *IEEE Transactions on Power Systems*, Vol. 6, No. 2, May 1991, pp. 576–583.

13. Wang, C., Shahidehpour, S. M., "Power Generation Scheduling for Multi-Area Hydro-Thermal Systems with Tie Line Constraints, Cascaded Reservoirs and Uncertain Data," *IEEE Transactions on Power Systems*, Vol. 8, No. 3, August 1993, pp. 1333–1340.

14. Li, C., Jap, P. J., Streiffert, D. L., "Implementation of Network Flow Programming to the Hydrothermal Coordination in an Energy Management System," *IEEE Transactions on Power Systems*, Vol. 8, No. 3, August 1993, pp. 1045–1054.

15. Nabona, N., "Multicomodity Network Flow Model for Long-Term Hydro Generation Optimization," *IEEE Transactions on Power Systems*, Vol. 8, No. 2, May 1993, pp. 395–404.

16. Piekutowski, M. R., Litwinowicz, T., Frowd, R. J., "Optimal Short Term Scheduling for a Large-Scale Cascaded Hydro System," 1993 *Power Industry Computer Application Conference*, Phoenix. AZ, pp. 292–298.

8 Production Cost Models

8.1 INTRODUCTION

Production cost models are computational models designed to calculate a generation system's production costs, requirements for energy imports, availability of energy for sales to other systems, and fuel consumption. They are widely used throughout the electric utility industry as an aid in long-range system planning, in fuel budgeting, and in system operation. The primary function of computing future system energy costs is accomplished by using computer models of expected load patterns and simulating the operation of the generation system to meet these loads. Since generating units are not perfectly reliable and future load levels cannot be forecast with certainty, many production cost programs are based on probabilistic models and are used to compute the statistically expected need for emergency energy and capacity supplies or the need for controlled load demand reductions.

The digital simulation of the generation system involves representation of:

1. Generating unit efficiency characteristics (input–output curves, etc.).
2. Fuel costs per unit of energy supplied.
3. System operating policies with regard to scheduling of unit operation and the economic dispatching of groups of units that are on-line.
4. Contracts for the purchases and sales of both energy and power capability.

When hydroelectric plants are a part of the power system, the production cost simulation will involve models of the policies used to operate these plants. The first production cost models were deterministic, in that the status of all units and energy resources was assumed to be known and the load is a single estimate.

Production cost programs involve modeling all of the generation characteristics and many of the controls discussed previously, including fuel costs and supply, economic dispatch, unit commitment and hydrothermal coordination. They also involve modeling the effects of transactions, a subject to be considered in a later chapter. Deterministic programs incorporate the generation scheduling techniques in some sort of simulation model. In the most detailed of these, the on-line unit commitment program might be used in an off-line study mode. These are used in studying issues that are related to system operations such as purchase and sale decisions, transmission access issues and near-term decisions regarding operator-controlled demand management.

Stochastic production cost models are usually used for longer-range studies that do not involve near-term operational considerations. In these problem areas, the risk of sudden, random, generating unit failures and random deviations of the load from the mean forecast are considered as probability distributions. This chapter describes the basic ideas used in the probabilistic production cost models.

It is not possible to delve into all the details involved in a typical modern computer program since these programs may be quite large, with tens of thousands of lines of code and thousands of items of data. Any such discussion would be almost instantly out of date since new problems keep arising. For example, the original purpose of these production cost programs was primarily computation of future system operating costs. In recent years, these models have been used to study such diverse areas as the possible effects of load management, the impact of fuel shortages, issues related to nonutility generation, and the reliability of future systems.

The "universal" block diagram in Figure 8.1 shows the organization of a "typical" energy production cost program. The computation simulates the system operation on a chronological basis with system data input being altered at the start of each interval. These programs must be able to recognize and take into account, in some fashion, the need for scheduled maintenance outages. Logic may be incorporated in this type of program to simulate the maintenance outage allocation procedure actually used, as well as to process maintenance schedules that are input to the program.

Expansion planning and fuel budgeting production cost programs require load models that cover weeks, months, and/or years. The expected load patterns may be modeled by the use of typical, normalized hourly load curves for the various types of days expected in each subinterval (i.e., month or week) or else by the use of load duration or load distribution curves. Load models used in studying operational issues involve the next few hours, days or weeks and are usually chronological load cycles.

A *load duration curve* expresses the period of time (say number of hours) in a fixed interval (day, week, month, or year) that the load is expected to equal or exceed a given megawatt value. It is usually plotted with the load on the vertical axis and the time period on the horizontal axis.

The scheduling of unit maintenance outages may involve time intervals as short as a day or as long as a year. The requirements for economic data such as unit, plant, and system consumption and fuel costs, are usually on a monthly basis. When these time interval requirements conflict, as they often do, the load model must be created in the model for the smallest subinterval involved in the simulation.

Production cost programs may be found in many control centers as part of the overall "application program" structure. These production cost models are usually intended to produce shorter-term computations of production costs (i.e., a few hours to the entire week) in order to facilitate negotiations for energy (or power) interchange or to compute cost savings in order to allocate economic

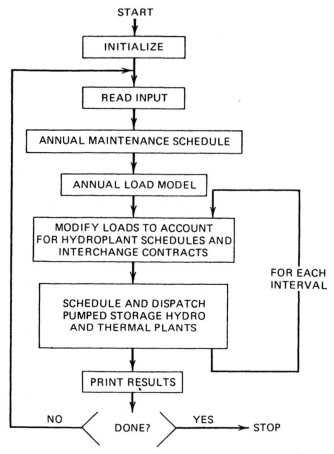

FIG. 8.1 Block diagram for a typical, single area energy production cost program used for planning.

benefits among pooled companies. In either application, the production cost simulation is used to evaluate costs under two or more assumptions. For example, in interchange negotiations, the system operators can evaluate the cost of producing the energy on the system versus the costs of purchasing it.

In U.S. power pools where units owned by several different utilities are dispatched by the control center, it is usually necessary to compute the production cost "savings" due to pooled operation. That is, each seller of energy is paid for the cost of producing the energy sold and may be given one-half the production cost "savings" of the system receiving the energy. One way of determining these savings is to simulate the production costs of each system supplying just its own load. In fact, in at least one U.S. pool this is called "own-load dispatch." These computed production costs can be compared with actual costs to arrive at the charges for transferring energy. The models used are deterministic and typically use the actual load patterns that occurred during the period under study. Scheduling computations frequently are performed with models that are similar to those used for real-time operational control.

Production cost computations are also needed in fuel budgeting. This involves making computations to forecast the needs for future fuel supplies at specific plant sites. Arrangements for fuel supplies vary greatly among utilities. In some instances, the utility may control the mining of coal or the production and transportation of natural gas; in others it may contract for fuel to be delivered to the plant. In many cases, the utility will have made a long-term arrangement with a fuel supplier for the fuel needed for a specific plant. (Examples are "mine-mouth" coal plants or nuclear units.) In still other cases, the utility may have to obtain fuel supplies on the open (i.e., "spot") market at whatever prices are prevailing at that time. In any case, it is necessary to make a computation of the expected fuel supply requirements so that proper arrangements can be made sufficiently in advance of the requirements. This requires a forecast of specific quantities (and large quantities) of fuel at given future dates.

Fuel budgeting models are usually very detailed. Deterministic or probabilistic production cost simulations may be used for this application. In some cases, where the emphasis is on the scheduling of fuel resources, transportation and fuel storage, the production cost computations might be one part of a large linear programming model. In these cases, the loads might be modeled by the expected energy demand in a day, week, month or season. Scheduling of generation would be done using a linear model of the input–output characteristics.

The operating center production cost needs may have a 7-day time horizon. The fuel budgeting time span may encompass 1 to 5 years and might, in the case of the mine-mouth plant studies, extend out to the expected life of the plant. System expansion studies usually encompass a minimum of 10 years and in many cases extend to 30 years into the future. It is this difference in time horizon that makes different models and approaches suitable for different problems.

8.2   USES AND TYPES OF PRODUCTION COST PROGRAMS

Table 8.1 lists the major features that may vary from program to program and indicates, along the horizontal axis, the major program uses of:

1. Long-range planning.
2. Fuel budgeting.
3. Operations planning.
4. Weekly schedules.
5. Allocation of pool savings.

TABLE 8.1 Energy Production Cost Programs

Load Model	Interval Considered	Economic Dispatch Procedure for Thermal Units	Long-Range Planning	Fuel Budgeting	Operations Planning	Weekly Schedules	Allocation of "Pool Savings"
Total energy or load duration	Seasons or years	Block loading[a]	×				
Load duration or load cycles	Months or weeks	Incremental loading	×	×	×		
Load duration or load cycles	Months, weeks or days	Incremental loading with forced outages considered	×	×	×		
Load cycle	Weeks or days	Incremental loading (losses)	×	×	×	×	×

[a] The term "block loading" refers to the scheduling of complete units in economic order without regard to incremental cost. The procedure is illustrated in this section.

Also indicated are the types of programs that have been found useful, so far, in each application. The type of load model used will determine, in part, the suitability of each program type for a given application.

The types of production cost programs shown in Table 8.1, which utilize chronological load patterns (i.e., load cycles) and deterministic scheduling methods, are computer implementations of the economic dispatching techniques and unit commitment methods explored previously. That is, production costs and fuel consumption are computed repetitively, assuming that the load cycles are known for an extended period into the future and that the availability of every unit can be predicted with 100% certainty for each subinterval of that future period.

In models using probabilistic representations of the future loads and generating unit availabilities, the expected values of production costs and fuel consumption are computed without the assumption of a perfectly known future.

There are other types of production cost programs that are known by various names. Some include different ways of categorizing the program, models, or computational methods that are used. For example there are "Monte Carlo," probabilistic simulations that are detailed, deterministic programs with the added feature that unit outages and deviations of loads from those forecast are incorporated by the use of synthetic sampling techniques. Random numbers are generated at regular time intervals and used to develop sample results from the appropriate probability distributions. These numbers determine the status of a unit; operating at full capability, on forced outage, or coming back into a state where it is available, if it was previously unavailable. The magnitude of the load deviation from the magnitude forecast may also be determined by a random number using a "forecast error" probability density. Other programs might combine some of the approximate generation scheduling techniques with load models that separate the week into weekdays and weekend days and consider only 4 wks per year, one for each season. In these (so-called "quick-and-dirty") models, the weekly cost and fuel consumption are multiplied by appropriate scaling factors to compute total seasonal values. On the other end of the complexity scale, there are programs which consider the dispatch of several interconnected areas and utilize power flow constraints caused by the transmission interconnections to restrict interarea interchange levels. Optimal power flow programs could be used in the same fashion.

So far, networks have only been represented in production cost programs by simplified models, such as using penalty factors, using a DC power flow (or equivalent distribution factors based on a DC model) or using a transportation network. AC power flows are useful for security-constrained economic dispatch, unit commitment and purchase–sale analyses. Optimal power flows may be used to study transmission power and VAR flow patterns to develop prices for the use of transmission systems.

In the complex, deterministic programs, the loads may be represented by chronologically arranged load cycle patterns. These patterns consist of hourly (or bi-hourly) loads that might be calculated using typical, daily load cycle patterns for workdays, weekend days and holidays throughout the period. The development of these typical patterns from historical data is an art; using them to develop forecasts of future load cycles is straightforward once the overall load forecast is developed. The earlier load models were load-duration curves and we shall utilize them to explore the various techniques.

8.2.1 Production Costing Using Load-Duration Curves

In representing future loads, sometimes it is satisfactory to specify only the total energy generation for a period. This is satisfactory if only total fuel consumption and production costs are of interest and neither capacity limitations nor chronological effects are important.

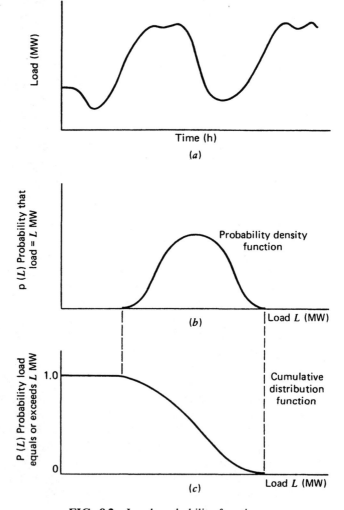

FIG. 8.2 Load probability functions.

Where capacity limitations are of more concern, a *load-duration curve* might be used. Figure 8.2 shows an expected load pattern in (a), a histogram of load for a given time period in (b), and the load-duration curve constructed from it in (c). In practical developments, the density and distribution functions may be developed as histograms where each load level, L, denotes a range of loads. These last two curves are expressed in both hours and per unit probability versus the megawatts of load. Figure 8.3 shows the more conventional representation of a load-duration curve where the probability has been multiplied by the period length to show the number of hours that the load equals, or exceeds, a given level, L (MW). It is conventional in deterministic production cost analyses to show this curve with the load on the vertical axis. In the probabilistic calculations, the form shown on Figure 8.2c is used.

In the simulation of the economic dispatch procedures with this type of load model, thermal units may be *block-loaded*. This means the units (or major segments of a unit) on the system are ordered in some fashion (usually cost) and are assumed to be fully loaded, or loaded up to the limitations of the load-duration curve. Figure 8.4 shows this procedure for a system where the internal peak load is 1700 MW. The units are considered to be loaded in a sequence determined by their average cost at full load in ₹/MWh. The amount of energy generated by each unit is equal to the area under the load-duration curve between the load levels in megawatts supplied by each unit.

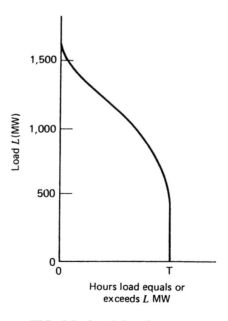

FIG. 8.3 Load-duration curve.

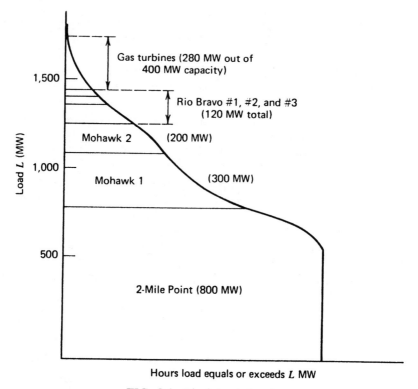

FIG. 8.4   Block-loaded units.

This system consists of three plants plus an array of gas-turbine generating units. These are:

Unit	Maximum capability (MW)
2-Mile Point	800
Mohawk 1	300
Mohawk 2	200
Rio Bravo 1	75
Rio Bravo 2	25
Rio Bravo 3	20
Eight gas turbines (each 50 MW)	400
	Total 1820

Note that in this system, the gas turbines are not used appreciably since the peak load is only 1700 MW and each unit is assumed to be available all the time during the interval.

Besides representing the thermal generating plants, the various production cost programs must also simulate the effects of hydroelectric plants with and without water storage, contracts for energy and capacity purchases and sales, and pumped-storage hydroelectric plants. The action of all these results in a modified load to be served by the array of thermal units. The scheduling of the thermal plants should be simulated to consider the security practices and policies of the power system as well as to simulate, to some appropriate degree, the economic dispatch procedures used on the system to control the unit output levels.

More complex production cost programs used to cover shorter time periods may duplicate the logic and procedures used in the control of the units. The most complex involve the procedures discussed in the previous three chapters on unit commitment and hydrothermal scheduling. These programs will usually use hourly forecasts of energy (i.e., the "hourly, integrated load" forecast) and thermal-generating-unit models that include incremental cost functions, start-up costs, and various other operating constraints.

EXAMPLE 8A

Let us consider the load-duration curve technique for a system of two units. Initially, the random forced outages of the generating units will be neglected. Then, we will incorporate consideration of these outages in order to show their effects on production costs and the ability of the small sample system to serve the load pattern expected. The load consists of the following:

x-Load (MW)	Duration (h)	Energy (MWh)
100	20	2000
80	60	4800
40	20	800
	Total = 100	7600

From these data, we may construct a load-duration curve in tabular and graphic form. The load-duration curve shows the number of hours that the load equals or exceeds a given value.

x-Load (MW)	Exact Duration, $Tp(x)$	$TP_n(k)$ Hours that Load Equals or Exceeds x
0	0	100
20	0	100
40	20	100
60	0	80
80	60	80
100	20	20
100+		0

In this table, p(x) is the load density function: the probability that the load is exactly x MW and $P_n(x)$ is the load distribution function; the probability that the load is equal to, or exceeds, x MW.

The table has been created for uniform load-level steps of 20 MW each. The table also introduces the notation that is useful in regarding the load-duration curve as a form of probability distribution. The load density and distribution functions, p(x) and $P_n(x)$, respectively, are probabilities. Thus, p(20) = 0, p(40) = 20/100 = 0.2, p(60) = 0, and so forth, and $P_n(20) = P_n(40) = 1.0$, $P_n(60) =$ 0.8, and so forth. The distribution function, $P_n(x)$, and the density, p(x), are related as follows.

$$P_n(x) = 1 - \int_x^{\infty} p(x)\, dx \tag{8.1}$$

For discrete-density functions (or histograms) in tabular form, it is easiest to construct the distribution by cumulating the probability densities from the highest to the lowest values of the argument (the load levels).

The load-duration curve is shown in Figure 8.5 in a way that is convenient to use for the development of the probabilistic scheduling methods.

The two units of the generating system have the following characteristics.

Unit	Power Output (MW)	Fuel Input (10^6 Btu/h)	Fuel Cost (R/10^6 Btu)	Fuel Cost Rate (R/h)	Incremental Fuel Cost (R/MWh)	Unit Forced Outage Rate (per unit)
1	0	160	1	160	—	
	80	800	1	800	8	0.05
2	0	80	2	160	—	
	40	400	2	800	16	0.10

In this table the fuel cost rate for each unit is a linear function of the power output, P. That is,

F(P) = fuel cost at zero output + incremental cost rate × P.

In addition to the usual input–output characteristics, forced outage rates are assumed. This rate represents the fraction of time that the unit is not available, due to a failure of some sort, out of the total time that the unit should be available for service. In computing forced outage rates, periods where a unit is on scheduled outage for maintenance are excluded. The unit forced outage rates are initially neglected, and the two units are assumed to be available 100% of the time.

Units are "block-loaded," with unit 1 being used first because of its lower average cost per MWh. The load-duration curve itself may be used to visualize the unit loadings. Figure 8.6 shows the two units block-loaded.

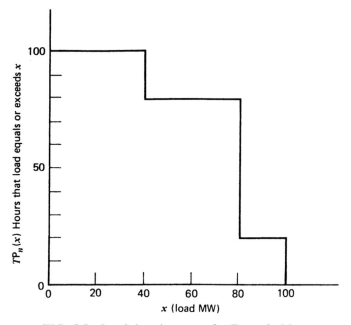

FIG. 8.5 Load-duration curve for Example 8A.

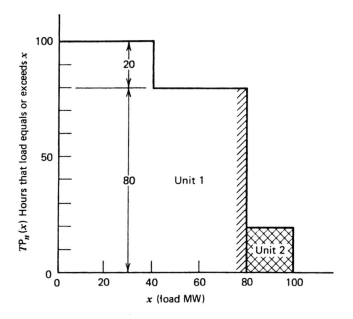

FIG. 8.6 Load-duration curve with block-loaded units.

Unit 1 is on-line for 100 h and is generating at an output level of 80 MW for 80 h and 40 MW for 20 h. Therefore, the production costs for unit 1 for this period:

$$= \text{hours on line} \times \text{no load fuel cost rate}$$
$$+ \text{energy generated} \times \text{incremental fuel cost rate}$$

$$= 100 \text{ h} \times 160 \text{ R/h} + (6400 + 800) \text{ MWh} \times 8 \text{ R/MWh}$$

$$= 16,000 \text{ R} + 57,600 \text{ R} = 73,600 \text{ R}$$

Similarly, unit 2 is required only 20 h in the interval and generates 400 MWh at a constant output level of 20 MW. Therefore, its production costs for this period:

$$= 20 \text{ h} \times 160 \text{ R/h} + 400 \text{ MWh} \times 16 \text{ R/MWh} = 9600 \text{ R}$$

These data are summarized as follows.

Unit	Load (MW)	Duration (h)	Energy (MWh)	Fuel Used (10^6 Btu)	Fuel Cost (R)
1	40	20	800	9600	9600
	80	80	6400	64000	64000
			7200	73600	73600
2	20	20	400	4800	9600
			7600	78400	83200

Note that these two units can easily supply the expected loads. If a third unit were available it would not be used, except as standby reserve.

This same basic approach to compute the production cost of a particular unit is used in most production cost models that represent individual unit characteristics. The simulation will determine the hours that the unit is on-line and the total duration or each of the unit's MW output levels. If the incremental cost is allowed to vary with loading level, the unit cost can be calculated as:

$$= \text{hours on line} \times \text{no load fuel cost}$$

$$+ \sum (\text{power generated} \times \text{hours at this level} \times \text{incremental fuel cost rate at this power level})$$

summed over the period. When nonzero, minimum loading levels are considered, this has to be modified to:

$$= \text{hours on-line} \times \text{fuel cost rate at minimum load}$$

$$+ \sum [(\text{power level} - \text{minimum power}) \times \text{incremental fuel cost rate} \times \text{hours at this level}]$$

It gets more involved when continuous functions (polynomials, for example) are used to model input–output cost curves.

8.2.2 Outages Considered

Next, let us consider the effects of the random forced outages of these units and compute the expected production costs. This is a situation that contains relatively few possible events so that the expected operation of each unit may be determined by enumeration of all the possible outcomes. For this procedure, it is easiest at this point to utilize the load density rather than the load-distribution function.

EXAMPLE 8B

Load level by load level, the operation and generation of the two units are as follows.

1. Load = 40 MW; duration 20 h

Unit 1:	On-line	20 h
	Operates	$0.95 \times 20 = 19$ h
	Output	40 MW
	Energy	$19 \times 40 = 760$ MWh
Unit 2:	On-line	1 h
	Operates	$0.9 \times 1 = 0.9$ h
	Output	40 MW
	Energy	$0.9 \times 40 = 36$ MWh

$$
\begin{aligned}
\text{Load energy} &= 800 \text{ MWh} \\
\text{Generation} &= 796 \text{ MWh} \\
\text{Unserved energy} &= 4 \text{ MWh} \\
\text{Shortages} \quad & \quad 40 \text{ MW for } 0.1 \text{ h}
\end{aligned}
$$

2. Load = 80 MW; duration 60 h

Unit 1:	On-line	60 h
	Operates	$0.95 \times 60 = 57$ h
	Output	80 MW
	Energy	$57 \times 80 = 4560$ MWh
Unit 2:	On-line	60 h total
	Operates	$0.9 \times 3 = 2.7$ h
	Output	40 MW
	Energy	$2.7 \times 40 = 108$ MWh

$$
\begin{aligned}
\text{Load energy} &= 4800 \text{ MWh} \\
\text{Generation} &= 4668 \text{ MWh} \\
\text{Unserved energy} &= 132 \text{ MWh}
\end{aligned}
$$

Shortages   80 MW for 0.3 h = 24 MWh

       40 MW for 2.7 h = 108 MWh

             132 MWh

3. Load = 100 MW; duration 20 h

Unit 1:

On-line	20 h
Operates	$0.95 \times 20 = 19$ h
Output	80 MW
Energy	$19 \times 80 = 1520$ MWh

Unit 2:

On-line	20 h
Operates	as follows:

a. Unit 1 on-line and operating 19 h
Unit 2 operates $0.9 \times 19 = 17.1$ h

Output	20 MW
Energy	$17.1 \times 20 = 342$ MWh
Shortage	20 MW for 1.9 h

b. Unit 1 supposedly on-line, but not operating 1 h
Unit 2 operates $0.9 \times 1 = 0.9$ h,

Output	40 MW
Energy	$0.9 \times 40 = 36$ MWh
Shortages	100 MW for 0.1 h
	60 MW for 0.9 h

$$
\begin{aligned}
\text{Load energy} &= 2000 \text{ MWh} \\
\text{Generation} &= 1898 \text{ MWh} \\
\text{Unserved energy} &= \;102 \text{ MWh}
\end{aligned}
$$

Shortages   100 MW for 0.1 h = 10 MWh

       60 MW for 0.9 h = 54 MWh

       20 MW for 1.9 h = 38 MWh

              102 MWh

Because this example is so small, it has been necessary to make an arbitrary assumption concerning the commitment of the second unit. The assumption made is that the second unit will be on-line for any load level that equals or exceeds the capacity of the first unit. Thus, the second unit is on-line for the 60-h duration of the 80 MW load. This assumption agrees with the algorithm developed later in the chapter.

The enumeration of the possible states is not quite complete. We have accounted for the periods when the load is satisfied and the times when there will be a real shortage of capacity. In addition, we need to separate the periods when the load is satisfied into periods where there is excess capability (more generation than load) and periods when the available capacity exactly matches the load (generation equals load). The latter periods are called *zero MW shortage* because there is no reserve capacity in that period. This information is needed in case an additional unit becomes available or emergency capacity needs to be purchased. This additional capacity would need to be operated during the entire period of a zero MW shortage because the occurrence of a real shortage is a random event depending on the failure of an operating generator.

For this example there are two such periods, one during the 40-MW load period and the other during the 80-MW load period. That is, the additional "zero MW shortage" conditions occur during those periods when the load is supplied precisely with no additional available capacity. Therefore, to the shortage events presented previously, we add the following.

Load (MW)	Duration (h)	Unit 1	Unit 2	Zero Reserve Expected Duration
1. 40	20	Out	In	$0.05 \times 0.9 \times 20 = 0.9$
2. 80	60	In	Out	$0.95 \times 0.1 \times 60 = \underline{5.7}$
				6.6 h

These "zero MW shortage" events are of significance, since their total expected duration determines the number of hours that any additional units will be required.

All these events may be presented in an orderly fashion. Since each unit may be either on or off and there are three loads, the total number of possible events is $3 \times 2 \times 2 = 12$. These are summarized along with the consequence of each event in Table 8.2.

Now, having enumerated all the possible operating events, it is possible to compute the expected production costs and shortages. Recall from Example 8A that the operating cost characteristics of the two units are

$$F_1 = 160 + 8P_1, \text{ R/h}$$

and

$$F_2 = 160 + 16P_2, \text{ R/h}$$

and the fuel costs are 1 and 2 R/10^6 Btu, respectively. The calculated operating costs considering forced ourages are computed using the data from Table 8.2.

TABLE 8.2 Summary of All Possible States

Load (MW)	Duration (h)	Event No.	Unit 1 Status[a]	Unit 1 Power (MW)	Unit 2 Status[a]	Unit 2 Power (MW)	Combined Event Duration (h)	Combined Event Consequence
40	20	1	1	40	1	0	17.1	Load satisfied; unit 2 not required
		2	1	40	0	0	1.9	Same as event no. 1
		3	0	0	1	40	0.9	Load satisfied; 0 MW shortage 0.9 h
		4	0	0	0	0	0.1	40 MW shortage 0.1 h
80	60	5	1	80	1	0	51.3	Load satisfied; unit 2 not required
		6	1	80	0	0	5.7	Load satisfied; 0 MW shortage 5.7 h
		7	0	0	1	40	2.7	40 MW shortage 2.7 h
		8	0	0	0	0	0.3	80 MW shortage 0.3 h
100	20	9	1	80	1	20	17.1	Load satisfied
		10	1	80	0	0	1.9	20 MW shortage 1.9 h
		11	0	0	1	40	0.9	60 MW shortage 0.9 h
		12	0	0	0	0	0.1	100 MW shortage 0.1 h

[a] 1 denotes available and 0 denotes unavailable.

These are:

Unit	Hours On-line	Total Expected Operating Hours	Expected Energy Generation (MWh)	Expected Fuel Use (10^6 Btu)	Expected Production Cost (R)
1	100	95.0	6840	69920	69920
2	81	72.9	522	10008	20016
			Totals 7362	79928	89936

The expected energy generated by unit 1 is the summation over the load levels of the product of the probability that the unit is available, $p = 0.95$, times the load level in MW, times the hours duration of the load level. The expected production costs for unit 1

$$= 95 \text{ h} \times 160 \text{ R/h} + 6840 \text{ MWh} \times 8 \text{ R/MWh}$$

and for unit 2

$$= 72.9 \text{ h} \times 160 \text{ R/h} + 522 \text{ MWh} \times 16 \text{ R/MWh}$$

Compared to the results of Example 8A, the fuel consumption has increased 1.95% over that found neglecting random forced outages, and the total cost has increased 8.1%. This cost would be increased even more if the unserved energy, 238 MWh, were to be supplied by some high-cost emergency source.

The expected unserved demands and energy may be summarized from the preceding data as shown in Table 8.3. The last column is the distribution of the need for additional capacity, $TP_n(x)$, referred to previously, computed after the two units have been scheduled. Data such as these are computed in probabilistic production cost programs to provide probabilistic measures of the generation system adequacy (i.e., reliability). If costs are assigned to the unsupplied demand and energy (representing replacement costs for emergency purchases of capacity and energy or the economic loss to society as a whole),

TABLE 8.3 Unserved Load

Unserved Demand (MW)	Duration of Shortage (h)	Unserved Energy (MWh)	Duration of Given Shortages or More (h)
0	6.6	0	12.6
20	1.9	38	6.0
40	2.8	112	4.1
60	0.9	54	1.3
80	0.3	24	0.4
100	0.1	10	0.1
Totals	12.6	238	

these data will provide an additional economic measure of the generation system.

This relatively simple example leads to a lengthy series of computations. The results point out the importance of considering random forced outages of generating units when production costs are being computed for prolonged future periods. The small size of this example tends to magnify the expected unserved demand distribution. In order to supply, reliably, a peak demand of 100 MW with a small number of units, the total capacity would be somewhere in the neighborhood of 200 MW. On the other hand, the relatively low forced outage rates of the units used in Example 8B tend to minimize the effects of outages on fuel consumption. Large steam turbine generators of 600 MW capacity, or more, frequently exhibit forced outage rates in excess of 10%.

It should also be fairly obvious at this point that the process of enumerating each possible state in order to compute expected operation, energy generation, and unserved demands, cannot be carried much further without an organized and efficient scheduling method. For N_L load levels and N units, each of which may be on or off, there are $N_L \times 2^N$ possible events to enumerate. The next section will develop the types of procedures that are found in many probabilistic production cost programs.

8.3   PROBABILISTIC PRODUCTION COST PROGRAMS

Until the 1970s, production cost estimates were usually computed on the basis that the total generating capacity is always available, except for scheduled maintenance outages. Operating experience indicates that the forced outage rate of thermal-generating units tends to increase with the unit size. Power system energy production costs are adversely affected by this phenomenon. The frequent long-duration outages of the more efficient base-load units require running the less efficient, more expensive plants at higher than expected capacity factors* and the importation of emergency energy. Some utility systems report the operation of peaking units for more than 150 h each month, when these same units were originally justified under the assumption that they would be run over a few hours per month, if at all.

Two measures of system unreliability (i.e., generation system inadequacy to serve the expected demands) due to random, forced generator failures are:

* *Capacity factor* is defined as follows.

$$\frac{\text{MWh generated by the unit}}{(\text{Number of hours in the period of interest})(\text{unit full-load MW capacity})}$$

Thus, a higher value (close to unity) indicates that a unit was run most of the time at full load. A lower value indicates the unit was loaded below full capacity most of the time or was shut down part of the time.

1. The period of time when the load is greater than available generation capacity.
2. The expected levels of power and energy that must be imported to satisfy the load.

The maximum emergency import power and total energy imported are different dimensions of the same measure. These quantities and the expected shortage duration are useful as sensitive indicators of the need for additional capacity or interconnection capability. Some of these ideas are discussed further in the Appendix.

8.3.1   Probabilistic Production Cost Computations

Production cost programs that recognize unit forced outages and compute the statistically expected energy production cost have been developed and used widely. Mathematical methods based on probability methods make use of probabilistic models of both the load to be served and the energy and capacity resources. The models of the generation need to represent the unavailability of basic energy resources (i.e., hydro-availability), the random forced outages of units, and the effects of contracts for energy sales and/or purchases. The computation may also include the expected cost of emergency energy over the tie lines, which is sometimes referred to as the *cost of unsupplied energy*.

The basic difficulties that were noted when using deterministic approaches to the calculation of systems production cost were:

1. The base-load units of a system are loaded in the models for nearly 100% of an interval.
2. The midrange, or "cycling," units are loaded for periods that depend on their priority rank and the shape of the load-duration curve.
3. For any system with reasonably adequate reserve level, the peaking units have nearly zero capacity factors.

These conditions are, in fact, all violated to a greater or lesser extent whenever random-unit forced outages occur on a real system. The unavailability of thermal-generating units due to unexpected, randomly occurring outages is fairly high for large-sized units. Values of 10% are common for full forced outages. That is, for a full forced outage rate of q, per unit, the particular generating unit is completely unavailable for $100q\%$ of the time it is supposed to be available. Generating units also suffer partial outages where the units must be derated (i.e., run at less than full capacity) for some period of time, due to the forced outage of some system component (e.g., a boiler feed pump or a fan motor). These partial forced outages may reach very significant levels. It is not uncommon to see data reflecting a 25% forced reduction in maximum generating unit capability for 20% of the time it is supposed to be available.

Data on unit outage rates are collected and processed in the United States by the National Electric Reliability Council (NERC). The collection and processing of these data are important and difficult tasks. Performance data of this nature are essential if rational projections of component and system unavailability are to be made.

There are two techniques that have been used to handle the convolution of the load distributions with the capacity–probability density functions of the units: numerical convolutions where discrete values are used to model all of the distributions, and analytical methods that use continuous functional representations. Both techniques may be further divided into approaches that perform the convolutions in different orders. In what will be referred to here as the *unserved load distribution method*, the individual unit probability–capacity densities are convolved with the load distribution in a sequence determined by a fixed economic loading criterion to develop a series of *unserved load distributions*. Unit energy production is the difference between the unserved load energy before the unit is scheduled (i.e., convolved with the previous unserved load distribution) and after it has been scheduled. The load forecast is the initial unserved load distribution. In the *expected cost method*, the unit probability–capacity densities are first convolved with each other in sequence to develop distributions of available capacity and the expected cost curve as a function of the total power generated. This expected cost curve may then be used with the load distribution to produce the expected value of the production cost to serve the given load forecast distribution. We shall explore the numerical convolution techniques.

The analytical methods use orthogonal functions to represent both the load and capacity–probability densities of the units. These are the methods based on the use of *cumulants*. The merit of this analytical method is that it is usually a much more rapid computation. The drawback appears to be the concern over accuracy (as compared with numerical convolution results). The references at the end of this chapter provide a convenient starting point for a further exploration of this approach. The discussions of the numerical convolution techniques which follow should provide a sufficient basis for appreciating the approach, its utility, and its difficulties.

8.3.2 Simulating Economic Scheduling with the Unserved Load Method

In the developments that follow, it is assumed that data are available that describe generating units in the following format.

Maximum Power Output Available (MW)	Probability Unit Is Available to Load to this Power (per unit)	Cost of Generating Maximum Available ($\mathbb{R}$/h)
$C(1) = 0$	$p(1)$	$F(1)$ = minimum cost
$C(2)$	$p(2)$	$F(2)$
$C(3)$	$p(3)$	$F(3)$
⋮	⋮	⋮
$C(n)$ = maximum	$p(n)$	$F(n)$

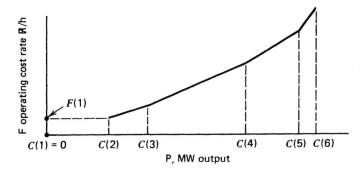

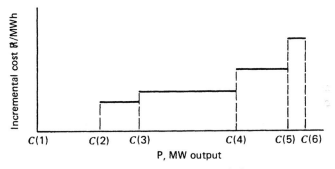

FIG. 8.7   Unit characteristics.

Pictorially, the unit characteristics needed are shown in Figure 8.7.

The probabilistic production cost procedure uses thermal-unit heat rate characteristics (i.e., heat input rate versus electric power output) that are linear segments. This type of heat rate characteristic is essential to the development of an efficient probabilistic computational algorithm since it results in stepped incremental cost curves. This simples the economic scheduling algorithm since any segment is fully loaded before the next is required. These unit input–output characteristics may have any number of segments so that a unit may be represented with as much detail as is desired. Unit thermal data are converted to cost per hour using fuel costs and other operating costs, as is the case with any economic dispatching technique.

The probabilistic production cost model simulates economic loading procedures and constraints. Fuel budgeting and planning studies utilize suitable approximations in order to permit the probabilistic computation of expected future costs. For instance, unit commitment will usually be approximated using a priority order. The priority list might be computed on the basis of average cost per megawatt-hour at full load with units grouped in blocks by minimum downtime requirements. Within each block of units with similar downtimes, units could be ordered economically by average cost per megawatt-hour at full load.

With unit commitment order established, the various available loading segments can be placed in sequence, in order of increasing incremental costs. The loading of units in this fashion is identical to using equal incremental cost scheduling where input–output curves are made up of straight-line segments. Finally, emergency sources (i.e., tie lines or pseudo tie lines) are placed last on the loading order list. The essential difference between the results of the probabilistic procedure and the usual economic dispatch computations is that all the units will be required if generator forced outages are considered.

"Must-run" units are usually designated in these computations by requiring minimum downtimes equal to or greater than a week (i.e., $7 \times 24 = 168$ h), or more. These base-load units are committed first. After the must-run units are committed, they must supply their minimum power. The next lowest-cost block of capacity may be either a subsequent loading segment on a committed unit or a new unit to be committed. (Remember that units must be committed before they are loaded further.) Following this or a similar procedure results in a list of unit loading segments, arranged in economic loading order, which is then convenient and efficient to use in the probabilistic production cost calculations and to modify for each scheduling interval.

Storage hydro-units and system sales/purchase contracts for interconnected systems must also be simulated in production cost programs. The exact treatment of each depends on the constraints and costs involved. For example, a monthly load model might be modified to account for storage hydro by *peak shaving*. In the peak-shaving approach, the hydro-unit production is scheduled to serve the peak load levels, ignoring hydraulic constraints (but not the capacity limit) and assuming a single incremental cost curve for the thermal system for the entire scheduling interval. This can be done taking into account both hydro-unit forced outages and hydro-energy availability (i.e., amount of interval energy available versus the probability of its being available). System purchases and sales are often simulated as if they were stored energy systems. Sales (or purchases) from specific units are more difficult to model, and the modeling depends on the details of the contract. For instance, a "pure" unit transaction is made only when the unit is available. Other "less pure" contracts might be made where the transaction might still take place using energy produced by other units under specified conditions.

In the probabilistic production cost approach, the load is modeled in the same way as it was in the previously illustrated load-duration curve approach; as a probability distribution expressed in terms of hours that the load is expected to equal or exceed the value on the horizontal axis. This is a monotonically decreasing function with increasing load and could be converted to a "pure" probability distribution by dividing by the number of hours in the load interval being modeled. This model is illustrated in Figures 8.2, 8.3, 8.5, and 8.6. Therefore, each load-duration curve is treated either as a cumulative probability distribution,

$$P_n(x) \text{ versus } x$$

where P_n = probability of needing x MW, or more; or when expressed in hours, it is $TP_n(x)$, where T is the duration of the particular time interval. Also,

$$P_n(x) = 1 \quad \text{for } x \leq 0$$

The load distribution is usually expressed in a table, $TP_n(x)$, which may be fairly short. The table needs to be only as long as the maximum load divided by the uniform MW interval size used in constructing the table. In applying this approach to a digital computer, it is both convenient and computationally efficient to think in terms of regular discrete steps and recursive algorithms. Various load-duration curves for the entire interval to be studied are arranged in the sequence to be used in the scheduling logic. There is no requirement that a single distribution $P_n(x)$ be used for all time periods. In developing the unit commitment schedule, it is necessary to verify not only that the maximum load plus spinning reserve is equal to or less than the sum of the capacities of the committed units, but also that the sum of the minimum loading levels of the committed units is not greater than the minimum load to be served.

A number of different descriptions have been used in the literature to explain this probabilistic procedure of thermal unit scheduling. The following has been found to be the easiest to grasp by someone unfamiliar with this procedure, and is theoretically sound. If there is a segment of capacity with a total of C MW available for scheduling, and if we denote:

$q = $ the probability that C MW are unavailable (i.e., its unavailability)

and

$p = 1 - q$

$= $ the probability or "availability" of this segment

then after this segment has been scheduled, the probability of needing x MW or more is now $P'_n(x)$. Since the occurrence of loads and unexpected unit outages are statistically independent events, the new probability distribution is a combination of mutually exclusive events with the same measure of need for additional capacity. That is,

$$P'_n(x) = qP_n(x) + pP_n(x + C) \tag{8.2}$$

In words, $qP_n(x)$ is the probability that new capacity C is unavailable times the probability of needing x, or more, MW, and $pP_n(x + C)$ is the probability C is available times the probability $(x + C)$, or more, is needed. These two terms represent two mutually exclusive events, each representing combined events where x MW, or more, remain to be served by the generation system.

This is a recursive computational algorithm, similar to the one used to develop the capacity outage distribution in the Appendix, and will be used in sequence to convolve each unit or loading segment with the distribution of load not served. It should be recognized that the argument of the probability distribution can be negative after load has been supplied and that $P_n(x)$ is zero for x greater than the peak load. Initially, when only the load distribution is used to develop $T P_n(x)$, $P_n(x) = 1$ for all $x \leq 0$.

Example 8B provides an introduction to the complexities involved in an enumerative approach to the problem at hand. By extending some of the ideas presented briefly in the Appendix to this chapter, a recursive technique (i.e., algorithm) may be developed to organize the probabilistic production cost calculations.

First, we note that the generation requirements for any generating segment are determined by the knowledge of the distribution $T P_n(x)$ that exists prior to the dispatch (i.e., scheduling) of the particular generating segment. That is, the value of $T P_n(0)$ determines the required hours of operation of a new unit. The area under the distribution $T P_n(x)$ for x between zero and the rating of the unit loading segment determines the requirements for energy production. Assuming the particular generation segment being dispatched is not perfectly reliable (i.e., that it is unavailable for some fraction of the time it is required), there will be a residual distribution of demands that cannot be served by this particular segment because of its forced outage.

Let us represent the forced outage (i.e., unavailability) rate for a generation segment of C MW, and $T P_n(x)$, the distribution of unserved load prior to scheduling the unit. Assume the unit segment to be scheduled is a complete generating unit with an input–output cost characteristic

$$F = F_0 + F_1 P, \qquad R/h$$

for $0 \leq P \leq C$ MW. The unit will be required $T P_n(0)$ hours, but on average it will be available only $(1 - q)T P_n(0)$ hours. The energy required by the load distribution that could be served by the unit is

$$E = T \int_{x=0}^{x=C} P_n(x)\, dx$$

or

$$= T \sum_{x=0+}^{x=C} P_n(x)\, \Delta x$$

for discrete distributions. The unit can only generate $(1 - q)E$ because of its expected unavailability.

These data are sufficient to compute the expected production costs. These costs for this period

$$= F_0 \times (1 - q)T P_n(0) + (1 - q)EF_1, \qquad R$$

Having scheduled the unit, there is a residual of unserved demands due to the forced outages of the unit. The recursive algorithm for the distribution of the probabilities of unserved load may be used to develop the new distribution of unserved load after the unit is scheduled. That is,

$$T P_n'(x) = q\, T P_n(x) + (1 - q)T P_n(x + C) \tag{8.3}$$

The process may be repeated until all units have been scheduled and a residual distribution remains that gives the final distribution of unserved demand.

Refer to the unit data described in Figure 8.7 and the accompanying text. The minimum load cost, $F(1)$, shown on this figure is associated only with the first loading segment, $C(2)$ to $C(3)$, since the demands of this portion of the unit will determine the maximum hours of operation of the unit.

A general scheduling algorithm may be developed based on these conditions. In this development, we temporarily put aside until the next section some of the practical and theoretical problems associated with scheduling units with multiple steps and nonzero minimum load restrictions. The procedure shown in flowchart form on Figure 8.8 is a method for computing the expected production costs for a single time period, T hours in duration.

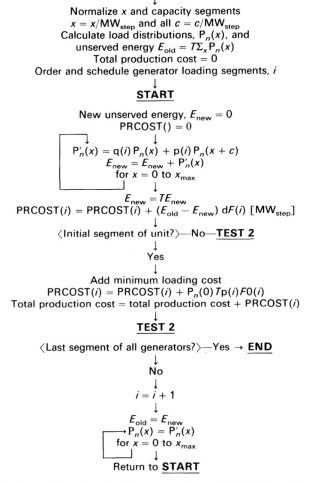

$\downarrow$
Normalize x and capacity segments
$x = x/MW_{step}$ and all $c = c/MW_{step}$
Calculate load distributions, $P_n(x)$, and
unserved energy $E_{old} = T\Sigma_x P_n(x)$
Total production cost = 0
Order and schedule generator loading segments, i
$\downarrow$
START

New unserved energy, $E_{new} = 0$
PRCOST() = 0
$\downarrow$

$P'_n(x) = q(i)P_n(x) + p(i)P_n(x + c)$
$E_{new} = E_{new} + P'_n(x)$
for $x = 0$ to x_{max}

$\downarrow$
$E_{new} = TE_{new}$
PRCOST(i) = PRCOST(i) + ($E_{old} - E_{new}$) $dF(i)$ [MW$_{step}$]
$\downarrow$
⟨Initial segment of unit?⟩—No—**TEST 2**
$\downarrow$
Yes
$\downarrow$
Add minimum loading cost
PRCOST(i) = PRCOST(i) + $P_n(0)$ $Tp(i)F0(i)$
Total production cost = total production cost + PRCOST(i)
$\downarrow$
TEST 2

⟨Last segment of all generators?⟩—Yes → **END**
$\downarrow$
No
$\downarrow$
$i = i + 1$
$\downarrow$
$E_{old} = E_{new}$
$P_n(x) = P'_n(x)$
for $x = 0$ to x_{max}
$\downarrow$
Return to **START**

FIG. 8.8   Unserved load method for computing probabilistic production costs.

Besides the terms defined on Figure 8.7 we require the following nomenclature and definitions:

$$i = 1, 2, \ldots, i_{max}, \text{ ordered capacity segments to be scheduled}$$

$$c(i) = C(i + 1) - C(i), \text{ capacity of the } i^{th} \text{ segment (MW)}$$

$$dF(i) = \frac{[F(i + 1) - F(i)]}{c(i)} \quad \text{incremental cost rate for the } i^{th} \text{ segment}$$
$$(\text{R/MWh})$$

$$F0(i) = \text{minimum load cost rate for } i^{th} \text{ segment of unit (R/h)}$$

$$p(i) = \text{availability of segment } i \text{ (per unit)}$$

$$q(i) = 1 - p(i), \text{ unavailability of segment } i \text{ (per unit)}$$

$$x = 0, 1, 2, \ldots, x_{max}, \text{ equally spaced load levels}$$

$$MW_{step} = \text{uniform interval for representing load distribution (MW)}$$

$$PRCOST(i) = \text{production costs for } i^{th} \text{ segment (R)}$$

$$E, E', E''' \ldots = \text{remaining unserved load energy}$$

In this algorithm, the energy generated by any particular loading segment of a generator is computed as the difference in unserved energy before and after the segment is scheduled. Since the incremental cost $[dF(i)]$ of any segment is constant, this is sufficient to determine the added costs due to loading of the unit above its minimum. For initial portions of a unit, $TP_n(0)$ determines the number of hours of operation required of the unit and is used to add the minimum load operating costs. We will illustrate the application of this procedure to the system described in Examples 8A and 8B.

EXAMPLE 8C

The computation of the expected production costs using the method shown in Figure 8.8 and the procedures involved can be illustrated with the data in Example 8A. Initially, we will ignore the forced outage of the two units and then follow this with an extension to incorporate the inclusion of forced outages.

With zero forced outage rates, the analysis of Example 8A is merely repeated in a different format where the load-duration curve is treated as a probability distribution. Figure 8.9 shows the initial load-duration curve in part (a); the modified curve after unit 1 is loaded is shown in part (b), and the final curve after both units are loaded is shown in part (c). Negative values of x represent load that has been served.

The computations involved in the convolutions may be illustrated in tabular

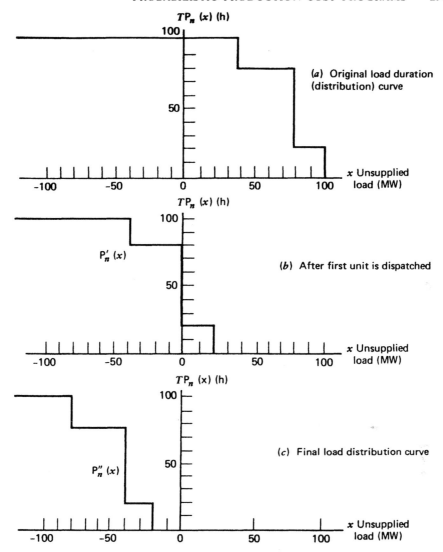

FIG. 8.9 Load-distribution curves redrawn as load probability distributions.

format. In general, in going from the jth distribution to the $(j + 1)^{st}$,

$$P_n^{j+1}(x) = qP_n^j(x) + pP_n^j(x + c)$$

where

p = 1 − q = "innage rate" of unit or segment being loaded
x + c, x = unsupplied load variables (MW)
c = capacity of unit (MW)
$P_n^j(x)$ = probability of needing to supply x or more MW at j^{th} stage

Both sides of the recursive relationship above may be multiplied by the interval duration, T, to convert it to the format illustrated in Figure 8.9. Recall that unit 1 was rated at 80 MW and unit 2 at 40 MW, and for Example 8A all $q = 0$ and all $p = 1$.

Table 8.4 shows the load probability for unserved loads of 0 to $100+$ MW. The range of valid MW values need not extend beyond the maximum load nor be less than zero. If you wish to consider the distribution extended to show the served load, $TP_n(x)$ may be extended to negative values. Only the energy for the positive x portion of this distribution represents real load energy. A negative unsupplied energy is, of course, an energy that has been supplied.

The remaining unsupplied energy levels at each step are denoted on the bottom of each column in Table 8.4 and are computed as follows.

$$E = 100 \times 20 + 80(80 - 20) + 40 \times (100 - 80) \text{ MWh}$$
$$= 20 \text{ h} \times (100 + 100 + 80 + 80 + 20) \text{ MW}$$
$$= 7600 \text{ MWh}$$
$$E' = 20 \times (20) = 400 \text{ MWh}$$
$$E'' = 0$$

Unit 1 was on-line for 100 h and generated $7600 - 400 = 7200$ MWh. Unit 2 was on-line for 80 h and generated 400 MWh. The unit loadings, loading levels, durations at those levels, fuel consumption, and production costs can easily be determined using these data. The numerical results are the same as shown in Example 8A. You should be able to duplicate those results using the distributions $P_n(x)$, $P_n'(x)$ and $P_n''(x)$.

Next let us consider forced outage rates for each unit. Let

$$q_1 = 0.05 \text{ per unit}$$

TABLE 8.4 Load Probability for Unserved Loads after Scheduling Two Units

x (MW)	$TP_n(x)$ (h)	$TP_n'(x) = TP_n(x + 80)$ (h)	$TP_n''(x) = TP_n'(x + 40)$ (h)
0	100	80	0
20	100	20	
40	100	0	
60	80		
80	80		
100	20		
100+	0		
Energy (for $x \geq 0$) $= E$		$= E'$	$= E''$

and

$$q_2 = 0.10 \text{ per unit}$$

be the forced outage rates of units 1 and 2, respectively. The recursive equation to obtain $P'_n(x)$ from the original load distribution, omitting the common factor T, is now

$$P'_n(x) = 0.05\, P_n(x) + 0.95\, P_n(x + 80)$$

The original and resultant unserved load distributions are now as follows (Figure 8.10 shows these distributions).

x (MW)	$T P_n(x)$ (h)	$T P'_n(x)$ (h)
0	100	$76 + 5 = 81$
20	100	$19 + 5 = 24$
40	100	$0 + 5 = 5$
60	80	$0 + 4 = 4$
80	80	$0 + 4 = 4$
100	20	$0 + 1 = 1$
100+	0	0
Energy	7600 MWh	760 MWh

These data may be used to compute the loadings, durations, energy produced, fuel consumption, and production cost for unit 1. Unit 1 may be loaded to 80 MW for 80 h and 40 MW for a maximum of 20 h according to the distribution $T P_n(x)$ shown in Figure 8.10. The unit is available only 95% of the time on the average. The loadings, generation, fuel consumption, and

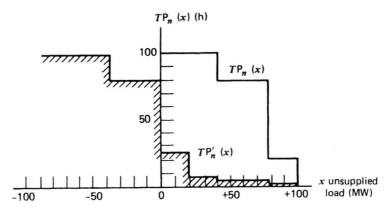

FIG. 8.10 Original and convolved load probability distributions.

fuel cost data for unit 1 are as follows and are identical with those from Example 8A.

Unit 1 Load (MW)	Duration (h)	Energy (MWh)	Fuel Used (10^6 Btu)	Fuel Cost (ℝ)
40	$0.95 \times 20 = 19$	760	9120	9120
80	$0.95 \times 80 = 76$	6080	60800	60800
		6840	69920	69920

If only production cost and/or fuel consumption are required, without detailed loading profiles, the production costs may be computed using the algorithm developed. That is, the production cost of unit 1:

$$= 160 \text{ ℝ/h} \times 0.95 \times 100 \text{ h} + 8 \text{ ℝ/MWh} \times (7600 - 760) \text{ MWh}$$

$$= 69,920 \text{ ℝ}$$

The detailed loadings and durations for unit 2 may also be computed using the distribution of unserved energy after the unit has been scheduled, $T P'_n(x)$. The unit is required 81 h, is required at zero load for $81 - 24 = 57$ h, may generate 40 MW for 5 h and 20 MW for $24 - 5 = 19$ h. The resulting generation and fuel costs are as follows.

Unit 2 Load (MW)	Duration (h)	Energy (MWh)	Fuel Used (10^6 Btu)	Fuel Cost (ℝ)
0	51.3	0	4104	8208
20	17.1	342	4104	8208
40	4.5	180	1800	3600
	72.9	522	10008	20016

However, the fuel consumption and production costs may be easily computed using the scheduling algorithm developed. The convolution of the second unit is done in accord with

$$P''_n(x) = 0.1 P'_n(x) + 0.9 P'_n(x + 40)$$

where the factor T has again been omitted.

The results are shown in Table 8.5. With these data, the production costs for unit 2 are simply

$$= 160 \text{ ℝ/h} \times 0.90 \times 81 \text{ h} + 16 \text{ ℝ/MWh} \times (760 - 238) \text{ MWh}$$

$$= 20,016 \text{ ℝ}$$

TABLE 8.5 Load Probability for Unserved Loads after Scheduling Unit 1 and Unit 2

x (MW)	$TP'_n(x)$ (h)	$TP''_n(x)$ (h)
0	81	12.6
20	24	6.0
40	5	4.1
60	4	1.3
80	4	0.4
100	1	0.1
100+	0	0
Energy	760 MWh	238 MWh

The final, unserved energy distribution is shown in Figure 8.11. Note that there is still an expected requirement to supply 100 MW. The probability of needing this much capacity is 0.001 per unit (or 0.1%), which is not insignificant.

In order to complete the example, we may compute the cost of supplying the remaining 238 MWh of unsupplied load energy. This must be based on an estimate of the cost of emergency energy supply or the value of unsupplied energy. For this example, let us assume that emergency energy may be purchased (or generated) from a unit with a net heat rate of 12,000 Btu/KWh and a fuel cost of 2 R/MBtu. These are equal to the heat rate and cost associated with unit 2 and are not too far out of line with the costs for energy from the two units previously scheduled. The cost of supplying this 238 MWh is then

$$238 \text{ MWh} \times 12 \text{ MBtu/MWh} \times 2 \text{ R/MBtu} = 5,712 \text{ R}$$

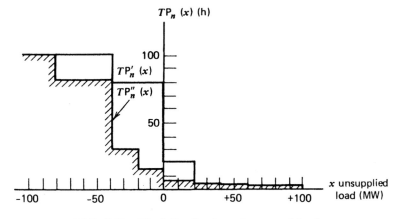

FIG. 8.11 Final distribution of unserved load.

TABLE 8.6   Results of Examples 8A and 8C Compared

	Fuel Used $(10^6$ Btu)	Fuel Cost (R)	Unsupplied Energy (MWh)	Cost of Emergency Energy (R)	Total Cost (R)
Example 8A	78400	83200	0	0	83200
Example 8D	79928	89936	238	5712	95648
Difference	1528	6736	—	—	12448
% Difference	1.95%	8.1%	—	—	15%

In summary, we may compare the results of Example 8A (computed with forced outages neglected) with the results from Example 8C, where they have been included and an allowance has been made for purchasing emergency energy (see Table 8.6). Ignoring forced outages results in a 1.95% underestimate of fuel consumption, a complete neglect of the need for and costs of emergency energy supplies, and an 8.1% underestimate of the total production costs.

The final unsupplied energy distribution may also be used to provide indexes for the need for additional transmission and/or generation capacity. This is an entire new area, however, and will not be explored here since the primary concern of this text is the operation, scheduling, and cost for power generation.

8.3.3   The Expected Cost Method

The expected cost technique is both an extension of an idea explored earlier in the discussion of hydrothermal scheduling, the system composite cost characteristic, and a variation in the convolution process used in the probabilistic approach. Using a composite system cost characteristic simplifies the computation of the total system production cost to serve a given load pattern. The expected cost per hour is given by the composite cost characteristic as a function of the power level. Calculating the production cost merely involves looking up the cost rates determined by the various load levels in the load model.

The unserved load technique of the previous section starts the convolution procedure with the probability distribution of the load pattern, and successively convolves the generation segments in an order determined by economics in order to compute successive distributions of unserved loads. Energy generation and costs of each segment were determined as a step in the procedure. In the expected cost method, the order of convolution is reversed; we start by convolving the generation probability densities and calculating expected costs to serve various levels of power generated by the system. Total costs are then computed by summing the costs to serve each load level in the forecast load model.

The expected cost method develops two functions in tabular form:

1. The probability density function of a capacity outage of x MW, $P_e(x)$.
2. The expected cost for serving a load of k (MW).

In this method, the function $P_e(x)$ represents the probability that the on-line generating units have an outage of *exactly* x MW. Keep in mind that the variables x and k, defined above, refer to the outage and load magnitudes, respectively. The expected cost rate for serving k MW of load demand is identical in its nature to the composite cost characteristic discussed in an earlier chapter, *except that it is a statistical expectation* that is computed in a fashion that recognizes the probability of random outages of the generation capacity. Thus, any generation being scheduled must serve the load demand, including any capacity shortages due to both random outages of previously scheduled capacity and demand levels in excess of the previously scheduled capacity. Therefore, we require the probability density function of the generation capacity. This function may be computed in a recursive manner, similar to those explored in the appendix of this chapter.

The recursive algorithm for developing a new capacity outage density, $P'_e(x)$, when adding a unit of "c" (MW) is:

$$P'_e(x) = qP_e(x - c) + pP_e(x) \tag{8.4}$$

where

$P_e(x)$ = prior probability of a capacity outage of x MW

c = capacity of generation segment

q = forced outage rate

$p = 1 - q$

and x ranges from zero to the total capacity, s, previously convolved. We need the initial values of this density function (i.e., for $s = 0$) in order to start the recursive computations. With no capacity scheduled these are:

$$P_e(x) = 1.0 \quad \text{for } x = 0$$

and

$$P_e(x) = 0 \quad \text{for all nonzero values of } x$$

We may develop the algorithm for recursive computation of the expected cost function by considering a simplified case where generators are represented by a single straight-line cost characteristic where minimum power level is zero and maximum is given by $c(i)$ MW. The index "i" represents the ith unit, as previously. Let $p(i) = 1 - q(i)$ represent the availability of this unit and $F_i(L)$ the cost rate (R/h) when the unit is generating a power of L MW. When all units have been scheduled, the maximum generation is the value $S = \sum_i c(i)$, the sum of all generator capacities. The load that may be supplied is denoted by k MW, and ranges from zero to S. (Note that there is a significant difference between s, the capacity scheduled previously as part of this computational process, and S, the total capacity of the system.)

Assume that we are in the midst of computing of the expected cost function, $EC(k)$. The capacity scheduled to this point is s MW. The new unit to be scheduled, unit "i," has a capacity of $c(i)$ MW. For any load level below the total capacity previously scheduled, s; that is for,

$$k \leq s$$

the new segment will supply the loads that were not served because of the outages of the previously scheduled segments *within the range of its capabilities*. The generation to be scheduled can only be loaded between zero and the maximum, c. For a given load level, k, the loading of the new segment is:

$$
\begin{aligned}
L &= k - (s - x), \quad \text{for } 0 \leq [k - (s - x)] \leq c \\
&= 0 \quad \text{for } [k - (s - x)] < 0 \\
&= c \quad \text{for } [k - (s - x)] > c
\end{aligned}
\tag{8.5}
$$

There will be a feasible set of outages $\{x\}$ that must be considered. The increase in the expected cost to serve load level, k, is then,

$$\Delta EC(k) = p(i) \sum_{\{x\}} P_e(x) F(L) \quad \text{for } 0 \leq k \leq s \tag{8.6}$$

When the load level k exceeds s,

$$EC(k) = EC(s)$$

EXAMPLE 8D

The previous 2-unit case of Examples 8A, 8B, and 8C can be used to illustrate the procedure. Load levels and capacity steps will be taken at 20-MW intervals so that the initial capacity–probability density is:

x (MW)	$P_e(x)$
0	1.0
Nonzero	0

The first unit is an 80-MW unit with $p(1) = 0.95$ and $F_1 = 160 + 8P_1$. The unit loading is

$$L = k - (s - x) = k + x, \quad \text{since } s = 0$$

The first expected cost table is then:

k (MW)	ΔEC(k)(R/h)	EC(k)(R/h)
0	$0.95[P_e(0)F_1(0)]$ = 152	152
20	$0.95[P_e(0)F_1(20)]$ = 304	304
40	$0.95[P_e(0)F_1(40)]$ = 456	456
60	$0.95[P_e(0)F_1(60)]$ = 608	608
80	$0.95[P_e(0)F_1(80)]$ = 760	760
100		760
⋮		⋮

The new value of the dispatched capacity is $s = 80$ and the new outage-probability table is:

x (MW)	$P_e(x)$
0	0.95
20	0
40	0
60	0
80	0.05
100	0
	1.00

The second unit's data are:

$$c = 40 \text{ MW}, \qquad q = 0.10, \qquad p = 0.90, \qquad \text{and} \qquad F_2 = 160 + 16P_2$$

Therefore, $L = k - (s - x) = k + x - 80$, and the second expected cost table is:

k (MW)	ΔEC(k)(R/h)		EC(k)(R/h)
0	$0.9[0.5F_2(0)]$	= 7.2	152 + 7.2 = 159.2
20	$0.9[0.05F_2(20)]$	= 21.6	325.6
40	$0.9[0.05F_2(40)]$	= 36	492
60	$0.9[0.05F_2(40)]$	= 36	644
80	$0.9[0.05F_2(40) + 0.95F_2(0)]$	= 172.8	932.8
100	$0.9[0.05F_2(40) + 0.95F_2(20)]$	= 446.4	1206.4
120	$0.9[F_2(40)]$	= 720	1480
140	.	.	1480
160	.	.	.

The new value of s is 120 MW and the new outage-probability table is:

x (MW)	$P_e(x)$
0	0.855
20	0
40	0.095
60	0
80	0.045
100	0
120	0.005
140	0
	1.000

We could stop at this point. Instead let's add an emergency source (an interconnection, perhaps) that will supply emergency power at a rate of 24 R/MW or energy at 24 R/MWh. We assume the source to be perfectly reliable, so that we may represent this source by a large unit with

$$c \geq 120 \text{ MW}, \qquad q = 0, \qquad p = 1.0, \qquad \text{and} \qquad F = 24(L)$$

where L represents the emergency load. Then

$$L = k + x - S = k + x - 120$$

The final expected cost function computations are:

k (MW)	$\Delta EC(k)$(R/h)			$EC(k)$(R/h)
0	0.005[24(0)]	=	0	159.2
20	0.005[24(20)]	=	2.4	328
40	0.005[24(40)] + 0.045[24(0)]	=	4.8	496.8
60	0.005[24(60)] + 0.045[24(20)]	=	28.8	672.8
80	0.005[24(80)] + 0.045[24(40)] + 0.095[24(0)]	=	52.8	985.6
100	0.005[24(100)] + 0.045[24(60)] + 0.095[24(20)]	=	122.4	1328.8
120	0.005[24(120)] + 0.045[24(80)] + 0.095[24(40)] + 0.855[24(0)]	=	192	1672
140	.		= 672	2152
160	.		= 1152	2632
$\vdots$	$\vdots$		$\vdots$	$\vdots$

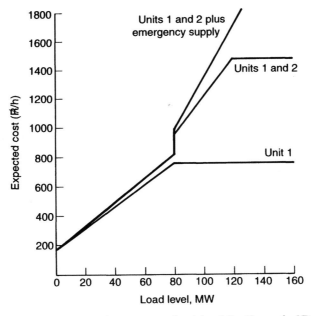

FIG. 8.12   Expected cost versus load level for Example 8D.

Figure 8.12 illustrates the expected cost versus the load level for this simple example.

The calculation of the production cost involves the determination of the expected cost at each load level in R/h and the duration of that load level. This duration is the probability density function of the load multiplied by the period length in hours, so that we are, in effect, performing the final step in convolving the load and capacity–cost distributions. For Example 8D we may develop the following table. This value agrees with that obtained in Example 8B when the cost of the 238 MWh of emergency energy required is included.

Load (MW)	Duration (h)	Expected Cost Rate (R/h)	Expected Cost (R)
40	20	496.8	9936
80	60	985.6	59136
100	20	1328.8	26576
		Total production cost =	95648

A computational flow chart similar to Figure 8.9 could be developed. (We leave this as a potential exercise.) The expected cost method has the merit that the cost rate data remain fixed with a fixed generation system and may be used to compute thermal-unit costs for different load patterns and energy purchases or sales without recomputation. As presented here, the expected cost method suffers from the lack of readily available data concerning the costs and fuel consumption of individual units. These data may be obtained when care is taken in the computational process to save the appropriate information. This involves more sophisticated programming techniques rather than new engineering applications. The same comment applies to the utilization of more realistic generation models with nonzero minimum loads and with partial outage states. All these complications can be, and have been, incorporated in various computer models that implement the expected cost method.

Similar comments apply to the unserved load method presented previously. The flowchart in Figure 8.9 offers clues to a number of programming techniques that have been applied in various instances to create more efficient computational procedures. For instance, one could replace the *unserved load* distribution by an *unserved energy* distribution as a function of the load level. This saves a step or two in the computation and would speed things up quite a bit. But these "tricks of the trade" have a way of becoming less important with the availability of ever-more-rapid small computers.

8.3.4   A Discussion of Some Practical Problems

The examples illustrate the simplicity of the basic computation of the scheduling technique used in this type of probabilistic production cost program where the load is modeled using a discrete tabular format. There are detailed complications, extensions, and exceptions that arise in the practical implementation of any production cost technique. This section reviews the procedures used previously, in the unserved load method, to point out some of these considerations. No attempt is made to describe a complete, detailed program. The intent is to point out some of the practical considerations and discuss some of the approaches that may be used.

First, consider Figure 8.13, which shows the *cumulative load distribution* (i.e.,

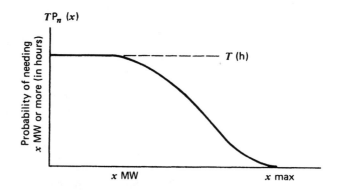

FIG. 8.13   Load probability distribution.

TABLE 8.7   Sample Subinterval Loading Data: Segment Data

Unit Number (i)	No. (j)	P_{min} (MW)	P_{max} (MW)	Cost	Outage Rate (q_{ij}) per Unit	Innage or Availability Rate
3	1	0	20	₽/h	0.05	0.95
1	1	0	20	₽/h	0.02	0.98
4	1	0	40	₽/h	0.02	0.98
1	2	20+	60	₽/MHh	0.05	0.95
3	2	20+	50	₽/MWh	0.05	0.95
4	2	40+	50	₽/MWh	0.05	
⋮	⋮	⋮	⋮	⋮	⋮	⋮

a load-duration curve treated as a cumulative probability distribution) for an interval of T hours. Next, assume an ordered list of loading segments as shown in Table 8.7. Units 3, 1, and 4 are to be committed initially, so that the sum of their capacities at full output equals or exceeds the peak load plus capacity required for spinning reserves. If we assume that two segments for each of these three units, this commitment totals 160 MW. Assume such a table includes all the units available in that subinterval. The cost data for the first three loading segments are the total costs per hour at the minimum loading levels of 20, 20, and 40 MW, respectively, and the remaining cost data are the incremental costs in ₽ per MWh for the particular segment. Table 8.7 is the ordered list of loading segments where each segment is loaded, generation and cost are computed, and the cumulative load distribution function is convolved with the segment.

There are two problems presented by these data that have not been discussed previously. First, the minimum loading sections of the initially committed units must be loaded at their minimum load points. For instance, the minimum load for unit 4 is 40 MW, which means it cannot satisfy loads less than 40 MW. Second, each unit has more than one loading segment. The loading of a unit's second loading segment, by considering the probability distribution of unserved load after the first segment of a unit has been scheduled, would violate the combinatorial probability rules that have been used to develop the scheduling algorithm, since the unserved load distribution includes events where the first unit was out of service. That is, the loading of a second or later section is not statistically independent of the availability of the previously scheduled sections of the particular unit. Both these concerns require further exploration in order to avoid the commitment of known errors in the procedure.

The situation with block-loaded units (or a nonzero minimum loading limit) is easily handled. Suppose the unserved load distribution prior to loading such a block-loaded segment is $T P_n(x)$ and the unit data are

$$q = \text{unavailability rate, per unit}$$
$$p = 1 - q$$
$$c = \text{capacity of segment}$$

By *block-loading* it is meant that the output of this particular segment is limited to exactly c MW. The nonzero minimum loading limit may be handled in a similar fashion.

The convolution of this segment with $T P_n(x)$ now must be handled in parts. For load demands below the minimum output, c, the unit is completely unavailable. For $x \geq c$, the unit may be loaded to c MW output. The algorithm for combining the mutually exclusive events where x, or more, MW of load remain unserved must now be performed in segments, depending on the load. For load levels, x, such that

$$x \geq c$$

the new unserved load distribution is

$$P'_x(x) = q \, P_n(x) + p \, P_n(x + c) \tag{8.7}$$

where the period length, T, has been omitted. For some loads, $x < c$, the unit cannot operate to supply the load. Let $p_n(x)$ denote the probability density of load x. In discrete form,

$$p_n(x) = P_n(x) - P_n(x + \text{MW}_{\text{step}}) \tag{8.8}$$

where $\text{MW}_{\text{step}} = $ uniform interval in tabulation of $P_n(x)$. For loads equal to or greater than c, the probability of exactly x MW after the unit has been scheduled is

$$p'_n(x) = q \, p_n(x) + p \, p_n(x + c) \tag{8.9}$$

For loads less than c (i.e., $0 \leq x \leq c$),

$$p'_n(x) = p_n(x) + p \, p_n(x + c) \tag{8.10}$$

For convenience in computation, let

$$p_n(x) = (q + p) p_n(x) \tag{8.11}$$

for $0 \leq x < c$. Then for this same load range,

$$p'_n(x) = q p_n(x) + p \, p_n(x + c) + p \, p_n(x) \tag{8.12}$$

Next, the new unserved load energy distribution may be found by integration of the density function from the maximum load to the load in question. For discrete representations and for $x \geq c$,

$$P'_n(x) = q\,P_n(x) + pP_n(x + c) \tag{8.13}$$

For loads less than c; that is, $0 \leq x \leq c$,

$$P'_n(x) = q\,P_n(x) + p\,P_n(x + c) + p[P_n(x) - P_n(c)] \tag{8.14}$$

The last term represents those events for loads between x and c, wherein the unit cannot operate. The term $[P_n(x) - P_n(c)]$ is the probability density of those loads taken as a whole. The first term, $q\,P_n(x)$, resulted from assuming that the unit could supply any load below its maximum.

This format for the block-loaded unit makes it easy to modify the unserved load scheduling algorithm presented previously. The effects of restriction to block-loading a unit may be illustrated using the data from Example 8C.

EXAMPLE 8E

The two-unit system and load distribution of Example 8C will be used with one modification. Instead of allowing the second unit to operate anywhere between 0 and 40 MW output, we will assume its operation is restricted to 40 MW only. The cost of this unit was

$$F_2 = 160 + 16P_2, \qquad \text{R/h}$$

so that for $P_2 = 40$ MW, $F_2 = 800$ R/h.

Recall (see Table 8.5) that after the first 80 MW unit was scheduled, the unserved load distribution was

x (MW)	$T P'_n(x)$ (h)
0	81
20	24
40	5
60	4
80	4
100	1

With an unserved load energy of 760 MWh. With a restriction to block-loading, the unit is on-line only 5 h. The energy it generates is therefore $5 \times 40 \times 0.9 = 180$ MWh. The new distribution of unserved load after the unit is scheduled is as follows.

x (MW)	$TP'_n(x)$ (h)	$q\,TP'(x) + p\,TP'(x + c)$	$p\,T[P_n(x) - P_n(c)]$	$TP''_n(x)$ (h)
0	81	12.6	$0.9[81 - 5]$	81.0
20	24	6.0	$0.9[24 - 5]$	23.1
40	5	4.1	—	4.1
60	4	1.3	—	1.3
80	4	0.4	—	0.4
100	1	0.1	—	0.1

The unserved load energy is now 580 MWh.

The quantitative significance of the precise treatment of block-loaded units has been magnified by the smallness of this example. In studies of practical-sized systems, block-loading restrictions are frequently ignored by removing the restriction on minimum loadings or are treated in some satisfactory, approximate fashion. For long-range studies, these restrictions usually have minor impact on overall production costs.

The analysis of the effects of the statistical dependence of the multiple-loading segments of a unit is somewhat more complicated. The distribution of unserved load probabilities, $TP_n(x)$, at any point in the scheduling algorithm is independent of the order in which various units are scheduled. Only the generation and hours of operation are dependent on the scheduling order. This may easily be verified by a simple numerical example, or it may be deduced from the recursive relationship presented for $TP_n(x)$.

Suppose we have a second section to be incrementally loaded for some machine at a point in the computations where the distribution of unserved load is $TP_n(x)$. The outage of this second incremental loading section is obviously not statistically independent of the outage of the unit as a whole. Therefore, the effect of the first section must be removed from $TP_n(x)$, prior to determining the loading of the second segment. This is known as *deconvolution*.

For this illustration of one method for handling multiple segments, we will assume:

1. The capacity of the segment extends from C_1 to C_2 where $C_2 > C_1$.
2. The first segment had a capacity of C_1.
3. The outage rates of both segments are equal to q per unit.

In the process of arriving at the distribution $TP_n(x)$, the initial segment of

C_1 MW was convolved in the usual fashion. That is,

$$TP_n(x) = q\,TP'_n(x) + p\,TP'_n(x + C_1) \tag{8.15}$$

The distribution $TP_n(x)$ is independent of the order in which segments are convolved. Only the loading of each segment depends on this order.

Therefore, we may consider that $TP'_n(x)$ represents an artificial distribution of load probabilities with the initial segment of the unit removed. This pseudo-distribution, $TP'_n(x)$, must be determined in order to evaluate the loading on the segment between C_1 and C_2. Several techniques may be used to recover $TP'_n(x)$ from $TP_n(x)$. The convolution equation may be solved for either $TP'_n(x)$ or $TP'_n(x + c)$. The deconvolution process is started at the maximum load if the equation is solved for $TP'_n(x)$. That is,

$$TP'_n(x) = \frac{1}{q}\,TP_n(x) - \frac{p}{q}\,TP'_n(x + c)$$

and $\hfill(8.16)$

$$TP'_n(x) = 0 \quad \text{for } x > \text{maximum load}$$

We will use this procedure to illustrate the method because the procedures and algorithms discussed have not preserved the distributions for negative values of unserved load (i.e., already-served loads). As a practical computational matter, it would be better practice to preserve the entire distribution $TP_n(x)$ and solve the convolution equation for $TP_n(x + c)$. That is,

$$TP_n(x + c) = \frac{1}{p}\,TP_n(x) - \frac{q}{p}\,TP'_n(x) \tag{8.17}$$

or shifting arguments, by letting $y = x + c$,

$$TP_n(y) = \frac{1}{p}\,TP_n(y - c) - \frac{q}{p}\,TP'_n(y - c) \tag{8.18}$$

In this case, the deconvolution is started at the point at which

$$-y = \text{sum of dispatched generation}$$

since

$$TP_n(y) = T$$

$$\text{for all } y < -\text{sum of dispatched generation}$$

Even though we will use the first deconvolution equation for illustration, the second should be used in any computer implementation where repeated deconvolutions are to take place. Since $q \ll p$, the factors $1/q$ and p/q in the first formulation will amplify any numerical errors that occur in computing the successive distributions. We use this potentially, numerically treacherous formulation here only as a convenience in illustration.

To return, we obtain the deconvolved distribution $TP'_n(x)$ by removing the effects of the first loading segment. Then the loading of the second segment from C_1 to C_2 is determined using $TP'_n(x)$, and the new, remaining distribution of unserved load is obtained by adding the total unit of C_2 MW to the distribution so that

$$TP''_n(x) = q\,TP'_n(x) + p\,TP'_n(x + C_2) \tag{8.19}$$

EXAMPLE 8F

Assume that in our previous examples, the first unit had a total capacity of 100 MW instead of 80. This last segment might have an incremental cost rate of 20 R/MWh so that it would not be dispatched until after the second unit had been used. Assume the outage rate of 0.05 per unit applies to the entire unit. Let us determine the loading on this second section and the final distribution of unserved load.

The distribution of unserved load from the previous examples is

x (MW)	$TP''_n(x)$ (h)
0	12.6
20	6.0
40	4.1
60	1.3
80	0.4
100	0.1

The deconvolved distribution may be computed starting at $x = 100$ MW using Eq. 8.16 and working up the table. The table was constructed with $c = 80$ MW for the capacity of this unit. The deconvolved distribution is

$$TP'_n(100) = \frac{0.1}{0.05} = 2$$

$$TP'_n(80) = \frac{0.4}{0.05} = 8$$

$$\vdots$$

The new distribution, adding the entire 100 MW unit, is determined using $c = 100$ MW and is

$$TP''_x(x) = 0.05\,TP'_x(x) + 0.95\,TP'_n(x + 100)$$

The results are as follows.

x (MW)	$T P_n(x)$ (h)	$T P'_n(x)$ (h)	$T P''_n(x)$ (h)
0	12.6	100	6.9
20	6.0	82	4.1
40	4.1	82	4.1
60	1.3	26	1.3
80	0.4	8	0.4
100	0.1	2	0.1
Energy	238 MWh		200 MWh

Thus, the second section of the first unit generates 38 MHh.

This computation may be verified by examining the detailed results of Example 8B, where the various load and outage combination events were enumerated. At a load of 100 MW, the second segment of unit 2 would have been loaded to the extent shown by this example. You should be able to identify two periods where the second section would have reduced previous shortages of 0 and 20 MW. This procedure and the example are theoretically correct but computationally tedious. Furthermore, the repeated deconvolution process may lead to numerical round-off errors unless care is taken in any practical implementation.

Approximations are frequently made in treating sequential loading segments. These are usually based on the assumption that the subsequent loading sections are independent of the previously loaded segments. That is, that they are equivalent to new, independent units with ratings that are equal to the capacity increment of the segment. When these types of approximations are made, they are justified on the basis of numerical tests. They generally perform more than adequately for larger systems but should be avoided for small systems.

The two extensions discussed here are only examples of the many extensions and modifications that may be made. When the computations of expected production costs are made as a function of the load to be served, these characteristics may be used as pseudogenerators in scheduling hydroelectric plants, pumped-storage units, or units with limited fuel supplies.

There have been further extensions in the theoretical development as well. It is quite feasible to represent the distribution of available capacity by the use of suitable orthogonal polynomials. Gram–Charlier series are frequently used to model probabilistic phenomena. They are most useful with a reasonably uniform set of generator capacities and outage rates. By representing the expected load distribution also as an analytic function it is possible to develop analytical expressions for unserved energy distributions and expected production costs. Care must be exercised in using these approximations when one or two very large generators are added to systems previously composed of a uniform array of capacities. We will not delve further into this area in this text. The remainder of this chapter is devoted to a further example and problems.

8.4   SAMPLE COMPUTATION AND EXERCISE

The discussion of the probabilistic techniques is more difficult than their performance. We will illustrate the unserved load method further using a three-unit system. The three generating units each may be loaded from 0 MW to their respective ratings. For ease of computation, we assume linear input–output cost curves and only full-forced outage rates (that is, the unit is either completely available or completely unavailable). The unit data are as follows.

Unit No.	Maximum Rating (MW)	Input–Output Cost Curve (R/h)	Full-Forced Outage Rate (per unit)
1	60	$60 + 3P_1$	0.2
2	50	$70 + 3.5P_2$	0.1
3	20	$80 + 4P_3$	0.1

In these cost curves P_i are in MW. In addition, the system is served over a tie line. Emergency energy is available without limit (MW or MWh) at a cost rate of 5 R/MWh.

The load model is a distribution curve for a 4-week interval (a 672-h period). That is, the expected load is as shown in Table 8.8. The total load energy is 43,680 MWh.

8.4.1   No Forced Outages

The economic dispatch of these units for each load level is straightforward. The units are to be loaded in the order shown. The sum of the peak load demand

TABLE 8.8   Load Distribution

Load Level (MW)	Hours of Existence	Probability	Hours Load Equals or Exceeds	Probability of Needing Load or More (pu)
30	134.4	0.2	672.0	1.00
50	134.4	0.2	537.6	0.80
70	134.4	0.2	403.2	0.60
80	168.0	0.25	268.8	0.40
100	100.8	0.15	100.8	0.15
	672.0			

(100 MW) and the total capability (130 MW) is 230 MW. Therefore, the probability table of needing capacity will extend eventually from -130 MW to $+100$ MW. It is convenient in digital computer implementation to work in uniform MW steps. For this example, we will use 10 MW.

As each unit is dispatched, the probability distribution of needing x or more MW [i.e., $P_n(x)$] is modified (i.e., convolved) using the following:

$$T P'_n(x) = T P_n(x + c)$$

where

$P'_n(x)$ and $P_n(x) =$ new and old distributions, respectively

$T =$ time period, 672 h in this instance

$c =$ capability of unit or segment when it is in state j

Table 8.9 shows initial distribution in the second column. The load energy to be served is

$$E = 672 \sum_{x=0}^{100} P_n(x)\, \Delta x = 43{,}680 \text{ MWh}$$

With zero-forced outage rate, the 60-MW unit loading results in the $P_n(x)$ distribution shown in the third column. The resultant load energy to be served is now:

$$E' = (0.15 \times 20 + 0.4 \times 10 + 0.6 \times 10) \times 672 = 8736 \text{ MWh}$$

which means unit 1 generated

$$43{,}680 - 8736 = 34{,}944 \text{ MWh}$$

The unit was on-line for 672 h, and the incremental cost rate was 3 R/h. Therefore, the cost for unit 1 is

$$\text{Total cost} = \sum_T F(P_t) \times \Delta t = \sum_T (60 + 3P_t)\,\Delta t = \sum_T (60\,\Delta t + 3P_t\,\Delta t)$$

$$= 60T + 3 \text{ (MWh generated), since MWh} = \sum_T P_t\,\Delta t$$

$$= 60 \text{ R/h} \times 672 \text{ h} + 34{,}944 \text{ MWh} \times 3 \text{ R/MWh} = 145{,}152 \text{ R}$$

Unit 2 serves the remaining load distribution (third column) and results in the distribution shown in the fourth column. This unit is only on-line for 60% of the interval, so that its cost is

$$0.6 \times 70 \text{ R/h} \times 672 \text{ h} + 8736 \text{ MWh} \times 3.5 \text{ R/MWh} = 58{,}800 \text{ R}$$

The total system cost is 203,952 R, and unit 3 is not used at all. These results are summarized in Table 8.10.

TABLE 8.9   Three-Unit Example: Zero-Forced Outage Rates

x (MW)	$P_n(x)$ (pu)	$P_n'(x)$ (pu)	$P_n''(x)$ (pu)
−130			
−120			
−110			
−100			
−90			
−80			1.0
−70			0.8
−60			0.8
−50			0.6
−40			0.6
−30		1.0	0.4
−20		0.8	0.15
−10		0.8	0.15
0		0.6	0
10		0.6	
20		0.4	
30	1.0	0.15	
40	0.8	0.15	
50	0.8	0	
60	0.6		
70	0.6		
80	0.4		
90	0.15		
100	0.15		
110	0		
E/672	65	13	0
MWh	43680	8736	0

TABLE 8.10   Summary of Results: Zero-Forced Outage Rates

Unit Number	Capacity (MW)	Outage Rate (pu)	Hours On-Line	Energy Generated (MWh)	Cost (R)
1	60	0.000	672.0	34944.0	145152.0
2	50	0.000	403.0	8736.0	58800.0
3	20	0.000	0	0	0
4	100	0.000	0	0	0
Total	230			43680.0	203952.0

Average system cost = 4.6692 R/MWh.

8.4.2 Forced Outages Included

When the forced outage is included, the convolution of the probability distribution is accomplished by

$$P'_n(x) = q\,P_n(x) + p\,P_n(x + c)$$

where

$$q = \text{forced outage rate (pu)}$$

$$p = 1 - q = \text{``innage'' rate}$$

Table 8.11 shows the computations for the first unit in the third and fourth columns.

The first unit is on-line $0.8 \times 672 = 537.6\,\text{h}$ and generates 27,955.2 MWh. (The initial load demand contains 43,680 MWh; the modified distribution in

TABLE 8.11 Three-Unit Example Including Forced Outage Rates

x (MW)	$P_n(x)$ (pu)	$P_n(x + 60)$ (pu)	$P'_n(x)$ (pu)	$P'_n(x + 50)$ (pu)	$P''_n(x)$ (pu)	$P'''_n(x)$ (pu)
−130	↑	↑	↑	↑		
−120						
−110						
−90						
−80				1.00		
−70				0.84		
−60				0.84		
−50				0.68		
−40				0.68		
−30		1.0	1.0	0.52		
−20		0.8	0.84	0.32		
−10		0.8	0.84	0.28		
0		0.6	0.68	0.16	0.212	
10		0.6	0.68	0.12	0.176	
20		0.4	0.52	0.12	0.160	
30	1.0	0.15	0.32	0.08	0.104	
40	0.8	0.15	0.28	0.03	0.055	
50	0.8	0	0.16	0.03	0.043	
60	0.6	↓	0.12	0	0.012	
70	0.6		0.12	↓	0.012	
80	0.4		0.08		0.008	
90	0.15		0.03		0.003	
100	0.15		0.03		0.003	
110	0	↓	0		0	
E/672	65		23.4		5.76	
MWh	43680		15724.8		3870.72	

column 4 contains 15,724.8 MWh.) Therefore, the first unit's cost is

$$60 \, \text{R/h} \times 537.6 \, \text{h} + 3 \, \text{R/MWh} \times 27,955.2 \, \text{MWh} = 116,121.6 \, \text{R}$$

The distribution of needed capacity is shown *partially* in the sixth column of Table 8.11. Sufficient data are shown to compute the load energy remaining. (*Load energy* is the portion of the distribution, $P_n^j(x)$, for $x \geq 0$). The unserved load energy after scheduling unit 2 is

$$0.576 \times 10 \times 672 = 3870.72 \, \text{MWh}$$

This means unit 2 generated an energy of $15,724.8 - 3870.72 = 11,854.08 \, \text{MWh}$ at an incremental cost of 3.5 R/MWh; or 41,489.28 R. The unit was on-line for 411.264 h at a cost rate of 70 R/h. This brings the total cost to 70,277.76 R for unit 2. Note that the operating time (i.e., the "hours on-line") is $0.9 \times 0.68 \times 672$ h. The first factor represents the probability that the unit is available, the second the fraction of the time interval that the load requires unit 2, and the 672-h factor is the length of the interval.

Table 8.12 shows a summary of the results for this three-unit plus tie-line sample exercise when outage rates are included. The third unit and tie line are utilized a substantial amount compared with ignoring forced outages. The total cost for the 4-wk interval increased by almost 5%.

The resulting successive convolutions are shown in Figure 8.14. After the entire 130 MW of generating capacity has been dispatched, the distribution of unserved load is represented by the portion of the lowest curve to the right of the zero MW point (it is shaded).

Table 8.13 shows the distribution of emergency energy delivery over the tie line.

This chapter has only provided an introduction to this area. Practical schemes exist to handle much more complex unit and load models, to incorporate limited energy and pumped-storage units, and to compute generation reliability indices. They are all based on techniques similar to those introduced here.

TABLE 8.12   Results

Unit No.	Capacity (MW)	Outage Rate (pu)	Hours On-Line	Energy Generated (MWh)	Cost (R)
1	60	0.200	538.0	27955.0	116122.0
2	50	0.100	411.0	11854.0	70278.0
3	20	0.100	128.0	2032.0	18386.0
4	100	0.000	111.0	1839.0	9193.0
Total	230			43680	213979.0

Average system cost = 4.8589 R/MWh.

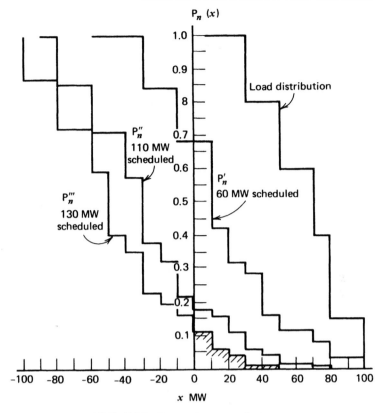

FIG. 8.14 Successive convolutions.

TABLE 8.13 Emergency Energy

Level No.	Loading (MW)	Hours
1	10.0	30.71
2	20.0	11.02
3	30.0	22.04
4	40.0	0.81
5	50.0	4.50
6	60.0	3.02
7	70.0	0.27
8	80.0	2.15
9	100.0	0.20
Total		74.72

APPENDIX
Probability Methods and Uses in Generation Planning

The major application of probability methods in power systems has been in the area of planning generating capacity requirements. This application, no matter what particular technique is used, assigns a probability to the generating capacity available, describes the load demands in some manner, and provides a numerical measure of the probability of failing to supply the expected power or energy demands. By defining a standard *risk level* (i.e., a standard or maximum probability of failure) and allowing system load demands to grow as a function of time, these probability methods may be utilized to calculate the time when new generating capacity will be required.

Three general categories of probability methods and measures have been developed and applied to the generation planning problem. These are:

1. The loss-of-load method.
2. The loss-of-energy method.
3. The frequency and duration method.

The first measures reliability as the probability of meeting peak loads (or its converse, the failure probability). The second uses the expected loss of energy as a reliability measure. The frequency and duration method is based on a somewhat different approach. It calculates the expected frequencies of outages of various amounts of capacity and their corresponding expected durations. These calculated values are then used with appropriate, forecasted loads and reliability standards to establish capacity reserve margins.

The mathematical techniques used are straightforward applications of probability methods. First, to review combined probabilities, let

$P(A)$ = probability that event A occurs

$P(B)$ = probability that event B occurs

$P(A \cap B)$ = joint probability that A and B occur together

$P(A \cup B)$ = probability that either A occurs by itself, or B occurs by itself, or A and B occur together.

Conditional probabilities will be omitted from this discussion. [A *conditional probability* is the probability that A will occur if B already has occurred and may be expressed $P(A/B)$].

A few needed rules from combinatorial probabilities are:

1. If A and B are independent events (i.e., whether A occurs or not has no bearing on B), then the joint probability that A and B occur together is $P(A \cap B) = P(A)P(B)$.

2. If the favorable result of an event is for A or B or both to occur, then the probability of this favorable result is $P(A \cup B) = P(A) + P(B) - P(A \cap B)$.

3. If, in rule 2, A and B are "mutually exclusive" events (i.e., if one occurs, the other cannot), then $P(A \cap B) = 0$ and $P(A \cup B) = P(A) + P(B)$.

4. The number of combinations of n things taken r at a time is given by the formula

$$_nC_r = \frac{n!}{r!(n-r)!} \tag{8A.1}$$

5. In general, the probability of exactly r occurrences in n trials of an event that has a constant probability of occurrence p is

$$P_n(r) = {}_nC_r p^r q^{n-r} = \frac{n!}{r!(n-r)!} p^r q^{n-r} \tag{8A.2}$$

where $q = 1 - p$.

Rule 5 is a generalized form of the binomial expansion, applying to all terms of the binomial $(p + q)^n$. This distribution has had widespread use in generating-system probability studies. For example, assume that a generation system is composed of four identical units and that each of these units has a probability p of being in service at any randomly chosen time. The probability of being out of service is $q = 1 - p$. Assume that each machine's behavior is independent of the others. Then, a table may be constructed showing the probability of having 4, 3, 2, 1, and none in service.

Number in Service	Probability of Occurrence
4	$P(4) = {}_4C_4 p^4 q^{4-4} = \dfrac{4!}{4!(4-4)!} p^4 = p^4$
3	$P(3) = {}_4C_3 p^3 q^{4-3} = \dfrac{4!}{3!(4-3)!} p^3 q = 4p^3 q$
2	$P(2) = {}_4C_2 p^2 q^{4-2} = \dfrac{4!}{2!(4-2)!} p^2 q^2 = 6p^2 q^2$
1	$P(1) = {}_4C_1 p^1 q^{4-1} = \dfrac{4!}{1!(4-1)!} pq^3 = 4pq^3$
0	$P(0) = {}_4C_0 p^0 q^{4-0} = \dfrac{4!}{0!(4-0)!} q^4 = q^4$

In this table, each of the probabilities is a term of the binomial expansion of the form:

$$_4C_n p^n q^{4-n}$$

where n is the number of units in service.

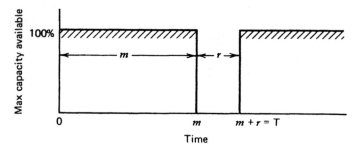

FIG. 8.15 Average availability cycle for a unit with two states.

These relationships assume a long-term average availability cycle, as shown in Figure 8.15 for a given unit. In this long-term average cycle,

$$m = \text{average time available before failures}$$

$$r = \text{average repair time}$$

$$T = m + r = \text{mean time between failures}$$

Using these definitions for the generator taken as a binary state device,

$$p = \frac{m}{T} = \text{"innage rate" (per unit)}$$

$$q = 1 - p = \frac{r}{T} = \text{"outage rate" (per unit)}$$

Generating units may also be considered to be multistate devices when each state is characterized by the maximum available capacity and the probability of existence of that particular state. For instance, a large unit may have a forced reduction in output of, say, 20% of its rating when one boiler feed pump is out of service. This may happen 25% of the total time the unit is supposed to be available. In this case, each unit state (j) can be characterized by

$$C(j) = \text{maximum capacity available in state } (j)$$

$$p(j) = \text{probability that the unit is in state } (j)$$

where

$$\sum_{j=1}^{n} p(j) = 1.0$$

$$C(1) = 0 \text{ (unit down)}$$

$$C(n) = 100\% \text{ capability (unit at full capacity)}$$

In the probabilistic production cost calculations we attach other parameters to a state, such as the incremental cost for loading the unit between $C(j-1)$ and $C(j)$ MW.

The use of reliability techniques based on probability mathematics for generation planning frequently involves the construction of tables that show capacity on outage and the corresponding probability of that much, or more, capacity being on outage. The binomial probability distribution is cumbersome to use in practical computations. We will illustrate the simple numerical convolution using recursive techniques that are useful and efficient in handling units of various capacities and outage rates. The model of the generating capacity to be developed in this case is a table such as the following.

k	O_k Generating Capacity Outage (MW)	Probability of Occurrence of O_k or greater $= P_0(O_k)$
1	0	1.000000
2	15	0.950000
3	25	0.813000
4	35	0.095261
$\vdots$	$\vdots$	$\vdots$

On this table

$$k = \text{index showing the entry number}$$

$$O_k = \text{generating capacity outage (MW)}$$

$$P_0(O_k) = \text{cumulative probability} = \text{probability of the occurrence}$$
$$\text{of an outage of } O_k, \text{ or larger}$$

This probability is a distribution rather than the density described with the binomial probability. It is a cumulative value rather than an exact probability (i.e., "exact" means probability density function).

Let each machine of the previously discussed hypothetical four-machine system be rated 10 MW, and let $p(k)$ be the exact probability of occurrence of a particular event characterized by a given outage value. The table started previously may be expanded into Table 8.14. The function $P(O_k)$ is monotonic, and it should be obvious that the probability of having a zero or larger capacity outage is 1.0.

Since all generators do not have the same capacity or outage rate, the simple relationship for the binomial distribution in Table 8.14 does not hold in the general case. Beside the unit capability, the only other parameter associated with a generator in this technique is the average outage existence rate, q.

A simple recursive algorithm exists to add a unit to an existing outage probability table. Suppose an outage probability table exists that gives

$$P_0(x) \text{ versus } x$$

TABLE 8.14   Outage Probabilities

k	No. of Machines in Service	MW Outage O_k	$p(k)$ = Exact Probability of Outage O_k	$P(O_k)$ = Probability of Outage O_k, or Larger
1	4	0	p^4	$p^4 + 4p^3q + 6p^2q^2 + 4pq^3 + q^4 \equiv 1.0$
2	3	10	$4p^3q$	$4p^3q + 6p^2q^2 + 4pq^3 + q^4$
3	2	20	$6p^2q^2$	$6p^2q^2 + pq^3 + q^4$
4	1	30	$4pq^3$	$4pq^3 + q^4$
5	0	40	q^4	q^4

Installed capacity = 40 MW.

where

$$P_0(x) = \text{probability of } x \text{ MW or more on outage}$$

$$x = \text{MW outage state}$$

Now suppose you wish to add an "n-state" unit to the table that is described by

$$p(j) = \text{probability unit is in state } j$$

$$C(j) = \text{maximum capacity of state } j$$

$$C(n) = \text{capacity of unit}$$

$$O_j = C(n) - C(j) = \text{MW outage for state } j$$

Then the new table of outage probabilities may be found by a numerical convolution:

$$P_0'(x) = \sum_{j=1}^{n} p(j)P_0(x - O_j) \tag{8A.3}$$

where

$$P_0(\leq 0) = 1.0$$

This algorithm is an application of the combinational rules for independent, mutually exclusive "events." Each term of the algorithm is made up of (1) the event that the new unit is in state j with an outage O_j MW, and (2) the event that the "old" system has an outage of $(x - O_j)$ MW. The combined event, therefore, has an outage of x MW, or more.

EXAMPLE 8G

Assume we have a generating system consisting of the following machines with their associated outage rate.

MW	Outage Rate
MW	0.02
10	0.02
10	0.02
10	0.02
10	0.02
5	0.02

The exact probability outage table for the first four units could be calculated using the binomial distribution directly and would result in the following table.

MW Outage x	Exact Probability $p(x)$	Cumulative Probability $P_0(x)$
0	0.922368	1.000000
10	0.075295	0.077632
20	0.002305	0.002337
30	0.000032	0.000032
40	0	0

Now, the fifth machine can exist in either of two states: (1) it is in service with a probability of $p = 1 - q = 0.98$ and no additional system capacity is out, or (2) it is out of service with a probability of being in that state of $q = 0.02$, and 5 MW additional capacity is out of service.

The resulting outage-probability table will have additional outages because of the new combinations that have been added. This can be easily overcome by expanding the table developed for four machines to include these new outages. This is shown in Table 8.15, along with an example where the fifth, 5 MW, unit is added to the table.

TABLE 8.15   Adding Fifth Unit

x (MW)	$P_0(x)$	$0.98\,P_0(x)$	$0.02\,P_0(x - 5)$	$P_0'(x)$
0	1.000000	0.980000	0.020000	1.000000
5	0.077632	0.076079	0.020000	0.096079
10	0.077632	0.076079	0.001553	0.077632
15	0.002337	0.002290	0.001553	0.003843
20	0.002337	0.002290	0.000047	0.002337
25	0.000032	0.000031	0.000047	0.000078
30	0.000032	0.000031	0	0.000031
35	0	0	0	0
40	0	0	0	0
45	0	0	0	0

The correctness of this approach and the resulting table may be seen by calculating the exact state probabilities for all possible combinations. That is,

MW Out x	Exact Probability p(x)
	New machine in service
0 + 0	$0.922368 \times 0.98 = 0.903921$
10 + 0	$0.075295 \times 0.98 = 0.073789$
20 + 0	$0.002305 \times 0.98 = 0.002258$
30 + 0	$0.000032 \times 0.98 = 0.000031$
40 + 0	$0 \times 0.98 = 0$
	New machine out of service
0 + 5 = 5	$0.922368 \times 0.02 = 0.018447$
10 + 5 = 15	$0.075295 \times 0.02 = 0.001506$
20 + 5 = 25	$0.002305 \times 0.02 = 0.000047$
30 + 5 = 35	$0.000032 \times 0.02 = 0$
40 + 5 = 45	$0 \times 0.02 = 0$

The exact state probabilities are combined by adding the probabilities for the mutually exclusive events that have identical outages; the results are shown in Table 8.16. Table 8.16 is the capacity model for the five-unit system and is usually assumed to be fixed until new machines are added or a machine is retired, or the model is altered to reflect scheduled maintenance outage.

This model was constructed using maximum capacities and calculating capacity outage probability distributions. Similar techniques may be used to construct available capacity distributions. A similar convolution is used in the probabilistic production cost computations. The form of the distribution is different because we are dealing with a scheduling problem rather than with

TABLE 8.16 Table of Combined Probabilities

MW Outage x	Exact Probability p(x)	Cumulative Probability $P'_0(x)$
0	0.903921	1.000000
5	0.018447	0.096079
10	0.073789	0.077632
15	0.001506	0.003843
20	0.002259	0.002337
25	0.000047	0.000078
30	0.000031	0.000031
35	0	0
40	0	0
45	0	0

the static, long-range planning problem. In the present case, we are interested in a distribution of capacity outage probabilities; in the scheduling problem, we require a distribution of unserved load probabilities.

PROBLEMS

8.1 Add another unit to Example 8G (in the Appendix). The new unit should have a capacity of 10 MW and an availability of 90%. That is, its outage rate is 0.10 per unit. Use the recursive algorithm illustrated in the Appendix. How far must the MW outage table be extended?

8.2 If the probability density function of unsupplied load power for a 1-h interval is $p_n(x)$ and the cumulative distribution is

$$P_n(x) = 1 - \int_0^x p_n(y)\, dy$$

demonstrate, using ordinary calculus, that the unsupplied energy is

$$\int_0^{x_{max}} P_n(y)\, dy$$

where

$$x_{max} = \text{maximum load in the 1-h interval}$$

$$y = \text{dummy variable used to represent the load}$$

Hint: $p_n(x)$ is the probability, or normalized duration, that a load of x MW exists. The energy represented by this load is then $x p_n(x)$. Find the total energy represented by the entire load distribution.

8.3 Complete Table 8.11 for the second unit (i.e., complete the sixth column). Convolve the third unit and determine the data for column 7 $[P_n'''(x)]$ and the energy generation of the third unit and its total cost. Find the distribution of energy to be served over the tie line. If this energy costs 5 R/MWh, what is the cost of this emergency supply and the total cost of production for this 4-wk interval?

8.4 Repeat Example 8C to find the minimum cost dispatch assuming that the fuel for unit 2 has been obtained under a take-or-pay contract and is limited to 4500 MBtu. Emergency energy will be purchased at 50 R/MWh. Find the minimum expected system cost including the cost of emergency energy.

8.5   Repeat the calculation of the system in Section 8.4 using the *expected cost method*. Show the development of the characteristic as each unit is scheduled. Plot the expected cost versus the power output. Check the total cost against the results in Section 8.4.

8.6   Repeat the sample computation of Section 8.4, except assume the input–output characteristic of unit 2 with its ratings have changed to the following.

Output (MW)		Input–Output Cost Curve (R/h)	Forced Outage Rate
Section 1	0–50	$70 + 3.5P_2$	0.1
Section 2	50–60	$245 + 4.5(P_2 - 50)$	0.1

Schedule section 2 of unit 2 after unit 3 and before the emergency energy. Use the techniques of Example 8F and deconvolve section 1 of unit 2 prior to determining the loading on section 2. Repeat the analysis, ignoring the statistical dependence of section 2 on section 1. (That is, schedule a 10-MW "unit" to represent section 2 without deconvolving section 1.)

FURTHER READING

The literature concerning production cost simulations is profuse. A survey of various types of model is contained in reference 1. References 2–4 describe deterministic models designed for long-range planning. Reference 5 provides an entry into the literature of Monte Carlo simulation methods applied to generation planning and production cost computations.

The two texts referred to in references 6 and 7 provide an introduction to the use of probabilistic models and methods for power-generation planning. Reference 8 illustrates the application to a single area. These methods have been extended to consider the effects of transmission interconnections on generation system reliability in references 9–12.

The original probabilistic production cost technique was presented by E. Jamoulle and his associates in a difficult-to-locate Belgian publication (reference 13). The basic methodology has been discussed and illustrated in a number of IEEE papers; references 14–16 are examples.

In many of these articles, the presentation of the probabilistic methodology is couched in a sometimes confusing manner. Where authors such as R. R. Booth and others discuss an "equivalent load distribution," they are referring to the same distribution, $TP_n(x)$, discussed in this chapter. These authors allow the distribution to grow from zero load to some maximum value equal to the sum of the maximum load plus the sum of the capacity on forced outage. We have found this concept difficult to impart and prefer the present presentation. The practical results are identical to those found more commonly in the literature.

The models of approximation using orthogonal expansions to represent capacity distributions have been presented by Stremel and his associates. Reference 17 provides an entry into this literature.

References 15 and 18 lead into the development of the expected production cost method.

References 19–26 contain examples of different approaches to computing probabilistic data and the extension of the methods to different problem areas and generation plant configurations. The last two references are extensions of these techniques to incorporate transmission network. Reference 28 is concerned with unit commitment, but it represents the type of technique that would be useful in shorter-term production cost applications involving transmission-constrained scheduling.

1. Wood, A. J., "Energy Production Cost Models," Symposium on Modeling and Simulation, University of Pittsburgh, April 1972. Published in the Conference Proceedings.

2. Bailey, E. S., Jr., Galloway, C. D., Hawkins, E. S., Wood, A. J., "Generation Planning Programs for Interconnected Systems: Part II, Production Costing Programs," *AIEE Special Supplement*, 1963, pp. 775–788.

3. Brennan, M. K., Galloway, C. D., Kirchmayer, L. K., "Digital Computer Aids Economic-Probabilistic Study of Generation Systems—I," *AIEE Transactions* (Power Apparatus and Systems), Vol. 77, August 1958, pp. 564–571.

4. Galloway, C. D., Kirchmayer, L. K., "Digital Computer Aids Economic-Probabilistic Study of Generation Systems—II," *AIEE Transactions* (Power Apparatus and Systems), Vol. 77, August 1958, pp. 571–577.

5. Dillard, J. K., Sels, H. K., "An Introduction to the Study of System Planning by Operational Gaming Models," *AIEE Transactions* (Power Apparatus and Systems), Vol. 78, Part III, December 1959, pp. 1284–1290.

6. Billinton, R., *Power System Reliability Evaluation*, Gordon and Breach, New York, 1970.

7. Billinton, R., Binglee, R. J., Wood, A. J., *Power System Reliability Calculations*, MIT Press, Cambridge, MA., 1973.

8. Garver, L. L., "Reserve Planning using Outage Probabilities and Load Uncertainties," *IEEE Transactions on Power Apparatus and Systems*, PAS-89, April 1970, pp. 514–521.

9. Cook, V. M., Galloway, C. D., Steinberg, M. J., Wood, A. J., "Determination of Reserve Requirements of Two Interconnected Systems," *AIEE Transactions*, (Power Apparatus and Systems), Vol. 82, Part III, April 1963, pp. 18–33.

10. Bailey, E. S., Jr., Galloway, C. D., Hawkins, E. S., Wood, A. J., "Generation Planning Programs for Interconnected Systems: Part I, Expansion Programs," *AIEE Special Supplement*, 1963, pp. 761–764.

11. Spears, H. T., Hicks, K. L., Lee, S. T., "Probability of Loss of Load for Three Areas," *IEEE Transactions on Power Apparatus and Systems*, Vol. PAS-89, April 1970, pp. 521–527.

12. Pang, C. K., Wood, A. J., "Multi-Area Generation System Reliability Calculations,"

IEEE Transactions on Power Apparatus and Systems, Vol. PAS-94, March/April 1975, pp. 508–517.

13. Baleriaux, H., Jamoulle, E., F. Linard de Guertechin, F., "Simulation de l'Exploitation d'un Parc de Machines Thermiques de Production d'Electricité Couple à des Station de Pompage," *Review E (Edition SRBE)*, Vol. 5, No. 7, 1967, pp. 3–24.

14. Booth, R. R., "Power System Simulation Model Based on Probability Analysis," *IEEE Transactions on Power Apparatus and Systems*, Vol. PAS-91, January/February 1972, pp. 62–69.

15. Sager, M. A., Ringlee, R. J., Wood, A. J., "A New Generation Production Cost Program to Recognize Forced Outages," *IEEE Transactions on Power Apparatus and Systems*, Vol. PAS-91, September/October 1972, pp. 2114–2124.

16. Sager, M. A., Wood, A. J., "Power System Production Cost Calculations—Sample Studies Recognizing Forced Outages," *IEEE Transactions on Power Apparatus and Systems*, PAS-92, January/February 1973, pp. 154–158.

17. Stremel, J. P., Jenkins, R. T., Babb, R. A. Bayless, W. D., "Production Costing using the Cumulant Method of Representing the Equivalent Load Curve," *IEEE Transactions on Power Apparatus and Systems*, Vol. PAS-99, September/October 1980, pp. 1947–1956.

18. Sidenblad, K. M., Lee, S. T. Y., "A Probabilistic Production Costing Methodology for Systems with Storage," *IEEE Transactions on Power Apparatus and Systems*, Vol. PAS-100, June 1981, pp. 3116–3124.

19. Caramania, M., Stremel, J., Fleck, W., Daniel, S., "Probabilistic Production Costing: An Investigation of Alternative Algorithms," *International Journal of Electrical Power and Energy Systems*, Vol. 5, No. 2, 1983, pp. 75–86.

20. Durán, H., "A Recursive Approach to the Cumulant Method of Calculating Reliability and Production Cost," *IEEE Transactions on Power Apparatus and Systems*, Vol. PAS-104, No. 1, January 1985, pp. 82–90.

21. Lee, F. N., "New Multi-Area Production Costing Method," *IEEE Transactions on Power Systems*, Vol. 3, No. 3, August 1988, pp. 915–922.

22. Ansari, S. H., Patton, A. D., "A New Markov Model for Base-Loaded Units for Use in Production Costing," *IEEE Transactions on Power Systems*, Vol. 5, No. 3, August 1990, pp. 797–804.

23. Dohner, C. V., Sager, M. A., Wood, A. J., "Operating Economy Benefits of Improved Gas Turbine Reliability," *IEEE Transactions on Power Systems*, Vol. 4, No. 1, February 1989, pp. 257–263.

24. Lin, M.-Y., Breiphol, A. M., Lee, F. N., "Comparison of Probabilistic Production Cost Simulation Methods," *IEEE Transactions on Power Systems*, Vol. 4, No. 4, November 1989, pp. 1326–1334.

25. Sutanto, D., Outhred, H. R., Lee, Y. B., "Probabilistic Power System Production Cost and Reliability Calculation by the Z-Transform Method," *IEEE Transactions on Energy Conversion*, Vol. 4, No. 4, December 1989, pp. 559–566.

26. Fockens, S., van Wijk, A. J. M., Turkenburg, W. C., Singh, C., "A Concise Method for Calculating Expected Unserved Energy in Generating System Reliability Analysis," *IEEE Transactions on Power Systems*, Vol. 6, No. 3, August 1991, pp. 1085–1091.

27. Pereira, M. V. F., Gorenstin, B. G., Morozowski, F. M., dos Santos, S. J. B.,

"Chronological Probabilistic Production Costing and Wheeling Calculations with Transmission Network Modeling," *IEEE Transactions on Power Systems*, Vol. 7, No. 2, May 1992, pp. 885–891.

28. Shaw, J. J., "A Direct Method for Security-Constrained Unit Commitment," IEEE Paper 94 SM 591-8 PWRS presented at the IEEE Power Engineering Society Meeting, San Francisco, CA, July 24–28, 1994, to be published in the *IEEE Transactions on Power Systems*.

9 Control of Generation

9.1 INTRODUCTION

So far, this text has concentrated on methods of establishing optimum dispatch and scheduling of generating units. It is important to realize, however, that such optimized dispatching would be useless without a method of control over the generator units. Indeed, the control of generator units was the first problem faced in early power-system design. The methods developed for control of individual generators and eventually control of large interconnections play a vital role in modern energy control centers.

A generator driven by a steam turbine can be represented as a large rotating mass with two opposing torques acting on the rotation. As shown in Figure 9.1, T_{mech}, the mechanical torque, acts to increase rotational speed whereas T_{elec}, the electrical torque, acts to slow it down. When T_{mech} and T_{elec} are equal in magnitude, the rotational speed, ω, will be constant. If the electrical load is increased so that T_{elec} is larger than T_{mech}, the entire rotating system will begin to slow down. Since it would be damaging to let the equipment slow down too far, something must be done to increase the mechanical torque T_{mech} to restore equilibrium; that is, to bring the rotational speed back to an acceptable value and the torques to equality so that the speed is again held constant.

This process must be repeated constantly on a power system because the loads change constantly. Furthermore, because there are many generators supplying power into the transmission system, some means must be provided to allocate the load changes to the gnerators. To accomplish this, a series of control systems are connected to the generator units. A governor on each unit maintains its speed while supplementary control, usually originating at a remote control center, acts to allocate generation. Figure 9.2 shows an overview of the generation control problem.

9.2. GENERATOR MODEL

Before starting, it will be useful for us to define our terms.

ω = rotational speed (rad/sec)

α = rotational acceleration

δ = phase angle of a rotating machine

328

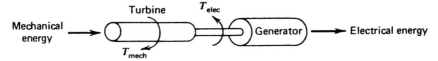

FIG. 9.1   Mechanical and electrical torques in a generating unit.

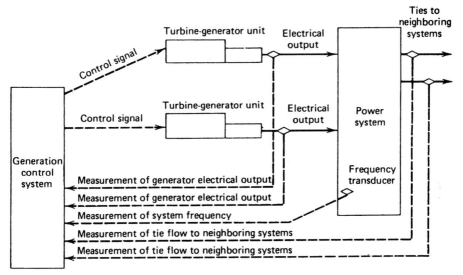

FIG. 9.2   Overview of generation control problem.

T_{net} = net accelerating torque in a machine

T_{mech} = mechanical torque exerted on the machine by the turbine

T_{elec} = electrical torque exerted on the machine by the generator

P_{net} = net accelerating power

P_{mech} = mechanical power input

P_{elec} = electrical power output

I = moment of inertia for the machine

M = angular momentum of the machine

where all quantities (except phase angle) will be in per unit on the machine base, or, in the case of ω, on the standard system frequency base. Thus, for example, M is in per unit power/per unit frequency/sec.

In the development to follow, we are interested in deviations of quantities about steady-state values. All steady-state or nominal values will have a "0" subscript (e.g., ω_0, T_{net_0}), and all deviations from nominal will be designated by a "Δ" (e.g., $\Delta\omega$, ΔT_{net}). Some basic relationships are

$$I\alpha = T_{\text{net}} \tag{9.1}$$

$$M = \omega I \tag{9.2}$$

$$P_{\text{net}} = \omega T_{\text{net}} = \omega(I\alpha) = M\alpha \tag{9.3}$$

To start, we will focus our attention on a single rotating machine. Assume that the machine has a steady speed of ω_0 and phase angle δ_0. Due to various electrical or mechanical disturbances, the machine will be subjected to differences in mechanical and electrical torque, causing it to accelerate or decelerate. We are chiefly interested in the deviations of speed, $\Delta\omega$, and deviations in phase angle, $\Delta\delta$, from nominal.

The phase angle deviation, $\Delta\delta$, is equal to the difference in phase angle between the machine as subjected to an acceleration of α and a reference axis rotating at exactly ω_0. If the speed of the machine under acceleration is

$$\omega = \omega_0 + \alpha t \tag{9.4}$$

then

$$\Delta\delta = \underbrace{\int (\omega_0 + \alpha t)\, dt}_{\substack{\text{Machine absolute} \\ \text{phase angle}}} - \underbrace{\int \omega_0\, dt}_{\substack{\text{Phase angle of} \\ \text{reference axis}}}$$

$$= \omega_0 t + \tfrac{1}{2}\alpha t^2 - \omega_0 t$$

$$= \tfrac{1}{2}\alpha t^2 \tag{9.5}$$

The deviation from nominal speed, $\Delta\omega$, may then be expressed as

$$\Delta\omega = \alpha t = \frac{d}{dt}(\Delta\delta) \tag{9.6}$$

The relationship between phase angle deviation, speed deviation, and net accelerating torque is

$$T_{\text{net}} = I\alpha = I\,\frac{d}{dt}(\Delta\omega) = I\,\frac{d^2}{dt^2}(\Delta\delta) \tag{9.7}$$

Next, we will relate the deviations in mechanical and electrical power to the deviations in rotating speed and mechanical torques. The relationship between net accelerating power and the electrical and mechanical powers is

$$P_{net} = P_{mech} - P_{elec} \qquad (9.8)$$

which is written as the sum of the steady-state value and the deviation term,

$$P_{net} = P_{net_0} + \Delta P_{net} \qquad (9.9)$$

where

$$P_{net_0} = P_{mech_0} - P_{elec_0}$$
$$\Delta P_{net} = \Delta P_{mech} - \Delta P_{elec}$$

Then

$$P_{net} = (P_{mech_0} - P_{elec_0}) + (\Delta P_{mech} - \Delta P_{elec}) \qquad (9.10)$$

Similarly for torques,

$$T_{net} = (T_{mech_0} - T_{elec_0}) + (\Delta T_{mech} - \Delta T_{elec}) \qquad (9.11)$$

Using Eq. 9.3, we can see that

$$P_{net} = P_{net_0} + \Delta P_{net} = (\omega_0 + \Delta \omega)(T_{net_0} + \Delta T_{net}) \qquad (9.12)$$

Substituting Eqs. 9.10 and 9.11, we obtain

$$(P_{mech_0} - P_{elec_0}) + (\Delta P_{mech} - \Delta P_{elec}) = (\omega_0 + \Delta \omega)[(T_{mech_0} - T_{elec_0})$$
$$+ (\Delta T_{mech} - \Delta T_{elec})] \qquad (9.13)$$

Assume that the steady-state quantities can be factored out since

$$P_{mech_0} = P_{elec_0}$$

and

$$T_{mech_0} = T_{elec_0}$$

and further assume that the second-order terms involving products of $\Delta \omega$ with ΔT_{mech} and ΔT_{elec} can be neglected. Then

$$\Delta P_{mech} - \Delta P_{elec} = \omega_0(\Delta T_{mech} - \Delta T_{elec}) \qquad (9.14)$$

As shown in Eq. 9.7, the net torque is related to the speed change as follows:

$$(T_{mech_0} - T_{elec_0}) + (\Delta T_{mech} - \Delta T_{elec}) = I \frac{d}{dt}(\Delta \omega) \qquad (9.15)$$

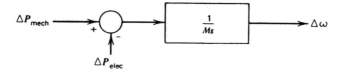

FIG. 9.3 Relationship between mechanical and electrical power and speed change.

then since $T_{mech_0} = T_{elec_0}$, we can combine Eqs. 9.14 and 9.15 to get

$$\Delta P_{mech} - \Delta P_{elec} = \omega_0 I \frac{d}{dt}(\Delta\omega)$$

$$= M \frac{d}{dt}(\Delta\omega) \qquad (9.16)$$

This can be expressed in Laplace transform operator notation as

$$\Delta P_{mech} - \Delta P_{elec} = Ms\,\Delta\omega \qquad (9.17)$$

This is shown in block diagram form in Figure 9.3.

The units for M are watts per radian per second per second. We will always use per unit power over per unit speed per second where the per unit refers to the machine rating as the base (see Example 9A).

9.3 LOAD MODEL

The loads on a power system consist of a variety of electrical devices. Some of them are purely resistive, some are motor loads with variable power–frequency characteristics, and others exhibit quite different characteristics. Since motor loads are a dominant part of the electrical load, there is a need to model the effect of a change in frequency on the net load drawn by the system. The relationship between the change in load due to the change in frequency is given by

$$\Delta P_{L(freq)} = D\,\Delta\omega \quad \text{or} \quad D = \frac{\Delta P_{L(freq)}}{\Delta\omega}$$

where D is expressed as percent change in load divided by percent change in frequency. For example, if load changed by 1.5% for a 1% change in frequency, then D would equal 1.5. However, the value of D used in solving for system dynamic response must be changed if the system base MVA is different from the nominal value of load. Suppose the D referred to here was for a net

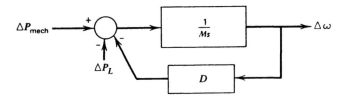

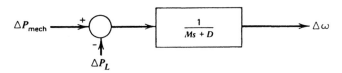

FIG. 9.4   Block diagram of rotating mass and load as seen by prime-mover output.

connected load of 1200 MVA and the entire dynamics problem were to be set up for a 1000-MVA system base. Note that $D = 1.5$ tells us that the load would change by 1.5 pu for 1 pu change in frequency. That is, the load would change by 1.5×1200 MVA or 1800 MVA for a 1 pu change in frequency. When expressed on a 1000-MVA base, D becomes

$$D_{1000\text{-MVA base}} = 1.5 \times \left(\frac{1200}{1000}\right) = 1.8$$

The net change in P_{elec} in Figure 9.3 (Eq. 9.15) is

$$\Delta P_{\text{elec}} = \underbrace{\Delta P_L}_{\substack{\text{Nonfrequency-}\\\text{sensitive load}\\\text{change}}} + \underbrace{D\,\Delta\omega}_{\substack{\text{Frequency-sensitive}\\\text{load change}}} \qquad (9.18)$$

Including this in the block diagram results in the new block diagram shown in Figure 9.4.

EXAMPLE 9A

We are given an isolated power system with a 600-MVA generating unit having an M of 7.6 pu MW/pu frequency/sec on a machine base. The unit is supplying a load of 400 MVA. The load changes by 2% for a 1% change in frequency.

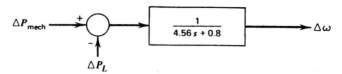

FIG. 9.5   Block diagram for system in Example 9A.

First, we will set up the block diagram of the equivalent generator load system. Everything will be referenced to a 100 MVA base.

$$M = 7.6 \times \frac{600}{1000} = 4.56 \text{ on a 1000-MVA base}$$

$$D = 2 \times \frac{400}{1000} = 0.8 \text{ on a 1000-MVA base}$$

Then the block diagram is as shown in Figure 9.5.

Suppose the load suddenly increases by 10 MVA (or 0.01 pu); that is,

$$\Delta P_L(s) = \frac{0.01}{s}$$

then

$$\Delta\omega(s) = -\frac{0.01}{s}\left(\frac{1}{4.56s + 0.8}\right)$$

or taking the inverse Laplace transform,

$$\Delta\omega(t) = (0.01/0.8)e^{-(0.8/4.56)t} - (0.01/0.8)$$

$$= 0.0125e^{-0.175t} - 0.0125$$

The final value of $\Delta\omega$ is -0.0125 pu, which is a drop of 0.75 Hz on a 60-Hz system.

When two or more generators are connected to a transmission system network , we must take account of the phase angle difference across the network in analyzing frequency changes. However, for the sake of governor analysis, which we are interested in here, we can assume that frequency will be constant over those parts of the network that are tightly interconnected. When making such an assumption, we can then lump the rotating mass of the turbine generators together into an equivalent that is driven by the sum of the individual turbine mechanical outputs. This is illustrated in Figure 9.6 where all turbine generators were lumped into a single equivalent rotating mass, M_{equiv}.

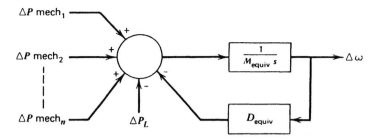

FIG. 9.6 Multi-turbine-generator system equivalent.

Similarly, all individual system loads were lumped into an equivalent load with damping coefficient, D_{equiv}.

9.4 PRIME-MOVER MODEL

The prime mover driving a generator unit may be a steam turbine or a hydroturbine. The models for the prime mover must take account of the steam supply and boiler control system characteristics in the case of a steam turbine, or the penstock characteristics for a hydro turbine. Throughout the remainder of this chapter, only the simplest prime-mover model, the nonreheat turbine, will be used. The models for other more complex prime movers, including hydro turbines, are developed in the references (see Further Reading).

The model for a nonreheat turbine, shown in Figure 9.7, relates the position of the valve that controls emission of steam into the turbine to the power output of the turbine, where

$$T_{CH} = \text{"charging time" time constant}$$

$$\Delta P_{valve} = \text{per unit change in valve position from nominal}$$

The combined prime-mover–generator-load model for a single generating unit can be built by combining Figure 9.4 and 9.7, as shown in Figure 9.8.

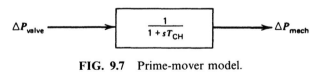

FIG. 9.7 Prime-mover model.

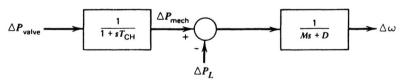

FIG. 9.8 Prime-mover–generator-load model.

9.5 GOVERNOR MODEL

Suppose a generating unit is operated with fixed mechanical power output from the turbine. The result of any load change would be a speed change sufficient to cause the frequency-sensitive load to exactly compensate for the load change (as in Example 9A). This condition would allow system frequency to drift far outside acceptable limits. This is overcome by adding a governing mechanism that senses the machine speed, and adjusts the input valve to change the mechanical power output to compensate for load changes and to restore frequency to nominal value. The earliest such mechanism used rotating "flyballs" to sense speed and to provide mechanical motion in response to speed changes. Modern governors use electronic means to sense speed changes and often use a combination of electronic, mechanical, and hydraulic means to effect the required valve position changes. The simplest governor, called the *isochronous governor*, adjusts the input valve to a point that brings frequency back to nominal value. If we simply connect the output of the speed-sensing mechanism to the valve through a direct linkage, it would never bring the frequency to nominal. To force the frequency error to zero, one must provide what control engineers call reset action. Reset action is accomplished by integrating the frequency (or speed) error, which is the difference between actual speed and desired or reference speed.

We will illustrate such a speed-governing mechanism with the diagram shown in Figure 9.9. The speed-measurement device's output, ω, is compared with a reference, ω_{ref}, to produce an error signal, $\Delta\omega$. The error, $\Delta\omega$, is negated and then amplified by a gain K_G and integrated to produce a control signal, ΔP_{valve}, which causes the main steam supply valve to open (ΔP_{valve} position) when $\Delta\omega$ is negative. If, for example, the machine is running at reference speed and the electrical load increases, ω will fall below ω_{ref} and $\Delta\omega$ will be negative. The action of the gain and integrator will be to open the steam valve, causing the turbine to increase its mechanical output, thereby increasing the electrical

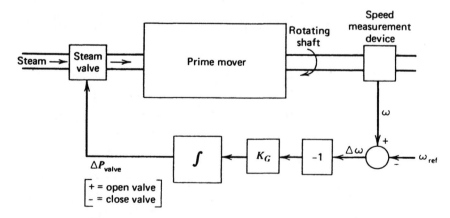

FIG. 9.9 Isochronous governor.

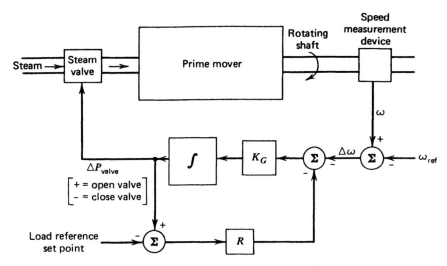

FIG. 9.10   Governor with speed-droop feedback loop.

output of the generator and increasing the speed ω. When ω exactly equals ω_{ref}, the steam valve stays at the new position (further opened) to allow the turbine generator to meet the increased electrical load.

The isochronous (constant speed) governor of Figure 9.9 cannot be used if two or more generators are electrically connected to the same system since each generator would have to have precisely the same speed setting or they would "fight" each other, each trying to pull the system's speed (or frequency) to its own setting. To be able to run two or more generating units in parallel on a generating system, the governors are provided with a feedback signal that causes the speed error to go to zero at different values of generator output.

This can be accomplished by adding a feedback loop around the integrator as shown in Figure 9.10. Note that we have also inserted a new input, called the *load reference*, that we will discuss shortly. The block diagram for this governor is shown in Figure 9.11, where the governor now has a net gain of $1/R$ and a time constant T_G.

The result of adding the feedback loop with gain R is a governor characteristic as shown in Fig. 9.12. The value of R determines the slope of the characteristic. That is, R determines the change on the unit's output for a given change in frequency. Common practice is to set R on each generating unit so that a change from 0 to 100% (i.e., rated) output will result in the same frequency change for each unit. As a result, a change in electrical load on a system will be compensated by generator unit output changes proportional to each unit's rated output.

If two generators with drooping governor characteristics are connected to a power system, there will always be a unique frequency, at which they will share a load change between them. This is illustrated in Figure 9.13, showing two units with drooping characteristics connected to a common load.

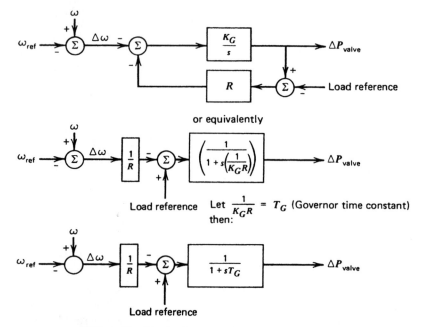

FIG. 9.11 Block diagram of governor with droop.

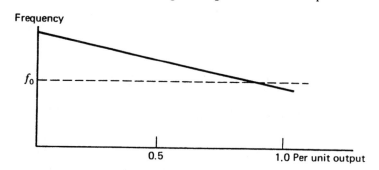

FIG. 9.12 Speed-droop characteristic.

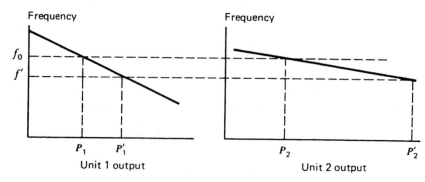

FIG. 9.13 Allocation of unit outputs with governor droop.

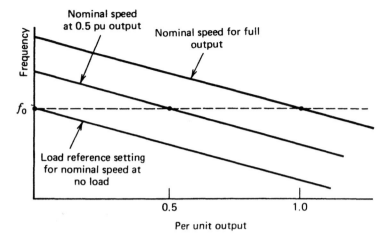

FIG. 9.14 Speed-changer settings.

As shown in Figure 9.13, the two units start at a nominal frequency of f_0. When a load increase, ΔP_L, causes the units to slow down, the governors increase output until the units seek a new, common operating frequency, f'. The amount of load pickup on each unit is proportional to the slope of its droop characteristic. Unit 1 increases its output from P_1 to P'_1, unit 2 increases its output from P_2 to P'_2 such that the net generation increase, $P'_1 - P_1 + P'_2 - P_2$, is equal to ΔP_L. Note that the actual frequency sought also depends on the load's frequency characteristic as well.

Figure 9.10 shows an input labeled "load reference set point." By changing the load reference, the generator's governor characteristic can be set to give reference frequency at any desired unit output. This is illustrated in Figure 9.14. *The basic control input to a generating unit as far as generation control is concerned is the load reference set point.* By adjusting this set point on each unit, a desired unit dispatch can be maintained while holding system frequency close to the desired nominal value.

Note that a steady-state change in ΔP_{valve} of 1.0 pu requires a value of R pu change in frequency, $\Delta\omega$. One often hears unit regulation referred to in percent. For instance, a 3% regulation for a unit would indicate that a 100% (1.0 pu) change in valve position (or equivalently a 100% change in unit output) requies a 3% change in frequency. Therefore, R is equal to pu change in frequency divided by pu change in unit output. That is,

$$R = \frac{\Delta\omega}{\Delta P}\ \text{pu}$$

At this point, we can construct a block diagram of a governor–prime–mover–rotating mass/load model as shown in Figure 9.15. Suppose that this generator

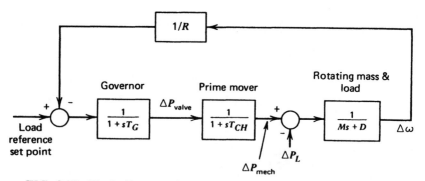

FIG. 9.15   Block diagram of governor, prime mover, and rotating mass.

experiences a step increase in load,

$$\Delta P_L(s) = \frac{\Delta P_L}{s} \tag{9.19}$$

The transfer function relating the load change, ΔP_L, to the frequency change, $\Delta \omega$, is

$$\Delta \omega(s) = \Delta P_L(s) \left[\cfrac{\cfrac{-1}{Ms + D}}{1 + \cfrac{1}{R}\left(\cfrac{1}{1 + sT_G}\right)\left(\cfrac{1}{1 + sT_{CH}}\right)\left(\cfrac{1}{Ms + D}\right)} \right] \tag{9.20}$$

The steady-state value of $\Delta \omega(s)$ may be found by

$$\Delta \omega \text{ steady state} = \lim_{s \to 0} \left[s \, \Delta \omega(s) \right]$$

$$= \cfrac{-\Delta P_L\left(\cfrac{1}{D}\right)}{1 + \left(\cfrac{1}{R}\right)\left(\cfrac{1}{D}\right)} = \cfrac{-\Delta P_L}{\cfrac{1}{R} + D} \tag{9-21}$$

Note that if D were zero, the change in speed would simply be

$$\Delta \omega = -R \, \Delta P_L \tag{9.22}$$

If several generators (each having its own governor and prime mover) were connected to the system, the frequency change would be

$$\Delta \omega = \cfrac{-\Delta P_L}{\cfrac{1}{R_1} + \cfrac{1}{R_2} + \cdots + \cfrac{1}{R_n} + D} \tag{9.23}$$

9.6   TIE-LINE MODEL

The power flowing across a transmission line can be modeled using the DC load flow method shown in Chapter 4.

$$P_{\text{tie flow}} = \frac{1}{X_{\text{tie}}} (\theta_1 - \theta_2) \tag{9.24}$$

This tie flow is a steady-state quantity. For purposes of analysis here, we will perturb Eq. 9.24 to obtain deviations from nominal flow as a function of deviations in phase angle from nominal.

$$P_{\text{tie flow}} + \Delta P_{\text{tie flow}} = \frac{1}{X_{\text{tie}}} [(\theta_1 + \Delta\theta_1) - (\theta_2 + \Delta\theta_2)]$$

$$= \frac{1}{X_{\text{tie}}} (\theta_1 - \theta_2) + \frac{1}{X_{\text{tie}}} (\Delta\theta_1 - \Delta\theta_2) \tag{9.25}$$

Then

$$\Delta P_{\text{tie flow}} = \frac{1}{X_{\text{tie}}} (\Delta\theta_1 - \Delta\theta_2) \tag{9.26}$$

where $\Delta\theta_1$ and $\Delta\theta_2$ are equivalent to $\Delta\delta_1$ and $\Delta\delta_2$ as defined in Eq. 9.6. Then, using the relationship of Eq. 9.6,

$$\Delta P_{\text{tie flow}} = \frac{T}{s} (\Delta\omega_1 - \Delta\omega_2) \tag{9.27}$$

where $T = 377 \times 1/X_{\text{tie}}$ (for a 60-Hz system).

Note that $\Delta\theta$ must be in radians for ΔP_{tie} to be in per unit megawatts, but $\Delta\omega$ is in per unit speed change. Therefore, we must multiply $\Delta\omega$ by 377 rad/sec (the base frequency in rad/sec at 60 Hz). T may be thought of as the "tie-line stiffness" coefficient.

Suppose now that we have an interconnected power system broken into two areas each having one generator. The areas are connected by a single transmission line. The power flow over the transmission line will appear as a a positive load to one area and an equal but negative load to the other, or vice versa, depending on the direction of flow. The direction of flow will be dictated by the relative phase angle between the areas, which is determined by the relative speed deviations in the areas. A block diagram representing this interconnection can be drawn as in Figure 9.16. Note that the tie power flow was defined as going from area 1 to area 2; therefore, the flow appears as a load to area 1 and a power source (negative load) to area 2. If one assumes that mechanical powers are constant, the rotating masses and tie line exhibit damped oscillatory characteristics known as synchronizing oscillations. (See problem 9.3 at the end of this chapter.)

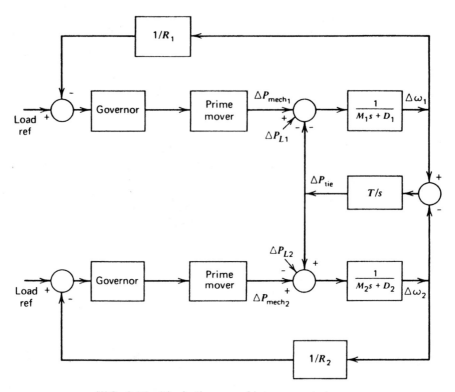

FIG. 9.16   Block diagram of interconnected areas.

It is quite important to analyze the steady-state frequency deviation, tie-flow deviation, and generator outputs for an interconnected area after a load change occurs. Let there be a load change ΔP_{L_1} in area 1. In the steady state, after all synchronizing oscillations have damped out, the frequency will be constant and equal to the same value on both areas. Then

$$\Delta \omega_1 = \Delta \omega_2 = \Delta \omega \quad \text{and} \quad \frac{d(\Delta \omega_1)}{dt} = \frac{d(\Delta \omega_2)}{dt} = 0 \qquad (9.28)$$

and

$$\Delta P_{mech_1} - \Delta P_{tie} - \Delta P_{L_1} = \Delta \omega D_1$$

$$\Delta P_{mech_2} + \Delta P_{tie} = \Delta \omega D_2$$

$$\Delta P_{mech_1} = \frac{-\Delta \omega}{R_1} \qquad (9.29)$$

$$\Delta P_{mech_2} = \frac{-\Delta \omega}{R_2}$$

By making appropriate substitutions in Eq. 9.29,

$$-\Delta P_{\text{tie}} - \Delta P_{L_1} = \Delta\omega\left(\frac{1}{R_1} + D_1\right)$$
$$+\Delta P_{\text{tie}} = \Delta\omega\left(\frac{1}{R_2} + D_2\right) \tag{9.30}$$

or, finally

$$\Delta\omega = \frac{-\Delta P_{L_1}}{\dfrac{1}{R_1} + \dfrac{1}{R_2} + D_1 + D_2} \tag{9.31}$$

from which we can derive the change in tie flow:

$$\Delta P_{\text{tie}} = \frac{-\Delta P_{L_1}\left(\dfrac{1}{R_2} + D_2\right)}{\dfrac{1}{R_1} + \dfrac{1}{R_2} + D_1 + D_2} \tag{9.32}$$

Note that the conditions described in Eqs. 9.28 through 9.32 are for the new steady-state conditions after the load change. The new tie flow is determined by the net change in load and generation in each area. We do not need to know the tie stiffness to determine this new tie flow, although the tie stiffness will determine how much difference in phase angle across the tie will result from the new tie flow.

EXAMPLE 9B

You are given two system areas connected by a tie line with the following characteristics.

Area 1	Area 2
$R = 0.01$ pu	$R = 0.02$ pu
$D = 0.8$ pu	$D = 1.0$ pu
Base MVA $= 500$	Base MVA $= 500$

A load change of 100 MW (0.2 pu) occurs in area 1. What is the new steady-state frequency and what is the change in tie flow? Assume both areas were at nominal frequency (60 Hz) to begin.

$$\Delta\omega = \frac{-\Delta P_{L_1}}{\dfrac{1}{R_1} + \dfrac{1}{R_2} + D_1 + D_2} = \frac{-0.2}{\dfrac{1}{0.01} + \dfrac{1}{0.02} + 0.8 + 1} = -0.00131752 \text{ pu}$$

$$f_{new} = 60 - 0.00132(60) = 59.92 \text{ Hz}$$

$$\Delta P_{tie} = \Delta\omega\left(\frac{1}{R_2} + D_2\right) = -0.00131752\left(\frac{1}{0.02} + 1\right) = -0.06719368 \text{ pu}$$

$$= -33.6 \text{ MW}$$

The change in prime-mover power would be

$$\Delta P_{mech_1} = \frac{-\Delta\omega}{R_1} = -\left(\frac{-0.00131752}{0.01}\right) = 0.13175231 \text{ pu} = 65.876 \text{ MW}$$

$$\Delta P_{mech_2} = \frac{-\Delta\omega}{R_2} = -\left(\frac{-0.00131752}{0.02}\right) = 0.06587615 \text{ pu} = 32.938 \text{ MW}$$

$$= 98.814 \text{ MW}$$

The total changes in generation is 98.814 MA, which is 1.186 MW short of the 100 MW load change. The change in total area load due to frequency drop would be

For area $1 = \Delta\omega D_1 = -0.0010540 \text{ pu} = -0.527 \text{ MW}$

For area $2 = \Delta\omega D_2 = -0.00131752 \text{ pu} = -0.6588 \text{ MW}$

Therefore, the total load change is $= 1.186 \text{ MW}$, which accounts for the difference in total generation change and total load change. (See Problem 9.2 for further variations on this problem.)

If we were to analyze the dynamics of the two-area systems, we would find that a step change in load would always result in a frequency error. This is illustrated in Figure 9.17, which shows the frequency response of the system to a step-load change. Note that Figure 9.17 only shows the average frequency (omitting any high-frequency oscillations).

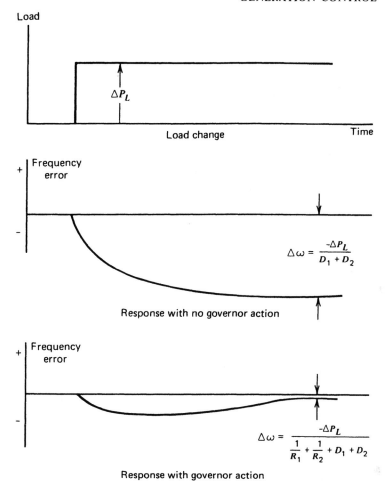

FIG. 9.17   Frequency response to load change.

9.7   GENERATION CONTROL

Automatic generation control (AGC) is the name given to a control system having three major objectives:

1. To hold system frequency at or very close to a specified nominal value (e.g., 60 Hz).
2. To maintain the correct value of interchange power between control areas.
3. To maintain each unit's generation at the most economic value.

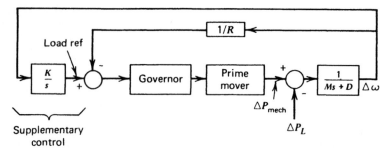

FIG. 9.18 Supplementary control added to generating init.

9.7.1 Supplementary Control Action

To understand each of the three objectives just listed, we may start out assuming that we are studying a single generating unit supplying load to an isolated power system. As shown in Section 9.5, a load change will produce a frequency change with a magnitude that depends on the droop characteristics of the governor and the frequency characteristics of the system load. Once a load change has occurred, a supplementary control must act to restore the frequency to nominal value. This can be accomplished by adding a reset (integral) control to the governor, as shown in Figure 9.18.

The reset control action of the supplementary control will force the frequency error to zero by adjustment of the speed reference set point. For example, the error shown in the bottom diagram of Figure 9.17 would be forced to zero.

9.7.2 Tie-Line Control

When two utilities interconnect their systems, they do so for several reasons. One is to be able to buy and sell power with neighboring systems whose operating costs make such transactions profitable. Further, even if no power is being transmitted over ties to neighboring systems, if one system has a sudden loss of a generating unit, the units throughout all the interconnection will experience a frequency change and can help in restoring frequency.

Interconnections present a very interesting control problem with respect to allocation of generation to meet load. The hypothetical situation in Figure 9.19 will be used to illustrate this problem. Assume both systems in Figure 9.19 have equal generation and load characteristics ($R_1 = R_2$, $D_1 = D_2$) and, further, assume system 1 was sending 100 MW to system 2 under an interchange agreement made between the operators of each system. Now, let system 2 experience a sudden load increase of 30 MW. Since both units have equal generation characteristics, they will both experience a 15 MW increase, and the tie line will experience an increase in flow from 100 MW to 115 MW. Thus, the 30 MW load increase in system 2 will have been satisfied by a 15 MW increase in generation in system 2, plus a 15 MW increase in tie flow into system 2. This

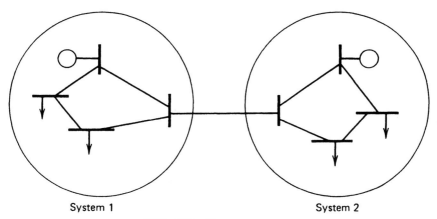

FIG. 9.19   Two-area system.

would be fine, except that system 1 contracted to sell only 100 MW, not 115 MW, and its generating costs have just gone up without anyone to bill the extra cost to. What is needed at this point is a control scheme that recognizes the fact that the 30 MW load increase occurred in system 2 and, therefore, would increase generation in system 2 by 30 MW while restoring frequency to nominal value. It would also restore generation in system 1 to its output before the load increase occurred.

Such a control system must use two pieces of information: the system frequency and the net power flowing in or out over the tie lines. Such a control scheme would, of necessity, have to recognize the following.

1. If frequency decreased and net interchange power leaving the system increased, a load increase has occurred outside the system.
2. If frequency decreased and net interchange power leaving the system decreased, a load increase has occurred inside the system.

This can be extended to cases where frequency increases. We will make the following definitions.

$$P_{\text{net int}} = \text{total actual net interchange}$$
$$(+ \text{ for power leaving the system}; - \text{ for power entering})$$

$$P_{\text{net int sched}} = \text{scheduled or desired value of interchange} \qquad (9.33)$$

$$\Delta P_{\text{net int}} = P_{\text{net int}} - P_{\text{net int sched}}$$

Then, a summary of the tie-line frequency control scheme can be given as in the table in Figure 9.20.

$\Delta\omega$	$\Delta P_{\text{net int}}$	Load change		Resulting control action
$-$	$-$	ΔP_{L_1}	$+$	Increase P_{gen} in system 1
		ΔP_{L_2}	0	
$+$	$+$	ΔP_{L_1}	$-$	Decrease P_{gen} in system 1
		ΔP_{L_2}	0	
$-$	$+$	ΔP_{L_1}	0	Increase P_{gen} in system 2
		ΔP_{L_2}	$+$	
$+$	$-$	ΔP_{L_1}	0	Decrease P_{gen} in system 2
		ΔP_{L_2}	$-$	

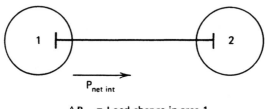

ΔP_{L_1} = Load change in area 1

ΔP_{L_2} = Load change in area 2

FIG. 9.20   Tie-line frequency control actions for two-area system.

We define a *control area* to be a part of an interconnected system within which the load and generation will be controlled as per the rules in Figure 9.20. The control area's boundary is simply the tie-line points where power flow is metered. All tie lines crossing the boundary must be metered so that total control area net interchange power can be calculated.

The rules set forth in Figure 9.20 can be implemented by a control mechanism that weighs frequency deviation, $\Delta\omega$, and net interchange power, $\Delta P_{\text{net int}}$. The frequency response and tie flows resulting from a load change, ΔP_{L_1}, in the two-area system of Figure 9.16 are derived in Eqs. 9.28 through 9.32. These results are repeated here.

Load Change	Frequency Change	Change in Net Interchange
ΔP_{L_1}	$\Delta\omega = \dfrac{-\Delta P_{L_1}}{\dfrac{1}{R_1} + \dfrac{1}{R_2} + D_1 + D_2}$	$\Delta P_{\text{net int}_1} = \dfrac{-\Delta P_{L_1}\left(\dfrac{1}{R_2} + D_2\right)}{\dfrac{1}{R_1} + \dfrac{1}{R_2} + D_1 + D_2}$

$$(9.34)$$

This corresponds to the first row of the table in Figure 9.20; we would therefore require that

$$\Delta P_{\text{gen}_1} = \Delta P_{L_1}$$
$$\Delta P_{\text{gen}_2} = 0$$

The required change in generation, historically called the *area control error* or ACE, represents the shift in the area's generation required to restore frequency and net interchange to their desired values. The equaions for ACE for each area are

$$ACE_1 = -\Delta P_{\text{net int}_1} - B_1 \, \Delta\omega$$
$$ACE_2 = -\Delta P_{\text{net int}_2} - B_2 \, \Delta\omega$$

(9.35)

where B_1 and B_2 are called *frequency bias factors*. We can see from Eq. 9.34 that setting bias factors as follows:

$$B_1 = \left(\frac{1}{R_1} + D_1 \right)$$
$$B_2 = \left(\frac{1}{R_2} + D_2 \right)$$

(9.36)

results in

$$ACE_1 = \left(\frac{+\Delta P_{L_1}\left(\dfrac{1}{R_2} + D_2\right)}{\dfrac{1}{R_1} + \dfrac{1}{R_2} + D_1 + D_2} \right) - \left(\frac{1}{R_1} + D_1 \right)\left(\frac{-\Delta P_{L_1}}{\dfrac{1}{R_1} + \dfrac{1}{R_2} + D_1 + D_2} \right) = \Delta P_{L_1}$$

$$ACE_2 = \left(\frac{-\Delta P_{L_1}\left(\dfrac{1}{R_2} + D_2\right)}{\dfrac{1}{R_1} + \dfrac{1}{R_2} + D_1 + D_2} \right) - \left(\frac{1}{R_2} + D_2 \right)\left(\frac{-\Delta P_{L_1}}{\dfrac{1}{R_1} + \dfrac{1}{R_2} + D_1 + D_2} \right) = 0$$

This control can be carried out using the scheme outlined in Figure 9.21. Note that the values of B_1 and B_2 would have to change each time a unit was committed or decommitted, in order to have the exact values as given in Eq. 9.36. Actually, the integral action of the supplementary controller will guarantee a reset of ACE to zero even when B_1 and B_2 are in error.

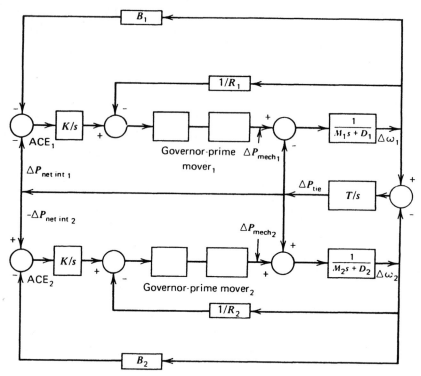

FIG. 9.21 Tie-line bias supplementary control for two areas.

9.7.3 Generation Allocation

If each control area in an interconnected system had a single generating unit, the control system of Figure 9.21 would suffice to provide stable frequency and tie-line interchange. However, power systems consist of control areas with many generating units with outputs that must be set according to economics. That is, we must couple an economic dispatch calculation to the control mechanism so it will know how much of each area's total generation is required from each individual unit.

One must remember that a particular total generation value will not usually exist for a very long time, since the load on a power system varies continually as people and industries use individual electric loads. Therefore, it is impossible to simply specify a total generation, calculate the economic dispatch for each unit, and then give the control mechanism the values of megawatt output for each unit—unless such a calculation can be made very quickly. Until the widespread use of digital computer-based control systems, it was common practice to construct control mechanisms such as we have been describing using analog computers. Although analog computers are not generally proposed for new control-center installations today, there are some in active use. An analog computer can provide the economic dispatch and allocation of generation in an area on an instantaneous basis through the use of function generators set to equal the units' incremental heat rate curves. B matrix loss formulas were also incorporated into analog schemes by setting the matrix coefficients on precision potentiometers.

When using digital computers, it is desirable to be able to carry out the economic-dispatch calculations at intervals of one to several minutes. Either the output of the economic dispatch calculation is fed to an analog computer (i.e., a "digitally directed analog" control system) or the output is fed to another program in the computer that executes the control functions (i.e., a "direct digital" control system). Whether the control is analog or digital, the allocation of generation must be made instantly when the required area total generation changes. Since the economic-dispatch calculation is to be executed every few minutes, a means must be provided to indicate how the generation is to be allocated for values of total generation other than that used in the economic-dispatch calculation.

The allocation of individual generator output over a range of total generation values is accomplished using base points and participation factors. The economic-dispatch calculation is executed with a total generation equal to the sum of the present values of unit generation as measured. The result of this calculation is a set of base-point generations, $P_{i_{\text{base}}}$, which is equal to the most economic output for each generator unit. The rate of change of each unit's output with respect to a change in total generation is called the unit's *participation factor*, *pf* (see Section 3.8 and Example 3I in Chapter 3). The base point and participation factors are used as follows

$$P_{i_{\text{des}}} = P_{i_{\text{base}}} + pf_i \times \Delta P_{\text{total}} \tag{9.37}$$

where

$$\Delta P_{\text{total}} = P_{\text{new total}} - \sum_{\substack{\text{all} \\ \text{gen}}} P_{i_{\text{base}}} \tag{9.38}$$

and

$$P_{i_{\text{des}}} = \text{new desired output from unit } i$$

$$P_{i_{\text{base}}} = \text{base-point generation for unit } i$$

$$pf_i = \text{participation factor for unit } i$$

$$\Delta P_{\text{total}} = \text{change in total generation}$$

$$P_{\text{new total}} = \text{new total generation}$$

Note that by definition (e.g., see Eq. 3.35) the participation factors must sum to unity. In a direct digital control scheme, the generation allocation would be made by running a computer code that was programmed to execute according to Eqs. 9.37 and 9.38.

9.7.4 Automatic Generation Control (AGC) Implementation

Modern implementation of automatic generation control (AGC) schemes usually consists of a central location where information pertaining to the system is telemetered. Control actions are determined in a digital computer and then transmitted to the generation units via the same telemetry channels. To implement an AGC system, one would require the following information at the control center.

1. Unit megawatt output for each committed unit.
2. Megawatt flow over each tie line to neighboring systems.
3. System frequency.

The output of the execution of an AGC program must be transmitted to each of the generating units. Present practice is to transmit raised or lower pulses of varying lengths to the unit. Control equipment then changes the unit's load reference set point up or down in proportion to the pulse length. The "length" of the control pulse may be encoded in the bits of a digital word that is transmitted over a digital telemetry channel. The use of digital telemetry is becoming commonplace in modern systems wherein supervisory control (opening and closing substation breakers), telemetry information (measurements of MW, MVAR, MVA voltage, etc.) and control information (unit raise/lower) is all sent via the same channels.

The basic reset control loop for a unit consists of an integrator with gain K as shown in Figure 9.22. The control loop is implemented as shown in Figure 9.23. The P_{des} control input used in Figures 9.22 and 9.23 is a function of system frequency deviation, net interchange error, and each unit's deviation from its scheduled economic output.

The overall control scheme we are going to develop starts with ACE, which is a measure of the error in total generation from total desired generation. ACE is calculated according to Figure 9.24. ACE serves to indicate when total generation must be raised or lowered in a control area. However, ACE is not the only error signal that must "drive" our controller. The individual units

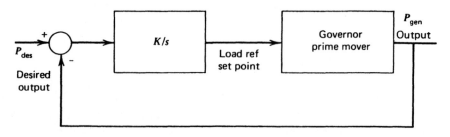

FIG. 9.22 Basic generation control loop.

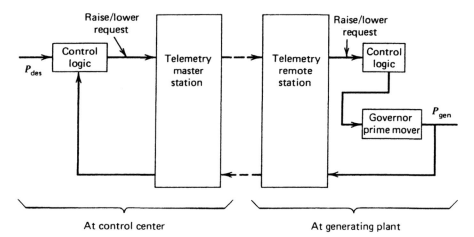

FIG. 9.23   Basic generation control loop via telemetry.

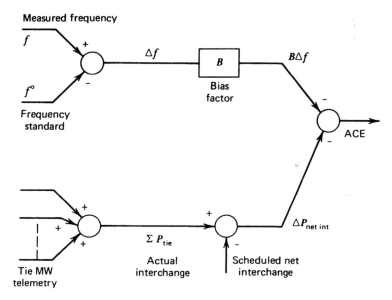

FIG. 9.24   ACE calculation.

may deviate from the economic output as determined by the base point and participation-factor calculation.

The AGC control logic must also be driven by the errors in unit output so as to force the units to obey the economic dispatch. To do this, the sum of the unit output errors is added to ACE to form a composite error signal that drives the entire control system. Such a control system is shown schematically in Figure 9.25, where we have combined the ACE calculation, the generation allocation calculation, and the unit control loop.

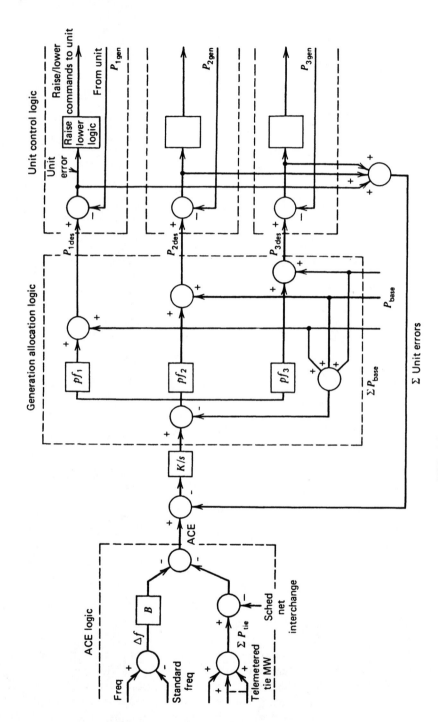

FIG. 9.25 Overview of AGC logic.

Investigation of Figure 9.25 shows an overall control system that will try to drive ACE to zero as well as driving each unit's output to its required economic value. Readers are cautioned that there are many variations to the control execution shown in Figure 9.25. This is especially true of digital implementations of AGC where great sophistication can be programmed into an AGC computer code.

Often the question is asked as to what constitutes "good" AGC design. This is difficult to answer, other than in a general way, since what is "good" for one system may be different in another. Three general criteria can be given.

1. The ACE signal should ideally be kept from becoming too large. Since ACE is directly influenced by random load variations, this criterion can be treated statistically by saying that the standard deviation of ACE should be small.

2. ACE should not be allowed to "drift." This means that the integral of ACE over an appropriate time should be small. "Drift" in ACE has the effect of creating system time errors or what are termed *inadvertent interchange errors*.

3. The amount of control action called for by the AGC should be kept to a minimum. Many of the errors in ACE, for example, are simply random load changes that need not cause control action. Trying to "chase" these random load variations will only wear out the unit speed-changing hardware.

To achieve the objectives of good AGC, many features are added, as described briefly in the next section.

9.7.5   AGC Features

This section will serve as a simple catalog of some of the features that can be found in most AGC systems.

Assist action: Often the incremental heat rate curves for generating units will give trouble to an AGC when an excessive ACE occurs. If one unit's participation factor is dominant, it will take most of the control action and the other units will remain relatively fixed. Although it is the proper thing to do as far as economics are concerned, the one unit that is taking all the action will not be able to change its output fast enough when a large ACE calls for a large change in generation. The assist logic then comes into action by moving more of the units to correct ACE. When the ACE is corrected, the AGC then restores the units back to economic output.

Filtering of ACE: As indicated earlier, much of the change in ACE may be random noise that need not be "chased" by the generating units. Most

AGC programs use elaborate, adaptive nonlinear filtering schemes to try to filter out random noise from true ACE deviations that need control action.

Telemetry failure logic: Logic must be provided to insure that the AGC will not take wrong action when a telemetered value it is using fails. The usual design is to suspend all AGC action when this condition happens.

Unit control detection: Sometimes a generating unit will not respond to raised lower pulses. For the sake of overall control, the AGC ought to take this into account. Such logic will detect a unit that is not following raised/lower pulses and suspend control to it, thereby causing the AGC to reallocate control action among the other units on control.

Ramp control: Special logic allows the AGC to ramp a unit form one output to another at a specified rate of change in output. This is most useful in bringing units on-line and up to full output.

Rate limiting: All AGC designs must account for the fact that units cannot change their output too rapidly. This is especially true of thermal units where mechanical and thermal stresses are limiting. The AGC must limit the rate of change such units will be called on to undergo during fast load changes.

Unit control modes: Many units in power systems are not under full AGC control. Various special control modes must be provided such as manual, base load, and base load and regulating. For example, base load and regulating units are held at their base load value—but are allowed to move as assist action dictates, and are then restored to base-load value.

PROBLEMS

9.1 Suppose that you are given a single area with three generating units as shown in Figure 9.26.

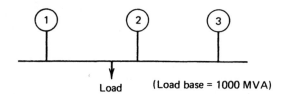

Load (Load base = 1000 MVA)

FIG. 9.26 Three-generator system for Problem 9.1.

Unit	Rating (MVA)	Speed Droop R (per unit on unit base)
1	100	0.01
2	500	0.015
3	500	0.015

The units are initially loaded as follows:

$$P_1 = \ \ 80 \text{ MW}$$

$$P_2 = 300 \text{ MW}$$

$$P_3 = 400 \text{ MW}$$

Assume $D = 0$; what is the new generation on each unit for a 50-MW load increase? Repeat with $D = 1.0$ pu (i.e., 1.0 pu on load base). Be careful to convert all quantities to a common base when solving.

9.2 Using the values of R and D in each area, for Example 9B, resolve for the 100-MW load change in area 1 under the following conditions:

$$\text{Area 1:} \quad \text{base MVA} = 2000 \text{ MVA}$$

$$\text{Area 2:} \quad \text{base MVA} = \ \ 500 \text{ MVA}$$

Then solve for a load change of 100 MW occurring in area 2 with R values and D values as in Example 9B and base MVA for each area as before.

9.3 Given the block diagram of two interconnected areas shown in Figure 9.27 (consider the prime-mover output to be constant, i.e., a "blocked" governor):

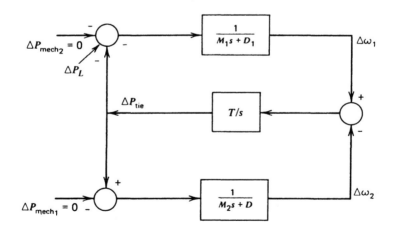

FIG. 9.27   Two-area system for Problem 9.3

a. Derive the transfer functions that relate $\Delta\omega_1(s)$ and $\Delta\omega_2(s)$ to a load change $\Delta P_L(s)$.

b. For the following data (all quantities refer to a 1000-MVA base),

$$M_1 = 3.5 \text{ pu} \qquad D_1 = 1.00$$
$$M_2 = 4.0 \text{ pu} \qquad D_2 = 0.75$$
$$T = 377 \times 0.02 \text{ pu} = 7.54 \text{ pu}$$

calculate the final frequency for load-step change in area 1 of 0.2 pu (i.e., 200 MW). Assume frequency was at nominal and tie flow was 0 pu.

c. Derive the transfer function relating tie flow, $\Delta P_{\text{tie}}(s)$ to $\Delta P_L(s)$. For the data of part b calculate the frequency of oscillation of the tie power flow. What happens to this frequency as tie stiffness increases (i.e. $T \to \infty$)?

9.4 Given two generating units with data as follows.

Unit 1: Fuel cost: $F_1 = 1.0 \text{ R/MBtu}$

$$H_1(P_1) = 500 + 7P_1 + 0.002P_1^2 \text{ MBtu/h}$$

$$150 < P_1 < 600 \quad \text{Rate limit} = 2 \text{ MW/min}$$

Unit 2: Fuel cost: $F_2 = 0.98 \text{ R/MBtu}$

$$H_2(P_2) = 200 + 8P_2 + 0.0025P_2^2 \text{ MBtu/h}$$

$$125 \le P_2 \le 500 \text{ MW} \quad \text{Rate limit} = 2 \text{ MW/min}$$

a. Calculate the economic base points and participation factors for these two units supplying 500 MW total. Use Eq. 3.35 to calculate participation factors.

b. Assume a load change of 10 MW occurs and that we wish to clear the ACE to 0 in 5 min. Is this possible if the units are to be allocated by base points and participation factors?

c. Assume the same load change as in part b, but assume that the rate limit on unit 1 is now 0.5 MW/min.

This problem demonstrates the flaw in using Eq. 3.35 to calculate the participation factors. An alternate procedure would generate participation factors as follows.

Let t be the time in minutes between economic-dispatch calculation executions. Each unit will be assigned a range that must be obeyed in performing the economic dispatch.

$$P_i^{\max} = P_i^0 + t \times \text{rate limit}_i$$
$$P_i^{\min} = P_i^0 - t \times \text{rate limit}_i \tag{9.39}$$

The range thus defined is simply the maximum and minimum excursion the unit could undergo within t minutes. If one of the limits described is outside the unit's normal economic limits, the economic limit would be used. Participation factors can then be calculated by resolving the economic dispatch at a higher value and enforcing the new limits described previously.

d. Assume $T = 5$ min and that the perturbed economic dispatch is to be resolved for 510 MW. Calculate the new participation factors as

$$pf_i = \frac{P_i^\Delta - P_{i_{\text{base pt}}}}{\Delta P_{\text{total}}}$$

where

$$P_{i_{\text{base pt}}} = \text{base economic solution}$$

$$P_{1_{\text{base}}} + P_{2_{\text{base}}} = 500 \text{ MW}$$

$$P_i^\Delta = \text{perturbed solution}$$

$$P_1^\Delta + P_2^\Delta = 510 \text{ MW}$$

with limits as calculated in Eq. 9.35.

Assume the initial unit generations P_i^0 were the same as the base points found in part a. And assume the rate limits were as in part c (i.e., unit 1 rate lim = 0.5 MW/ min, unit 2 rate lim = 2 MW/min). Now check to see if 1part c gives a different result.

9.5 The interconnected systems in the eastern United States and Canada have a total capacity of about 5×10^5 MW. The equivalent inertia and damping constants are approximately

$$M = 8 \text{ pu MW/pu frequency/sec}$$

and

$$D = 1.5$$

both on the system capacity base. It is necessary to correct for time errors every so often. The electrical energy involved is not insignificant.

a. Assume that a time error of 1 sec is to be corrected by deliberately supplying a power unbalance of a constant amount for a period of 1 h. Find the power unbalance required. Express the amount in MWH.

b. Is this energy requirement a function of the power unbalance? Assume a power unbalance is applied to the system of a duration "delta T". During this period, the unbalance of power is constant; after the period it is zero. Does it make any difference if the length of time is long or short? Show the response of the system. The time deviation is the integral of the frequency deviation.

9.6 In Fig. 9.16 assume that system 2 represents a system so large that it is effectively an "infinite bus." M_2 is much greater than M_1 and the frequency deviation in system 2 is zero.

    a. Draw the block diagram including the tie line between areas 1 and 2. What is the transfer function for a load change in area 1 and the tie flow?

    b. The reactance of the tie is 1 pu on a 1000-MW base. Initially, the tie flow is zero. System 1 has an inertia constant (M_1) of 10 on the same base. Load damping and governor action are neglected. Determine the equation for the tie-line power flow swings for a sudden short in area 1 that causes an instantaneous power drop of 0.02 pu (2%), which is restored instantly. Assume that $\Delta P_{L_i}(s) = -0.02$, and find the frequency of oscillation and maximum angular deviation between areas 1 and 2.

FURTHER READING

The reader should be familiar with the basics of control theory before attempting to read many of the references cited here. A good introduction to automatic generation control is the book *Control of Generation and Power Flow on Interconnected Systems*, by Nathan Cohn (reference 4 in Chapter 1). Other sources of introductory material are contained in references 1–3.

Descriptions of how steam turbine generators are modeled are found in references 4 and 5; reference 6 shows how hydro-units can be modeled. Reference 7 shows the effects to be expected from various prime-mover and governing systems. References 8–10 are representative of advances made in AGC techniques through the late 1960s and early 1970s. Other special interests in AGC design include special-purpose optimal filters (see references 10 and 11), direct digital control schemes (see references 12–15), and control of jointly owned generating units (see reference 16).

Research in control theory toward "optimal control" techniques was used in several papers presented in the late 1960s and early 1970s. As far as is known to the authors, optimal control techniques have not, as of the writing of this text, been utilized successfully in a working AGC system. Reference 17 is representative of the papers using optimal control theory.

Recent research has included an approach that takes the short-term load forecast, economic dispatch, and AGC problems, and approaches them as one overall control problem. References 18 and 19 illustrate this approach. References 20–22 are excellent overviews of more recent work in AGC.

1. Friedlander, G. D., "Computer-Controlled Power Systems, Part I—Boiler-Turbine Unit Controls," *IEEE Spectrum*, April 1965, pp. 60–81.

2. Friedlander, G. D., "Computer-Controlled Power Systems, Part II—Area Controls and Load Dispatch," *IEEE Spectrum*, May 1965, pp. 72–91.

3. Ewart, D. N., "Automatic Generation Control—Performance Under Normal Conditions," *Systems Engineering for Power: Status and Prospects*, U. S. Government Document, CONF-750867, 1975, pp. 1–14.

4. Anderson, P. M., *Modeling Thermal Power Plants for Dynamic Stability Studies*, Cyclone Copy Center, Ames, IA, 1974.

5. IEEE Committee Report, "Dynamic Models for Steam and Hydro Turbines in Power System Studies," *IEEE Transactions on Power Apparatus and Systems*, Vol. PAS-92, November/December 1973, pp. 1904–1915.

6. Undril, J. M., Woodward, J. L., "Nonlinear Hydro Governing Model and Improved Calculation for Determining Temporary Droop," *IEEE Transactions on Power Apparatus and Systems*, Vol. PAS-86, April 1967, pp, 443–453.

7. Concordia, C., Kirchmayer, L. K., de Mello, F. P., Schulz, R. P., "Effect of Prime-Mover Response and Governing Characteristics on System Dynamic Performance," *Proceedings of the American Power Conference*, 1966.

8. Cohn, N., "Considerations in the Regulation of Interconnected Areas," *IEEE Transactions on Power Apparatus and Systems*, Vol. PAS-86, December 1967, pp. 1527–1538.

9. Cohn N. "Techniques for Improving the Control of Bulk Power Transfers on Interconnected Systems," *IEEE Transactions on Power Apparatus and Systems*, Vol. PAS-90, November/December 1971, pp. 2409–2419.

10. Cooke, J. L., "Analysis of Power System's Power-Density Spectra," *IEEE Transactions on Power Apparatus and Systems*, Vol. PAS-83, January 1964, pp. 34–41.

11. Ross, C. W., "Error Adaptive Control Computer for Interconnected Power Systems," *IEEE Transactions on Power Apparatus and Systems*, Vol. PAS-85, July 1966, pp. 742–749.

12. Ross, C. W., "A Comprehensive Direct Digital Load-Frequency Controller," *IEEE Power Industry Computer Applications Conference Proceedings*, 1967.

13. Ross, C. W., Green, T. A., "Dynamic Performance Evaluation of a Computer-Controlled Electric Power System" *IEEE Transactions on Power Apparatus and Systems*, Vol. PAS-91, May/June 1972, pp. 1158–1165.

14. de Mello, F. P., Mills, R. J., B'Rells, W. F., "Automatic Generation Control—Part I: Process Modeling," *IEEE Transactions on Power Apparatus and Systems*, Vol. PAS-92, March/April 1973, pp. 710–715.

15. de Mello, F. P., Mills, R. J., B'Rells, W. F., "Automatic Generation Control—Part II: Digital Control Techniques," *IEEE Transactions on Powder Apparatus and Systems*, Vol. PAS-92, March/April 1973, pp. 716–724.

16. Podmore, R., Gibbard, M. J., Ross, D. W., Anderson, K. R., Page, R. G., Argo, K., Coons, K., "Automatic Generation Control of Jointly Owned Generating Unit," *IEEE Transactions on Power Apparatus and Systems*, Vol. PAS-98, January/February 1979, pp. 207–218.

17. Elgerd, O. I., Fosha, C. E., "The Megawatt-Frequency Control Problem: A New Approach Via Optimal Control Theory," *Proceedings, Power Industry Computer Applications Conference*, 1969.

18. Zaborszky, J., Singh, J., "A Reevaluation of the Normal Operating State Control of the Power Systems using Computer Control and System Theory: Estimation," *Power Industry Computer Applications Conference Proceedings*, 1979.

19. Mukai, H., Singh, J., Spare, J., Zaborszky, J. "A Reevaluation of the Normal Operating State Control of the Power System using Computer Control and System

Theory—Part II: Dispatch Targeting," *IEEE Transactions on Power Apparatus and Systems*, Vol. PAS-100, January 1981, pp. 309–317.

20. Van Slyck, L. S., Jaleeli, N., Kelley, W. R., "Comprehensive Shakedown of an Automatic Generation Control Process," *IEEE Transactions on Power Systems*, Vol. 4, No. 2, May 1989, pp. 771–781.

21. Jaleeli, N., Van Slyck, L. S., Ewart, D. N., Fink, L. H., Hoffmann, A. G., "Understanding Automatic Generation Control," *IEEE Transactions on Power Systems*, Vol. 7, No. 3, August 1992, pp. 1106–1122.

22. Douglas, L. D., Green, T. A., Kramer, R. A., "New Approaches to the AGC Non-Conforming Load Problem," 1933 *IEEE Power Industry Computer Applications Conference*, pp. 48–57.

10 Interchange of Power and Energy

10.1 INTRODUCTION

This chapter reviews the interchange of power and energy, primarily the practices in Canada and the United States where there are numerous, major electric utilities operating in parallel in three large AC interconnections. In many other parts of the world, simpler commercial structures of the electric power industry exist. Many countries have one to two major generation–transmission utilities with local distribution utilities. The industry structure is important in discussing the interchange of power and energy since the purchase and sale of power and energy is a commercial business where the parties to any transaction expect to enhance their own economic positions under nonemergency situations. In North America, the "market place" is large, geographically widespread, and the transmission networks in the major interconnections are owned and operated by multiple entities. This has led to the development of a number of common practices in the interchange of power and energy between electric utilities. Where the transmission network is (or was) owned by a single entity, the past and developing practices regarding transactions may be different than those in the United States and Canada. We will confine the discussions of the commercial aspects of the electric energy markets to the practices in North America, circa early 1995.

The market structures for electric energy and power are changing. In the past, interconnected electric utility systems dealt only with each other to buy and sell power and energy. Only occasionally did nonutility entities become involved, and these were usually large industrial organizations with their own generation. Many of these industrial firms had a need for process heat or steam and developed internal generation (i.e., *cogeneration plants*) to supply steam and electric power. Some developed electric power beyond the internal needs of the plant so that they could arrange for sale of the excess to the local utility system. The earlier markets only involved "wholesale transactions", the sale and purchase of electric energy to utilities for ultimate delivery to the consumer. With the exception of industrial cogenerators, all aspects of the interchange arrangements were made between interconnected utilities.

In more recent times, there has been an opening of the market to facilitate the involvement of more nonutility organizations, consumers as well as generators. Throughout the world there has been a movement towards deregulation of the electric utility industry and an opening of the market to nonutility entities, mainly nonutility generating firms. There is agitation to open the use of the transmission system to all utilities and nonutility generators by providing "open transmission access." Because of the multiple ownership of the transmission systems in North America and the absence of a single entity charged with the control of the entire (or even regional) bulk power system, there are many unresolved issues (as of July 1995). These concern generation control, control of flows on the transmission system circuits, and establishment of schemes for setting "fair and equitable" rates for the use of the transmission network by parties beyond the utility owner of the local network. This last factor is an important issue since it is the very transmission interconnections that make the commercial market physically feasible. The discussions involve concerns over monopoly practices by, and the property rights of, the owners of the various parts of the network.

Nevertheless, the movement towards more nonutility participation continues and more entities are becoming involved in the operation of the interconnected systems. Most all of the nonutility participants are involved in supplying power and energy to utilities or large industrial firms. The use of a transmission system by parties other than its owner may involve "wheeling" arrangements (that is, an arrangement to use the transmission system owned by another party to deliver power). There have been wheeling arrangements as long as there have been interconnections between more than two utilities. In most cases, the development of transmission service (i.e., wheeling) rates has been based on simplified physical models designed to facilitate commercial arrangements. As long as the market was restricted to a few parties, these arrangements were usually mutually satisfactory. With the introduction of nonutility participants, there is a need for the development of rate structures based on more realistic models of the power system.

The growth of the number and size of energy transactions has emphasized the need for intersystem agreements on power flows over "parallel" transmission circuits. Two neighboring utilities may engage in the purchase–sale of a large block of power. They may have more than enough unused transmission capacity in the direct interconnections between the systems to carry the power. But, since the systems are interconnected in an AC network that includes a large number of utilities, when the transaction takes place, a large portion of the power may actually flow over circuits owned by other systems. The flow pattern is determined by physical laws, not commercial arrangements. The problems caused by these parallel path flows have been handled (at least in North America) by mutual agreements between interconnected utility systems. In the past, there was a general, if unspoken, agreement to attempt to accommodate the transactions. But, as the numbers and sizes of the transactions have increased, there have been more incidences of local circuit overloads caused by remote transactions.

We emphasize these points because in other parts of the world they do not exist in the same form. Many of the problems associated with transmission system use, transmission access, and parallel path issues, are a consequence of multiple ownership of the transmission network. They are *structural problems*, not physical problems. On the other hand, when a formerly nationalized grid is deregulated and turned into a single, privatized network there are problems, but they are not the problems that arise from the need to treat multiple transmission owners on a fair and equitable basis.

Interutility transfers of energy are easily accomplished. Recall the computation of the *area control error*, ACE, in the chapter on generation control. A major component of ACE is the *scheduled net interchange*. To arrange for the sale of energy between two interconnected systems, the seller increases its net interchange by the amount of the sale, and the purchaser decreases its net interchange by a similar amount. (We ignore losses.) The AGC systems in the two utilities will adjust the total generation accordingly and the energy will be transferred from the selling system to the purchaser. With normal controls, the power will flow over the transmission network in a pattern determined by the loads, generation, control settings, and network impedances and configuration. (Notice that *network ownership* is not a factor.)

The AGC scheme of Chapter 9 develops an autonomous, local control based upon ACE. It is predicated (implicitly, at least) on the existence of a well-defined control area that usually corresponds to the geographical and electrical boundaries of one or more utilities. Interchanges are presumed to be scheduled between utility control centers so that the net interchange schedule is well defined and relatively stable over time. With many participants engaged in transactions and, perhaps, private generators selling power to entities beyond the local control area, the interchange schedule may be subject to more frequent changes and some local loads may no longer be the primary responsibility of the local utility. AGC systems may have to become more complex with more information being supplied in real time on all local generation, load substations, and all transactions. New arrangements may be needed to assign responsibility for control actions and frequency regulation. Utilities have done these tasks in the past out of their own self-interest. A new incentive may be needed as the need for frequency and tie-line control becomes a marketplace concern; not just the concern of the utility.

This chapter reviews the practices that have evolved in all-utility interchange arrangements. This leads to a brief discussion of power pools and other commercial arrangements designed to facilitate economic interchange. Many of the issues raised by the use of the transmission system are unresolved issues that await the full and mature development of new patterns for coordinating bulk power system operations and defining, packaging and pricing transmission services. We can only discuss possible outcomes.

There are evolving market structures that include nonutility participants. These may include organizations that have generation resources, distributing utilities, and consumers, usually larger industrial firms. In these areas, we must venture into questions involving *price*. No transactions take place without involving prices, even those between utilities. Disputes naturally arise over what are *fair price levels.* (Price and fairness, like beauty, are in the eye of the beholder. The price level wanted for an older automobile may seem very fair to me as the seller and outrageous to the purchaser. We may both be correct and no sale will take place. Or one, or both, of us may be willing to change our views so that we do consummate a sale; in which case, the price agreed upon is "fair," by definition.)

In areas where there is regulation of utility charges to consumers, prices are usually based on costs. (In most markets in capitalistic economies, prices are based on market action rather than being administered by governments.) There is usually a stated principal that utilities may recover no more than a given margin above "cost." There may be some dispute over what costs should be included and how they should be allocated to each consumer class, but, generally, the notion of cost-based pricing is firmly established. Where utilities are dealing with each other or with nonutility entities, there may, or may not, be an obligation to base prices on costs. In many situations, market forces will set price levels. Transactions will be negotiated when both parties can agree upon terms that each considers advantageous, or at least satisfactory.

This chapter also introduces the concept of *wheeling*, the delivery of power and energy over a transmission system (or systems) not owned by controlled by the generating entity or the purchasing entity. At the center of the idea of selling transmission capacity to others is the definition and measurement of the available transmission capacity for transferring power. This is not an easy quantity to define since it depends upon acceptable notions of reliable, or *secure* system operating practices, a very subjective issue. In the communication network areas such as telephone systems, data transmission networks, and so on, the path capacities are more readily definable. Signals may be rerouted when a channel is fully loaded and the party desiring communication service will receive a "busy signal" if there is no capacity currently available. This does not carry over into interconnected AC power systems. Certainly, there are definable physical limits to the current that may be carried by each portion of the system without causing permanent physical damage. There is a need to reduce these absolute limits to provide some margin for the inability to predict the loading levels with certainty. There must also be some margin, or reserve, retained to permit the system to survive forced outages of circuit elements and generators. Voltage magnitudes in the system must be kept within controllable ranges. It is here where art, experience, and opinion enter and make the exact definition of available transmission capacity difficult. Thus, in any commercial arrangement for energy transactions, the question of available transmission capacity may arise and need to be settled.

Outside of North America, a major shift in the structure of the electric utility industries that has taken place in the past decade is that of splitting up formerly integrated, government utility organizations. This has usually involved the *privatization* of governmentally sponsored utilities and the separation of the original utility into separate and independent, private organizations owned by shareholders. Some of the resuting entities may be generation companies, others distribution utilities with the responsibility for the distribution of power to the ultimate consumer, and one organization that has control of the transmission network and is responsible for establishing a market for, and scheduling of, generation. Where this has happened, it has led to the development of a market structure involving a few large organizations that were formerly part of the state system, plus nonutility generators. These are markets that tend to be dominated by a few large participants.

In the United States, the electric utility industry is very diverse, with 200 to 400 major utilities (depending upon the precise definition used), plus a few thousand other organizations that are also classified as utilities. Many are investors-owned. Some are governmentally sponsored organizations at both state and federal levels. Still others are consumer-owned utilities. Given this diversity, the new market structures that may evolve under *deregulation* in the United States are apt to be different than those in countries where state systems have been privatized.

The discussions of these issues and their resolutions in this text has to be tentative, and, we trust, unbiased. Any change in a long-standing industry naturally meets with opposition, objections, and controversy, as well as enthusiastic advocacy.

10.2 ECONOMY INTERCHANGE BETWEEN INTERCONNECTED UTILITIES

Electric power systems interconnect because the interconnected system is more reliable, it is a better system to operate, and it may be operated at less cost than if left as separate parts. We saw in a previous chapter that interconnected systems have better regulating characteristics. A load change in any of the sytems is taken care of by all units in the interconnection, not just the units in the control area where the load change occurred. This fact also makes interconnections more reliable since the loss of a generating unit in one of them can be made up from spinning reserve among units throughout the interconnection. Thus, if a unit is lost in one control area, governing action from units in all connected areas will increase generation outputs to make up the deficit until standby units can be brought on-line. If a power system were to run in isolation and lose a large unit, the chance of the other units in that isolated system being able to make up the deficit are greatly reduced. Extra units would have to be run as spinning reserve, and this would mean less-economic operation. Furthermore, a generation system will generally require a smaller installed generation capacity reserve if it is planned as part of an interconnected system.

One of the most important reasons for interconnecting with neighboring systems centers on the better economics of operation that can be attained when utilities are interconnected. This opportunity to improve the operating economics arises any time two power systems are operating with different incremental costs. As Example 10A will show, if there is a sufficient difference in the incremental cost between the systems, it will pay both systems to exchange power at an equitable price. To see how this can happen, one need merely reason as follows. Given the following situation:

- Utility A is generating at a lower incremental cost than utility B.
- If utility B were to buy the next megawatt of power for its load from utility A at a price less than if it generated that megawatt from its own generation, it would save money in supplying that increment of load.
- Utility A would benefit economically from selling power to utility B, as long as utility B is willing to pay a price that is greater than utility A's cost of generating that block of power.

The key to achieving a mutually beneficial transaction is in establishing a "fair" price for the economy interchange sale.

There are other, longer-term interchange transactions that are economically advantageous to interconnected utilities. One system may have a surplus of power and energy and may wish to sell it to an interconnected company on a long-term firm-supply basis. It may, in other circumstances, wish to arrange to see this excess only on a "when, and if available" basis. The purchaser would probably agree to pay more for a firm supply (the first case) than for the interruptable supply of the second case.

In all these transactions, the question of a "fair and equitable price" enters into the arrangement. The economy interchange examples that follow are all based on an equal division of the operating costs that are saved by the utilities involved in the interchange. This is not always the case since "fair and equitable" is a very subjective concept; what is fair and equitable to one party may appear as grossly unfair and inequitable to the other. The 50–50 split of savings in the examples in this chapter should not be taken as advocacy of this particular price schedule. It is used since it has been quite common in interchange practices in the United States. Pricing arrangements for long-term interchange between vary widely and may include "take-or-pay" contracts, split savings, or fixed price schedules.

Before we look at the pricing of interchange power, we will present an example showing how the interchange power affects production costs.

EXAMPLE 10A

Two utility operating areas are shown in Figure 10.1. Data giving the heat rates and fuel costs for each unit in both areas are given here.

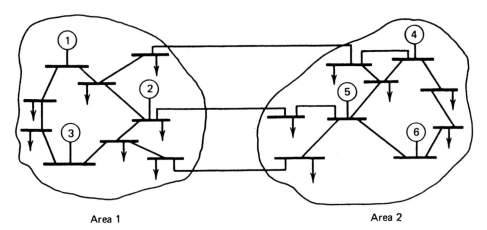

FIG. 10.1   Interconnected areas for Example 10A.

Unit data:

$$F_i(P_i) = f_i(a_i + b_i P_i + c_i P_i^2)$$

$$P_i^{min} \le P_i \le P_i^{max}$$

Unit No.	Fuel Cost f_i (R/MBtu)	Cost Coefficients			Unit Limits	
		a_i	b_i	c_i	P_i^{min} (MW)	P_i^{max} (MW)
1	2.0	561	7.92	0.001562	150	600
2	2.0	310	7.85	0.00194	100	400
3	2.0	78	7.97	0.00482	50	200
4	1.9	500	7.06	0.00139	140	590
5	1.9	295	7.46	0.00184	110	440
6	1.9	295	7.46	0.00184	110	440

Area 1:   Load =   700 MW

Max total generation = 1200 MW

Min total generation =   300 MW

Area 2:   Load = 1100 MW

Max total generation = 1470 MW

Min total generation =   360 MW

First, we will assume that each area operates independently; that is, each
will supply its own load from its own generation. This will necessitate
performing a separate economic dispatch calculation for each area. The results
of an independent economic dispatch are given here.

Area 1:
$$P_1 = 322.7 \text{ MW}$$
$$P_2 = 277.9 \text{ MW} \qquad \text{Total generation} = 700 \text{ MW}$$
$$P_3 = 99.4 \text{ MW}$$
$$\lambda = 17.856 \text{ R/MWh}$$

Operating cost, area 1 = 13,677.21 R/h

Area 1:
$$P_4 = 524.7 \text{ MW}$$
$$P_5 = 287.7 \text{ MW} \qquad \text{Total generation} = 1100 \text{ MW}$$
$$P_6 = 287.7 \text{ MW}$$
$$\lambda = 16.185 \text{ R/MWh}$$

Operating cost, area 2 = 18,569.23 R/h

Total operating cost for both areas = 13,677.21 + 18,569.23
= 32,246.44 R/h

Now suppose the two areas are interconnected by several transmission circuits such that the two areas may be thought of, and operated, as one system. If we now dispatch them as one system, considering the loads in each area to be the same as just shown, we get a different dispatch for the units.

$$P_1 = 184.0 \text{ MW}$$
$$P_2 = 166.2 \text{ MW} \qquad \text{Total generation in area 1} = 404.6 \text{ MW}$$
$$P_3 = 54.4 \text{ MW}$$

$$P_4 = 590.0 \text{ MW}$$
$$P_5 = 402.7 \text{ MW} \qquad \text{Total generation in area 2} = \underline{1395.4 \text{ MW}}$$
$$P_6 = 402.7 \text{ MW}$$

Total generation for
entire system $= 1800.0 \text{ MW}$

$$\lambda = 16.990 \text{ R/h}$$

Operating cost, area 1 = 8530.93 R/h

Operating cost, area 2 = $\underline{23,453.89 \text{ R/h}}$

Total operating cost = 31,984.82 R/h

Interchange power = 295.4 MW from area 2 to area 1

Note that area 1 is now generating less than when it was isolated, and area 2 is generating more. If we ignore losses, we can see that the change in generation in each area corresponds to the net power flow over the interconnecting circuits. This is called the *interchange power*. Note also that the overall cost of operating both systems is now less than the sum of the costs to operate the areas when each supplied its own load.

Example 10A has shown that interconnecting two power systems can have a marked economic advantage when power can be interchanged. If we look at the net change in operating cost for each area, we will discover that area 1 had a decrease in operating cost while area 2 had an increase. Obviously, area 1 should pay area 2 for the power transmitted over the interconnection, but how much should be paid? This question can be, and is, approached differently by each party.

Assume that the two systems did interchange the 295.4 MW for 1 h. Analyzing the effects of this interchange gives the following.

Area 1 costs:	without the interchange	13,677.21 ℞
	with the interchange	8530.93
	Savings	5146.28 ℞
Area 2 costs:	without interchange	18,569.23 ℞
	with interchange	23,453.89
	Increased cost	4884.66 ℞
	Combined, net savings	261.62 ℞

Area 1: Area 1 can argue that area 2 had a net operating cost increase of 4884.66 ℞ and therefore area 1 ought to pay area 2 this 4884.66 ℞. Note that if this were agreed to, area 1 should reduce its net operating cost by $13,677.21 - (8530.93 + 4884.66) = 261.62$ ℞ when the cost of the purchase is included.

Area 2: Area 2 can argue that area 1 had a net decrease in operating cost of 5146.28 ℞ and therefore area 1 ought to pay area 2 this 5146.28 ℞. Note that if this were agreed to, area 2 would have a net decrease in its operating costs when the revenues from the sale are included: $18,569.23 - 23,453.89 + 5146.28 = 261.62$ ℞.

The problem with each of these approaches is, of course, that there is no agreement concerning a mutually acceptable "fair" price. In both cases, one party to the transaction gets all the economic benefits while the other gains nothing. A common practice in such cases is to price the sale at the cost of generation plus one-half the savings in operating costs of the purchaser. This splits the savings equally between the two operating areas. This means that area 1 would pay area 2 the amount of 5015.47 ℞ and each area would have 130.81 ℞ reduction in operating costs.

Such transactions are usually not carried out if the net savings are very small. In such a case, the errors in measuring interchange flows might cause the transaction to be uneconomic. The transaction may also appear to be uneconomic to a potential seller if the utility is concerned with conserving its fuel resources to serve its own customers.

10.3 INTERUTILITY ECONOMY ENERGY EVALUATION

In example 10A, we saw how two power systems could operate interconnected for less money than if they operated separately. We obtained a dispatch of the interconnected systems by assuming that we had all the information necessary (input–output curves, fuel costs, unit limits, on-line status, etc.) in one location and could calculate the overall dispatch as if the areas were part of the same system. However, unless the two power systems have formed a power pool or transmit this information to each other, or a third party, who will arrange the transaction; this assumption is incorrect. The most common situation involves system operations personnel, located in offices within each of the control areas, who can talk to each other by telephone. We can assume that each office has the data and computation equipment needed to perform an economic dispatch calculation for its own power system and that all information about the neighboring system must come over the telephone (or some other communications network). How should the two operations offices coordinate their operations to obtain best economic operation of both systems?

The simplest way to coordinate the operations of the two power systems is to note that if someone were performing an economic dispatch for both systems combined, the most economic way to operate would require the incremental cost to be the same at each generating plant, assuming that losses are ignored. The two operations offices can achieve the same result by taking the following steps.

1. Assume there is no interchange power being transmitted between the two systems.
2. Each system operations office runs an economic dispatch calculation for its own system.
3. By talking over the telephone, the offices can determine which system has the lower incremental cost. The operations office in the system with lower incremental cost then runs a series of economic dispatch calculations, each one having a greater total demand (that is, the total load is increased at each step). Similarly, the operations office in the system having higher incremental cost runs a series of economic dispatch calculations, each having a lower total demand.
4. Each increase in total demand on the system with lower incremental cost will tend to raise its incremental cost, and each decrease in demand on the high incremental cost system will tend to lower its incremental cost. By running the economic dispatch steps and conversing over the telephone, the two operations offices can determine the level of interchange energy that will bring the two systems toward most economic operation.

Under idealized "free market" conditions where both utilities are attempting to minimize their respective operating costs, and assuming no physical limitations on the transfer, their power negotiations (or bartering) will lead to the same economic results as a pool dispatch performed on a single area basis. These assumptions, however, are critical. In many practical situations, there are both physical and institutional constraints that prevent interconnected utility systems from achieving optimum economic dispatch.

EXAMPLE 10B

Starting from the "no interchange" conditions of Example 10A, we will find the most economic operation by carrying out the steps outlined earlier. Since area 2 has a lower incremental cost before the transaction, we will run a series of economic dispatch calculations with increasing load steps of 50 MW, and an identical series on area 1 with decreasing load steps of 50 MW.

Area 1:

Step	Demand (MW)	Area 1 Incremental Cost (ℝMWh)	Assumed Interchange from Area 2 (MW)
1	700	17.856	0
2	650	17.710	50
3	600	17.563	100
4	550	17.416	150
5	500	17.270	200
6	450	17.123	250
7	400	16.976	300
8	350	16.816	350

Area 2:

Step	Demand (MW)	Area 2 Incremental Cost (ℝMWh)	Assumed Interchange from Area 1 (MW)
1	1100	16.185	0
2	1150	16.291	50
3	1200	16.395	100
4	1250	16.501	150
5	1300	16.656	200
6	1350	16.831	250
7	1400	17.006	300
8	1450	17.181	350

Note that at step 6, area 1's incremental cost is just slightly above area 2's incremental cost, but that the relationship then changes at step 7. Thus, for minimum total operating costs, the two systems ought to be interchanging between 250 and 300 MW interchange.

This procedure can be repeated with smaller steps between 250 and 300 MW, if desired.

10.4 INTERCHANGE EVALUATION WITH UNIT COMMITMENT

In Examples 10A and 10B, there was an implicit assumption that conditions remained constant on the two power systems as the interchange was evaluated. Usually, this assumption is a good one if the interchange is to take place for a period of up to 1 h. However, there may be good economic reasons to transmit interchange power for periods extending from several hours to several days. Obviously, when studying such extended periods, we will have to take into account many more factors than just the relative incremental costs of the two systems.

Extended interchange transactions require that a model of the load to be served in each system (i.e., the expected load levels as a function of time) be included, as well as the unit commitment schedule for each. The procedure for studying interchange of power over extended periods of time is as follows.

1. Each system must run a base-unit commitment study extending over the length of the period in question. These base-unit commitment studies are run without the interchange, each system serving its own load as given by a load forecast extending over the entire time period.

2. Each system then runs another unit commitment, one system having an increase in load, the other a decrease in load over the time the interchange is to take place.

3. Each system then calculates a total production cost for the base-unit commitment and for the unit commitment reflecting the effect of the interchange. The difference in cost for each system represents the cost of the interchange power (a positive change in cost for the selling sytem and a negative change in cost for the buying system). The price for the interchange can then be negotiated. If the agreed-on pricing policy is to "split the savings," the price will be set by splitting the savings of the purchaser and adding the change in the cost for the selling system. If the savings are negative, it obviously would not pay to carry out the interchange contract.

The unit commitment calculation allows the system to adjust for the start-up and shut-down times to take more effective advantage of the interchange power.

It may pay for one system to leave an uneconomical unit off-line entirely during a peak in load and buy the necessary interchange power instead.

10.5 MULTIPLE-UTILITY INTERCHANGE TRANSACTIONS

Most power systems are interconnected with all their immediate neighboring systems. This may mean that one system will have interchange power being bought and sold simultaneously with several neighbors. In this case, the price for the interchange must be set while taking account of the other interchanges. For example, if one system were to sell interchange power to two neighbouring systems in sequence, it would probably quote a higher price for the second sale, since the first sale would have raised its incremental cost. On the other hand, if the selling utility was a member of a power pool, the sale price might be set by the power and energy pricing portions of the pool agreement to be at a level such that the seller receives the cost of the generation for the sale plus one-half the total savings of all the purchasers. In this case, assuming that a pool control center exists, the sale prices would be computed by this center and would differ from the prices under multiple interchange contracts. The order in which the interchange transaction agreements are made is very important in costing the interchange where there is no central pool dispatching office.

Another phenomenon that can take place with multiple neighbors is called "wheeling." This occurs when a system's transmission system is simply being used to transmit power from one neighbor, through an intermediate system, to a third system. The intermediate system's AGC will keep net interchange to a specified value, regardless of the power being passed through it. The power being passed through will change the transmission losses incurred in the intermediate system. When the losses are increased, this can represent an unfair burden on the intermediate system, since if it is not part of the interchange agreement, the increased losses will be supplied by the intermediate system's generation. As a result, systems often assess a "wheeling" charge for such power passed through its transmission network.

The determination of an appropriate (i.e., "acceptable") wheeling charge involves both engineering and economics. Utilities providing a wheeling service to other utilities are enlarging the scope of the market for interchange transactions. Past practices amongst utilities have been established by mutual agreement amongst interconnected systems in a region. A transaction between two utilities that are not directly interconnected may also be arranged by having *each intermediate utility* purchase and resell the power until it goes from the original generator of the sale power to the utility ultimately purchasing it. This is known (in the United States, at least) as *displacement*.

For example, consider a three-party transaction. A locates power and energy in C and makes an arrangement with an intervening system B for transmission. Then C sells to B and B sells to A. The price level to A may be set as the cost of C's generation plus the wheeling charges of B plus one-half of A's savings. It may also be set at B's net costs plus one-half of A's savings. Price is a matter of negotiation in this type of transaction, when prior agreements on pricing policies are absent.

Often, utility companies will enter into interchange agreements that give the amount and schedule of the interchange power but leave the final price out. Instead of agreeing on the price, the contract specifies that the systems will operate with the interchange and then decide on its cost after it has taken place. By doing so, the systems can use the actual load on the systems and the actual unit commitment schedules rather than the predicted load and commitment schedules. Even when the price has been negotiated prior to the interchange, utilities will many times wish to verify the economic gains projected by performing after-the-fact production costs.

Power systems are often interconnected with many neighboring systems and interchange may be carried out with each one. When carrying out the after-the-face production costs, the operations offices must be careful to duplicate the order of the interchange agreements. This is illustrated in Example 10C.

EXAMPLE 10C

Suppose area 1 of Example 10A was interconnected with a third system, here designated area 3, and that interchange agreements were entered into as follows.

Interchange agreement A: area 1 buys 300 MW from area 2

Interchange agreement B: area 1 sells 100 MW to area 3

Data for area 1 and area 2 will be the same as in Example 10A. For this example, we assume that area 3 will not reduce its own generation below 450 MW for reasons that might include unit commitment or spinning-reserve requirements. The area 3 cost characteristics are as follows.

Total Demand (MW)	Area 3 Incremental Cost (R/MWh)	Area 3 Total Production Cost (R/h)
450	18.125	8220.00
550	18.400	10042.00

First, let us see what the cost would be under a split-savings pricing policy if the interchange agreements were made with agreement A first, then agreement B.

	Area 1 Gen. (MW)	Area 1 Cost (R/h)	Area 2 Gen. (MW)	Area 2 Cost (R/h)	Area 3 Gen. (MW)	Area 3 Cost (R/h)
Start	700	13677.21	1100	18569.23	550	10042.00
After agreement A	400	8452.27	1400	23532.25	550	10042.00
After agreement B	500	10164.57	1400	23532.25	450	8220.00

Agreement A: Saves area 1   5224.94 R
              Costs area 2   4963.02 R
              After splitting savings, area 1 pays area 2   5093.98 R

Agreement B: Costs area 1   1712.30 R
              Saves area 3   1822.00 R
              After splitting savings, area 3 pays area 1   1767.15 R

Summary of payments:

        Area 1 pays a net   3326.83 R
        Area 2 receives   5093.98 R
        Area 3 pays   1767.15 R

Now let the transactions be costed assuming the same split-savings pricing policy but with the interchange agreements made with agreement B first, then agreement A.

	Area 1 Gen. (MW)	Area 1 Cost (R/h)	Area 2 Gen. (MW)	Area 2 Cost (R/h)	Area 3 Gen. (MW)	Area 3 Cost (R/h)
Start	700	13677.21	1100	18569.23	550	10042.00
After agreement B:	800	15477.55	1100	18569.23	450	8220.00
After agreement A:	500	10164.57	1400	23532.25	450	8220.00

Agreement B: Costs area 1   1800.34 R
              Saves area 3   1822.00 R
              After splitting savings, area 3 pays area 1   1811.17 R

Agreement A: Saves area 1   5312.98 R
              Costs area 2   4963.02 R
              After splitting savings, area 1 pays area 2   5138.00 R

Summary of payments:

        Area 1 pays a net   3326.83 R
        Area 2 receives   5138.00 R
        Area 3 pays   1811.17 R

Except for area 1, the payments for the interchanged power are different, depending on the order in which the agreements were carried out. If agreement A were carried out first, area 2 would be selling power to area 1 at a lower incremental cost than if agreement B were carried out first. Obviously, it would be to a seller's (area 2 in this case) advantage to sell when the buyer's (area 1) incremental cost is high, and, conversely, it is to a buyer's (area 3) advantage to buy from a seller (area 1) whose incremental cost is low.

When several two-party interchange agreements are made, the pricing must follow the proper sequence. In this example, the utility supplying the energy receives more than its incremental production costs no matter which transaction is costed initially. The rate that the other two areas pay per MWh are different and depend on the order of evaluation. These differences may be summarized as follows in terms of R/MWh.

	Cost Rates (R/MWh)	
Area	A Costed First	B Costed First
1 pays	16.634	16.634
2 receives	16.980	17.127
3 pays	17.673	18.112

The central dispatch of a pool can avoid this problem by developing a single cost rate for every transaction that takes place in a given interval.

10.6 OTHER TYPES OF INTERCHANGE

There are other reasons for interchanging power than simply obtaining economic benefits. Arrangements are usually made between power companies to interconnect for a variety of reasons. Ultimately, of course, economics plays the dominant role.

10.6.1 Capacity Interchange

Normally, a power system will add generation to make sure that the available capacity of the units it has equals its predicted peak load plus a reserve to cover unit outages. If for some reason this criterion cannot be met, the system may enter into a capacity agreement with a neighboring system, provided that neighboring system has surplus capacity beyond what it needs to supply its own peak load and maintain its own reserves. In selling capacity, the system that has a surplus agrees to cover the reserve needs of the other system. This may require running an extra unit during certain hours, which represents a cost to the selling system. The advantage of such agreements is to let each system schedule generation additions at longer intervals by buying capacity when it is short and selling capacity when a large unit has just been brought on-line and it has a surplus. Pure capacity reserve interchange agreements do not entitle the purchaser to any energy other than emergency energy requirements.

10.6.2   Diversity Interchange

Daily diversity interchange arrangements may be made between two large systems covering operating areas that span different time zones. Under such circumstances, one system may experience its peak load at a different time of the day than the other system simply because the second system is 1 h behind. If the two systems experience such a phenomenon, they can help each other by interchanging power during the peak. The system that peaked first would buy power from the other and then pay it back when the other system reached its peak load.

This type of interchange can also occur between systems that peak at different seasons of the year. Typically, one system will peak in the summer due to air-conditioning load and the other will peak in winter due to heating load. The winter-peaking system would buy power during the winter months from the summer-peaking system whose system load is presumably lower at that time of year. Then in the summer, the situation is reversed and the summer-peaking system buys power from the winter-peaking system.

10.6.3   Energy Banking

Energy-banking agreements usually occur when a predominantly hydro system is interconnected to a predominantly thermal system. During high water runoff periods, the hydro system may have energy to spare and will sell it to the thermal system. Conversely, the hydro system may also need to import energy during periods of low runoff. The prices for such arrangements are usually set by negotiations between the specific systems involved in the agreement.

Instead of accounting for the interchange and charging each other for the transactions on the basis of hour-by-hour operating costs, it is common practice in some areas for utilities to agree to a banking arrangement, whereby one of the systems acts as a bank and the other acts as a depositor. The depositor would "deposit" energy whenever it had a surplus and only the MWh "deposited" would be accounted for. Then, whenever the depositor needed energy, it would simply withdraw the energy up to MWh it had in the account with the other system. Which system is "banker" or "depositor" depends on the exchange contract. It may be that the roles are reversed as a function of the time of year.

10.6.4   Emergency Power Interchange

It is very likely that at some future time a power system will have a series of generation failures that require it to import power or shed load. Under such emergencies, it is useful to have agreements with neighboring systems that commit them to supply power so that there will be time to shed load. This may occur at times that are not convenient or economical from an incremental cost point of view. Therefore, such agreements often stipulate that emergency power be priced very high.

10.6.5   Inadvertent Power Exchange

The AGC systems of utilities are not perfect devices with the result that there are regularly occurring instances where the error in controlling interchange results in a significant, accumulated amount of energy. This is known as *inadvertent interchange.* Under normal circumstances, system operators will "pay back" the accumulated inadvertent interchange energy megawatt-hour for megawatt-hour, usually during similar time periods in the next week. Differences in cost rates are ignored.

Occasionally, utilities will suffer prolonged shortages of fuel or water, and the inadvertent interchange energy may grow beyond normal practice. If done deliberately, this is known as "leaning on the ties." When this occurs, systems will normally agree to pay back the inadvertent energy at the same time of day that the errors occurred. This tends to equalize the economic transfer. In severe fuel shortage situations, interconnected utilities may agree to compensate each other by paying for the inadvertent interchange at price levels that reflect the real cost of generating the exchange energy.

10.7   POWER POOLS

Interchange of power between systems can be economically advantageous, as has been demonstrated previously. However, when a system is interconnected with many neighbors, the process of setting up one transaction at a time with each neighbor can become very time consuming and will rarely result in the optimum production cost. To overcome this burden, several utilities may form a power pool that incorporates a central dispatch office. The power pool is administered from a central location that has responsibility for setting up interchange between members, as well as other administrative tasks. The pool members relinquish certain responsibilities to the pool operating office in return for greater economies in operation.

The agreement the pool members sign is usually very complex. The complexity arises because the members of the pool are attempting to gain greater benefits from the pool operation and to allocate these benefits equitably among the members. In addition to maximizing the economic benefits of interchange between the pool members, pools help member companies by coordinating unit commitment and maintenance scheduling, providing a centralized assessment of system security at the pool office, calculating better hydro-schedules for member companies, and so forth. Pools provide increased reliability by allowing members to draw energy from the pool transmission grid during emergencies as well as covering each others' reserves when units are down for maintenance or on forced outage.

Some of the difficulties in setting up a power pool involving nonaffiliated companies or systems arise because the member companies are independently owned and for the most part independently operated. Therefore, one cannot just make the assumption that the pool is exactly the same entity as a system under one ownership. If one member's transmission system is heavily loaded with power flows that chiefly benefit that member's neighbors, then the system that owns the transmission is entitled to a reimbursement for the use of the transmission facilities. If one member is directed to commit a unit to cover a reserve deficiency in a neighboring system, that system is also likewise entitled to a reimbursement.

These reimbursement arrangements are built into the agreement that the members sign when forming the pool. The more the members try to push for maximum economic operation, the more complicated such agreements become. Nevertheless, the savings obtainable are quite significant and have led many interconnected utility systems throughout the world to form centrally dispatched power pools when feasible.

A list of operating advantages for centrally dispatched power pools, ordered by greatest expected economic advantage, might look as follows:

1. Minimize operating costs (maximize operating efficiency).
2. Perform a system-wide unit commitment.
3. Minimize the reserves being carried throughout the system.
4. Coordinate maintenance scheduling to minimize costs and maximize reliability by sharing reserves during maintenance periods.
5. Maximize the benefits of emergency procedures.

There are disadvantages that must be weighed against these operating and economic advantages. Although it is generally true that power pools with centralized dispatch offices will reduce overall operating costs, some of the individual utilities may perceive the pool requirements and disciplines as disadvantageous. Factors that have been cited include.

1. The complexity of the pool agreement and the continuing costs of supporting the interutility structure required to manage and administer the pool.
2. The operating and investment costs associated with the central dispatch office and the needed communication and computation facilities.
3. The relinquishing of the right to engage in independent transactions outside of the pool by the individual companies to the pool office and the requirement that any outside transactions be priced on a split-saving basis based on pool members' costs.

4. The additional complexity that may result in dealing with regulatory agencies if the pool operates in more than one state.

5. The feeling on the part of the management of some utilities that the pool structure is displacing some of an individual system's management responsibilities and restricting some of the freedom of independent action possible to serve the needs of its own customers.

Power pools without central dispatch control centers can be administered through a central office that simply acts as a brokerage house to arrange transactions among members. In the opposite extreme, the pool can have a fully staffed central office with real-time data telemetered to central computers that calculate the best pool-wide economic dispatch and provide control signals to the member companies.

By far the most difficult task of pool operation is to decide who will pay what to whom for all the economic transactions and special reimbursements built into the pool agreement. There are several ways to solve this problem, and some will be illustrated in Section 10.7.2.

10.7.1   The Energy-Broker System

As with sales and purchases of various commodities or financial instruments (e.g., stock), it is often advantageous for interconnected power systems to deal through a broker who sets up sales and purchases of energy instead of dealing directly with each other. The advantage of this arrangement is that the broker can observe all the buy and sell offers at one time and achieve better economy of operation. When utilities negotiate exchanges of power and energy in pairs, the "market place" is somewhat haphazard like a bazaar. The introduction of a central broker to accept quotations to sell and quotations to purchase creates an orderly marketplace where supply, demand, and prices are known simultaneously.

The simplest form of "broker" scheme is the "bulletin board." In this type of scheme, the utility members post offers to buy or sell power and energy at regular, frequent intervals. Members are free to access the bulletin board (via some sort of data exchange network) at all times. Members finding attractive offers are free to contact those posting the offers and make direct arrangements for the transaction. Like any such informally structured market, many transactions will be made outside the marketplace. More complex brokers are those set up to arrange the matching of buyers and sellers directly, and, perhaps, to set transaction prices.

In one power broker scheme in use, the companies that are members of the broker system send hourly buy and sell offers for energy to the broker who matches them according to certain rules. Hourly, each member transmits an incremental cost and the number of megawatt-hours it is willing to sell or its decremental cost and the number of megawatt-hours it will buy. The broker sets up the transactions by matching the lowest cost seller with the highest cost buyer, proceeding in this manner until all offers are processed. The matched buyers and sellers will price the transaction on the basis of rules established in setting up the power broker scheme. A common arrangement is to compensate the seller for this incremental generation costs and split the savings of the buyer equally with the seller. The pricing formula for this arrangement is as follows. Let

$$F_s' = \text{incremental cost of the selling utility (R/MWh)}$$

$$F_b' = \text{decremental cost of the buying utility (R/MWh)}$$

$$F_c = \text{cost rate of the transaction (R/MWh)}$$

Then,

$$F_c = F_s' + \frac{1}{2}(F_b' - F_s')$$

$$= \frac{1}{2}(F_s' + F_b')$$

(10.1)

In words, the transaction's cost rate is the average of the seller's incremental cost and the purchaser's decremental cost. In this text, decremental cost is the reduction in operating cost when the generation is reduced a small amount. Example 10D illustrates the power brokerage process.

EXAMPLE 10D

In this example, four power systems have sent their buy/sell offers to the broker. In the table that follows, these are tabulated and the maximum pool savings possible is calculated.

Utilities Selling Energy	Incremental Cost (R/MWh)	MWh for Sale	Seller's Total Increase in Cost (R)
A	25	100	2500
B	30	100	3000

Utilities Buying Energy	Decremental Cost (R/MWh)	MWh for Purchase	Buyer's Total Decrease in Cost (R)
C	35	50	1750
D	45	150	6750

$$\text{Net pool savings} = (1750\ R + 6750\ R) - (2500\ R + 3000\ R)$$

$$= 8500\ R - 5500\ R = 3000\ R$$

The broker sets up transactions as shown in the following table.

Transaction	Savings Computation		Total Transaction Savings (R)
1. A sells 100 MWh to D	100 MWh (45 − 25) R/MWh	=	2000
2. B sells 50 MWh to D	50 MWh (45 − 30) R/MWh	=	750
3. B sells 50 MWh to C	50 MWh (35 − 30) R/MWh	=	250
		Total	3000

The rates and total payments are easily computed under the split-savings arrangement. These are shown in the following table.

Transaction	Price (R/MWh)	Total cost (R)
1. A sells 100 MWh to D	35.0	3500
2. B sells 50 MWh to D	37.5	1875
3. B sells 50 MWh to C	32.5	1625
	Total	7000

A receives 3500 R from D; B receives 3500 R from D and C. Note that each participant benefits: A receives 1000 R above its costs; B receives 500 R above its costs; C saves 125 R; and D saves 1375 R.

The chief advantage of a broker system is its simplicity. All that is required to get a broker system into operation is a communications circuit to each member's operations office and some means of setting up the transactions. The transactions can be set up manually or, in the case of more modern brokerage arrangements, by a computer program that is given all the buy/sell offers and automatically sets up the transactions. With this type of broker, the quoting systems are commonly only informed of the "match" suggested by the broker and are free to enter into the transaction or not as each see fit.

Economists have sometimes argued that the broker pricing scheme should set one single "clearing price" for energy each time period. The logic behind this is that the market-determined price level should be based on the participants' needs and willingness to buy or sell. This removes the absolute need for quoting cost-based prices. Utilities would be free to quote offers at whatever price level they wished, but would be (under most rules that have been suggested) obligated to deliver or purchase the energy quoted at the market clearing price. The transactions market would be similar to the stock exchange. Objections raised have been that in times of shortage, price levels could rise dramatically and uncontrollably.

Power broker schemes can be extended to handle long-term economy interchange and to arrange capacity sales. This enables brokers to assist in minimizing costs for spinning reserves and coordinate unit commitments in interconnected systems.

10.7.2   Allocating Pool Savings

All methods of allocating the savings achieved by a central pool dispatch are based on the premise that no pool member should have higher generation production expenses than it could achieve by dispatching its own generation to meet its own load.

We saw previously in the pool broker system that one of the ways to allocate pool savings is simply to split them in proportion each system's net interchange during the interval. In the broker method of matching buyers and sellers based on their incremental and decremental costs, calculations of savings are relatively easy to make since the agreed incremental costs and amounts of energy must be transmitted to the broker at the start. When a central economic dispatch is used, it is easier to act as if the power were sold to the pool by the selling systems and then bought from the pool by the buying systems. In addition, allowances may be made for the fact that one system's transmission system is being used more than others in carrying out the pool transactions.

There are two general types of allocation schemes that have been used in U.S. pool control centers. One, illustrated in Example 10E, may be performed in a real-time mode with cost and savings allocations made periodically using the incremental and decremental costs of the systems. In this scheme, power is sold to and purchased from the pool and participants' accounts are updated currently. In the other approach, illustrated in Example 10F, the allocation of costs and savings is done after the fact using total production costs. Example 10E shows a scheme using incremental costs similar to one used by a U.S. pool made up of several member systems.

EXAMPLE 10E

Assume that the same four systems as given in Example 10D were scheduled to transact energy by a central dispatching scheme. Also, assume that 10% of the gross system's savings was to be set aside to compensate those systems that provided transmission facilities to the pool. The first table shows the calculation of the net system savings.

Utilities Selling Energy	Incremental Cost (R/MWh)	MWh for Sale	Seller's Total Increase in Cost (R)
A	25	100	2500
B	30	100	3000

Utilities Buying Energy	Decremental Cost (₨/MWh)	MWh for Purchase	Buyer's Total Decrease in Cost (₨)
C	35	50	1750
D	45	150	6750
		Pool savings	3000
Savings withheld for transmission compensation[a]			300
		Net savings	2700

[a] 10% savings withheld for transmission compensation.

Next, the weighted average incremental costs for selling and buying power are calculated.

Seller's weighted average incremental cost

$$= \left[\frac{(25 \text{ ₨/MWh} \times 100 \text{ MWh}) + (30 \text{ ₨/MWh} \times 100 \text{ MWh})}{100 \text{ MWh} + 100 \text{ MWh}} \right] = 27.50 \text{ ₨/MWh}$$

Buyer's weighted average decremental cost

$$= \left[\frac{(35 \text{ ₨/MWh} \times 50 \text{ MWh}) + (45 \text{ ₨/MWh} \times 150 \text{ MWh})}{50 \text{ MWh} + 150 \text{ MWh}} \right] = 42.50 \text{ ₨/MWh}$$

Finally, the individual transactions savings are calculated.

1. A sells 100 MWh to pool:

$$100 \text{ MWh} \frac{42.50 - 25 \text{ ₨/MWh}}{2} \times 0.9 = 787.50 \text{ ₨}$$

2. B sells 100 MWh to pool:

$$100 \text{ MWh} \frac{42.50 - 30 \text{ ₨/MWh}}{2} \times 0.9 = 562.50 \text{ ₨}$$

3. C buys 50 MWh from pool:

$$50 \text{ MWh} \frac{35 - 27.50 \text{ ₨/MWh}}{2} \times 0.9 = 168.75 \text{ ₨}$$

4. D buys 150 MWh from pool:

$$150 \text{ MWh} \frac{45 - 27.50 \text{ ₨/MWh}}{2} \times 0.9 = 1181.25 \text{ ₨}$$

$$\overline{2700.00 \text{ ₨ Net savings}}$$

The total transfers for this hour are then:

$$
\begin{aligned}
\text{C buys 50 MWh for } 42.5 \times 50 - 168.75 &= 1956.23 \text{ R} \\
\text{D buys 150 MWh for } 42.5 \times 150 - 1181.25 &= 5193.75 \text{ R} \\
\text{Total} \quad &\; 7150.00 \text{ R}
\end{aligned}
$$

$$
\begin{aligned}
\text{A sells 100 MWh for } 27.5 \times 100 + 787.5 &= 3537.50 \text{ R} \\
\text{B sells 100 MWh for } 27.5 \times 100 + 562.5 &= 3312.50 \text{ R} \\
&\; 6850.00 \text{ R}
\end{aligned}
$$

$$
\begin{aligned}
\text{Total transmission charge} \quad & 300.00 \text{ R} \\
\text{Total} \quad & 7150.00 \text{ R}
\end{aligned}
$$

The 300 R that was set aside for transmission compensation would be split up among the four systems according to some agreed-upon rule reflecting each system's contribution to the pool transmission network.

The second type of savings allocation method is based on after-the-fact computations of individual pool member costs as if each were operating strictly so as to serve their own individual load. In this type of calculation, the unit commitment, hydro-schedules, and economic dispatch of each individual pool member are recomputed for an interval after the pool load has been served. This "own load dispatch" is performed with each individual system's generating capacity, including any portions of jointly owned units, to achieve maximum operating economy for the individual system.

The costs for these computed individual production costs are then summed and the total pool savings are computed as the difference between this cost and the actual cost determined by the central pool dispatch.

These savings are then allocated among the members of the pool according to the specific rules established in the pool agreement. One method could be based on rules similar to those illustrated previously. That is, any interval for which savings are being distributed, buyers and sellers will split the savings equally.

Specific computational procedures may vary from pool to pool. Those members of the pool supplying energy in excess of the needs of their own loads will be compensated for their increased production expenses and receive a portion of the overall savings due to a pool-wide dispatch. The process is complicated because of the need to perform individual system production cost calculations. Pool agreements may contain provisions for compensation to members supplying capacity reserves as well as energy to the pool. A logical questions that requires resolution by the pool members involves the fairness of comparing an after-the-fact production cost analysis that utilizes a known load pattern with a pool dispatch that was forced to use load forecasts. With the load pattern known with certainty, the internal unit commitment may be optimized to a greater extent that was feasible by the pool control center. Example 10F illustrates this type of procedure for the three systems of Example 10C for one period. In this example, only the effects of the economic dispatch are shown since the unit commitment process is not involved.

EXAMPLE 10F

The three areas and load levels are identical to those in Example 10C. (Generation data are in Examples 10A and 10B as well.) In this case, the three areas are assumed to be members of a centrally dispatched power pool. The pool's rules for pricing pool interchange are as follows.

1. Each area delivering power and energy to the pool in excess of its own load will receive compensation for its increased production costs.
2. The total pool savings will be computed as the difference between the sum of the production costs of the individual areas (each computed on the basis that it supplied its own load) and the pool-wide production cost.
3. These savings will be split equally between the suppliers of pool capacity or energy and the areas receiving pool-supplied capacity or energy.
4. In each interval where savings are allocated (usually a week, but in this example only 1 h), the cost rate for pricing the interchange will be one-half the sum of the total pool savings plus the cost of generating the pool energy divided by the total pool energy. The total pool energy is the sum of the energies in the interval supplied by all areas, each generating energy in excess of its own load.

The pool production costs are as follows.

Area	Area Load (MW or MWh)	Own-Load Production Cost (R/h)
1	700	13677.21
2	1100	18569.23
3	550	10042.00
Total	2350	42288.44

Under the pool dispatch, areas 1 and 2 are dispatched at an incremental cost of 17.149 R/MWh to generate a total of 1900 MW. Area 3 is limited to supplying 450 MW of its own load at an incremental cost of 18.125 R/MWh. The generation and costs of the three areas and the pool under pool dispatch are given in the following table.

Area	Area Generation (MW or MWh)	Production Cost (R/h)	Incremental Cost (R/MWh)
1	458.9	9458.74	17.149
2	1441.1	24232.66	17.149
3	450.0	8220.00	18.125
Pool	2350.0	41911.40	17.149

Therefore, the total savings due to the pool dispatch for this 1 h are

$$42{,}288.44 \text{ R} - 41{,}911.40 \text{ R} = 377.04 \text{ R}$$

In this example, area 2 is supplying 341.1 MWh in excess of its own load to the pool. This is the total pool energy. Therefore, the price rate for allocating savings is computed as follows.

Cost of pool energy:

Cost of energy supplied to the pool by area 2
$$= 24{,}232.66 \text{ R} - 18{,}569.23 \text{ R} = 5663.43 \text{ R}$$
$$+ 1/2 \text{ pool savings} = \underline{188.52 \text{ R}}$$
$$\text{Total} \quad 5851.95 \text{ R}$$

$$\text{Interchange price rate} = \frac{5851.95}{341.1} = 17.156 \text{ R/MWh}$$

The final outcome for each area is shown in the following table.

Area	Pool Energy Received (MWh)	Interchange Cost (R)	Production Cost (R)	Net Cost (R)
1	+241.1	4136.34	9458.74	13595.08
2	−341.1	−5851.95	24232.66	18380.71
3	+100	1715.61	8220.00	9935.61
Pool		0	41911.40	41911.40

Note that each area's net production costs are reduced as compared with what they would have been under isolated dispatch. Furthermore, the ambiguity involved in pricing different transactions in alternative sequences has been avoided.

Example 10F is based on only a single load level so that after-the-fact unit commitment and production costing is not required. It could have been done on a real-time basis, in fact. This example also illustrates the complete transaction allocation that must be done for savings allocation schemes.

Complete own-load dispatch computations for cost and savings allocations are usually performed for a weekly period. The implementation may be complex since hourly loads and unit status data are required. An on-line, real-time allocation scheme avoids these complications.

No matter how these savings allocations are performed, you should appreciate that any estimates of "savings" involves finding the difference between actual, known costs and *costs as they might have been*. There is a great deal of room for disagreement about how to estimate these second, hypothetical costs.

10.8 TRANSMISSION EFFECTS AND ISSUES

This topic involves both technical and structural considerations. There are some technical issues that transcend the organizational market structure issues, but many of these arise only because of the multiple ownership of interconnected power transmission networks. There are basic technical issues of defining a network's capability to transfer power that involve physical capacity to handle power flows reliably (or *securely*). Even here (or is it *especially here?*), nontechnical matters are involved in defining *acceptable levels of network unreliability*. In an economic environment where capital and financing is available to develop multiple parallel paths in a transmission network, transmission capability may be restricted by the desire of the utilities and involved governmental agencies to insure very high levels of system security. Widespread blackouts and prolonged power shortages are to be avoided. Networks are designed with large capacity margins so that elements tend to be loaded conservatively. Normal failures of single major elements will not cause loss of load. Even simultaneous occurrences of two failures of major elements will not cause load curtailment. In most foreseeable circumstances, there will not be cascading outages that spread across the interconnected system. *Cascading outages* can occur where the loss of a transmission circuit, due to a prolonged fault, would result in the overloading of parallel circuits. These, in turn, might be opened in time by the action of protective relaying systems. Thus, the single event could cascade into a regional series of events that could result in a blackout.

In economic climates where capital and financing are difficult to obtain, and in areas where environmental restrictions prevent adding transmission capacity, power transmission networks may be designed using less-stringent reliability standards and operated in a fashion such that loads are expected to be curtailed when major transmission elements suffer outages. Security and reliability standards may be similar to the previous situation, with the exception that controlled load disconnection is not considered to be a "failure" event. Even in systems where "defensive operational scheduling" practices are normally followed (i.e., loss of single or two major system elements does not result in cascading outages), there are occasions where it is more economic to resort to using special system controls. These might drop load automatically when a remote generation source loses one of its transmission links to the system. This is a simple example; there are more complex arrangements that have been used. When a variety of specialized system control schemes are used, it is necessary to keep track of the various systems and keep every interconnected system abreast of changes and new developments.

In any interconnected system, there is a need to define in quantitative terms the maximum amount of power that may be transferred without violating whatever system reliability and security criteria are in place. Therefore, it is necessary to consider the types of operating limitations that exist in AC power networks. These include thermal limits sets by the capability of the lines and apparatus to absorb and dissipate the heat created by the current flowing in the various elements. These limits are usually expressed as a maximum allowable temperature rise above specified ambient conditions. The intent is to prevent the extreme, sustained temperatures that might cause lines to sag and equipment to be damaged. Even with these straightforward thermal limitations, there are variable ambient conditions that make actual danger points occur in the summer at lower power transfers than in the colder months. Next are limits set by the interplay of system limitations, equipment limitations, economics, and service reliability ("security") standards. These include voltage–VAR-related conditions and stability considerations.

Voltage and VAR conditions arise because voltage magnitudes within the system must remain within a bandwidth that is set by the voltage tolerances of both system and consumer equipment. Large high-voltage equipment and consumer equipment (motors, transformers, etc.) are generally limited to excursions of about $\pm 5\%$ of their rated voltage. The voltage magnitude bandwidth tolerance on the system is affected (and generally enlarged) by the ability of various voltage-correcting devices to restore voltages to a bandwidth acceptable to the apparatus. Key control devices include tapchanging transformers and various types of VAR-supply devices. At shorter transmission distances (say 50 miles or 100 km), the thermal limits and voltage–VAR limitations generally are the restricting system conditions. Of course, it is theoretically possible to add additional circuits and VAR-support equipment, but economic considerations generally set a practical limit on what is done to increase transmission capacity.

Transmission capability limits can be imposed by voltage instability, steady-state stability, and transient stability. In all cases, the network has to be able to survive possible conditions that can lead to unstable situations. These instability-inducing conditions usually become more intense as the system loading increases. The need to avoid these operating regimes then places a practical limit on the power that can be transmitted. At longer distances it is usually transient stability that sets the limits. The various limits are found by testing the network under increasingly heavy loading conditions and seeking ways to alleviate or prevent the instabilities. At some point, it becomes impossible or uneconomic to increase the limits further. Besides economic considerations, the actual power transfer limitations found will depend upon the testing criteria utilized. Is it sufficient to test the network's ability to survive a single-phase fault that is successfully cleared and the line reclosed, or should the network be tested using a bolted three-phase fault that requires switching a line segment?

10.8.1 Transfer Limitations

The operators in an interconnected AC system are interested in the limits to the amount of power that may be transferred between various systems or buses. The amount of power transfer capability available at any given time is a function of the system-wide pattern of loads, generation and circuit availability. This has led the United States systems to establish definitions of "incremental transfer capability." These definitions depend upon testing the network to meet selected security constraints (one or two simultaneous outages) under various sets of operating conditions to determine the added ("incremental") power that maybe transferred safely. This requires the cooperative efforts of a number of utilities in a region and only provides general guidelines concerning the transfer capability limits.

All of these tests and limitations depend a great deal upon the use of subjective criteria, definitions, and procedures that are a result of mutual agreement amongst the utilities. Practices differ. As an example, take the matter of determining the ability of an interconnected system to transmit an additional block of 500 MW between two systems separated by one or more intervening systems. If the operators test the systems' capability under the existing and planned optimal generation schedules, the network's loading criteria are violated. However, by shifting generation by a fairly small amount, the transfer would satisfy all of the systems' criteria. Should the transfer take place? In the systems in North America the answer would generally be "yes," with the added proviso that the cost for the transfer would include the recovery of the added generation cost of the systems that shifted generation off of an optimal economic dispatch.

Transfer limits can be determined for relatively simple interconnections where DC approximations are satisfactory to establish network flows. Sometimes these techniques may be used to study incremental flows. But, in most cases, it requires an AC power flow of some sort to investigate transfer limits and answer questions similar to the one in the previous paragraph.

This leads to what has been termed the "busy-signal problem." When I attempt to place a call that would require the use of an already-loaded communication channel, the system controls attempt to reroute my call, and if they are unsuccessful, I receive a busy signal. In present AC power systems, if a request is made in initiate a new transaction over a transmission system that is loaded to near maximum capability, it is feasible to do a moderate amount of "rerouting" of power flows by shifting generation and perhaps some switching of circuits. But if these measures are unsuccessful, or precluded by current operating practices, I will only find out some time after the request has been made, and, unless I am conversant in power system operating practices, I may not understand why the particular answer was given.

This is the point in the discussion where institutional problems become quite important. As long as the parties that are interested are interconnected electric utilities and other technically competent organizations that all can agree with each other about the operating rules, definitions of transfer capability, and the various assumptions used in establishing limitations, there is not a serious problem. Suppose, however, that all these parties do not agree. Suppose that some are satisfied with the present arrangements while others are eager to expand the network capability for the marketing, or purchasing, of power over a wider geographic area. They would like a concrete definition of network transfer capacity that did incorporate so many variable and ambiguous factors. The lack of a simple "busy signal" becomes even more pressing when nonutility entities are permitted access to the transmission system to make sales and purchases.

The situation is similar when measures to relieve local constraints are required in order to facilitate the use of the interconnected system by nonlocal parties. Who should pay for these measures? How should the costs be allocated? These are all real concerns when the interconnected system is owned by many individual utilities and serves the needs of even more individual organizations.

10.8.2   Wheeling

The term "wheeling" has a number of definitions; we will stay with a simple one. Wheeling is the use of some party's (or parties') transmission system(s) for the benefit of other parties. Wheeling occurs on an AC interconnection that contains more than two utilities (more properly, two parties) whenever a transaction takes place. (If there are only two parties, there is no third party to perform wheeling.) As used here, the term "parties" includes both utility and nonutility organizations.

Consider the six interconnected control areas shown in Figure 10.2. Suppose areas A and C negotiate the sale of 100 MW by A to C. Area A will increase its

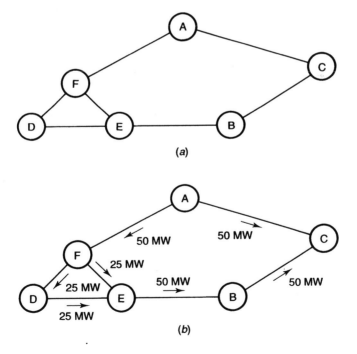

FIG. 10.2   Six interconnected control areas. (*a*) Configuration; (*b*) Incremental power flows when area A sells 100 MW to area C.

scheduled net interchange by 100 MW and C will reduce its net interchange schedule by the same amount. (We ignore losses.) The generation in A will increase by the 100 MW sale and that in C will decrease by the 100 MW purchase. Figure 10.2b shows the resulting *changes* in power flows, obtained by finding the difference between power flows before and after the transaction. Note that not all of the transaction flows over the direct interconnections between the two systems. The other systems are all wheeling some amount of the transaction. (In the United States, these are called "*parallel path or loop flows.*")

The number of possibilities for transactions is very large, and the power flow pattern that results depends on the configuration and the purchase–sale combination plus the schedules in all of the systems. In the United States, various arrangements have been worked out between the utilities in different regions to facilitate interutility transactions that involve wheeling. These past arrangements would generally ignore flows over parallel paths were the two systems contiguous and owned sufficient transmission capacity to permit the transfer. (This capacity is usually calculated on the basis of nominal or nameplate ratings) In that case, wheeling was *not* taking place, by mutual agreement. The extension of this arrangement to noncontiguous utilities led to the artifice known as the "contract path." In making arrangements for wheeling, the two utilities would rent the capability needed on any path that would interconnect the two utilities. Thus, on Figure 10.2, a 100-MW transaction between systems A and D might involve arranging a "contract path" between them that would have 100 MW available. Flows over any parallel paths are ignored. As artificial as these concepts may appear, they are commercial arrangements that have the merit of facilitating mutually beneficial transactions between systems.

Difficulties arise when wheeling increases power losses in the intervening systems and when the parallel path flows utilize capacity that is needed by a wheeling utility. Increased transmission losses may be supplied by the seller so that the purchaser in a transaction receives the net power that was purchased. In other cases, the transaction cost may include a payment to the wheeling utility to compensate it for the incremental losses. The relief of third-party network element loading caused by wheeling is a more difficult problem to resolve. If it is a situation that involves overloading a third party's system on a recurring basis, the utilities engaged in the transaction may be required to cease the transfer or pay for additional equipment in someone else's system. Both approaches have been used in the past.

Loop flows and arrangements for parallel path compensation become more important as the demand for transmission capacity increases at a faster rate than actual capability does. This is the situation in most developed countries. New, high-voltage transmission facilities are becoming more difficult to construct. Another unresolved issue has to do with the participation of organizations that are basically consumers. Should they be allowed access to the power transmission network so that they may arrange for energy supplies from nonlocal resources? In the deregulated natural gas industry in the United States, this has been done.

10.8.3   Rates for Transmission Services in Multiparty Utility Transactions

Rates for transmission service have a great deal of influence on transactions when wheeling is involved. We have previously considered energy transaction prices based on split-savings concepts. Where wheeling services are involved, this same idea might be carried over so that the selling and wheeling utilities would split the savings with the purchaser on some agreed-upon basis. Both the seller and wheeling systems would want to recover their costs and would wish to receive a profit by splitting the savings of the purchaser. Some (many economists) would argue that transmission services should be offered on the basis of a "cost plus" price. A split-savings arrangement involving four or five utility systems might lose its economic attractiveness to the buyer by the time the potential savings were redistributed.

The notion of selling transmission service is not new. A number of different pricing schemes have been proposed and used. Most are based upon simplified models that allow such fictions as the "contract path." Some are based on an attempt to mimic a power flow, in that they would base prices on incremental power flows determined in some cases by using DC power flow models. The very simplest rates are a charge per MWh transferred, and ignore any path considerations.

More complex schemes are based on the "marginal cost" of transmission which is based on the use of bus incremental costs (BIC). The numerical evaluation of BIC is straightforward for a system in economic dispatch. In that case, the bus penalty factor times the incremental cost of power at the bus is equal to the system λ, except for generator buses that are at upper or lower limits. This is true for load buses as well as generator buses. (We will treat this situation in more detail in Chapter 13 on the optimal power flow.)

Consider any power system in economic dispatch.

1. If we have a single generator, then the cost to deliver an additional small increment of power at the generator bus is equal to the incremental cost of power for that generator.

2. If we have more than one generator attached to a bus and this is the only source of power, and the generators have been dispatched economically (i.e., equal λ), then the cost to deliver an additional small increment of power at this bus is equal to λ.

3. If there are multiple generators at different buses throughout the power system, and they have been dispatched economically, i.e., accurate penalty factors have been calculated and used in the economic dispatch—then the cost of delivery of an addition small increment of power at any individual generator bus will be that generator's own incremental cost. This cost will

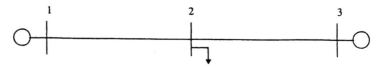

FIG. 10.3 Three-bus system.

not be equal across the system due to the fact that each generator's incremental cost is multiplied by its penalty factor.

It is important to stress that we are talking of an "additional small increment" of power at a bus and not a large increment. If the power increase is very small, the three statements above hold. If we are talking of a large increment in power delivered anywhere, the optimal dispatch must be recalculated and the cost is not equal to the incremental cost in any of the three cases above.

If we have the case shown in Figure 10.3, the power is all delivered to a load bus that is separated from either generator by a transmission line. In this case, the incremental cost of delivery of power to the load is not equal to the incremental cost of delivery at either generator bus. The exact value of the incremental cost at the load bus can be calculated, however, using the techniques developed in Chapter 13 (see Section 13.7). The incremental cost to deliver power at a bus is called the bus incremental cost (BIC) and plays a very important role in the operation of modern power systems. For a power system without any transmission limitations, the BIC at any bus in the system will usually be fairly close to the BIC at other buses. However, when there is a transmission constraint, this no longer holds.

Suppose the following situation were to arise in the system in Figure 10.3.

1. Generator 1 has high incremental cost and is at its low limit.
2. Generator 3 has low incremental cost and is not at either limit.

In such a case, the BIC at the load bus will be very close to the low incremental cost of the generator at bus 3.

Now let there be a limit to the power flowing on the transmission line from bus 3 to bus 2 so that no further power can be generated at bus 3. When the load is increased at bus 2, the increase must come entirely from the generator at bus 1 and its BIC will be much higher, reflecting the incremental cost of the bus 1 generator. Thus, the BICs are very useful to show when loading of the transmission system shifts the cost of delivery at certain buses in the network.

Next, let us consider how the bus incremental costs can be used to calculate the *short run marginal costs* (SRMC) of wheeling. Figure 10.4 shows three systems, A, B, and C, with A selling P_w MW to system C and system B wheeling that amount. The figure shows a single point for injecting the power (bus 1) and a single point for delivery to system C (bus 2). The operators of the wheeling

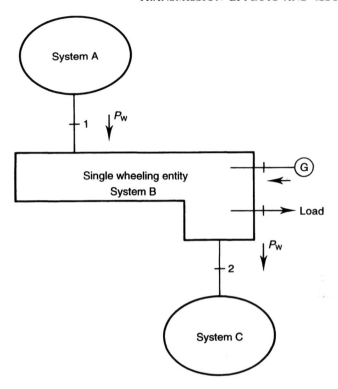

FIG. 10.4 Simple wheeling example.

system, B, can determine the incremental cost of power at both buses by using an optimal power flow (OPF). If these operators were to purchase the block of wheeled power at bus 1 at the incremental cost there, and sell it to system C at the incremental cost of power at bus 2, they would recover their (short run) marginal cost of transmission. Many engineers and economists have suggested that transmission service prices should be based upon these marginal costs since they include the cost of incremental transmission losses and network constraints. The equation to determine this marginal cost is,

$$\Delta F = (\partial F/\partial P_W)\Delta P_W = [\partial F/\partial P_i - \partial F/\partial P_j]\Delta P_W \qquad (10.2)$$

where the power ΔP_W is injected at bus i and withdrawn at bus j. Various implementations of the OPF may be programmed to determine the rate-of-change of the objective function with respect to independent variables and constraints. These computations may be used to evaluate the marginal transmission cost directly.

The six-bus case used previously in Chapter 4 may be used to illustrate these ideas. Two separate wheeling examples were run. In both examples, 50 MW were injected at bus 3 and withdrawn at bus 6. In the first case, no flow limits were imposed on any circuit element. Figure 10.5 shows the power flow that results when the OPF is used to schedule the base case using the generation cost data given in Example 4E. In the second case, a 100-MVA limit was imposed on the circuit connecting bus 3 and bus 6. Figure 10.6 shows the OPF results for this case. Note the redistribution of flows and the new generation schedule.

The short-run marginal transmission cost rates (in R/MWh) found were 0.522 for the unconstrained case and 3.096 for the constrained case. In the unconstrained example, the marginal cost reflects the effects of the incremental losses. The system dispatch is altered a slight amount to accommodate the additional losses caused by the 50-MW wheeling transfer. No major generation shifts are required. When the flow on the direct line, 3 to 6, is constrained, the generation pattern is shifted in the OPF solution to reduce the MVA flow on that circuit. In doing so, the marginal cost of wheeling is increased to reflect that change.

The effect of a constraint can be illustrated by considering the three-system wheeling situation shown on Figure 10.4. Suppose the transmission system is lossless. With no constraints on power flows, the marginal cost of power will be the same throughout the system. (It will be equal to the incremental cost of the next MWh generated in system B.) Now suppose that there is a constraint in system B such that *before the wheeled power is injected*, no more power may flow from the area near bus 1 to loads near bus 2. (See Figure 10.7 which shows a cut labeled "Transmission bottleneck.") Then, when the power to be wheeled is injected at bus 1 and withdrawn at bus 2, the schedule in system B will be adjusted so that the delivered power is absorbed near bus 1 and generated by units near bus 2. The difference in marginal costs will now increase, reflecting the marginal cost of the constraint. With no constraint violations, marginal costs of wheeling rise gradually to reflect incremental losses. When constraints are reached, the marginal wheeling costs are more volatile and change rapidly.

Marginal cost-based pricing for transmission services has a theoretical appeal. Not everyone is in agreement that transmission services should be priced this way. If the entire transaction is priced at the marginal cost rate after the transaction is in place, the wheeling utility may over- or under-recover its changes in operating costs. Perhaps more importantly, short-run marginal operating costs do not reflect the revenue required to pay the costs related to the investments in the wheeling system's facilities. These facilities make it possible to wheel the power. (It is quite possible that short-run marginal wheeling costs could be negative if a transaction were to result in incremental power flows that reduced the losses in the wheeling system.) Any pricing structure for transmission service needs to incorporate some means of generating the funds required to install and support any new facilities that are needed in order to accommodate growing demands for service. These are the long-run marginal costs. If the transmission network is to be treated as a separate entity, the price structure for transmission service needs to include the long-run costs as well as short-run operating costs.

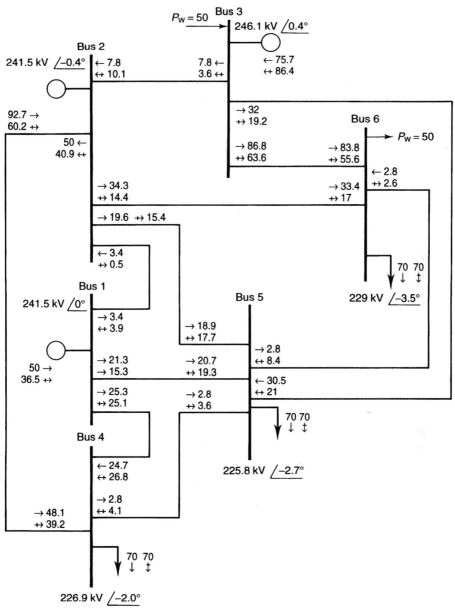

Six-bus network base case AC load flow

where → MW
       ↦ MVAR

FIG. 10.5 Six-bus case with 50 MW being wheeled between bus 3 and bus 6. OPF schedule with no line-flow constraint.

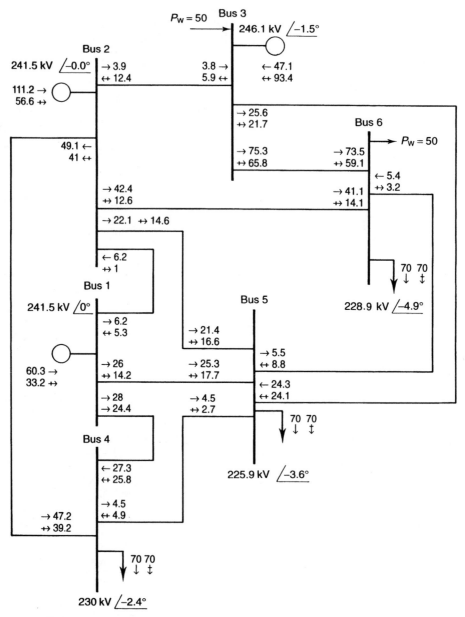

Six-bus network base case AC load flow

where   → MW
        ↦ MVAR

FIG. 10.6   Six-bus case with $P_W = 50$ MW and 100-MVA flow constraint imposed on lines 3–6.

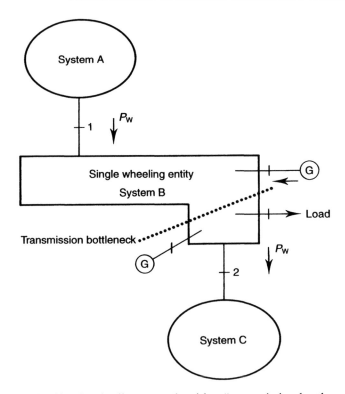

FIG. 10.7   Simple wheeling example with a "transmission bottleneck."

10.8.4   Some Observations

The nature of the electric utility business is changing. In the United Kingdom, the nationwide system was split into several generating companies and 11 regional distribution companies. The former state-owned system was privatized and a market set up to permit the introduction of independent generating companies. Similar developments have taken place in South America and the Philippines. In North America, these types of developments may result in changes in the scheduling and operation of electric power systems. It is conceivable that regional control centers may have the primary function of scheduling the use of the transmission system. Generation dispatch within any organization could still be based on minimizing operating costs, but constraints might be imposed by the transmission system dispatch and the scheduling of transactions could become the primary task of the regional control centers. It is too early (July 1995 at the time of writing) to tell if this will happen and exactly how it might happen.

10.9   TRANSACTIONS INVOLVING NONUTILITY PARTIES

Transactions involving nonutility organizations are increasing. A growing number of larger nonutility generators are being developed. Some of these are large industrial firms that have a need for process heat and steam and can generate electric energy for sale to others at very favorable costs. In some areas, nonutility generating companies have been created to supply some of the needs for new capacity in the region. These are established as profit-making organizations and not as regulated utilities. They must operate in parallel with the utility system and, therefore, there must be some coordination between the groups. The type of relationship and specific operating rules vary.

The customers of these nonutility generators may be utilities or retail consumers. Utilities may purchase the power for resale; this is classified as a "*wholesale market.*" Where sales are made directly to consumers (certain large industrial firms, for example), the transaction is a "*retail*" transaction. The distinction may be important from a commercial viewpoint because the transactions usually require the utilization of the interconnected utilities transmission systems, as well as the load's local utility transmission system. The same distinction between a wholesale and retail transaction would be made if the generating party to a transaction was an electric utility that was making a sale to a retail consumer located in the service territory of another, interconnected utility. When wheeling is involved, the distinction between wholesale and retail transactions tends to become more significant, particularly in the United States because of established practices.

Technical problems involving nonutility generators primarily involve coordination and scheduling issues. The scheduling of the nonutility generator's level of output may be handled in different fashions. It could have a fixed output contract, it might be scheduled by the local utility's control center, or it could be dispatched to meet the load(s) of the buyer of its power. In a market structured like the scheme developed in the United Kingdom, the schedule for some suppliers is determined by a posted price level.

The next four figures illustrate some of the control area configurations that can occur with nonutility parties involved in transactions. In each figure, the nonutility generator is denoted by "G". In Figure 10.8, the generator G is supplying power to the local utility, a wholesale transaction. The dispatch of G

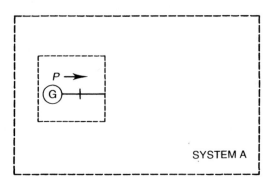

FIG. 10.8   Nonutility generator G delivering P MW to local system A.

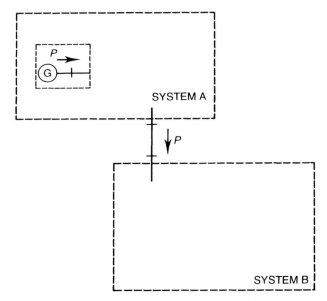

FIG. 10.9   Nonutility generator G delivering P MW to system B.

might be fixed, under control of the local utility, or be based upon a posted purchase price for energy and power. The utility AGC system could treat the generator as a local source or as part of scheduled interchange. In Figure 10.9, the generator is supplying power to a remote utility, and wholesale wheeling is involved. The output of G would be treated as scheduled interchange by both systems.

Suppose the generator G were to sell his output to a retail customer located within the service territory of the local utility. This is illustrated on Figure 10.10. This transaction requires retail wheeling by system A. The unit G could be scheduled in a variety of different fashions depending upon the agreement with system A. It might follow the load demands of the customer, in which case the

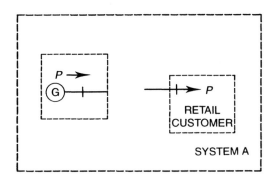

FIG. 10.10   Nonutility generator G delivering power to a retail customer in system A.

utility might treat the output and load as an interchange in its AGC system. If G were contracted to supply a fixed output level, utility A could treat it as a must-run unit and include both the load and the unit in its AGC system.

When this type of transaction involves a retail customer located in an interconnected system, such as shown in Figure 10.11, the situation is more complicated. One alternative would be for system A to treat the output of G as part of a scheduled interchange, with all of the output being delivered to B. System B could then treat the interchange as a schedule between A and the retail customer. The possible arrangements are many. The same type of arrangements would be required if the source were not the nonutility generator G but a third utility, say system C, that was supplying the customer in system B. Further, the "retail customer" could be a distribution utility; in which case "wholesale wheeling" is involved even though the physical situations are identical.

There has been a general movement towards the development of a nonutility generation industry. In many areas, the utilities (particularly those that face a shortage of generation capacity) encourage the installation of unregulated generation resources, and, in some instances, the utilities themselves have become involved in this industry. The movements towards privatization and deregulation encourage this trend. The situation with regard to allowing retail customers access to the transmission system is more contentious. There are a number of larger industrial firms where the cost of electric energy is a significant portion of their cost of production. Many of these organizations would like to obtain access to energy from sources other than the local utility. The issues are unresolved as yet.

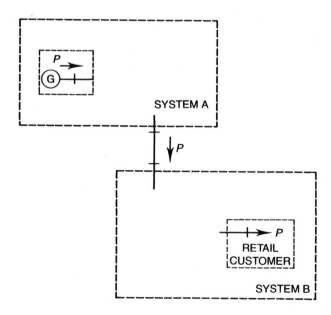

FIG. 10.11   Nonutility generator G delivering power to a retail customer in system B.

In countries where former integrated government power systems have been broken up and privatized, the industry structure seems to be headed for one where the bulk power transmission systems and central dispatch system remain as regulated monopolies. They have the responsibility to provide a market for the purchase and sale of generation and to schedule the operation of the power system to accommodate the generating utilities, the private generating organizations, and the distribution utilities. Furthermore, they may have to coordinate the operation of the system to facilitate the implementation of supply–purchase contracts made directly.

On the other hand, the trend in the United States seems to be less uniform. Some larger transmission-owning utilities favor a system based upon the centrally dispatched power pool. In this concept, the central dispatch office would be responsible for controlling generation and the transmission network. Contracts between buyers and sellers could be made separately, but the actual generation would be the result of an economic schedule of all of the units. Pool agreements would be structured similarly to existing power pool agreements, where no generating entity would have an operating cost higher than the one that it would have had, absent the pool control. This type of arrangement preserves the technical control of the system in the utility, while theoretically permitting any sort of transaction to take place. The "devil is in the details;" prices for transmission and generation services would require careful definition and, perhaps, continued regulatory supervision.

Other transmission-owning utilities appear to favor a more loosely structured market where transactions could be made between various parties, subject to the availability of transmission capacity. Transmission use would then become a separately priced item. This would, it is claimed, allow third-party brokers to make a more efficient (economic) marketplace. Here, the sticking points are apt to be the control and availability of transmission services, as well as their pricing. Technical problems may require more utility control than is deemed acceptable by "free marketers."

Utilities without extensive transmission want access to the networks of others in order to avail themselves of the generation markets. Large industrial concerns with significant electrical consumption are in the same camp. These groups advocate open transmission access with continued regulatory supervision of transmission rates and control, but market-determined pricing for power and energy.

PROBLEMS

10.1   Four areas are interconnected as shown in Figure 10.12. Each area has a total generation capacity of 700 MW currently on-line. The minimum loading of these units is 140 MW in each area. Area loads for a given hour are as shown in Figure 10.12. The transmission lines are each sufficient to transfer any amount of power required.

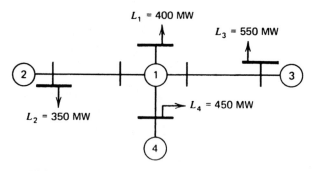

FIG. 10.12   Four-area system for Problem 10.1.

The composite input–output production cost characteristics of each area are as follows:

$$F_1 = 200 + 2P_1 + 0.005P_1^2 \qquad (\text{R}/\text{h})$$

$$F_2 = 325 + 3P_2 + 0.002143P_2^2 \quad (\text{R}/\text{h})$$

$$F_3 = 275 + 2.6P_3 + 0.003091P_3^2 \ (\text{R}/\text{h})$$

$$F_4 = 190 + 3.1P_4 + 0.00233P_4^2 \ (\text{R}/\text{h})$$

In all cases, $140 \le P_i \le 700\,\text{MW}$. Find the cost of each area if each independently supplies its own load, and the total cost for all four areas.

10.2   Assume that area 1 of Problem 10.1 engages in two transactions.

a. Area 1 buys 190 MW from area 2.

b. Area 1 sells 120 MW to area 3.

For each of these transactions, the price is based upon a 50–50 split-savings agreement. Find the price of each transaction, the net generation costs for each area including the sum it pays or receives under the split-savings agreement, with the order of the transactions (as given above) being as follows.

i. a then b.

ii. b then a.

In both instances, find the total cost for the four-area pool.

10.3   Assume that the four areas of Problem 10.1 are centrally dispatched by a pool control center.

a. Find the generation and production cost in each area.

b. Assume a split-savings pool agreement such that each area exporting receives its increased costs of production plus its proportionate share of 50% of the pool savings. Find the cost per MWh of transfer energy (i.e., "pool energy") and the net production cost of each area.

10.4 Assume that the four areas of Problem 10.1 are members of a "power broker." Previous to the hour shown in Problem 10.1, each area submits quotations to the broker to sell successive blocks of 25 or 50 MW and bids to purchase blocks of 25 or 50 MW. In furnishing these data to the broker, assume that the prices quoted are the average incremental costs for the block. The broker matching rules are as follows.

Rule 1. Quotations to sell and bids to buy are matched only wherever there is a direct connection between the quoting and bidding company.

Rule 2. Transactions are arranged in a priority order where the lowest remaining incremental cost for the quoting area is matched with the highest decremental cost for the bidding areas. [That is, lowest available incremental cost energy available for sale is matched with the area with the greatest available potential incremental cost savings (=decremental cost).]

Rule 3. "Matches" may be made for all or part of a block. The remainder of the block must be used, if possible, before the next block is utilized. Matching will cease when the absolute value of the difference between the incremental and decremental cost drops below 0.33 R/MWh.

Rule 4. No area may be both a buyer and a seller in any given hour.

Rule 5. The price per MWh for each matched transaction is one-half the sum of the absolute values of the incremental and decremental costs.

For this problem, assume that quotes and bids are supplied to the broker by each area as follows.

Area	Quotes to Sell	Quotes to Buy
1	100 MW in 25 MW blocks	100 MW in 25 MW blocks
2	200 MW in 50 MW blocks	None
3	None	200 MW in 50 MW blocks
4	25 MW	25 MW

a. Set up the power broker matching system and establish the transactions that can take place and the price of each.

b. Assume that all feasible transactions take place and find the net production cost to each area and the pool.

10.5 Repeat Problem 10.4 with the following assumptions simultaneously taken in place of those in Problem 10.4.

a. Each area is interconnected with every other area and transfers may take place directly between all pairs of areas.

b. The matched transactions will proceed until the difference between decremental costs is zero instead of 0.33 ℞/MWh.

10.6 Repeat Problem 10.5 with one "clearing price" that applies to all transactions and is equal to the price determined for the last matched transaction.

10.7 Use the cost data for the six-bus base case in Chapter 4, and the power flow and generator output data presented in Figures 10.5 and 10.6 that illustrate the wheeling of 50 MW between bus 3 and bus 6. We want to compute an estimate of the utility's net costs under all three cases. Let

Net cost = total production cost for all generators − charges for wheeling

Produce a table that shows the power generation for each unit and the total system operating cost in ℞/h for the three cases: the base case and the two wheeling cases. The generation data for an optimal power flow calculation of the base case to minimize operating costs with no line flow limits shows the following:

$$P_1 = 50.00 \text{ MW}$$
$$P_2 = 89.63 \text{ MW}$$
$$P_3 = 77.07 \text{ MW}$$
$$P_{loss} = 6.70 \text{ MW}$$

and a cost rate of 3126.36 ℞/h. For the two cases with 50 MW being wheeled, compute the charges for wheeling as (50 MW × the SRMC) for wheeling given in the chapter. These are 0.522 ℞/MWh and 3.096 ℞/MWh for the two wheeling cases. These charges represent income to the utility and reduce the total operating cost. (The question is really: "Does the use of the SRMC for wheeling only recover additional operating costs for the wheeling, or does it make an added profit for the utility?" Remember, this is only one example.)

FURTHER READING

References 1–3 provide a good historical look at the techniques that have gone into power pooling. Reference 4 is an excellent summary of the state-of-the-art (1980) of power brokering and pooling, and reviews the practices of most major U.S. power pools. The list of possible references dealing with the issues of utility deregulation, utility privatization and transmission access is very long. Only a few suggestions are given as a starting point. The references listed at the end of Chapter 13 are also relevant to the topics of this chapter; especially the treatment of SRMC for transmission. A great deal of original work was done by the late Professor Fred Schweppe and his associates at MIT and Harvard. This is summarized in the "*Spot Pricing of Electricity*" book, reference 5. Reference 6 is one example of an approach to establishing wheeling rates. The last three references discuss experiences in the United Kingdom and South America.

1. Mochon, H. H. Jr. "Practices of the New England Power Exchange," *Proceedings of the American Power Conference*, Vol. 34, 1972, pp. 911–925.
2. Happ, H. H. "Multi-Computer Configurations & Diakoptics: Dispatch of Real Power in Power Pools," *1967 Power Industry Computer Applications Conference Proceedings*, IEEE, pp. 95–107.
3. Roth, J. E., Ambrose, Z. C., Schappin, L. A., Gassert, J. D., Hunt, D. M., Williams, D. D., Wood, W., Matrin, L. W., "Economic Dispatch of Pennsylvania–New Jersey–Maryland Interconnection System Generation on a Multi-Area Basis," *1967 Power Industry Computer Applications Conference Proceedings*, IEEE, pp. 109–116.
4. "Power Pooling: Issues and Approaches," DOE/ERA/6385-1, U.S. Department of Energy, 1980.
5. Schweppe, F. C., Caramanis, M. C., Tabors, R. D., Bohn, R. E., "*Spot Pricing of Electricity*," Kluwer Academic, Boston, MA, 1987.
6. Clayton, J. S., Erwin, S. R., Gibson, C. A., "Interchange Costing and Wheeling Loss Evaluation by Means of Incrementals", *IEEE Transactions on Power Systems*, Vol. 5, No. 3, August 1989, pp. 1167–1175.
7. Rudnick, H., Palma, R., Fernandez, J. E., "Marginal Pricing and Supplement Cost Allocation in Transmission Open Access," *IEEE Transactions on Power Systems*, Vol. 10, No. 2, May 1995, pp. 1125–1142.
8. Perez-Arriaga, I. J., Rubio, F. J., Puerta, J. F., Arcleuz, J., Marin, J., "Marginal Pricing of Transmission Services: An Analysis of Cost Recovery," *IEEE Transactions on Power Systems*, Vol. 10, No. 1, February 1995, pp. 546–553.
9. Henney, A., "Challenging the Status Quo: Privatizing Electricity in England and Wales," *Public Utilities Fortnightly*, July 15, 1994, pp. 26–31.

11 Power System Security

11.1 INTRODUCTION

Up until now we have been mainly concerned with minimizing the cost of operating a power system. An overriding factor in the operation of a power system is the desire to maintain system security. System security involves practices designed to keep the system operating when components fail. For example, a generating unit may have to be taken off-line because of auxiliary equipment failure. By maintaining proper amounts of spinning reserve, the remaining units on the system can make up the deficit without too low a frequency drop or need to shed any load. Similarly, a transmission line may be damaged by a storm and taken out by automatic relaying. If, in committing and dispatching generation, proper regard for transmission flows is maintained, the remaining transmission lines can take the increased loading and still remain within limit.

Because the specific times at which initiating events that cause components to fail are unpredictable, the system must be operated at all times in such a way that the system will not be left in a dangerous condition should any credible initiating event occur. Since power system equipment is designed to be operated within certain limits, most pieces of equipment are protected by automatic devices that can cause equipment to be switched out of the system if these limits are violated. If any event occurs on a system that leaves it operating with limits violated, the event may be followed by a series of further actions that switch other equipment out of service. If this process of cascading failures continues, the entire system or large parts of it may completely collapse. This is usually referred to as a *system blackout*.

An example of the type of event sequence that can cause a blackout might start with a single line being opened due to an insulation failure; the remaining transmission circuits in the system will take up the flow that was flowing on the now-opened line. If one of the remaining lines is now too heavily loaded, it may open due to relay action, thereby causing even more load on the remaining lines. This type of process is often termed a *cascading outage*. Most power systems are operated such that any single initial failure event will not leave other components heavily overloaded, specifically to avoid cascading failures.

Most large power systems install equipment to allow operations personnel to monitor and operate the system in a reliable manner. This chapter will deal with the techniques and equipment used in these systems. We will lump these under the commonly used title *system security*.

Systems security can be broken down into three major functions that are carried out in an operations control center:

1. System monitoring.
2. Contingency analysis.
3. Security-constrained optimal power flow.

System monitoring provides the operators of the power system with pertinent up-to-date information on the conditions on the power system. Generally speaking, it is the most important function of the three. From the time that utilities went beyond systems of one unit supplying a group of loads, effective operation of the system required that critical quantities be measured and the values of the measurements be transmitted to a central location. Such systems of measurement and data transmission, called *telemetry systems*, have evolved to schemes that can monitor voltages, currents, power flows, and the status of circuit breakers, and switches in every substation in a power system transmission network. In addition, other critical information such as frequency, generator unit outputs and transformer tap positions can also be telemetered. With so much information telemetered simultaneously, no human operator could hope to check all of it in a reasonable time frame. For this reason, digital computers are usually installed in operations control centers to gather the telemetered data, process them, and place them in a data base from which operators can display information on large display monitors. More importantly, the computer can check incoming information against prestored limits and alarm the operators in the event of an overload or out-of-limit voltage.

State estimation is often used in such systems to combine telemetered system data with system models to produce the best estimate (in a statistical sense) of the current power system conditions or "state." We will discuss some of the highlights of these techniques in Chapter 12.

Such systems are usually combined with supervisory control systems that allow operators to control circuit breakers and disconnect switches and transformer taps remotely. Together, these systems are often referred to as *SCADA systems*, standing for supervisory control and data acquisition system. The SCADA system allows a few operators to monitor the generation and high-voltage transmission systems and to take action to correct overlords or out-of-limit voltages.

The second major security function is contingency analysis. The results of this type of analysis allow systems to be operated defensively. Many of the problems that occur on a power system can cause serious trouble within such a quick time period that the operator could not take action fast enough. This is often the case with cascading failures. Because of this aspect of systems operation, modern operations computers are equipped with contingency analysis programs that model possible systems troubles before they arise. These programs are based on a model of the power system and are used to study outage events and alarm the operators to any potential overlords or out-of-limit voltages. For example, the simplest form of contingency analysis can be put together with a standard power-flow program such as described in Chapter 4, together with procedures to set up the power-flow data for each outage to be studied by the power-flow program. Several variations of this type of contingency analysis scheme involve fast solution methods, automatic contingency event selection, and automatic initializing of the contingency power flows using actual system data and state estimation procedures.

The third major security function is security-constrained optimal power flow. In this function, a contingency analysis is combined with an optimal power flow which seeks to make changes to the optimal dispatch of generation, as well as other adjustments, so that when a security analysis is run, no contingencies result in violations. To show how this can be done, we shall divide the power system into four operating states.

- **Optimal dispatch:** this is the state that the power system is in prior to any contingency. It is optimal with respect to economic operation, but it may not be secure.
- **Post contingency:** is the state of the power system after a contingency has occurred. We shall assume here that this condition has a security violation (line or transformer beyond its flow limit, or a bus voltage outside the limit).
- **Secure dispatch:** is the state of the system with no contingency outages, but with corrections to the operating parameters to account for security violations.
- **Secure post-contingency:** is the state of the system when the contingency is applied to the base-operating condition—with corrections.

We shall illustrate the above with an example. Suppose the trivial power system consisting of two generators, a load, and a double circuit line, is to be operated with both generators supplying the load as shown below (ignore losses):

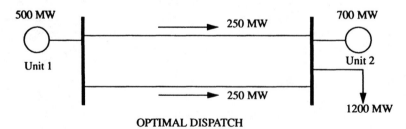

OPTIMAL DISPATCH

We assume that the system as shown is in economic dispatch, that is the 500 MW from unit 1 and the 700 MW from unit 2 is the optimum dispatch. Further, we assert that each circuit of the double circuit line can carry a maximum of 400 MW, so that there is no loading problem in the base-operating condition.

Now, we shall postulate that one of the two circuits making up the transmission line has been opened because of a failure. This results in

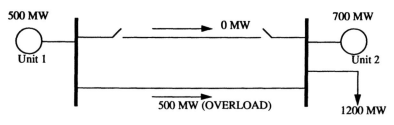

POST CONTINGENCY STATE

Now there is an overload on the remaining circuit. We shall assume for this example that we do not want this condition to arise and that we will correct the condition by lowering the generation on unit 1 to 400 MW. The secure dispatch is

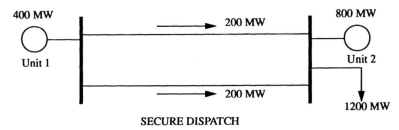

SECURE DISPATCH

Now, if the same contingency analysis is done, the post-contingency condition is

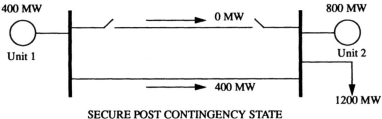

SECURE POST CONTINGENCY STATE

By adjusting the generation on unit 1 and unit 2, we have prevented the post-contingency operating state from having an overload. This is the essence of what is called "security corrections." Programs which can make control adjustments to the base or pre-contingency operation to prevent violations in the post-contingency conditions are called "security-constrained optimal power flows" or SCOPF. These programs can take account of many contingencies and calculate adjustments to generator MW, generator voltages, transformer taps, interchange, etc. We shall show how the SCOPF is formed in Chapter 13.

Together, the functions of system monitoring, contingency analysis, and corrective action analysis comprise a very complex set of tools that can aid in the secure operation of a power system. This chapter concentrates on contingency analysis.

11.2 FACTORS AFFECTING POWER SYSTEM SECURITY

As a consequence of many widespread blackouts in interconnected power systems, the priorities for operation of modern power systems have evolved to the following.

- Operate the system in such a way that power is delivered reliably.
- Within the constraints placed on the system operation by reliability considerations, the system will be operated most economically.

The greater part of this book is devoted to developing methods to operate a power system to gain maximum economy. But what factors affect its operation from a reliability standpoint? We will assume that the engineering groups who have designed the power system's transmission and generation systems have done so with reliability in mind. This means that adequate generation has been installed to meet the load and that adequate transmission has been installed to deliver the generated power to the load. If the operation of the system went on without sudden failures or without experiencing unanticipated operating states, we would probably have no reliability problems. However, any piece of equipment in the system can fail, either due to internal causes or due to external causes such as lightning strikes, objects hitting transmission towers, or human errors in setting relays. It is highly uneconomical, if not impossible, to build a power system with so much redundancy (i.e., extra transmission lines, reserve generation, etc.) that failures never cause load to be dropped on a system. Rather, systems are designed so that the probability of dropping load is acceptably small. Thus, most power systems are designed to have sufficient redundancy to withstand all major failure events, but this does not guarantee that the system will be 100% reliable.

Within the design and economic limitations, it is the job of the operators to try to maximize the reliability of the system they have at any given time. Usually, a power system is never operated with all equipment "in" (i.e., connected) since failures occur or maintenance may require taking equipment out of service. Thus, the operators play a considerable role in seeing that the system is reliable.

In this chapter, we will not be concerned with all the events that can cause trouble on a power system. Instead, we will concentrate on the possible consequences and remedial actions required by two major types of failure events—transmission-line outages and generation-unit failures.

Transmission-line failures cause changes in the flows and voltages on the transmission equipment remaining connected to the system. Therefore, the analysis of transmission failures requires methods to predict these flows and voltages so as to be sure they are within their respective limits. Generation failures can also cause flows and voltages to change in the transmission system, with the addition of dynamic problems involving system frequency and generator output.

11.3 CONTINGENCY ANALYSIS: DETECTION OF NETWORK PROBLEMS

We will briefly illustrate the kind of problems we have been describing by use of the six-bus network used in Chapter 4. The base-case power flow results for Example 4A are shown in Figure 11.1 and indicate a flow of 43.8 MW and 60.7 MVAR on the line from bus 3 to bus 6. The limit on this line can be expressed in MW or in MVA. For the purpose of this discussion, assume that we are only interested in the MW loading on the line. Now let us ask what will happen if the transmission line from bus 3 to bus 5 were to open. The resulting flows and voltages are shown in Figure 11.2. Note that the flow on the line from bus 3 to bus 6 has increased to 54.9 MW and that most of the other transmission lines also experienced changes in flow. Note also that the bus voltage magnitudes changed, particularly at bus 5, which is now almost 5% below nominal. Figures 11.3 and 11.4 are examples of generator outages and serve to illustrate the fact that generation outages can also result in changes in flows and voltages on a transmission network. In the example shown in Figure 11.3, all the generation lost from bus 3 is picked up on the generator at bus 1. Figure 11.4 shows the case when the loss of generation on bus 3 is made up by an increase in generation at buses 1 and 2. Clearly, the differences in flows and voltages show that how the lost generation is picked up by the remaining units is imporant.

If the system being modeled is part of a large interconnected network, the lost generation will be picked up by a large number of generating units outside the system's immediate control area. When this happens, the pickup in generation is seen as an increase in flow over the tie lines to the neighboring systems. To model this, we can build a network model of our own system plus an equivalent network of our neighbor's system and place the swing bus or reference bus in the equivalent system. A generator outage is then modeled so that all lost generation is picked up on the swing bus, which then appears as an increase on the tie flows, thus approximately modeling the generation loss when interconnected. If, however, the system of interest is not interconnected, then the loss of generation must be shown as a pickup in output on the other generation units within the system. An approximate method of doing this is shown in Section 11.3.2.

Operations personnel must know which line or generation outages will cause flows or voltages to fall outside limits. To predict the effects of outages,

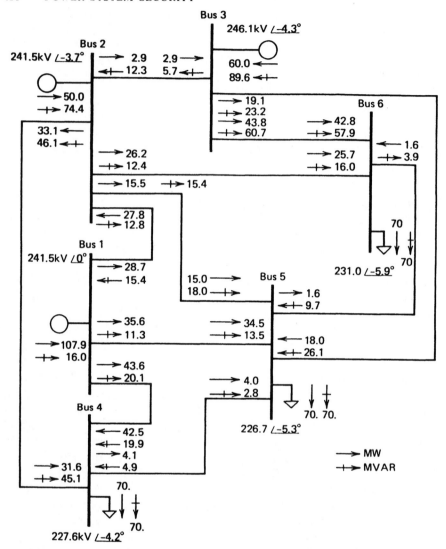

FIG. 11.1   Six-bus network base case AC power flow (see Example 4A).

contingency analysis techniques are used. Contingency analysis procedures model single failure events (i.e., one-line outage or one-generator outage) or multiple equipment failure events (i.e., two transmission lines, one transmission line plus one generator, etc.), one after another in sequence until "all credible outages" have been studied. For each outage tested, the contingency analysis procedure checks all lines and voltages in the network against their respective limits. The simplest form of such a contingency analysis technique is shown in Figure 11.5.

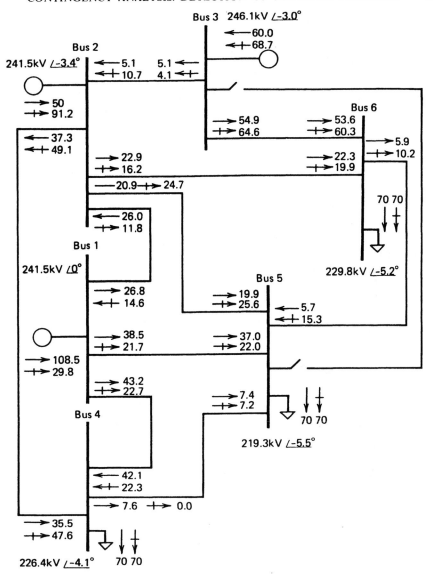

FIG. 11.2   Six-bus network line outage case; line from bus 3 to bus 5 opened.

The most difficult methodological problem to cope with in contingency analysis is the speed of solution of the model used. The most difficult logical problem is the selection of "all credible outages." If each outage case studied were to solve in 1 sec and several thousand outages were of concern, it would take close to 1 h before all cases could be reported. This would be useful if the system conditions did not change over that period of time. However, power

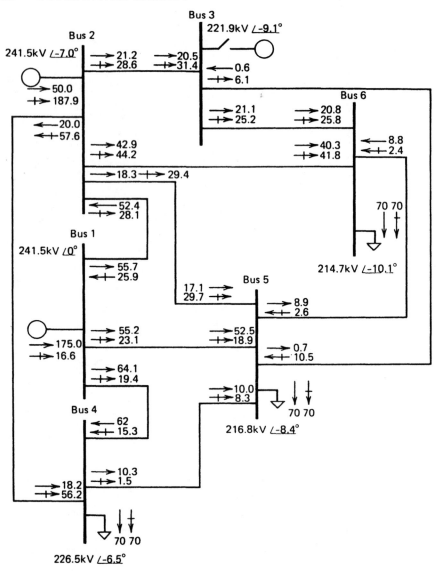

FIG. 11.3 Six-bus network generator outage case. Outage of generator on bus 3; lost generation picked up on generator 1.

systems are constantly undergoing changes and the operators usually need to know if the present operation of the system is safe, without waiting too long for the answer. Contingency analysis execution times of less than 1 min for several thousand outage cases are typical of computer and analytical technology as of 1995.

One way to gain speed of solution in a contingency analysis procedure is to

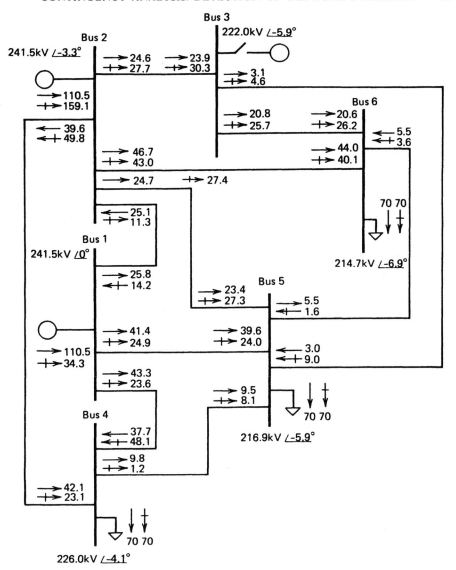

FIG. 11.4 Six-bus network generator outage case. Outage of generator on bus 3; lost generation picked up on generator 1 and generator 2.

use an approximate model of the power system. For many systems, the use of DC load flow models provides adequate capability. In such systems, the voltage magnitudes may not be of great concern and the DC load flow provides sufficient accuracy with respect to the megawatt flows. For other systems, voltage is a concern and full AC load flow analysis is required.

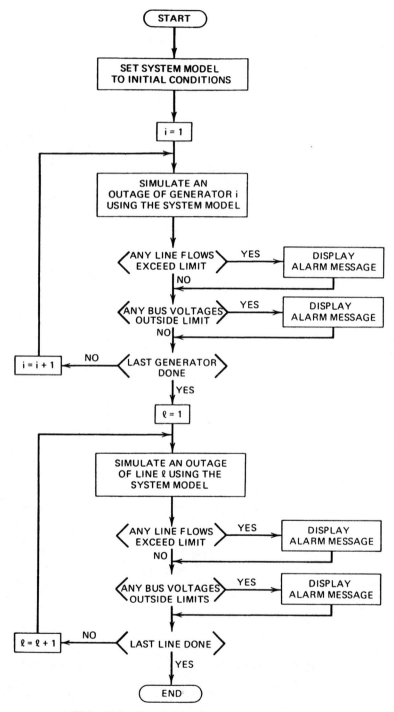

FIG. 11.5 Contingency analysis procedure.

11.3.1 An Overview of Security Analysis

A security analysis study which is run in an operations center must be executed very quickly in order to be of any use to operators. There are three basic ways to accomplish this.

- Study the power system with approximate but very fast algorithms.
- Select only the important cases for detailed analysis.
- Use a computer system made up of multiple processors or vector processors to gain speed.

The first method has been in use for many years and goes under various names such as "D factor methods," "linear sensitivity methods," "DC power flow methods," etc. This approach is useful if one only desires an approximate analysis of the effect of each outage. This text presents these methods under the name linear sensitivity factors and uses the same derivation as was presented in Chapter 4 under the DC power flow methods. It has all the limitations attributed to the DC power flow; that is, only branch MW flows are calculated and these are only within about 5% accuracy. There is no knowledge of MVAR flows or bus voltage magnitudes. Linear sensitivity factors are presented in Section 11.3.2.

If it is necessary to know a power system's MVA flows and bus voltage magnitudes after a contingency outage, then some form of complete AC power flow must be used. This presents a great deal of difficulty when thousands of cases must be checked. It is simply impossible, even on the fastest processors in existence today (1995) to execute thousands of complete AC power flows quickly enough. Fortunately, this need not be done as most of the cases result in power flow results which do not have flow or voltage limit violations. What is needed are ways to eliminate all or most of the nonviolation cases and only run complete power flows on the "critical" cases. These techniques go under the names of "contingency selection" or "contingency screening" and are introduced in Section 11.3.4.

Last of all, it must be mentioned that there are ways of running thousands of contingency power flows if special computing facilities are used. These facilities involve the use of many processors running separate cases in parallel, or vector processors which achieve parallel operation by "unwinding" the looping instruction sets in the computer code used. As of the writing of this edition (1995), such techniques are still in the research stage.

11.3.2 Linear Sensitivity Factors

The problem of studying thousands of possible outages becomes very difficult to solve if it is desired to present the results quickly. One of the easiest ways to provide a quick calculation of possible overloads is to use *linear sensitivity factors*. These factors show the approximate change in line flows for changes in generation on the network configuration and are derived from the DC load flow presented in Chapter 4. These factors can be derived in a variety of ways and basically come down to two types:

1. Generation shift factors.
2. Line outage distribution factors.

Here, we shall describe how these factors are used. The derivation of sensitivity factors is given in Appendix 11A.

The generation shift factors are designated $a_{\ell i}$ and have the following definition:

$$a_{\ell i} = \frac{\Delta f_\ell}{\Delta P_i} \tag{11.1}$$

where

ℓ = line index

i = bus index

Δf_ℓ = change in megawatt power flow on line ℓ when a change in generation, ΔP_i, occurs at bus i

ΔP_i = change in generation at bus i

It is assumed in this definition that the change in generation, ΔP_i, is exactly compensated by an opposite change in generation at the reference bus, and that all other generators remain fixed. The $a_{\ell i}$ factor then represents the sensitivity of the flow on line ℓ to a change in generation at bus i. Suppose one wanted to study the outage of a large generating unit and it was assumed that all the generation lost would be made up by the reference generation (we will deal with the case where the generation is picked up by many machines shortly). If the generator in question was generating P_i^0 MW and it was lost, we would represent ΔP_i as

$$\Delta P_i = -P_i^0 \tag{11.2}$$

and the new power flow on each line in the network could be calculated using a precalculated set of "a" factors as follows:

$$\hat{f}_\ell = f_\ell^0 + a_{\ell i}\Delta P_i \quad \text{for } \ell = 1 \dots L \tag{11.3}$$

where

$\hat{f}_\ell$ = flow on line ℓ after the generator on bus i fails

f_ℓ^0 = flow before the failure

The "outage flow," $\hat{f}_\ell$, on each line can be compared to its limit and those exceeding their limit flagged for alarming. This would tell the operations personnel that the loss of the generator on bus i would result in an overload on line ℓ.

The generation shift sensitivity factors are linear estimates of the change in flow with a change in power at a bus. Therefore, the effects of simultaneous changes on several generating buses can be calculated using superposition. Suppose, for example, that the loss of the generator on bus i were compensated by governor action on machines throughout the interconnected system. One frequently used method assumes that the remaining generators pick up in proportion to their maximum MW rating. Thus, the proportion of generation pickup from unit j ($j \neq i$) would be

$$\gamma_{ji} = \frac{P_j^{\max}}{\displaystyle\sum_{\substack{k \\ k \neq i}} P_k^{\max}} \tag{11.4}$$

where

$P_k^{\max}$ = maximum MW rating for generator k

γ_{ji} = proportionality factor for pickup on generating unit j when unit i fails

Then, to test for the flow on line ℓ, under the assumption that all the generators in the interconnection participate in making up the loss, use the following:

$$\hat{f}_\ell = f_\ell^0 + a_{\ell i}\Delta P_i - \sum_{j \neq i}[a_{\ell j}\gamma_{ji}\Delta P_i] \tag{11.5}$$

Note that this assumes that no unit will actually hit its maximum. If this is apt to be the case, a more detailed generation pickup algorithm that took account of generation limits would be required.

The line outage distribution factors are used in a similar manner, only they apply to the testing for overloads when transmission circuits are lost. By definition, the line outage distribution factor has the following meaning:

$$d_{\ell,k} = \frac{\Delta f_\ell}{f_k^0} \tag{11.6}$$

where

$d_{\ell,k}$ = line outage distribution factor when monitoring line ℓ after an outage on line k

Δf_ℓ = change in MW flow on line ℓ

f_k^0 = original flow on line k before it was outaged (opened)

If one knows the power on line ℓ and line k, the flow on line ℓ with line k out can be determined using "d" factors.

$$\hat{f}_\ell = f_\ell^0 + d_{\ell,k}f_k^0 \tag{11.7}$$

where

$$f_\ell^0, f_k^0 = \text{preoutage flows on lines } \ell \text{ and } k, \text{ respectively}$$

$$\hat{f}_\ell = \text{flow on line } \ell \text{ with line } k \text{ out}$$

By precalculating the line outage distribution factors, a very fast procedure can be set up to test all lines in the network for overload for the outage of a particular line. Furthermore, this procedure can be repeated for the outage of each line in turn, with overloads reported to the operations personnel in the form of alarm messages.

Using the generator and line outage procedures described earlier, one can program a digital computer to execute a contingency analysis study of the power system as shown in Figure 11.6. Note that a line flow can be positive or negative so that, as shown in Figure 11.6, we must check f against $-f_\ell^{max}$ as well as f_ℓ^{max}. This figure makes several assumptions; first, it assumes that the generator output for each of the generators in the system is available and that the line flow for each transmission line in the network is also available. Second, it assumes that the sensitivity factors have been calculated and stored, and that they are correct. The assumption that the generation and line flow MWs are available can be satisfied with telemetry systems or with state estimation techniques. The assumption that the sensitivity factors are correct is valid as long as the transmission network has not undergone any significant switching operations that would change its structure. For this reason, control systems that use sensitivity factors must have provision for updating the factors when the network is switched. A third assumption is that all generation pickup will be made on the reference bus. If this is not the case, substitute Eq. 11.5 in the generator outage loop.

EXAMPLE 11A

The $[X]$ matrix for our six-bus sample network is shown in Figure 11.7, together with the generation shift distribution factors and the line outage distribution factors.

The generation shift distribution factors that give the fraction of generation shift that is picked up on a transmission line are designated $a_{\ell i}$. The a factor is obtained by finding line ℓ along the rows and then finding the generator to be shifted along the columns. For instance, the shift factor for a change in the flow on line 1-4 when making a shift in generation on bus 3 is found in the second row, third column.

The line outage distribution factors are stored such that each row and column corresponds to one line in the network. The distribution factor $d_{\ell,k}$ is obtained by finding line ℓ along the rows and then finding line k along that row in the appropriate column. For instance, the line outage distribution factor that gives the fraction of flow picked up on line 3-5 for an outage on line 3-6

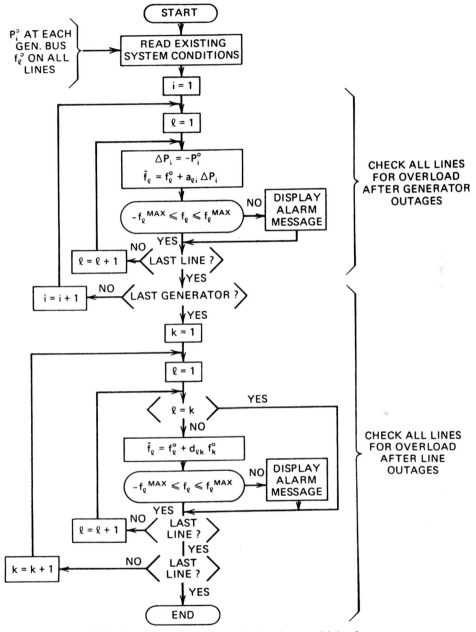

FIG. 11.6 Contingency analysis using sensitivity factors.

X Matrix for Six-Bus Sample System (Reference at Bus 1)

$$
\begin{bmatrix}
0 & 0 & 0 & 0 & 0 & 0 \\
0 & 0.09412 & 0.08051 & 0.06298 & 0.06435 & 0.08129 \\
0 & 0.08051 & 0.16590 & 0.05897 & 0.09077 & 0.12895 \\
0 & 0.06298 & 0.05897 & 0.10088 & 0.05422 & 0.05920 \\
0 & 0.06435 & 0.09077 & 0.05422 & 0.12215 & 0.08927 \\
0 & 0.08129 & 0.12895 & 0.05920 & 0.08927 & 0.16328
\end{bmatrix}
$$

Generation Shift Factors For Six-Bus Sample System

	Bus 1	Bus 2	Bus 3
$\ell = 1$ (line 1-2)	0	-0.47	-0.40
$\ell = 2$ (line 1-4)	0	-0.31	-0.29
$\ell = 3$ (line 1-5)	0	-0.21	-0.30
$\ell = 4$ (line 2-3)	0	0.05	-0.34
$\ell = 5$ (line 2-4)	0	0.31	0.22
$\ell = 6$ (line 2-5)	0	0.10	-0.03
$\ell = 7$ (line 2-6)	0	0.06	-0.24
$\ell = 8$ (line 3-5)	0	0.06	0.29
$\ell = 9$ (line 3-6)	0	-0.01	0.37
$\ell = 10$ (line 4-5)	0	0	-0.08
$\ell = 11$ (line 5-6)	0	-0.06	-0.13

Line Outage Distribution Factors for Six-Bus Sample System

	$k=1$ (Line 1-2)	$k=2$ (Line 1-4)	$k=3$ (Line 1-5)	$k=4$ (Line 2-3)	$k=5$ (Line 2-4)	$k=6$ (Line 2-5)	$k=7$ (Line 2-6)	$k=8$ (Line 3-5)	$k=9$ (Line 3-6)	$k=10$ (Line 4-5)	$k=11$ (Line 5-6)
$\ell = 1$ (line 1-2)		0.64	0.54	-0.11	-0.50	-0.21	-0.12	-0.14	0.01	0.01	0.13
$\ell = 2$ (line 1-4)	0.59		0.46	-0.03	0.61	-0.06	-0.04	-0.04	0	-0.33	0.04
$\ell = 3$ (line 1-5)	0.41	0.36		0.15	-0.11	0.27	0.16	0.18	-0.02	0.32	-0.17
$\ell = 4$ (line 2-3)	-0.10	-0.03	0.18		0.12	0.23	0.47	-0.40	-0.53	0.17	0.13
$\ell = 5$ (line 2-4)	-0.59	0.76	-0.17	0.16		0.30	0.17	0.19	-0.02	-0.67	-0.19
$\ell = 6$ (line 2-5)	-0.19	-0.06	0.33	0.22	0.23		0.24	0.27	-0.03	0.31	-0.26
$\ell = 7$ (line 2-6)	-0.12	-0.04	0.21	0.51	0.15	0.27		-0.20	0.58	0.20	0.44
$\ell = 8$ (line 3-5)	-0.12	-0.04	0.20	-0.38	0.14	0.27	-0.17		0.47	0.19	-0.42
$\ell = 9$ (line 3-6)	0.01	0	-0.03	-0.62	-0.02	-0.03	0.64	0.60		-0.02	0.56
$\ell = 10$ (line 4-5)	0.01	-0.24	0.29	0.13	-0.39	0.24	0.14	0.15	-0.02		-0.15
$\ell = 11$ (line 5-6)	0.11	0.03	-0.18	0.12	-0.13	-0.23	0.36	-0.40	0.42	-0.18	

FIG. 11.7   Outage factors for a six-bus system.

is found in the eighth row and ninth column. Figure 11.3 shows an outage of the generator on bus 3 with all pickup of lost generation coming on the generator at bus 1. To calculate the flow on line 1-4 after the outage of the generator on bus 3, we need (see Figure 11.1):

$$\text{Base-case flow on line 1-4} = 43.6 \text{ MW}$$

$$\text{Base-case generation on bus 3} = 60 \text{ MW}$$

$$\text{Generation shift distribution factor} = a_{1\text{-}4,3} = -0.29$$

Then the flow on line 1-4 after generator outage is = base-case $\text{flow}_{1\text{-}4}$ + $a_{1\text{-}4,3}\Delta P_{\text{gen}_3} = 43.6 + (-0.29)(-60 \text{ MW}) = 61 \text{ MW}$.

To show how the line outage and generation shift factors are used, calculate some flows for the outages shown in Figures 11.2 and 11.3. Figure 11.2 shows an outage of line 3-5. If we wish to calculate the power flowing on line 3-6 with line 3-5 opened, we would need the following.

$$\text{Base-case flow on line 3-5} = 19.1 \text{ MW}$$

$$\text{Base-case flow on line 3-6} = 43.8 \text{ MW}$$

$$\text{Line outage distribution factor: } d_{3\text{-}6,3\text{-}5} = 0.60$$

Then the flow on 3-6 after the outage is = base $\text{flow}_{3\text{-}6} + d_{3\text{-}6,3\text{-}5} \times$ base $\text{flow}_{3\text{-}5} = 43.8 + (0.60) \times (19.1) = 55.26 \text{ MW}$.

In both outage cases, the flows calculated by the sensitivity methods are reasonably close to the values calculated by the full AC load flows as shown in Figures 11.2 and 11.3.

11.3.3   AC Power Flow Methods

The calculations made by network sensitivity methods are faster than those made by AC power flow methods and therefore find wide use in operations control systems. However, there are many power systems where voltage magnitudes are the critical factor in assessing contingencies. In addition, there are some systems where VAR flows predominate on some circuits, such as underground cables, and an analysis of only the MW flows will not be adequate to indicate overloads. When such situations present themselves, the network sensitivity methods may not be adequate and the operations control system will have to incorporate a full AC power flow for contingency analysis.

When an AC power flow is to be used to study each contingency case, the speed of solution and the number of cases to be studied are critical. To repeat what was said before, if the contingency alarms come too late for operators to act, they are worthless. Most operations control centers that use an AC power flow program for contingency analysis use either a Newton–Raphson or the decoupled power flow. These solution algorithms are used because of their speed of solution and the fact that they are reasonably reliable in convergence when solving difficult cases. The decoupled load flow has the further advantage that a matrix alteration formula can be incorporated into it to simulate the outage of transmission lines without reinverting the system Jacobian matrix at each iteration.

The simplest AC security analysis procedure consists of running an AC power flow analysis for each possible generator, transmission line, and transformer outage as shown in Figure 11.8. This procedure will determine the overloads and voltage limit violations accurately (at least within the accuracy of the power flow program, the accuracy of the model data, and the accuracy with which we have obtained the initial conditions for the power flow). It does suffer a major drawback, however, and that concerns the time such a program takes to execute. If the list of outages has several thousand entries, then the total time to test for all of the outages can be too long.

We are thus confronted with a dilemma. Fast, but inaccurate, methods involving the a and d factors can be used to give rapid analysis of the system, but they cannot give information about MVAR flows and voltages. Slower, full AC power flow methods give full accuracy but take too long.

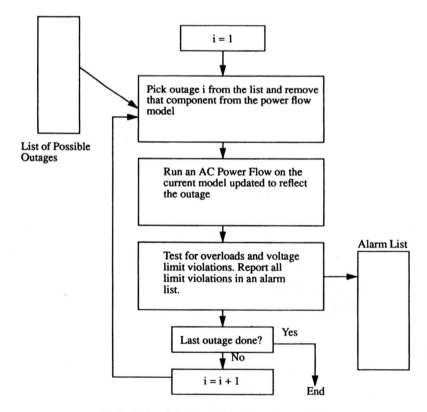

FIG. 11.8   AC power flow security analysis.

Fortunately, there is a way out of this dilemma. Because of the way the power system is designed and operated, very few of the outages will actually cause trouble. That is, most of the time spent running AC power flows will go for solutions of the power flow model that discover that there are no problems. Only a few of the power flow solutions will, in fact, conclude that an overload or voltage violation exists.

The solution to this dilemma is to find a way to select contingencies in such a way that only those that are likely to result in an overload or voltage limit violation will actually be studied in detail and the other cases will go unanalyzed. A flowchart for a process like this appears in Figure 11.9. Selecting

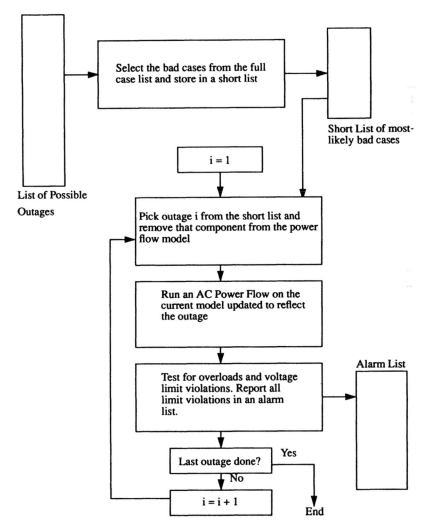

FIG. 11.9   AC power flow security analysis with contingency case selection.

the bad or likely trouble cases from the full outage case list is not an exact procedure and has been the subject of intense research for the past 15 years. Two sources of error can arise.

1. **Placing too many cases on the short list:** this is essentially the "conservative" approach and simply leads to longer run times for the security analysis procedure to execute.

2. **Skipping cases:** here, a case that would have shown a problem is not placed on the short list and results in possibly having that outage take place and cause trouble without the operators being warned.

11.3.4 Contingency Selection

We would like to get some measure as to how much a particular outage might affect the power system. The idea of a performance index seems to fulfill this need. The definition for the overload performance index (PI) is as follows:

$$\text{PI} = \sum_{\substack{\text{all branches} \\ l}} \left(\frac{P_{\text{flow } l}}{P_l^{\max}} \right)^{2n} \tag{11.8}$$

If n is a large number, the PI will be a small number if all flows are within limit, and it will be large if one or more lines are overloaded. The problem then is how to use this performance index.

Various techniques have been tried to obtain the value of PI when a branch is taken out. These calculations can be made exactly if $n = 1$; that is, a table of PI values, one for each line in the network, can be calculated quite quickly. The selection procedure then involves ordering the PI table from largest value to least. The lines corresponding to the top of the list are then the candidates for the short list. One procedure simply ordered the PI table and then picked the top N_c entries from this list and placed them on the short list (see reference 8).

However when $n = 1$, the PI does not snap from near zero to near infinity as the branch exceeds its limit. Instead, it rises as a quadratic function. A line that is just below its limit contributes to PI almost equal to one that is just over its limit. The result is a PI that may be large when many lines are loaded just below their limit. Thus the PI's ability to distinguish or detect bad cases is limited when $n = 1$. Ordering the PI values when $n = 1$ usually results in a list that is not at all representative of one with the truly bad cases at the top. Trying to develop an algorithm that can quickly calculate PI when $n = 2$ or larger has proven extremely difficult.

One way to perform an outage case selection is to perform what has been called the *1P1Q method* (see references 9 and 10). Here, a decoupled power flow is used. As shown in Figure 11.10, the solution procedure is interrupted after one iteration (one $P - \theta$ calculation and one $Q - V$ calculation; thus, the name 1P1Q). With this procedure, the PI can use as large an n value as desired, say $n = 5$. There appears to be sufficient information in the solution at the end of

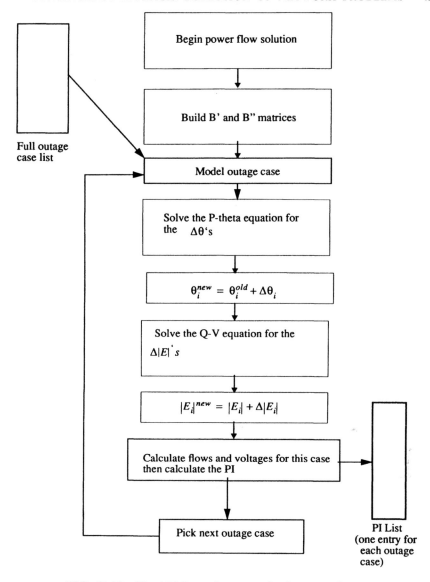

FIG. 11.10   The 1P1Q contingency selection procedure.

the first iteration of the decoupled power flow to give a reasonable PI. Another advantage to this procedure is the fact that the voltages can also be included in the PI. Thus, a different PI can be used, such as:

$$PI = \sum_{\substack{\text{all branches} \\ i}} \left(\frac{P_{\text{flow } l}}{P_l^{\max}} \right)^{2n} + \sum_{\substack{\text{all buses} \\ i}} \left(\frac{\Delta |E_i|}{\Delta |E|^{\max}} \right)^{2m} \tag{11.9}$$

where $\Delta|E_i|$ is the difference between the voltage magnitude as solved at the end of the 1P1Q procedure and the base-case voltage magnitude. $\Delta|E|^{max}$ is a value set by utility engineers indicating how much they wish to limit a bus voltage from changing on one outage case.

To complete the security analysis, the PI list is sorted so that the largest PI appears at the top. The security analysis can then start by executing full power flows with the case which is at the top of the list, then solve the case which is second, and so on down the list. This continues until either a fixed number of cases is solved, or until a predetermined number of cases are solved which do not have any alarms.

11.3.5   Concentric Relaxation

Another idea to enter the field of security analysis in power systems is that an outage only has a limited geographical effect. The loss of a transmission line does not cause much effect a thousand miles away; in fact, we might hope that it doesn't cause much trouble beyond 20 miles from the outage, although if the line were a heavily loaded, high-voltage line, its loss will most likely be felt more than 20 miles away.

To realize any benefit from the limited geographical effect of an outage, the power system must be divided into two parts: the affected part and the part that is unaffected. To make this division, the buses at the end of the outaged line are marked as layer zero. The buses that are one transmission line or transformer from layer zero are then labeled layer one. This same process can be carried out, layer by layer, until all the buses in the entire network are included. Some arbitrary number of layers is chosen and all buses included in that layer and lower-numbered layers are solved as a power flow with the outage in place. The buses in the higher-numbered layers are kept as constant voltage and phase angle (i.e., as reference buses).

This procedure can be used in two ways: either the solution of the layers included becomes the final solution of that case and all overloads and voltage violations are determined from this power flow, or the solution simply is used to form a performance index for that outage. Figure 11.11 illustrates this layering procedure.

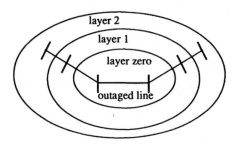

FIG. 11.11   Layering of outage effects.

The concentric relaxation procedure was originally proposed by Zaborsky (see reference 13). The trouble with the concentric relaxation technique is that it requires more layers for circuits whose influence is felt further from the outage.

11.3.6 Bounding

A paper by Brandwajn (reference 11) solves at least one of the problems in using the concentric relaxation method. Namely, it uses an adjustable region around the outage to solve for the outage case overloads. In reference 11, this is applied only to the linear (DC) power flow; it has subsequently been extended for AC network analysis.

To perform the analysis in the bounding technique we define three subsystems of the power system as follows:

$N1$ = the subsystem immediately surrounding the outaged line

$N2$ = the external subsystem that we shall not solve in detail

$N3$ = the set of boundary buses that separate N1 and N2

The subsystems appear as shown in Figure 11.12. The bounding method is based on the fact that we can make certain assumptions about the phase angle spread across the lines in N2, given the injections in N1 and the maximum phase angle appearing across any two buses in N3. In Appendix 11A of this chapter we show how to calculate the ΔP_k and the ΔP_m injections that will make the phase angles on buses k and m simulate the outage of line $k-m$.

If we are given a transmission line in N2 with flow f_{pq}^0, then there is a maximum amount that the flow on pq can shift. That is, it can increase from

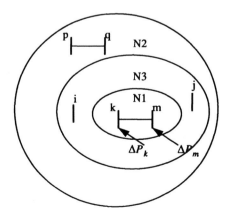

FIG. 11.12 Layers used in bounding analysis.

f^0_{pq} to its upper limit or it can decrease to its lower limit. Then,

$$\Delta f^{max}_{pq} = \text{smaller of } [(f^+_{pq} - f^0_{pq}), (f^0_{pq} - f^-_{pq})] \tag{11.10}$$

Further, we can translate this into a maximum change in phase angle difference as follows:

$$f_{pq} = \frac{1}{x_{pq}} (\theta_p - \theta_q) \tag{11.11}$$

or

$$\Delta f_{pq} = \frac{1}{x_{pq}} (\Delta\theta_p - \Delta\theta_q) \tag{11.12}$$

and finally:

$$(\Delta\theta_p - \Delta\theta_q)^{max} = \Delta f^{max}_{pq} x_{pq} \tag{11.13}$$

Thus, we can define the maximum change in the phase angle difference across pq. Reference 11 develops the theorem that:

$$|\Delta\theta_p - \Delta\theta_q| < |\Delta\theta_i - \Delta\theta_j| \tag{11.14}$$

where i and j are any pair of buses in N3, $\Delta\theta_i$ is the largest $\Delta\theta$ in N3, and $\Delta\theta_j$ is the smallest $\Delta\theta$ in N3 (see Appendix 11B).

Equation 11.14 is interpreted as follows: the right-hand side, $|\Delta\theta_i - \Delta\theta_j|$, provides an upper limit to the maximum change in angular spread across any circuit in N2. Thus, it provides us with a limit as to how far any of the N2 circuits can change their flow. By combining Eqs. 11.13 and 11.14 we obtain:

$$\Delta f^{max}_{pq} x_{pq} < |\Delta\theta_i - \Delta\theta_j| \tag{11.15}$$

Figure 11.13 shows a graphical interpretation of the bounding process. There are two cases represented in Figure 11.13: a circuit on the top of the figure that

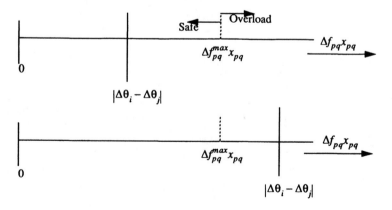

FIG. 11.13   Interpretation of bounding.

cannot go over limit, while that on the bottom could. In each case, the horizontal line represents the change in flow on circuit pq times its reactance, $\Delta f_{pq} x_{pq}$; the dotted line, labeled $\Delta f_{pq}^{max} x_{pq}$, represents the point where circuit pq will go into overload and is determined as explained previously. Any value of $\Delta f_{pq} x_{pq}$ to the right of the dotted line represents an overload.

The solid line labeled $|\Delta \theta_i - \Delta \theta_j|$ represents the upper limit on $\Delta f_{pq} x_{pq}$. Thus, if the solid line is below (to the left) of the dotted line, then the circuit theory upper limit predicts that the circuit cannot go into overload; if on the other hand, the solid line is above (to the right of) the dotted line, the circuit may be shifted in flow due to the outage so as to violate a limit.

A completely safe N2 region would be one in which the maximum $|\Delta \theta_i - \Delta \theta_j|$ upper limit is small enough to be less than all of the $\Delta f_{pq}^{max} x_{pq}$ limits. In fact, as the N1 region is enlarged, the value of $|\Delta \theta_i - \Delta \theta_j|$ will become smaller and smaller. Therefore, the test to determine whether the N1 region encompasses all possible overloaded circuits should be as follows:

All circuits in N2 are safe from overload if the value of $|\Delta \theta_i - \Delta \theta_j|$ is less than the smallest value of $\Delta f_{pq}^{max} x_{pq}$ over all pairs pq, where pq corresponds to the buses at the ends of circuits in N2

If this condition fails, then we have to expand N1, calculate a new $|\Delta \theta_i - \Delta \theta_j|$ in N3, and rerun the test over the newly defined N2 region circuits. When an N2 is found which passes the test, we are done and only region N1 need be studied in detail.

References 10 and 12 extend this concept to screening for AC contingency effects. Such contingency selection/screening techniques form the foundation for many real-time computer security analysis algorithms.

EXAMPLE 11B

In this example, we shall take the six-bus sample system used previously and show how the bounding technique works so that not all of the circuits in the system need be analyzed. Note that this is a small system so that the net savings in computer time may not be that great. Nonetheless, it demonstrates the principles used in the bounding technique quite well.

We shall study the outage of transmission line 3-6. The DC power flow will be used throughout and the initial conditions will be those shown in Figure 4.12. The MW limits on the transmission lines are shown in the table at the top of the next page.

Line	MW Limit
1-2	30
1-4	50
1-5	40
2-3	20
2-4	40
2-5	20
2-6	30
3-5	20
3-6	60
4-5	20
5-6	20

In this example, we shall proceed in steps. Step A will analyze the system as if the N1 and N3 regions consist of only line 3-6 itself, as shown in Figure 11.14. If the bounding criteria is met, no other analysis need be done as it will establish that no overloads exist anywhere in the system. If the bounding criteria fails, we still proceed to step B. Step B expands the bounded region from line 3-6 to include all buses which are once removed from buses 3 and 6; that is, it includes buses 2, 3, 5, and 6 as shown in Figure 11.15, and in this case the boundary of the region, N3, consists of buses 2 and 5.

To start, we need to calculate Δf_{pq}^{max} and then $\Delta f_{pq}^{max} x_{pq}$ as given in Eqs. 11.10 through 11.13. These values are given below where the flows and flow limits are all converted to per unit on a 100 MVA base. (The line reactances are found in the appendix to Chapter 4.)

Line	MW Limit (per unit)	f_{pq}^0 (per unit)	Δf_{pq}^{max}	x_{pq}	$\Delta f_{pq}^{max} x_{pq}$
1-2	0.30	0.253	0.047	0.20	0.0094
1-4	0.50	0.416	0.084	0.20	0.0168
1-5	0.40	0.331	0.069	0.30	0.0207
2-3	0.20	0.018	0.182	0.25	0.0455
2-4	0.40	0.325	0.075	0.10	0.0075
2-5	0.20	0.162	0.038	0.30	0.0114
2-6	0.30	0.248	0.052	0.20	0.0104
3-5	0.20	0.169	0.031	0.26	0.00806
3-6	0.60	0.449	—	—	—
4-5	0.20	0.041	0.159	0.40	0.0636
5-6	0.20	0.003	0.197	0.30	0.0591

For step A, we use Eq. 11A.13 from Appendix 11A to calculate $\delta_{3,36}$ and $\delta_{6,36}$ as

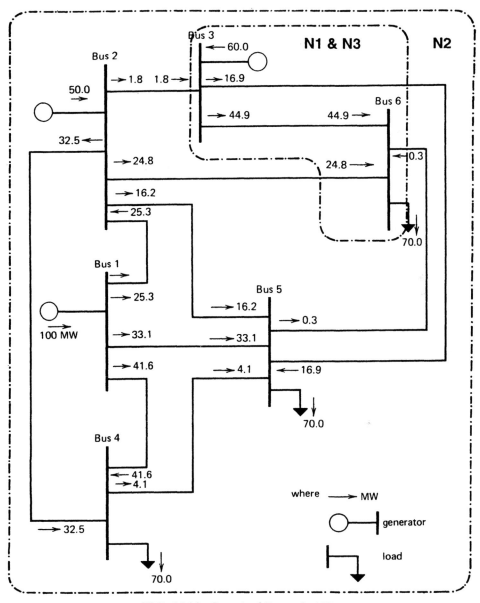

FIG. 11.14   Step A of Example 11B.

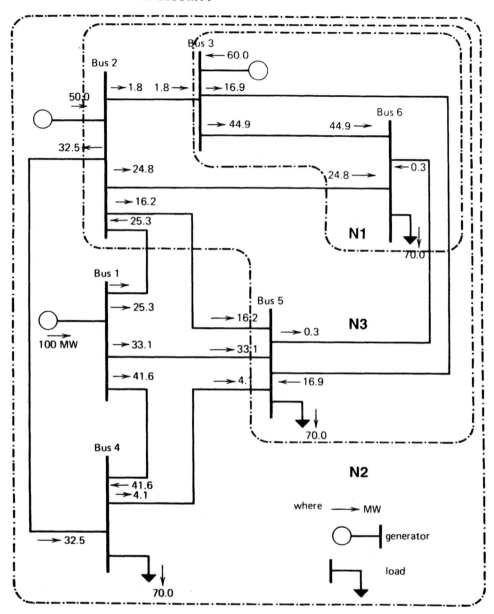

FIG. 11.15   Step B of Example 11B.

shown below.

$$\delta_{3,36} = \frac{(X_{33} - X_{36})x_{36}}{x_{36} - (X_{33} + X_{66} - 2X_{36})} = 0.12865$$

$$\delta_{6,36} = \frac{(X_{63} - X_{66})x_{36}}{x_{36} - (X_{ss} + X_{66} - X_{36})} = -0.11953$$

Then using Eq. 11A.11

$$|\Delta\theta_3 - \Delta\theta_6| = 0.111437$$

According to the criterion in Eq. 11.14, the value $|\Delta\theta_i - \Delta\theta_j|$ must be less that the smallest value of $|\Delta\theta_p - \Delta\theta_q|$ which equals $\Delta f_{pq}^{max} x_{pq}$ and is found in the table above to be at line 2-4. Since $|\Delta\theta_3 - \Delta\theta_6| = 0.111437$ and the minimum $|\Delta\theta_i - \Delta\theta_j|$ is $|\Delta\theta_2 - \Delta\theta_4|$ which has a value of 0.0075, the criteria fails. We must proceed to step B.

Step B requires that we calculate $|\Delta\theta_i - \Delta\theta_j|$ for buses 2 and 5. This value is 0.003564 and the bounding criteria is satisfied.

If we had used the d factors for the six-bus system as shown in Example 11A, we could simply find all the line flows for the 3-6 outage as shown in the table below.

Line	MW Limit (per unit)	f_{pq}^0 (per unit)	$f_{pq}^{3\text{-}6\ out}$	
1-2	0.30	0.253	0.257	
1-4	0.50	0.416	0.416	
1-5	0.40	0.331	0.322	
2-3	0.20	0.018	−0.220	overload
2-4	0.40	0.325	0.316	
2-5	0.20	0.162	0.148	
2-6	0.30	0.248	0.508	overload
3-5	0.20	0.169	0.380	overload
3-6	0.60	0.449	—	
4-5	0.20	0.041	0.320	
5-6	0.20	0.003	0.191	

Note that three overloads exist on lines 2-3, 2-6, and 3-5, which are all within the bounded region N1 + N3 in Figure 11.15.

APPENDIX 11A
Calculation of Network Sensitivity Factors

First, we show how to derive the generation-shift sensitivity factors. To start, repeat Eq. 4.36.

$$\mathbf{\theta} = [X]\mathbf{P} \tag{11A.1}$$

This is the standard matrix calculation for the DC load flow. Since the DC power-flow model is a linear model, we may calculate perturbations about a given set of system conditions by use of the same model. Thus, if we are interested in the changes in bus phase angles, $\Delta\theta$, for a given set of changes in the bus power injections, $\Delta\mathbf{P}$, we can use the following calculation.

$$\Delta\mathbf{\theta} = [X]\Delta\mathbf{P} \tag{11A.2}$$

In Eq. 11A.1, it is assumed that the power on the swing bus is equal to the sum of the injections of all the other buses. Similarly, the net perturbation of the swing bus in Eq. 11A.2 is the sum of the perturbations on all the other buses.

Suppose that we are interested in calculating the generation shift sensitivity factors for the generator on bus i. To do this, we will set the perturbation on bus i to $+1$ and the perturbation on all the other buses to zero. We can then solve for the change in bus phase angles using the matrix calculation in Eq. 11A.3.

$$\Delta\mathbf{\theta} = [X]\begin{bmatrix} +1 \\ -1 \end{bmatrix}\begin{matrix} \text{—row } i \\ \text{—ref row} \end{matrix} \tag{11A.3}$$

The vector of bus power injection perturbations in Eq. 11A.3 represents the situation when a 1 pu power increase is made at bus i and is compensated by a 1 pu decrease in power at the reference bus. The $\Delta\theta$ values in Eq. 11A.3 are thus equal to the derivative of the bus angles with respect to a change in power injection at bus i. Then, the required sensitivity factors are

$$a_{\ell i} = \frac{df_\ell}{dP_i} = \frac{d}{dP_i}\left[\frac{1}{x_\ell}(\theta_n - \theta_m)\right]$$

$$= \frac{1}{x_\ell}\left(\frac{d\theta_n}{dP_i} - \frac{d\theta_m}{dP_i}\right) = \frac{1}{x_\ell}(X_{ni} - X_{mi}) \tag{11A.4}$$

where

$$X_{ni} = \frac{d\theta_n}{dP_i} = n^{\text{th}} \text{ element from the } \Delta\mathbf{\theta} \text{ vector in Eq. 11A.3}$$

$$X_{mi} = \frac{d\theta_m}{dP_i} = m^{\text{th}} \text{ element from the } \Delta\mathbf{\theta} \text{ vector in Eq. 11A.3}$$

$$x_\ell = \text{line reactance for line } \ell$$

A line outage may be modeled by adding two power injections to a system, one at each end of the line to be dropped. The line is actually left in the system and the effects of its being dropped are modeled by injections. Suppose line k

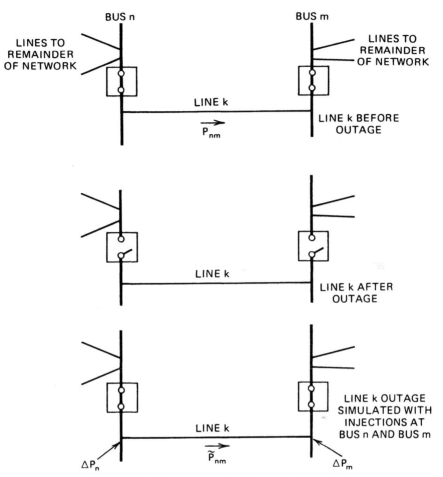

FIG. 11.16   Line outage modeling using injections.

from bus n to bus m were opened by circuit breakers as shown in Figure 11.16. Note that when the circuit breakers are opened, no current flows through them and the line is completely isolated from the remainder of the network. In the bottom part of Figure 11.16, the breakers are still closed but injections ΔP_n and ΔP_m have been added to bus n and bus m, respectively. If $\Delta P_n = \tilde{P}_{nm}$, where $\tilde{P}_{nm}$ is equal to the power flowing over the line, and $\Delta P_m = -\tilde{P}_{nm}$, we will still have no current flowing through the circuit breakers even though they are closed. As far as the remainder of the network is concerned, the line is disconnected.

Using Eq. 11A.2 relating to $\Delta\theta$ and ΔP, we have

$$\Delta\theta = [X]\Delta P$$

where

$$\Delta \mathbf{P} = \begin{bmatrix} \vdots \\ \Delta P_n \\ \vdots \\ \Delta P_m \end{bmatrix}$$

so that

$$\Delta \theta_n = X_{nn} \Delta P_n + X_{nm} \Delta P_m$$
$$\Delta \theta_m = X_{mn} \Delta P_n + X_{mm} \Delta P_m \tag{11A.5}$$

define

$\theta_n, \theta_m, P_{nm}$ to exist before the outage, where P_{nm} is the flow on line k from bus n to bus m

$\Delta \theta_n, \Delta \theta_m, \Delta P_{nm}$ to be the incremental changes resulting from the outage

$\tilde{\theta}_n, \tilde{\theta}_m, \tilde{P}_{nm}$ to exist after the outage

The outage modeling criteria requires that the incremental injections ΔP_n and ΔP_m equal the power flowing over the outaged line *after* the injections are imposed. Then, if we let the line reactance be x_k

$$\tilde{P}_{nm} = \Delta P_n = -\Delta P_m \tag{11A.6}$$

where

$$\tilde{P}_{nm} = \frac{1}{x_k} (\tilde{\theta}_n - \tilde{\theta}_m)$$

then

$$\Delta \theta_n = (X_{nn} - X_{nm}) \Delta P_n$$
$$\Delta \theta_m = (X_{mm} - X_{mn}) \Delta P_n \tag{11A.7}$$

and

$$\tilde{\theta}_n = \theta_n + \Delta \theta_n$$
$$\tilde{\theta}_m = \theta_m + \Delta \theta_m \tag{11A.8}$$

giving

$$\tilde{P}_{nm} = \frac{1}{x_k} (\tilde{\theta}_n - \tilde{\theta}_m) = \frac{1}{x_k} (\theta_n - \theta_m) + \frac{1}{x_k} (\Delta \theta_n - \Delta \theta_m)$$

or

$$\tilde{P}_{nm} = P_{nm} + \frac{1}{x_k} (X_{nn} + X_{mm} - 2X_{nm}) \Delta P_n \tag{11A.9}$$

Then (using the fact that $\tilde{P}_{nm}$ is set to ΔP_n)

$$\Delta P_n = \left[\frac{1}{1 - \dfrac{1}{x_k}(X_{nn} + X_{mm} - 2X_{nm})}\right] P_{nm} \qquad (11A.10)$$

Define a sensitivity factor δ as the ratio of the change in phase angle θ, anywhere in the system, to the original power P_{nm} flowing over a line nm before it was dropped. That is,

$$\delta_{i,nm} = \frac{\Delta\theta_i}{P_{nm}} \qquad (11A.11)$$

If neither n or m is the system reference bus, two injections, ΔP_n and ΔP_m, are imposed at buses n and m, respectively. This gives a change in phase angle at bus i equal to

$$\Delta\theta_i = X_{in}\Delta P_n + X_{im}\Delta P_m \qquad (11A.12)$$

Then using the relationship between ΔP_n and ΔP_m, the resulting δ factor is

$$\delta_{i,nm} = \frac{(X_{in} - X_{im})x_k}{x_k - (X_{nn} + X_{mm} - 2X_{nm})} \qquad (11A.13)$$

If either n or m is the reference bus, only one injection is made. The resulting δ factors are

$$\delta_{i,nm} = \frac{X_{in}x_k}{(x_k - X_{nn})} \qquad \text{for } m = \text{ref}$$

$$= \frac{-X_{im}x_k}{(x_k - X_{mm})} \qquad \text{for } n = \text{ref} \qquad (11A.14)$$

If bus i itself is the reference bus, then $\delta_{i,nm} = 0$ since the reference bus angle is constant.

The expression for $d_{\ell,k}$ is

$$d_{\ell,k} = \frac{\Delta f_\ell}{f_k^0} = \frac{\dfrac{1}{x_\ell}(\Delta\theta_i - \Delta\theta_j)}{f_k^0}$$

$$= \frac{1}{x_\ell}\left(\frac{\Delta\theta_i}{P_{nm}} - \frac{\Delta\theta_j}{P_{nm}}\right)$$

$$= \frac{1}{x_\ell}(\delta_{i,nm} - \delta_{j,nm}) \qquad (11A.15)$$

if neither i nor j is a reference bus

$$d_{\ell,k} = \frac{1}{x_\ell}\left(\frac{(X_{in} - X_{im})x_k - (X_{jn} - X_{jm})x_k}{x_k - (X_{nn} + X_{mm} - 2X_{nm})}\right)$$

$$= \frac{\dfrac{x_k}{x_\ell}(X_{in} - X_{jn} - X_{im} + X_{jm})}{x_k - (X_{nn} + X_{mm} - 2X_{nm})} \tag{11A.16}$$

The fact that the a and d factors are linear models of the power system allows us to use superposition to extend them. One very useful extension is to use the a and d factors to model the power system in its post-outage state; that is, to generate factors that model the system's sensitivity after a branch has been lost.

Suppose one desired to have the sensitivity factor between line ℓ and generator bus i when line k was opened. This is calculated by first assuming that the change in generation on bus i, ΔP_i, has a direct effect on line ℓ and an indirect effect through its influence on the power flowing on line k, which, in turn, influences line ℓ when line k is out. Then

$$\Delta f_\ell = a_{\ell i}\Delta P_i + d_{\ell,k}\Delta f_k \tag{11A.17}$$

However, we know that

$$\Delta f_k = a_{ki}\Delta P_i \tag{11A.18}$$

therefore,

$$\Delta f_\ell = a_{\ell i}\Delta P_i + d_{\ell,k}a_{ki}\Delta P_i = (a_{\ell i} + d_{\ell,k}a_{ki})\Delta P_i \tag{11A.19}$$

We can refer to $a_{\ell i} + d_{\ell,k}a_{ki}$ as the "compensated generation shift sensitivity."

The compensated sensitivity factors are useful in finding corrections to the generation dispatch that will make the post-contingency state of the system secure from overloads. This will be dealt with in Chapter 13 under the topic of "security-constrained optimal power flow."

APPENDIX 11B
Derivation of Equation 11.14

Equation 11.14, repeated here as Eq. 11B.1:

$$|\Delta\theta_p - \Delta\theta_q| < |\Delta\theta_i - \Delta\theta_j| \tag{11B.1}$$

is proved as shown in reference 11 (the proof is attributed to Moslehi).

Suppose that buses i and j have the highest and lowest values of $\Delta\theta$ in the N3 region. Then the following both hold:

$$\Delta\theta_i > \Delta\theta_f$$

and

$$\Delta\theta_j < \Delta\theta_f$$

for all buses f in N3. Taking any external bus in N2, call it bus e, we shall state that

$$\Delta\theta_e < \Delta\theta_i \qquad (11B.2)$$

and

$$\Delta\theta_e > \Delta\theta_j \qquad (11B.3)$$

Proof: Suppose Eq. 11B.2 is not true and there exists a bus e' such that

$$\Delta\theta_{e'} > \Delta\theta_i$$

and, further, suppose that

$$\Delta\theta_{e'} > \Delta\theta_e \qquad (11B.4)$$

for all the buses in N3. This implies that Eq. 11B.4 holds for the union of buses in N2 and N3. If we now look at the network as a DC power flow network, with no impedances to ground, and only the two injections at buses k and m, then all incremental power flows leaving node e' must be positive, since the incremental flows leaving node e' are found from

$$\Delta f_{e'e} = \frac{1}{x_{e'e}} (\Delta\theta_{e'} - \Delta\theta_e) \qquad (11B.5)$$

However, since the network in N2 and N3 is strictly passive, and there are no impedances to ground, this violates Kirchoff's current law, which requires all branch flows incident to a bus to sum to zero. The only way for this to be true would be if all flows were zero; that is, all incremental angle spreads were equal. We can continue this reasoning to the neighbor buses of e' until we reach node i and conclude that

$$\Delta\theta_{e'} = \Delta\theta_i \qquad (11B.6)$$

which contradicts Eq. 11B.4; thus, Eq. 11B.2 is proved. Equation 11B.3 is proved in a similar fashion. Then, as a result, Eq. 11B.1 is also proved.

PROBLEMS

11.1 Figure 11.17 shows a four-bus power system. Also given below are the impedance data for the transmission lines of the system as well as the generation and load values.

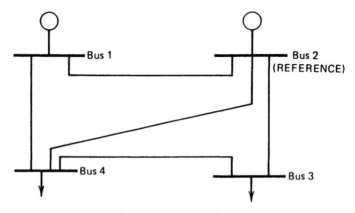

FIG. 11.17 Four-bus network for Problem 11.1.

Line	Line rectance (pu)
1-2	0.2
1-4	0.25
2-3	0.15
2-4	0.30
3-4	0.40

Bus	Load (MW)	Generation (MW)
1		150
2		350
3	220	
4	280	

a. Calculate the generation shift sensitivity coefficients for a shift in generation from bus 1 to bus 2.

b. Calculate the line outage sensitivity factors for outages on lines 1-2, 1-4, and 2-3.

11.2 In the system shown in Figure 11.18, three generators are serving a load of 1300 MW. The MW flow distribution, bus loads, and generator outputs are as shown. The generators have the following characteristics.

Generator No.	P_{min} (MW)	P_{max} (MW)
1	100	600
2	90	400
3	100	500

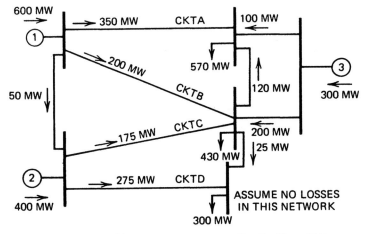

FIG. 11.18   Three-generator system for Problem 11.2.

The circuits have the following limits:

CKT A	600 MW max
CKT B	600 MW max
CKT C	450 MW max
CKT D	350 MW max

Throughout this problem we will only be concerned with flows on the circuit labeled A, B, C, and D. The generation shift sensitivity coefficients, $a_{\ell i}$, for circuits, A, B, C, and D are as follows.

CKT	Shift on Gen. 1	Shift on Gen. 2
A	0.7	0.08
B	0.2	0.02
C	0.06	0.54
D	0.04	0.36

Example: $\Delta P_{\text{flow}_\ell} = a_{\ell,i} \times \Delta P_i$

if

$$\ell = C \quad \text{and} \quad i = 2$$

$$\Delta P_{\text{flow}_c} = (0.54)\Delta P_2$$

Assume a shift on gen. 1 or gen. 2 will be compensated by an equal (opposite) shift on gen. 3. The line outage sensitivity factors $d_{\ell,k}$ are

	A	B	C	D
A	X	0.8	0.21	0.14
B	0.9	X	0.06	0.04
C	0.06	0.12	X	0.82
D	0.04	0.08	0.73	X

with ℓ as the row index (downward) and k as the column index (rightward).

As an example, suppose the loss of circuit k will increase the loading on circuit ℓ as follows.

$$P_{\text{flow}_\ell} = P_{\text{flow}_\ell} \text{ (before outage)} + d_{\ell,k} \times P_{\text{flow}_k} \text{ (before outage)}$$

if

$$\ell = A \quad \text{and} \quad k = B$$

The new flow on ℓ would be

$$P_{\text{flow}_A} = P_{\text{flow}_A} + (0.8)P_{\text{flow}_B}$$

a. Find the contingency (outage) flow distribution on circuits A, B, C, and D for an outage on circuit A. Repeat for an outage on B, then on C, then on D. (Only one circuit is lost at one time.) Are there any overloads?

b. Can you shift generation from gen. 1 to gen. 3, or from gen. 2 to gen. 3, so that no overloads occur? If so, how much shift?

11.3 Given the three-bus network shown in Figure 11.19 (see Example 4B), where

$$x_{12} = 0.2 \text{ pu}$$
$$x_{13} = 0.4 \text{ pu}$$
$$x_{23} = 0.25 \text{ pu}$$

the $[X]$ matrix is

$$\begin{bmatrix} 0.2118 & 0.1177 & 0 \\ 0.1177 & 0.1765 & 0 \\ 0 & 0 & 0 \end{bmatrix}$$

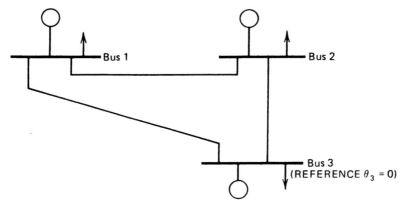

FIG. 11.19   Three-bus system for Problem 11.3.

Use a 100-MVA base. The base loads and generations are as follows.

Bus	Load (MW)	Gen. (MW)	Gen. min (MW)	Gen. max (MW)
1	100	150	50	250
2	300	180	60	250
3	100	170	60	300

a. Find base power flows on the transmission lines.

b. Calculate the generation shift factors for line 1-2. Calculate the shift in generation on bus 1 and 2 so as to force the flow on line 1-2 to zero MW. Assume for economic reasons that any shifts from base conditions are more expensive for shifts at the generator on bus 1 than for shifts on bus 2, and that the generator on bus 3 can be shifted without any penalty.

11.4   Using the system shown in Example 11B, find N1, N2 and N3 for the outage of the line from bus 2 to bus 4. Do you need to expand region N1? Where are the overloads, if any? (Use the same branch flow limits as shown in Example 11B.)

11.5   Using the data found in Figure 11.7, find the base-case bus phase angles and all line flows using the following bus loads and generators: all loads are 100 MW and all generators are also at 100 MW. Assume line flow limits as shown in the following table.

Line	MW Limit
1-2	70
1-4	90
1-5	70
2-3	20
2-4	50
2-5	40
2-6	60
3-5	30
3-6	70
4-5	30
5-6	20

For a line outage on line 1-4, find the change in phase angle across each of the remaining lines and see if the phase angle change across buses 1 and 4 meets the bounding criteria developed in the text.

11.6 Using the data from Problem 11.2, calculate the performance index, PI, for each outage case. Use a value of $n = 1$ and $n = 5$; that is for

$$ PI = \sum_{\text{all lines}} \left(\frac{\text{flow}_{ij}}{\text{flow max}_{ij}} \right)^{2n} $$

Which PI does a better job of predicting the case with the overload? Explain why.

FURTHER READING

The subject of power system security has received a great deal of attention in the engineering literature since the middle 1960s. The list of references presented here is therefore large but also quite limited nonetheless.

Reference 1 is a key paper on the topic of system security and energy control system philosophy. Reference 2 provides the basic theory for contingency assessment of power systems. Reference 3 covers contingency analysis using DC power flow methods. Reference 4 is a broad overview of security assessment and contains an excellent bibliography covering the literature on security assessment up to 1975.

The use of AC power flows in contingency analysis is possible with any AC load flow algorithm. However, the *fast-decoupled* power flow algorithm is generally recognized as the best for this purpose since its Jacobian matrix is constant and single-line outages can be modeled using the matrix inversion lemma. Reference 5 covers the fast-decoupled power flow algorithm and its application.

Correcting the generation dispatch by sensitivity methods is covered by reference 6. The use of linear programming to solve power systems problems is covered in reference 7.

References 8–12 cover some of the literature on contingency selection, and reference 13 gives a technique for solving the power flow using an approximation called concentric relaxation. References 14 and 15 give an indication of recent research on dynamic security assessment; that is, detecting fault cases that may cause dynamic or transient stability problems. Finally, reference 16 is concerned with the emerging area of voltage stability, which seeks to find contingencies which will cause such severe voltage problems as to bring on what is known as a "voltage collapse."

1. DyLiacco, T. E., "The Adaptive Reliability Control System," *IEEE Transactions on Power Apparatus and Systems,* Vol. PAS-86, May 1967, pp. 517–531.

2. El-Abiad, A. H., Stagg, G. W., "Automatic Evaluation of Power System Performance—Effects of Line and Transformer Outages," *AIEE Transactions on Power Apparatus and Systems,* Vol. PAS-81, February 1963, pp. 712–716.

3. Baughman, M., Schweppe, F. C., "Contingency Evaluation: Real Power Flows from a Linear Model," IEEE Summer Power Meeting, 1970, Paper CP 689-PWR.

4. Debs, A. S., Benson, A. R., "Security Assessment of Power Systems," *Systems Engineering For Power: Status and Prospects,* U.S. Government Document, CONF-750867, 1967, pp. 1–29.

5. Stott, B., Alsac, O., "Fast Decoupled Load Flow," *IEEE Transactions on Power Apparatus and Systems,* Vol. PAS-93, May/June 1974, pp. 859–869.

6. Thanikachalam, A., Tudor, J. R., "Optimal Rescheduling of Power for System Reliability," *IEEE Transactions on Power Apparatus and Systems,* Vol. PAS-90, July/August 1971, pp. 1776–1781.

7. Chan, S. M., Yip, E., "A Solution of the Transmission Limited Dispatch Problem by Sparse Linear Programming," *IEEE Transactions on Power Apparatus and Systems,* Vol. PAS-98, May/June 1979, pp. 1044–1053.

8. Ejebe, G. C., Wollenberg, B. F., "Automatic Contingency Selection," *IEEE Transactions on Power Apparatus and Systems,* Vol. PAS-98, January/February 1979, pp. 92–104.

9. Albuyeh, F., Bose, A., Heath, B., "Automatic Contingency Selection; Ranking Outages on the Basis of Real and Reactive Power Equations," *IEEE Transactions on Power Apparatus and Systems,* Vol. PAS-101, No. 1, January 1982, pp. 107–112.

10. Ejebe, G. C., VanMeeteren, H. P., Wollenberg, B. F. "Fast Contingency Screening and Evaluation for Voltage Security Analysis," *IEEE Transactions on Power Systems,* Vol. 3, No. 4, November 1988, pp. 1582–1590.

11. Brandwajn, V., "Efficient Bounding Method for Linear Contingency Analysis," *IEEE Transactions on Power Systems,* Vol. 3, No. 1, February 1988, pp. 38–43.

12. Brandwajn, V., Lauby, M. G., "Complete Bounding Method for AC Contingency Screening," *IEEE Transactions on Power Systems,* Vol. PWRS-4, May 1989, pp. 724–729.

13. Zaborsky, J., Whang, K. W., Prasad, K., "Fast Contingency Evaluation using Concentric Relaxation," *IEEE Transactions on Power Apparatus and Systems,* Vol. PAS-99, January/February 1980, pp. 28–36.

14. Fouad, A. A., "Dynamic Security Assessment Practices in North America," *IEEE Transactions on Power Systems*, Vol. 3, No. 3, August 1988, pp. 1310–1321.

15. El-Kady, M. A., Tang, C. K., Carvalho, V. F., Fouad, A. A., Vittal, V., "Dynamic Security Assessment Utilizing the Transient Energy Function Method," *IEEE Transactions on Power Systems*, Vol. PWRS-1, No. 3, August 1986, pp. 284–291.

16. Jasmon, G. B., Lee, L. H. C. C., "New Contingency Ranking Technique Incorporating a Voltage Stability Criterion," *IEE Proceedings, Part C: Generation, Transmission and Distribution*, Vol. 140, No. 2, March 1993, pp. 87–90.

12 An Introduction to State Estimation in Power Systems

12.1 INTRODUCTION

State estimation is the process of assigning a value to an unknown system state variable based on measurements from that system according to some criteria. Usually, the process involves imperfect measurements that are redundant and the process of estimating the system states is based on a statistical criterion that estimates the true value of the state variables to minimize or maximize the selected criterion. A commonly used and familiar criterion is that of minimizing the sum of the squares of the differences between the estimated and "true" (i.e., measured) values of a function.

The ideas of least-squares estimation have been known and used since the early part of the nineteenth century. The major developments in this area have taken place in the twentieth century in applications in the aerospace field. In these developments, the basic problems have involved the location of an aerospace vehicle (i.e., missile, airplane, or space vehicle) and the estimation of its trajectory given redundant and imperfect measurements of its position and velocity vector. In many applications, these measurements are based on optical observations and/or radar signals that may be contaminated with noise and may contain system measurement errors. State estimators may be both static and dynamic. Both types of estimators have been developed for power systems. This chapter will introduce the basic development of a static-state estimator.

In a power system, the state variables are the voltage magnitudes and relative phase angles at the system nodes. Measurements are required in order to estimate the system performance in real time for both system security control and constraints on economic dispatch. The inputs to an estimator are imperfect power system measurements of voltage magnitudes and power, VAR, or ampere-flow quantities. The estimator is designed to produce the "best estimate" of the system voltage and phase angles, recognizing that there are errors in the measured quantities and that there may be redundant measurements. The output data are then used in system control centers in the implementation of the security-constrained dispatch and control of the system as discussed in Chapters 11 and 13.

12.2   POWER SYSTEM STATE ESTIMATION

As introduced in Chapter 11, the problem of monitoring the power flows and voltages on a transmission system is very important in maintaining system security. By simply checking each measured value against its limit, the power system operators can tell where problems exist in the transmission system—and, it is hoped, they can take corrective actions to relieve overloaded lines or out-of-limit voltages.

Many problems are encountered in monitoring a transmission system. These problems come primarily from the nature of the measurement transducers and from communications problems in transmitting the measured values back to the operations control center.

Transducers from power system measurements, like any measurement device, will be subject to errors. If the errors are small, they may go undetected and can cause misinterpretation by those reading the measured values. In addition, transducers may have gross measurement errors that render their output useless. An example of such a gross error might involve having the transducer connected up backward; thus, giving the negative of the value being measured. Finally, the telemetry equipment often experiences periods when communications channels are completely out; thus, depriving the system operator of any information about some part of the power system network.

It is for these reasons that power system state estimation techniques have been developed. A state estimator, as we will see shortly, can "smooth out" small random errors in meter readings, detect and identify gross measurement errors, and "fill in" meter readings that have failed due to communications failures.

To begin, we will use a simple DC load flow example to illustrate the principles of state estimation. Suppose the three-bus DC load flow of Example 4B were operating with the load and generation shown in Figure 12.1. The only information we have about this system is provided by three MW power flow meters located as shown in Figure 12.2.

Only two of these meter readings are required to calculate the bus phase angles and all load and generation values fully. Suppose we use M_{13} and M_{32} and further suppose that M_{13} and M_{32} give us perfect readings of the flows on their respective transmission lines.

$$M_{13} = 5 \text{ MW} = 0.05 \text{ pu}$$

$$M_{32} = 40 \text{ MW} = 0.40 \text{ pu}$$

Then, the flows on lines 1-3 and 3-2 can be set equal to these meter readings.

$$f_{13} = \frac{1}{x_{13}} (\theta_1 - \theta_3) = M_{13} = 0.05 \text{ pu}$$

$$f_{32} = \frac{1}{x_{23}} (\theta_3 - \theta_2) = M_{32} = 0.40 \text{ pu}$$

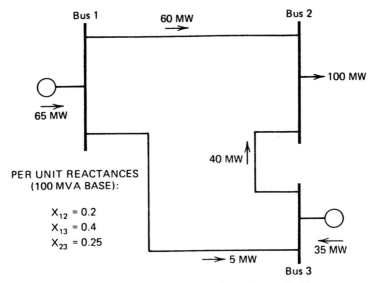

FIG. 12.1   Three-bus system from Example 4B.

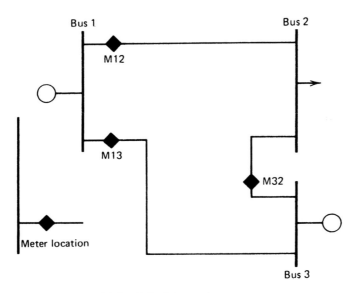

FIG. 12.2   Meter placement.

Since we know that $\theta_3 = 0$ rad, we can solve the f_{13} equation for θ_1, and the f_{32} equation for θ_2, resulting in

$$\theta_1 = 0.02 \text{ rad}$$
$$\theta_2 = -0.10 \text{ rad}$$

We will now investigate the case where all three meter readings have slight errors. Suppose the readings obtained are

$$M_{12} = 62 \text{ MW} = 0.62 \text{ pu}$$

$$M_{13} = 6 \text{ MW} = 0.06 \text{ pu}$$

$$M_{32} = 37 \text{ MW} = 0.37 \text{ pu}$$

If we use only the M_{13} and M_{32} readings, as before, we will calculate the phase angles as follows:

$$\theta_1 = 0.024 \text{ rad}$$

$$\theta_2 = -0.0925 \text{ rad}$$

$$\theta_3 = 0 \text{ rad (still assumed to equal zero)}$$

This results in the system flows as shown in Figure 12.3. Note that the predicted flows match at M_{13} and M_{32}, but the flow on line 1-2 does not match the reading of 62 MW from M_{12}. If we were to ignore the reading on M_{13} and use M_{12} and M_{32}, we could obtain the flows shown in Figure 12.4.

All we have accomplished is to match M_{12}, but at the expense of no longer matching M_{13}. What we need is a procedure that uses the information available from all three meters to produce the best estimate of the actual angles, line flows, and bus load and generations.

Before proceeding, let's discuss what we have been doing. Since the only thing we know about the power system comes to us from the measurements,

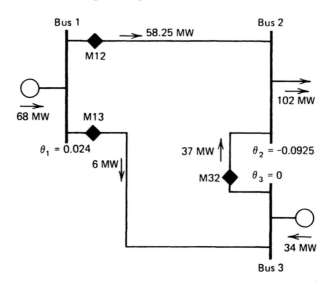

FIG. 12.3   Flows resulting from use of meters M_{13} and M_{32}.

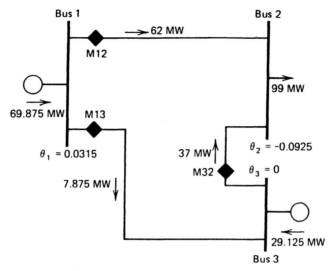

FIG. 12.4   Flows resulting from use of meters M_{12} and M_{32}.

we must use the measurements to estimate system conditions. Recall that in each instance the measurements were used to calculate the bus phase angles at bus 1 and 2. Once these phase angles were known, all unmeasured power flows, loads, and generations could be determined. We call θ_1 and θ_2 the *state variables* for the three-bus system since knowing them allows all other quantities to be calculated. In general, the state variables for a power system consist of the bus voltage magnitude at all buses and the phase angles at all but one bus. The swing or reference bus phase angle is usually assumed to be zero radians. Note that we could use real and imaginary components of bus voltage if desired. If we can use measurements to estimate the "states" (i.e., voltage magnitudes and phase angles) of the power system, then we can go on to calculate any power flows, generation, loads, and so forth that we desire. This presumes that the network configuration (i.e., breaker and disconnect switch statuses) is known and that the impedances in the network are also known. Automatic load tap changing autotransformers or phase angle regulators are often included in a network, and their tap positions may be telemetered to the control as a measured quantity. Strictly speaking, the transformer taps and phase angle regulator positions should also be considered as states since they must be known in order to calculate the flows through the transformers and regulators.

   To return to the three-bus DC power flow model, we have three meters providing us with a set of redundant readings with which to estimate the two states θ_1 and θ_2. We say that the readings are redundant since, as we saw earlier, only two readings are necessary to calculate θ_1 and θ_2, the other reading is always "extra." However, the "extra" reading does carry useful information and ought not to be discarded summarily.

This simple example serves to introduce the subject of *static-state estimation*, which is the art of estimating the exact system state given a set of imperfect measurements made on the power system. We will digress at this point to develop the theoretical background for static-state estimation. We will return to our three-bus system in Section 12.3.4.

12.3   MAXIMUM LIKELIHOOD WEIGHTED LEAST-SQUARES ESTIMATION

12.3.1   Introduction

Statistical estimation refers to a procedure where one uses samples to calculate the value of one or more unknown parameters in a system. Since the samples (or measurements) are inexact, the estimate obtained for the unknown parameter is also inexact. This leads to the problem of how to formulate a "best"estimate of the unknown parameters given the available measurements.

The development of the notions of state estimation may proceed along several lines, depending on the statistical criterion selected. Of the many criteria that have been examined and used in various applications, the following three are perhaps the most commonly encountered.

1. *The maximum likelihood criterion*, where the objective is to maximize the probability that the estimate of the state variable, $\hat{x}$, is the true value of the state variable vector, x (i.e., maximize $P(\hat{x}) = x$).
2. *The weighted least-squares criterion*, where the objective is to minimize the sum of the squares of the weighted deviations of the estimated measurements, $\hat{z}$, from the actual measurements, z.
3. *The minimum variance criterion*, where the object is to minimize the expected value of the sum of the squares of the deviations of the estimated components of the state variable vector from the corresponding components of the true state variable vector.

When normally distributed, unbiased meter error distributions are assumed, each of these approaches results in identical estimators. This chapter will utilize the maximum likelihood approach because the method introduces the measurement error weighting matrix $[R]$ in a straightforward manner.

The maximum likelihood procedure asks the following question: "What is the probability (or likelihood) that I will get the measurements I have obtained?" This probability depends on the random error in the measuring device (transducer) as well as the unknown parameters to be estimated. Therefore, a reasonable procedure would be one that simply chose the estimate as the value that maximizes this probability. As we will see shortly, the maximum likelihood estimator assumes that we know the probability density function (PDF) of the random errors in the measurement. Other estimation schemes could also be used. The "least-squares" estimator does not require that we know the probability density function for the sample or measurement errors. However, if we assume that the probability density function of sample or measurement error is a normal (Gaussian) distribution, we will end up with the same estimation formula. We will proceed to develop our estimation formula using the maximum likelihood criterion assuming normal distributions for measurement errors. The result will be a "least-squares" or more precisely a "weighted least-squares" estimation formula, even though we will develop the formulation using the maximum likelihood criteria. We will illustrate this method with a simple electrical circuit and show how the maximum likelihood estimate can be made.

First, we introduce the concept of *random measurement error*. Note that we have dropped the term "sample" since the concept of a measurement is much more appropriate to our discussion. The measurements are assumed to be in error: that is, the value obtained from the measurement device is close to the true value of the parameter being measured but differs by an unknown error. Mathematically, this can be modeled as follows.

Let z^{meas} be the value of a measurement as received from a measurement device. Let z^{true} be the true value of the quantity being measured. Finally, let η be the random measurement error. We can then represent our measured value as

$$z^{meas} = z^{true} + \eta \qquad (12.1)$$

The random number, η, serves to model the uncertainty in the measurements. If the measurement error is unbiased, the probability density function of η is usually chosen as a normal distribution with zero mean. Note that other measurement probability density functions will also work in the maximum likelihood method as well. The probability density function of η is

$$\text{PDF}(\eta) = \frac{1}{\sigma\sqrt{2\pi}} \exp(-\eta^2/2\sigma^2) \qquad (12.2)$$

where σ is called the standard deviation and σ^2 is called the variance of the random number. $\text{PDF}(\eta)$ describes the behavior of η. A plot of $\text{PDF}(\eta)$ is shown in Figure 12.5. Note that σ, the standard deviation, provides a way to model the seriousness of the random measurement error. If σ is large, the measurement is relatively inaccurate (i.e., a poor-quality measurement device), whereas a small value of σ denotes a small error spread (i.e., a higher-quality measurement device). The normal distribution is commonly used for modeling measurement errors since it is the distribution that will result when many factors contribute to the overall error.

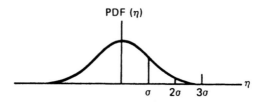

FIG. 12.5   The normal distribution.

12.3.2   Maximum Likelihood Concepts

The principle of maximum likelihood estimation is illustrated by using a simple DC circuit example as shown in Figure 12.6. In this example, we wish to estimate the value of the voltage source, x^{true}, using an ammeter with an error having a known standard deviation. The ammeter gives a reading of z_1^{meas}, which is equal to the sum of z_1^{true} (the true current flowing in our circuit) and η_1 (the error present in the ammeter). Then we can write

$$z_1^{\text{meas}} = z_1^{\text{true}} + \eta_1 \tag{12.3}$$

Since the mean value of η_1 is zero, we then know that the mean value of z_1^{meas} is equal to z_1^{true}. This allows us to write a probability density function for z_1^{meas} as

$$\text{PDF}(z_1^{\text{meas}}) = \frac{1}{\sigma_1\sqrt{2\pi}} \exp\left[\frac{-(z_1^{\text{meas}} - z_1^{\text{true}})^2}{2\sigma_1^2}\right] \tag{12.4}$$

where σ_1 is the standard deviation for the random error η_1. If we assume that the value of the resistance, r_1, in our circuit is known, then we can write

$$\text{PDF}(z_1^{\text{meas}}) = \frac{1}{\sigma_1\sqrt{2\pi}} \exp\left[\frac{-\left(z_1^{\text{meas}} - \frac{1}{r_1}x\right)^2}{2\sigma_1^2}\right] \tag{12.5}$$

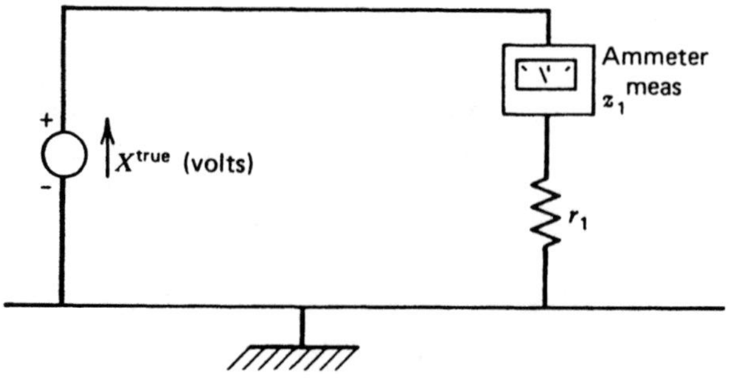

FIG. 12.6   Simple DC circuit with current measurement.

Coming back to our definition of a maximum likelihood estimator, we now wish to find an estimate of x (called x^{est}) that maximizes the probability that the observed measurement z_1^{meas} would occur. Since we have the probability density function of z_1^{meas}, we can write

$$\text{prob}(z_1^{meas}) = \int_{z_1^{meas}}^{z_1^{meas}+dz_1^{meas}} \text{PDF}(z_1^{meas})\,dz_1^{meas} \quad \text{as } dz_1^{meas} \to 0$$

$$= \text{PDF}(z_1^{meas})\,dz_1^{meas} \tag{12.6}$$

The maximum likelihood procedure then requires that we maximize the value of $\text{prob}(z_1^{meas})$, which is a function of x. That is,

$$\max_{x} \text{prob}(z_1^{meas}) = \max_{x} \text{PDF}(z_1^{meas})\,dz_1^{meas} \tag{12.7}$$

One convenient transformation that can be used at this point is to maximize the natural logarithm of $\text{PDF}(z_1^{meas})$ since maximizing the Ln of $\text{PDF}(z_1^{meas})$ will also maximize $\text{PDF}(z_1^{meas})$. Then we wish to find

$$\max_{x} \text{Ln}[\text{PDF}(z_1^{meas})]$$

or

$$\max_{x} \left[-\text{Ln}(\sigma_1\sqrt{2\pi}) - \frac{\left(z_1^{meas} - \dfrac{1}{r_1}x\right)^2}{2\sigma_1^2} \right]$$

Since the first term is constant, it can be ignored. We can maximize the function in brackets by minimizing the second term since it has a negative coefficient, that is,

$$\max_{x} \left[-\text{Ln}(\sigma_1\sqrt{2\pi}) - \frac{\left(z_1^{meas} - \dfrac{1}{r_1}x\right)^2}{2\sigma_1^2} \right]$$

is the same as

$$\min_{x} \left[\frac{\left(z_1^{meas} - \dfrac{1}{r_1}x\right)^2}{2\sigma_1^2} \right] \tag{12.8}$$

The value of x that minimizes the right-hand term is found by simply taking the first derivative and setting the result to zero:

$$\frac{d}{dx}\left[\frac{\left(z_1^{meas} - \dfrac{1}{r_1}x\right)^2}{2\sigma_1^2} \right] = \frac{-\left(z_1^{meas} - \dfrac{1}{r_1}x\right)}{r_1\sigma_1^2} = 0 \tag{12.9}$$

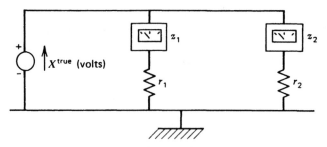

FIG. 12.7   DC circuit with two current measurements.

or

$$x^{est} = r_1 z_1^{meas}$$

To most readers this result was obvious from the beginning. All we have accomplished is to declare the maximum likelihood estimate of our voltage as simply the measured current times the known resistance. However, by adding a second measurement circuit, we have an entirely different situation in which the best estimate is not so obvious. Let us now add a second ammeter and resistance as shown in Figure 12.7.

Assume that both r_1 and r_2 are known. As before, model each meter reading as the sum of the true value and a random error:

$$z_1^{meas} = z_1^{true} + \eta_1$$
$$z_2^{meas} = z_2^{true} + \eta_2$$

(12.10)

where the errors will be represented as independent zero mean, normally distributed random variables with probability density functions:

$$\text{PDF}(\eta_1) = \frac{1}{\sigma_1\sqrt{2\pi}}\exp\left(\frac{-(\eta_1)^2}{2\sigma_1^2}\right)$$
$$\text{PDF}(\eta_2) = \frac{1}{\sigma_2\sqrt{2\pi}}\exp\left(\frac{-(\eta_2)^2}{2\sigma_2^2}\right)$$

(12.11)

and as before we can write the probability density functions of z_1^{meas} and z_2^{meas} as

$$\text{PDF}(z_1^{meas}) = \frac{1}{\sigma_1\sqrt{2\pi}}\exp\left[\frac{-\left(z_1^{meas} - \frac{1}{r_1}x\right)^2}{2\sigma_1^2}\right]$$
$$\text{PDF}(z_2^{meas}) = \frac{1}{\sigma_2\sqrt{2\pi}}\exp\left[\frac{-\left(z_2^{meas} - \frac{1}{r_2}x\right)^2}{2\sigma_2^2}\right]$$

(12.12)

The likelihood function must be the probability of obtaining the measurements z_1^{meas} and z_2^{meas}. Since we are assuming that the random errors η_1 and η_2 are independent random variables, the probability of obtaining z_1^{meas} and z_2^{meas} is simply the product of the probability of obtaining z_1^{meas} and the probability of obtaining z_2^{meas}.

$$
\text{prob}(z_1^{\text{meas}} \text{ and } z_2^{\text{meas}}) = \text{prob}(z_1^{\text{meas}}) \times (\text{prob}(z_2^{\text{meas}}))
$$

$$
= \text{PDF}(z_1^{\text{meas}}) \, \text{PDF}(z_2^{\text{meas}}) \, dz_1^{\text{meas}} \, dz_2^{\text{meas}}
$$

$$
= \left[\frac{1}{\sigma_1 \sqrt{2\pi}} \exp\left(\frac{-\left(z_1^{\text{meas}} - \dfrac{1}{r_1} x \right)^2}{2\sigma_1^2} \right) \right]
$$

$$
\times \left[\frac{1}{\sigma_2 \sqrt{2\pi}} \exp\left(\frac{-\left(z_2^{\text{meas}} - \dfrac{1}{r_2} x \right)^2}{2\sigma_2^2} \right) \right] dz_1^{\text{meas}} \, dz_2^{\text{meas}}
$$

$$
(12.13)
$$

To maximize the function we will again take its natural logarithm:

$$
\max_x \text{prob}(z_1^{\text{meas}} \text{ and } z_2^{\text{meas}})
$$

$$
= \max_x \left[-\text{Ln}(\sigma_1 \sqrt{2\pi}) - \frac{\left(z_1^{\text{meas}} - \dfrac{1}{r_1} x \right)^2}{2\sigma_1^2} - \text{Ln}(\sigma_2 \sqrt{2\pi}) - \frac{\left(z_2^{\text{meas}} - \dfrac{1}{r_2} x \right)^2}{2\sigma_2^2} \right]
$$

$$
= \min_x \left[\frac{\left(z_1^{\text{meas}} - \dfrac{1}{r_1} x \right)^2}{2\sigma_1^2} + \frac{\left(z_2^{\text{meas}} - \dfrac{1}{r_2} x \right)^2}{2\sigma_2^2} \right] \qquad (12.14)
$$

The minimum sought is found by

$$
\frac{d}{dx} \left[\frac{\left(z_1^{\text{meas}} - \dfrac{1}{r_1} x \right)^2}{2\sigma_1^2} + \frac{\left(z_2^{\text{meas}} - \dfrac{1}{r_2} x \right)^2}{2\sigma_2^2} \right]
$$

$$
= \frac{-\left(z_1^{\text{meas}} - \dfrac{1}{r_1} x \right)}{r_1 \sigma_1^2} - \frac{\left(z_2^{\text{meas}} - \dfrac{1}{r_2} x \right)}{r_2 \sigma_2^2} = 0
$$

giving

$$x^{\text{est}} = \frac{\left(\dfrac{z_1^{\text{meas}}}{r_1\sigma_1^2} + \dfrac{z_2^{\text{meas}}}{r_2\sigma_2^2}\right)}{\left(\dfrac{1}{r_1^2\sigma_1^2} + \dfrac{1}{r_2^2\sigma_2^2}\right)} \tag{12.15}$$

If one of the ammeters is of superior quality, its variance will be much smaller than that of the other meter. For example, if $\sigma_2^2 \ll \sigma_1^2$, then the equation for x^{est} becomes

$$x^{\text{est}} \simeq z_2^{\text{meas}} \times r_2$$

Thus, we see that the maximum likelihood method of estimating our unknown parameter gives us a way to weight the measurements properly according to their quality.

It should be obvious by now that we need not express our estimation problem as a maximum of the product of probability density functions. Instead, we can observe a direct way of writing what is needed by looking at Eqs. 12.8 and 12.14. In these equations, we see that the maximum likelihood estimate of our unknown parameter is always expressed as that value of the parameter that gives the minimum of the sum of the squares of the difference between each measured value and the true value being measured (expressed as a function of our unknown parameter) with each squared difference divided or "weighted" by the variance of the meter error. Thus, if we are estimating a single parameter, x, using N_m measurements, we would write the expression

$$\min_{x} J(x) = \sum_{i=1}^{N_m} \frac{[z_i^{\text{meas}} - f_i(x)]^2}{\sigma_i^2} \tag{12.16}$$

where

f_i = function that is used to calculate the value being measured by the ith measurement

σ_i^2 = variance for the ith measurement

$J(x)$ = measurement residual

N_m = number of independent measurements

z_i^{meas} = i^{th} measured quantity

Note that Eq. 12.16 may be expressed in per unit or in physical units such as MW, MVAR, or kV.

If we were to try to estimate N_s unknown parameters using N_m measurements, we would write

$$\min_{\{x_1, x_2, \ldots, x_{N_s}\}} J(x_1, x_2, \ldots, x_{N_s}) = \sum_{i=1}^{N_m} \frac{[z_i - f_i(x_1, x_2, \ldots, x_{N_s})]^2}{\sigma_i^2} \tag{12.17}$$

The estimation calculation shown in Eqs. 12.16 and 12.17 is known as a *weighted least-squares* estimator, which, as we have shown earlier, is equivalent to a maximum likelihood estimator if the measurement errors are modeled as random numbers having a normal distribution.

12.3.3 Matrix Formulation

If the functions $f_i(x_1, x_2, \ldots, x_{N_s})$ are linear functions, Eq. 12.17 has a closed-form solution. Let us write the function $f_i(x_1, x_2, \ldots, x_{N_s})$ as

$$f_i(x_1, x_2, \ldots, x_{N_s}) = f_i(\mathbf{x}) = h_{i1}x_1 + h_{i2}x_2 + \ldots + h_{iN_s}x_{N_s} \qquad (12.18)$$

Then, if we place all the f_i functions in a vector, we may write

$$\mathbf{f(x)} = \begin{bmatrix} f_1(\mathbf{x}) \\ f_2(\mathbf{x}) \\ \vdots \\ f_{N_m}(\mathbf{x}) \end{bmatrix} = [H]\mathbf{x} \qquad (12.19)$$

where

$[H]$ = an N_m by N_s matrix containing the coefficients of the linear functions $f_i(\mathbf{x})$

N_m = number of measurements

N_s = number of unknown parameters being estimated

Placing the measurements in a vector:

$$\mathbf{z}^{\text{meas}} = \begin{bmatrix} z_1^{\text{meas}} \\ z_2^{\text{meas}} \\ \vdots \\ z_{N_m}^{\text{meas}} \end{bmatrix} \qquad (12.20)$$

We may then write Eq. 12.17 in a very compact form.

$$\min_{\mathbf{x}} J(\mathbf{x}) = [\mathbf{z}^{\text{meas}} - \mathbf{f(x)}]^T [R^{-1}][\mathbf{z}^{\text{meas}} - \mathbf{f(x)}] \qquad (12.21)$$

where

$$[R] = \begin{bmatrix} \sigma_1^2 & & & \\ & \sigma_2^2 & & \\ & & \ddots & \\ & & & \sigma_{N_m}^2 \end{bmatrix}$$

$[R]$ is called the *covariance matrix of measurement errors*. To obtain the general expression for the minimum in Eq. 12.21, expand the expression and substitute $[H]\mathbf{x}$ for $\mathbf{f}(\mathbf{x})$ from Eq. 12.19.

$$\min_{\mathbf{x}} J(\mathbf{x}) = \{\mathbf{z}^{\text{meas}^T}[R^{-1}]\mathbf{z}^{\text{meas}} - \mathbf{x}^T[H]^T[R^{-1}]\mathbf{z}^{\text{meas}}$$

$$-\mathbf{z}^{\text{meas}^T}[R^{-1}][H]\mathbf{x} + \mathbf{x}^T[H]^T[R^{-1}][H]\mathbf{x}\} \qquad (12.22)$$

Similar to the procedures of Chapter 3, the minimum of $J(\mathbf{x})$ is found when $\partial J(\mathbf{x})/\partial x_i = 0$, for $i = 1, \ldots, N_s$; this is identical to stating that the gradient of $J(\mathbf{x})$, $\nabla J(\mathbf{x})$, is exactly zero.

The gradient of $J(\mathbf{x})$ is (see the appendix to this chapter)

$$\nabla J(\mathbf{x}) = -2[H]^T[R^{-1}]\mathbf{z}^{\text{meas}} + 2[H]^T[R^{-1}][H]\mathbf{x}$$

Then $\nabla J(\mathbf{x}) = \mathbf{0}$ gives

$$\mathbf{x}^{\text{est}} = [[H]^T[R^{-1}][H]]^{-1}[H]^T[R^{-1}]\mathbf{z}^{\text{meas}} \qquad (12.23)$$

Note that Eq. 12.23 holds when $N_s < N_m$; that is, when the number of parameters being estimated is less than the number of measurements being made.

When $N_s = N_m$, our estimation problem reduces to

$$\mathbf{x}^{\text{est}} = [H]^{-1}\mathbf{z}^{\text{meas}} \qquad (12.24)$$

There is also a closed-form solution to the problem when $N_s > N_m$, although in this case we are not estimating $\mathbf{x}$ to maximize a likelihood function since $N_s > N_m$ usually implies that many different values for $\mathbf{x}^{\text{est}}$ can be found that cause $f_i(\mathbf{x}^{\text{est}})$ to equal z_i^{meas} for all $i = 1, \ldots, N_m$ exactly. Rather, the objective is to find $\mathbf{x}^{\text{est}}$ such that the sum of the squares of x_i^{est} is minimized. That is,

$$\min_{\mathbf{x}} \sum_{i=1}^{N_s} x_i^2 = \mathbf{x}^T\mathbf{x} \qquad (12.25)$$

subject to the condition that $\mathbf{z}^{\text{meas}} = [H]\mathbf{x}$. The closed-form solution for this case is

$$\mathbf{x}^{\text{est}} = [H]^T[[H][H]^T]^{-1}\mathbf{z}^{\text{meas}} \qquad (12.26)$$

In power system state estimation, underdetermined problems (i.e., where $N_s > N_m$) are not solved, as shown in Eq. 12.26. Rather, "pseudo-measurements" are added to the measurement set to give a completely determined or overdetermined problem. We will discuss pseudo-measurements in Section 12.6.3. Table 12.1 summarizes the results for this section.

TABLE 12.1   Estimation Formulas

Case	Description	Solution	Comment
$N_s < N_m$	Overdetermined	$x^{est} = [[H]^T[R^{-1}][H]]^{-1}$ $\times \{H\}^T[R^{-1}]z^{meas}$	x^{est} is the maximum likelihood estimate of x given the measurements z^{meas}
$N_s = N_m$	Completely determined	$x^{est} = [H]^{-1}z^{meas}$	x^{est} fits the measured quantities to the measurements z^{meas} exactly
$N_s > N_m$	Underdetermined	$x^{est} = [H]^T[[H][H]^T]^{-1}z^{meas}$	x^{est} is the vector of minimum norm that fits the measured quantities to the measurements exactly. (The norm of a vector is equal to the sum of the squares of its components)

12.3.4   An Example of Weighted Least-Squares State Estimation

We now return to our three-bus example. Recall from Figure 12.2 that we have three measurements to determine θ_1 and θ_2, the phase angles at buses 1 and 2. From the development in the preceding section, we know that the states θ_1 and θ_2 can be estimated by minimizing a residual $J(\theta_1, \theta_2)$ where $J(\theta_1, \theta_2)$ is the sum of the squares of individual measurement residuals divided by the variance for each measurement.

To start, we will assume that all three meters have the following characteristics.

Meter full-scale value: 100 MW

Meter accuracy: ± 3 MW

This is interpreted to mean that the meters will give a reading within ± 3 MW of the true value being measured for approximately 99% of the time. Mathematically, we say that the errors are distributed according to a normal probability density function with a standard deviation, σ, as shown in Figure 12.8.

Notice that the probability of an error decreases as the error magnitude increases. By integrating the PDF between -3σ and $+3\sigma$ we come up with a value of approximately 0.99. We will assume that the meter's accuracy (in our case ± 3 MW) is being stated as equal to the 3σ points on the probability density function. Then ± 3 MW corresponds to a metering standard deviation of $\sigma = 1$ MW $= 0.01$ pu.

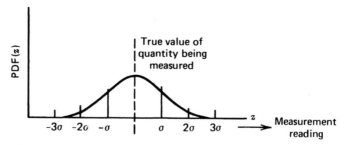

FIG. 12.8 Normal distribution of meter errors.

The formula developed in the last section for the weighted least-squares estimate is given in Eq. 12.23, which is repeated here.

$$\mathbf{x}^{est} = [[H]^T[R^{-1}][H]]^{-1}[H]^T[R^{-1}]\mathbf{z}^{meas}$$

where

$\mathbf{x}^{est}$ = vector of estimated state variables

$[H]$ = measurement function coefficient matrix

$[R]$ = measurement covariance matrix

$\mathbf{z}^{meas}$ = vector containing the measured values themselves

For the three-bus problem we have

$$\mathbf{x}^{est} = \begin{bmatrix} \theta_1^{est} \\ \theta_2^{est} \end{bmatrix} \tag{12.27}$$

To derive the $[H]$ matrix, we need to write the measurements as a function of the state variables θ_1 and θ_2. These functions are written in per unit as

$$M_{12} = f_{12} = \frac{1}{0.2}(\theta_1 - \theta_2) = 5\theta_1 - 5\theta_2$$

$$M_{13} = f_{13} = \frac{1}{0.4}(\theta_1 - \theta_3) = 2.5\theta_1 \tag{12.28}$$

$$M_{32} = f_{32} = \frac{1}{0.25}(\theta_3 - \theta_2) = -4\theta_2$$

The reference-bus phase angle, θ_3, is still assumed to be zero. Then

$$[H] = \begin{bmatrix} 5 & -5 \\ 2.5 & 0 \\ 0 & -4 \end{bmatrix}$$

The covariance matrix for the measurements, $[R]$, is

$$[R] = \begin{bmatrix} \sigma^2_{M12} & & \\ & \sigma^2_{M13} & \\ & & \sigma^2_{M32} \end{bmatrix} = \begin{bmatrix} 0.0001 & & \\ & 0.0001 & \\ & & 0.0001 \end{bmatrix}$$

Note that since the coefficients of $[H]$ are in per unit we must also write $[R]$ and z^{meas} in per unit.

Our least-squares "best" estimate of θ_1 and θ_2 is then calculated as

$$\begin{bmatrix} \theta^{est}_1 \\ \theta^{est}_2 \end{bmatrix} = \left[\begin{bmatrix} 5 & 2.5 & 0 \\ -5 & 0 & -4 \end{bmatrix} \begin{bmatrix} 0.0001 & & \\ & 0.0001 & \\ & & 0.0001 \end{bmatrix}^{-1} \begin{bmatrix} 5 & -5 \\ 2.5 & 0 \\ 0 & -4 \end{bmatrix} \right]^{-1}$$

$$\times \begin{bmatrix} 5 & 2.5 & 0 \\ -5 & 0 & -4 \end{bmatrix} \begin{bmatrix} 0.0001 & & \\ & 0.0001 & \\ & & 0.0001 \end{bmatrix}^{-1} \begin{bmatrix} 0.62 \\ 0.06 \\ 0.37 \end{bmatrix}$$

$$= \begin{bmatrix} 312500 & -250000 \\ -250000 & 410000 \end{bmatrix}^{-1} \begin{bmatrix} 32500 \\ -45800 \end{bmatrix}$$

$$= \begin{bmatrix} 0.028571 \\ -0.094286 \end{bmatrix}$$

where

$$z^{meas} = \begin{bmatrix} 0.62 \\ 0.06 \\ 0.37 \end{bmatrix}$$

From the estimated phase angles, we can calculate the power flowing in each transmission line and the net generation or load at each bus. The results are shown in Figure 12.9. If we calculate the value of $J(\theta_1, \theta_2)$, the residual, we get

$$J(\theta_1, \theta_2) = \frac{[z_{12} - f_{12}(\theta_1, \theta_2)]^2}{\sigma^2_{12}} + \frac{[z_{13} - f_{13}(\theta_1, \theta_2)]^2}{\sigma^2_{13}} + \frac{[z_{32} - f_{32}(\theta_1, \theta_2)]^2}{\sigma^2_{32}}$$

$$= \frac{[0.62 - (5\theta_1 - 5\theta_2)]^2}{0.0001} + \frac{[0.06 - (2.5\theta_1)]^2}{0.0001} + \frac{[0.37 + (4\theta_2)]^2}{0.0001}$$

$$= 2.14 \tag{12.29}$$

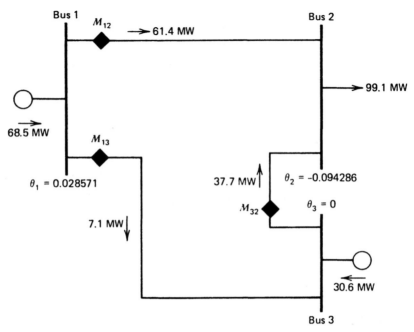

FIG. 12.9   Three-bus example with best estimates of θ_1 and θ_2.

Suppose the meter on the M_{13} transmission line was superior in quality to those on M_{12} and M_{32}. How will this affect the estimate of the states? Intuitively, we can reason that any measurement reading we get from M_{13} will be much closer to the true power flowing on line 1-3 than can be expected when comparing M_{12} and M_{32} to the flows on lines 1-2 and 3-2, respectively. Therefore, we would expect the results from the state estimator to reflect this if we set up the measurement data to reflect the fact that M_{13} is a superior measurement. To show this, we use the following metering data.

Meters M_{12} and M_{32}:   100 MW full scale
± 3 MW accuracy
($\sigma = 1$ MW $= 0.01$ pu)

Meter M_{13}:   100 MW full scale
± 0.3 MW accuracy
($\sigma = 0.1$ MW $= 0.001$ pu)

The covariance matrix to be used in the least-squares formula now becomes

$$[R] = \begin{bmatrix} \sigma_{M12}^2 & & \\ & \sigma_{M13}^2 & \\ & & \sigma_{M32}^2 \end{bmatrix} = \begin{bmatrix} 1 \times 10^{-4} & & \\ & 1 \times 10^{-6} & \\ & & 1 \times 10^{-4} \end{bmatrix}$$

We now solve Eq. 12.23 again with the new $[R]$ matrix.

$$
\begin{bmatrix} \theta_1 \\ \theta_2 \end{bmatrix} = \left[\begin{bmatrix} 5 & 2.5 & 0 \\ -5 & 0 & -4 \end{bmatrix} \begin{bmatrix} 1 \times 10^{-4} & & \\ & 1 \times 10^{-6} & \\ & & 1 \times 10^{-4} \end{bmatrix}^{-1} \begin{bmatrix} 5 & -5 \\ 2.5 & 0 \\ 0 & -4 \end{bmatrix} \right]^{-1}
$$

$$
\begin{bmatrix} 5 & 2.5 & 0 \\ -5 & 0 & -4 \end{bmatrix} \begin{bmatrix} 1 \times 10^{-4} & & \\ & 1 \times 10^{-6} & \\ & & 1 \times 10^{-4} \end{bmatrix}^{-1} \begin{bmatrix} 0.62 \\ 0.06 \\ 0.37 \end{bmatrix}
$$

$$
= \begin{bmatrix} 6.5 \times 10^6 & -2.5 \times 10^5 \\ -2.5 \times 10^5 & 4.1 \times 10^5 \end{bmatrix}^{-1} \begin{bmatrix} 1.81 \times 10^5 \\ -0.458 \times 10^5 \end{bmatrix}
$$

$$
= \begin{bmatrix} 0.024115 \\ -0.097003 \end{bmatrix}
$$

From these estimated phase angles, we obtain the network conditions shown in Figure 12.10. Compare the estimated flow on line 1-3, as just calculated, to the estimated flow calculated on line 1-3 in the previous least-squares estimate. Setting σ_{M13} to 0.1 MW has brought the estimated flow on line 1-3 much closer to the meter reading of 6.0 MW. Also, note that the estimates of flow on lines 1-2 and 3-2 are now further from the M_{12} and M_{32} meter readings, respectively, which is what we should have expected.

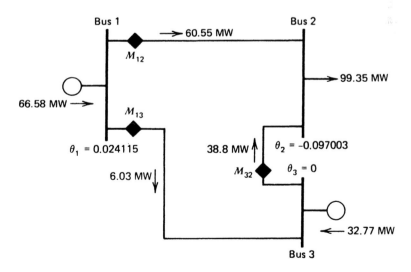

FIG. 12.10   Three-bus example with better meter at M_{13}.

12.4   STATE ESTIMATION OF AN AC NETWORK

12.4.1   Development of Method

We have demonstrated how the maximum likelihood estimation scheme developed in Section 12.3.2 led to a least-squares calculation for measurements from a linear system. In the least-squares calculation, we are trying to minimize the sum of measurement residuals:

$$\min_{\mathbf{x}} J(\mathbf{x}) = \sum_{i=1}^{N_m} \frac{[z_i - f_i(\mathbf{x})]^2}{\sigma_i^2} \tag{12.30}$$

In the case of a linear system, the $f_i(\mathbf{x})$ functions are themselves linear and we solve for the minimum of $J(\mathbf{x})$ directly. In an AC network, the measured quantities are MW, MVAR, MVA, amperes, transformer tap position, and voltage magnitude. The state variables are the voltage magnitude at each bus, the phase angles at all but the reference bus, and the transformer taps. The equation for power entering a bus is given in Eq. 4.21 and is clearly not a linear function of the voltage magnitude and phase angle at each bus. Therefore, the $f_i(\mathbf{x})$ functions will be nonlinear functions, except for a voltage magnitude measurement where $f_i(\mathbf{x})$ is simply unity times the particular x_i that corresponds to the voltage magnitude being measured. For MW and MVAR measurements on a transmission line from bus i to bus j we would have the following terms in $J(\mathbf{x})$:

$$\frac{\{MW_{ij}^{meas} - [|E_i|^2(G_{ij}) - |E_i||E_j|(\cos(\theta_i - \theta_j)G_{ij} + \sin(\theta_i - \theta_j)B_{ij})]\}^2}{\sigma_{MW_{ij}}^2} \tag{12.31}$$

and

$$\frac{\{MVAR_{ij}^{meas} - [-|E_i|^2(B_{cap_{ij}} + B_{ij}) - |E_i||E_j|(\sin(\theta_i - \theta_j)G_{ij} - \cos(\theta_i - \theta_j)B_{ij})]\}^2}{\sigma_{MVAR_{ij}}^2} \tag{12.32}$$

A voltage magnitude measurement would result in the following term in $J(\mathbf{x})$:

$$\frac{(|E_i|^{meas} - |E_i|)^2}{\sigma_{|E_i|}^2} \tag{12.33}$$

Similar functions can be derived for MVA or ampere measurements.

If we do not have a linear relationship between the states ($|E|$ values and θ values) and the power flows on a network, we will have to resort to an iterative technique to minimize $J(\mathbf{x})$. A commonly used technique for power system state estimation is to calculate the gradient of $J(\mathbf{x})$ and then force it to zero using Newton's method, as was done with the Newton load flow in Chapter 4. We will review how to use Newton's method on multidimensional problems before proceeding to the minimization of $J(\mathbf{x})$.

Given the functions $g_i(\mathbf{x})$, $i = 1, \ldots, n$, we wish to find $\mathbf{x}$ that gives $g_i(\mathbf{x}) = g_i^{\text{des}}$, for $i = 1, \ldots, n$. If we arrange the g_i functions in a vector we can write

$$\mathbf{g}^{\text{des}} - \mathbf{g}(\mathbf{x}) = 0 \tag{12.34}$$

by perturbing $\mathbf{x}$ we can write

$$\mathbf{g}^{\text{des}} - \mathbf{g}(\mathbf{x} + \Delta\mathbf{x}) = \mathbf{g}^{\text{des}} - \mathbf{g}(\mathbf{x}) - [\mathbf{g}'(\mathbf{x})]\Delta\mathbf{x} = 0 \tag{12.35}$$

where we have expanded $\mathbf{g}(\mathbf{x} + \Delta\mathbf{x})$ in a Taylor's series about $\mathbf{x}$ and ignored all higher-order terms. The $[\mathbf{g}'(\mathbf{x})]$ term is the Jacobian matrix of first derivatives of $\mathbf{g}(\mathbf{x})$. Then

$$\Delta\mathbf{x} = [\mathbf{g}'(\mathbf{x})]^{-1}[\mathbf{g}^{\text{des}} - \mathbf{g}(\mathbf{x})] \tag{12.36}$$

Note that if $\mathbf{g}^{\text{des}}$ is identically zero we have

$$\Delta\mathbf{x} = [\mathbf{g}'(\mathbf{x})]^{-1}[-\mathbf{g}(\mathbf{x})] \tag{12.37}$$

To solve for $\mathbf{g}^{\text{des}}$, we must solve for $\Delta\mathbf{x}$ using Eq. 12.36, then calculate $\mathbf{x}^{\text{new}} = \mathbf{x} + \Delta\mathbf{x}$ and reapply Eq. 12.36 until either $\Delta\mathbf{x}$ gets very small or $\mathbf{g}(\mathbf{x})$ comes close to $\mathbf{g}^{\text{des}}$.

Now let us return to the state estimation problem as given in Eq. 12.30:

$$\min_{\mathbf{x}} J(\mathbf{x}) = \sum_{i=1}^{N_m} \frac{[z_i - f_i(\mathbf{x})]^2}{\sigma_i^2}$$

We first form the gradient of $J(\mathbf{x})$ as

$$\nabla_x \mathbf{J}(\mathbf{x}) = \begin{bmatrix} \dfrac{\partial J(\mathbf{x})}{\partial x_1} \\[2mm] \dfrac{\partial J(\mathbf{x})}{\partial x_2} \\[2mm] \vdots \end{bmatrix}$$

$$= -2 \begin{bmatrix} \dfrac{\partial f_1}{\partial x_1} & \dfrac{\partial f_2}{\partial x_1} & \dfrac{\partial f_3}{\partial x_1} & \cdots \\[2mm] \dfrac{\partial f_1}{\partial x_2} & \dfrac{\partial f_2}{\partial x_2} & \dfrac{\partial f_3}{\partial x_2} & \cdots \\[2mm] \vdots & \vdots & \vdots & \end{bmatrix} \begin{bmatrix} \dfrac{1}{\sigma_1^2} & & \\ & \dfrac{1}{\sigma_2^2} & \\ & & \ddots \end{bmatrix} \begin{bmatrix} [z_1 - f_1(\mathbf{x})] \\[2mm] [z_2 - f_2(\mathbf{x})] \\[2mm] \vdots \end{bmatrix} \tag{12.38}$$

If we put the $f_i(x)$ functions in a vector form $\mathbf{f}(x)$ and calculate the Jacobian of $\mathbf{f}(x)$, we would obtain

$$\frac{\partial \mathbf{f}(x)}{\partial \mathbf{x}} = \begin{bmatrix} \dfrac{\partial f_1}{\partial x_1} & \dfrac{\partial f_1}{\partial x_2} & \dfrac{\partial f_1}{\partial x_3} & \cdots \\[2mm] \dfrac{\partial f_2}{\partial x_1} & \dfrac{\partial f_2}{\partial x_2} & \dfrac{\partial f_2}{\partial x_3} & \cdots \\[2mm] \vdots & \vdots & \vdots & \end{bmatrix} \tag{12.39}$$

We will call this matrix $[H]$. Then,

$$[H] = \begin{bmatrix} \dfrac{\partial f_1}{\partial x_1} & \dfrac{\partial f_1}{\partial x_2} & \dfrac{\partial f_1}{\partial x_3} & \cdots \\[2mm] \dfrac{\partial f_2}{\partial x_1} & \dfrac{\partial f_2}{\partial x_2} & \dfrac{\partial f_2}{\partial x_3} & \cdots \\[2mm] \vdots & \vdots & \vdots & \end{bmatrix} \tag{12.40}$$

And its transpose is

$$[H]^T = \begin{bmatrix} \dfrac{\partial f_1}{\partial x_1} & \dfrac{\partial f_2}{\partial x_1} & \dfrac{\partial f_3}{\partial x_1} & \cdots \\[2mm] \dfrac{\partial f_1}{\partial x_2} & \dfrac{\partial f_2}{\partial x_2} & \dfrac{\partial f_3}{\partial x_2} & \cdots \\[2mm] \vdots & \vdots & \vdots & \end{bmatrix} \tag{12.41}$$

Further, we write

$$\begin{bmatrix} \sigma_1^2 & & \\ & \sigma_2^2 & \\ & & \ddots \end{bmatrix} = [R] \tag{12.42}$$

Equation 12.38 can be written

$$\nabla_x J(x) = \left\{ -2[H]^T[R]^{-1} \begin{bmatrix} z_1 - f_1(x) \\ z_2 - f_2(x) \\ \vdots \end{bmatrix} \right\} \tag{12.43}$$

To make $\nabla_x J(x)$ equal zero, we will apply Newton's method as in Eq. 12.37, then

$$\Delta x = \left[\frac{\partial \nabla_x J(x)}{\partial \mathbf{x}} \right]^{-1} [-\nabla_x J(x)] \tag{12.44}$$

The Jacobian of $\nabla_x J(\mathbf{x})$ is calculated by treating $[H]$ as a constant matrix:

$$\frac{\partial \nabla_x J(\mathbf{x})}{\partial \mathbf{x}} = \frac{\partial}{\partial \mathbf{x}}\left\{ -2[H]^T[R]^{-1}\begin{bmatrix} z_1 - f_1(\mathbf{x}) \\ z_2 - f_2(\mathbf{x}) \\ \vdots \end{bmatrix}\right\}$$

$$= -2[H]^T[R]^{-1}[-H]$$

$$= 2[H]^T[R]^{-1}[H] \qquad (12.45)$$

Then

$$\Delta \mathbf{x} = \frac{1}{2}[[H]^T[R]^{-1}[H]]^{-1}\left\{2[H]^T[R]^{-1}\begin{bmatrix} z_1 - f_1(\mathbf{x}) \\ \vdots \end{bmatrix}\right\}$$

$$= [[H]^T[R]^{-1}[H]]^{-1}[H]^T[R]^{-1}\begin{bmatrix} z_1 - f_1(\mathbf{x}) \\ z_2 - f_2(\mathbf{x}) \\ \vdots \end{bmatrix} \qquad (12.46)$$

Equation 12.46 is obviously a close parallel to Eq. 12.23. To solve the AC state estimation problem, apply Eq. 12.46 iteratively as shown in Figure 12.11. Note that this is similar to the iterative process used in the Newton power flow solution.

12.4.2   Typical Results of State Estimation on an AC Network

Figure 12.12 shows our familiar six-bus system with $P + jQ$ measurements on each end of each transmission line and at each load and generator. Bus voltage is also measured at each system bus.

To demonstrate the use of state estimation on these measurements, the base-case conditions shown in Figure 11.1 were used together with a random number generating algorithm to produce measurements with random errors. The measurements were obtained by adding the random errors to the base-case flows, loads, generations, and bus-voltage magnitudes. The errors were generated so as to be representative of values drawn from a set of numbers having a normal probability density function with zero mean, and variance as specified for each measurement type. The measurement variances used were

$P + jQ$ measurements: $\sigma = 5$ MW for the P measurement

$\sigma = 5$ MVAR for the Q measurement

Voltage measurement: $\sigma = 3.83$ kV

The base conditions and the measurements are shown in Table 12.2. The state estimation algorithm shown in Figure 12.11 was run to obtain estimates

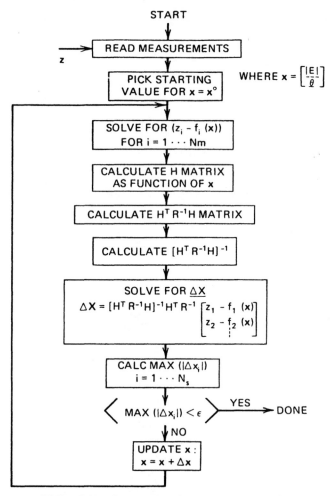

FIG. 12.11   State estimation solution algorithm.

for the bus-voltage magnitudes and phase angles given the measurements shown in Table 12.2. The procedure took three iterations with x^0 initially being set to 1.0 pu and 0 rad for the voltage magnitude and phase angle at each bus, respectively. At the beginning of each iteration, the sum of the measurement residuals, $J(x)$ (see Eq. 12.30), is calculated and displayed. At the end of each iteration, the maximum $\Delta|E|$ and the maximum $\Delta\theta$ are calculated and displayed. The iterative steps for the six-bus system used here produced the results given in Table 12.3.

The value of $J(x)$ at the end of the iterative procedure would be zero if all measurements were without error or if there were no redundancy in the measurements. When there are redundant measurements with errors, the value of $J(x)$ will not normally go to zero. Its value represents a measure of the overall

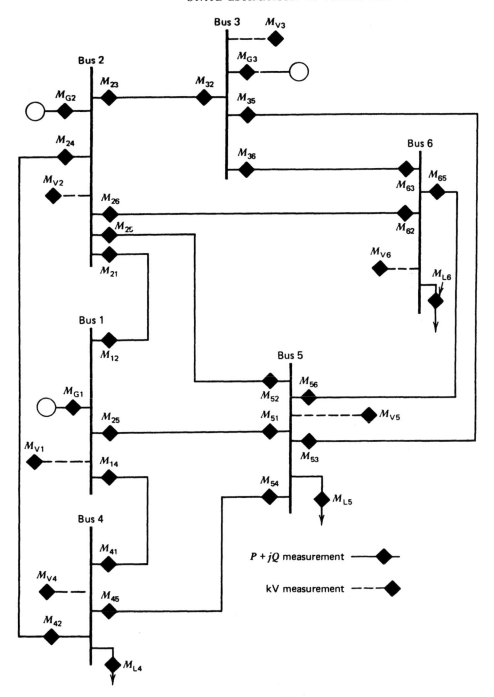

FIG. 12.12 Six-bus system with measurements.

TABLE 12.2   Base-Case Conditions

Measurement	Base-Case Value			Measured Value		
	kV	MW	MVAR	kV	MW	MVAR
M_{V1}	241.5			238.4		
M_{G1}		107.9	16.0		113.1	20.2
M_{12}		28.7	−15.4		31.5	−13.2
M_{14}		43.6	20.1		38.9	21.2
M_{15}		35.6	11.3		35.7	9.4
M_{V2}	241.5			237.8		
M_{G2}		50.0	74.4		48.4	71.9
M_{21}		−27.8	12.8		−34.9	9.7
M_{24}		33.1	46.1		32.8	38.3
M_{25}		15.5	15.4		17.4	22.0
M_{26}		26.2	12.4		22.3	15.0
M_{23}		2.9	−12.3		8.6	−11.9
M_{V3}	246.1			250.7		
M_{G3}		60.0	89.6		55.1	90.6
M_{32}		−2.9	5.7		−2.1	10.2
M_{35}		19.1	23.2		17.7	23.9
M_{36}		43.8	60.7		43.3	58.3
M_{V4}	227.6			225.7		
M_{L4}		70.0	70.0		71.8	71.9
M_{41}		−42.5	−19.9		−40.1	−14.3
M_{42}		−31.6	−45.1		−29.8	−44.3
M_{45}		4.1	−4.9		0.7	−17.4
M_{V5}	226.7			225.2		
M_{L5}		70.0	70.0		72.0	67.7
M_{54}		−4.0	−2.8		−2.1	−1.5
M_{51}		−34.5	−13.5		−36.6	−17.5
M_{52}		−15.0	−18.0		−11.7	−22.2
M_{53}		−18.0	−26.1		−25.1	−29.9
M_{56}		1.6	−9.7		−2.1	−0.8
M_{V6}	231.0			228.9		
M_{L6}		70.0	70.0		72.3	60.9
M_{65}		−1.6	3.9		1.0	2.9
M_{62}		−25.7	−16.0		−19.6	−22.3
M_{63}		−42.8	−57.9		−46.8	−51.1

TABLE 12.3   Iterative Results of State Estimator Solution

Iteration	$J(x)$ at Beginning of Iteration (pu)	Largest $\Delta\lvert E\rvert$ at End of Iteration (pu V)	Largest $\Delta\theta$ at End of Iteration (rad)
1	3696.86	0.1123	0.06422
2	43.67	0.004866	0.0017
3	40.33	0.0000146	0.0000227

fit of the estimated values to the measurement values. The value of $J(x)$ can, in fact, be used to detect the presence of bad measurements.

The estimated values from the state estimator are shown in Table 12.4, together with the base-case values and the measured values. Notice that, in general, the estimated values do a good job of calculating the true (base-case) conditions from which the measurements were made. For example, measurement M_{23} shows a P flow of 8.6 MW whereas the true flow is 2.9 MW and the estimator predicts a flow of 3.0 MW.

The example shown here started from a base case or "true" state that was shown in Table 12.2. In actual practice, we only have the measurements and the resulting estimate of the state, we never know the "true" state exactly and can only compare measurements with estimates. In the presentations to follow, however, we will leave the base-case or "true" conditions in our illustrations to aid the reader.

The results in Table 12.4 show one of the advantages of using a state estimation algorithm in that, even with measurement errors, the estimation algorithm calculates quantities that are the "best" possible estimates of the true bus voltages and generator, load, and transmission line MW and MVAR values.

There are, however, other advantages to using a state estimation algorithm. First, is the ability of the state estimator to detect and identify bad measurements, and, second, is the ability to estimate quantities that are not measured and telemetered. These are introduced later in the chapter.

12.5   STATE ESTIMATION BY ORTHOGONAL DECOMPOSITION

One problem with the standard least-squares method presented earlier in the chapter is the numerical difficulties encountered with some special state estimation problems. One of these comes about when we wish to drive a state estimator solution to match its measurement almost exactly. This is the case when we have a circuit such as shown in Figure 12.13. All of the actual flows and injections are shown in Figure 12.13 along with the values assumed for the measurements.

In this sample system, the measurement of power at bus 1 will be assumed to be zero MW. If the value of zero is dictated by the fact that the bus has no

TABLE 12.4   **State Estimation Solution**

Measurement	Base-Case Value			Measured Value			Estimated Value		
	kV	MW	MVAR	kV	MW	MVAR	kV	MW	MVAR
M_{V1}	241.5			238.4			240.6		
M_{G1}		107.9	16.0		113.1	20.2		111.9	18.7
M_{12}		28.7	−15.4		31.5	−13.2		30.4	−14.4
M_{14}		43.6	20.1		38.9	21.2		44.8	21.2
M_{15}		35.6	11.3		35.7	9.4		36.8	11.8
M_{V2}	241.5			237.8			239.9		
M_{G2}		50.0	74.4		48.4	71.9		47.5	70.3
M_{21}		−27.8	12.8		−34.9	9.7		−29.4	11.9
M_{24}		33.1	46.1		32.8	38.3		32.4	45.3
M_{25}		15.5	15.4		17.4	22.0		15.6	14.8
M_{26}		26.2	12.4		22.3	15.0		25.9	10.8
M_{23}		2.9	−12.3		8.6	−11.9		3.0	−12.6
M_{V3}	246.1			250.7			244.7		
M_{G3}		60.0	89.6		55.1	90.6		59.5	87.4
M_{32}		−2.9	5.7		−2.1	10.2		−3.0	6.2
M_{35}		19.1	23.2		17.7	23.9		19.2	22.9
M_{36}		43.8	60.7		43.3	58.3		43.3	58.3
M_{V4}	227.6			225.7			226.1		
M_{L4}		70.0	70.0		71.8	71.9		70.2	70.2
M_{41}		−42.5	−19.9		−40.1	−14.3		−43.6	−20.7
M_{42}		−31.6	−45.1		−29.8	−44.3		−30.9	−44.4
M_{45}		4.1	−4.9		0.7	−17.4		4.3	−5.1
M_{V5}	226.7			225.2			225.3		
M_{L5}		70.0	70.0		72.0	67.7		71.8	69.4
M_{54}		−4.0	−2.8		−2.1	−1.5		−4.2	−2.5
M_{51}		−34.5	−13.5		−36.6	−17.5		−35.6	−13.6
M_{52}		−15.0	−18.0		−11.7	−22.2		−15.1	−17.4
M_{53}		−18.0	−26.1		−25.1	−29.9		−18.1	−25.8
M_{56}		1.6	−9.7		−2.1	−0.8		1.3	−10.1
M_{V6}	231.0			228.9			230.1		
M_{L6}		70.0	70.0		72.3	60.9		68.9	65.8
M_{65}		−1.6	3.9		1.0	2.9		−1.2	4.4
M_{62}		−25.7	−16.0		−19.6	−22.3		−25.4	−14.5
M_{63}		−42.8	−57.9		−46.8	−51.1		−42.3	−55.7

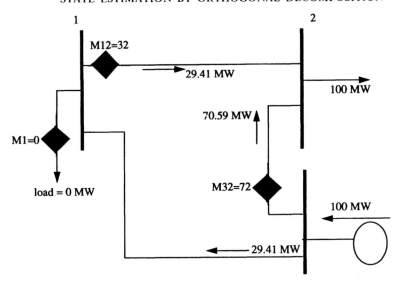

FIG. 12.13 Zero injection system example.

load or generation attached to it, then we know this value of zero MW with certainty and the concept of an error in its "measured" value is meaningless. Nonetheless, we proceed by setting up the standard state estimator equations and specifying the value of the measurement σ for M_1 as: $\sigma_{M1} = 10^{-2}$. This results in the following solution when using the state estimator equations as shown in Eq. 12.23:

$$P_{\text{flow}} \text{ estimate on line 1-2} = 30.76 \text{ MW}$$

$$P_{\text{flow}} \text{ estimate on line 3-2} = 72.52$$

$$\text{Injection estimate on bus 1} = 0.82$$

The estimator has not forced the bus injection to be exactly zero; instead, it reads 0.82 MW. This may not seem like such a big error. However, if there are many such buses (say 100) and they all have errors of this magnitude, then the estimator will have a large amount of load allocated to the buses that are known to be zero.

At first, the solution to this dilemma may seem to be simply forcing the σ value to a very small number for the zero injection buses and rerun the estimator. The problem with this is as follows. Suppose we had changed the zero injection σ to $\sigma_{M1} = 10^{-10}$. Hopefully, this would force the estimator to make the zero injection so dominant that it would result in the correct zero value coming out of the estimator calculation. In this case, the $[H^T R^{-1} H]$ matrix used in the standard least-squares method would look like this for the sample system:

$$[H] = \begin{bmatrix} 5.0 & -5.0 \\ 0 & -4.0 \\ 7.5 & -5.0 \end{bmatrix}$$

$$[R] = \begin{bmatrix} 10^{-4} & & \\ & 10^{-4} & \\ & & 10^{-20} \end{bmatrix}$$

then

$$[H^T R^{-1} H] = \begin{bmatrix} 56.25 \times 10^{20} & -37.5 \times 10^{20} \\ -37.5 \times 10^{20} & 25.0 \times 10^{20} \end{bmatrix}$$

Unfortunately, this matrix is very nearly singular. The reason is that the terms in the matrix are dominated by those terms which are multiplied by the 10^{20} terms from the inverse of the R matrix, and the other terms are so small by comparison that they are lost from the computer (unless one is using an extraordinarily long word length or extra double precision). When the above is presented to a standard matrix inversion routine or run into a Gaussian elimination solution routine, an error message results and garbage comes out of the estimator.

The solution to this dilemma is to use another algorithm for the least-squares solution. This algorithm is called the orthogonal decomposition algorithm and works as follows.

12.5.1   The Orthogonal Decomposition Algorithm

This algorithm goes under several different names in texts on linear algebra. It is often called the QR algorithm or the Gram–Schmidt decomposition. The idea is to take the state estimation least-squares equation, Eq. 12.23, and eliminate the R^{-1} matrix as follows: let

$$[R^{-1}] = R^{-1/2} R^{-1/2} \tag{12.47}$$

where

$$[R^{-1/2}] = \begin{bmatrix} \dfrac{1}{\sigma_{m1}} & & \\ & \dfrac{1}{\sigma_{m2}} & \\ & & \dfrac{1}{\sigma_{m3}} \end{bmatrix} \tag{12.48}$$

then

$$[H^T R^{-1} H]^{-1} = [H^T R^{-1/2} R^{-1/2} H]^{-1} = [H'^T H'] \tag{12.49}$$

with

$$[H'] = [R^{-1/2}][H] \qquad (12.50)$$

Finally, Eq. 12.23 becomes

$$\mathbf{x}^{est} = [H'^T H']^{-1}[H'^T]\mathbf{z}'^{meas} \qquad (12.51)$$

where

$$\mathbf{z}'^{meas} = [R^{-1/2}]\mathbf{z}^{meas} \qquad (12.52)$$

The idea of the orthogonal decomposition algorithm is to find a matrix $[Q]$ such that:

$$[H'] = [Q][U] \qquad (12.53)$$

(Note that in most linear algebra text books, this factorization would be written as $[H'] = [Q][R]$; however, we shall use $[Q][U]$ so as not to confuse the identity of the $[R]$ matrix.)

The matrix $[Q]$ has special properties. It is called an orthogonal matrix so that

$$[Q^T][Q] = [I] \qquad (12.54)$$

where $[I]$ is the identity matrix, which is to say that the transpose of $[Q]$ is its inverse. The matrix $[U]$ is now upper triangular in structure, although, since the $[H]$ matrix may not be square, $[U]$ will not be square either. Thus,

$$[H'] = \begin{bmatrix} h'_{11} & h'_{12} \\ h'_{21} & h'_{22} \\ h'_{31} & h'_{32} \end{bmatrix} = [Q][U] = \begin{bmatrix} q_{11} & q_{12} & q_{13} \\ q_{21} & q_{22} & q_{23} \\ q_{31} & q_{32} & q_{33} \end{bmatrix} \begin{bmatrix} u_{11} & u_{12} \\ 0 & u_{22} \\ 0 & 0 \end{bmatrix} \qquad (12.55)$$

Now, if we substitute $[Q][U]$ for $[H']$ in the state estimation equation:

$$\mathbf{x}^{est} = [U^T Q^T Q U]^{-1}[U^T][Q^T]\mathbf{z}' \qquad (12.56)$$

or

$$\mathbf{x}^{est} = [U^T U]^{-1} U^T \hat{\mathbf{z}} \qquad (12.57)$$

since

$$[Q^T Q] = I$$

and

$$\hat{\mathbf{z}} = [Q^T]\mathbf{z}' \qquad (12.58)$$

Then, by rearranging we get

$$[U^T U]\mathbf{x}^{est} = [U^T]\hat{\mathbf{z}} \qquad (12.59)$$

and we can eliminate U^T from both sides so that we are left with

$$[U]\mathbf{x}^{est} = \hat{\mathbf{z}} \tag{12.60}$$

or

$$\begin{bmatrix} u_{11} & u_{12} \\ 0 & u_{22} \\ 0 & 0 \end{bmatrix} \begin{bmatrix} x_1^{est} \\ x_2^{est} \end{bmatrix} = \begin{bmatrix} \hat{z}_1 \\ \hat{z}_2 \\ \hat{z}_3 \end{bmatrix} \tag{12.61}$$

This can be solved directly since U is upper triangular:

$$x_2^{est} = \frac{\hat{z}_2}{u_{22}} \tag{12.62}$$

and

$$x_1^{est} = \frac{1}{u_{11}} (\hat{z}_1 - u_{12} x_2^{est}) \tag{12.63}$$

The Q matrix and the U matrix are obtained, for our simple two-state–three-measurement problem here, using the Givens rotation method as explained in reference 15.

For the Givens rotation method, we start out to define the steps necessary to solve:

$$[Q^T][H] = [U] \tag{12.64}$$

where $[H]$ is a 2×2 matrix:

$$\begin{bmatrix} h_{11} & h_{12} \\ h_{21} & h_{22} \end{bmatrix}$$

and $[U]$ is

$$\begin{bmatrix} u_{11} & u_{12} \\ 0 & u_{22} \end{bmatrix}$$

The $[Q]$ matrix must be orthogonal, and when it is multiplied times $[H]$, it eliminates the h_{21} term. The terms in the $[Q]$ matrix are simply:

$$\begin{bmatrix} c & s \\ -s & c \end{bmatrix}$$

where

$$c = \frac{h_{11}}{\sqrt{h_{11}^2 + h_{21}^2}} \tag{12.65}$$

and

$$s = \frac{h_{21}}{\sqrt{h_{11}^2 + h_{21}^2}} \tag{12.66}$$

The reader can easily verify that the $[Q]$ matrix is indeed orthogonal and that:

$$\begin{bmatrix} u_{11} & u_{12} \\ 0 & u_{22} \end{bmatrix} = \begin{bmatrix} 1 & (ch_{12} + sh_{22}) \\ 0 & (-sh_{12} + ch_{22}) \end{bmatrix} \tag{12.67}$$

When we solve the 3×2 $[H]$ matrix in our three-measurement–two-state sample problem, we apply the Givens rotation three times to eliminate h_{21}, h_{31}, and h_{32}. That is, we need to solve

$$[Q^T] \begin{bmatrix} h_{11} & h_{12} \\ h_{21} & h_{22} \\ h_{31} & h_{32} \end{bmatrix} = \begin{bmatrix} u_{11} & u_{12} \\ 0 & u_{22} \\ 0 & 0 \end{bmatrix} \tag{12.68}$$

We will carry this out in three distinct steps, where each step can be represented as a Givens rotation. The result is that we represent $[Q^T]$ as the product of three matrices:

$$[Q^T] = [N_3][N_2][N_1] \tag{12.69}$$

These matrices are numbered as shown to indicate the order of application. In the case of the 3×2 $[H]$ matrix,

$$[N_1] = \begin{bmatrix} c & s & 0 \\ -s & c & 0 \\ 0 & 0 & 1 \end{bmatrix} \tag{12.70}$$

where c and s are defined exactly as before. Next, $[N_2]$ must be calculated so as to eliminate the 31 term which results from $[N_1][H]$. The actual procedure loads $[H]$ into $[U]$ and then determines each $[N]$ based on the current contents of $[U]$. The $[N_2]$ matrix will have terms like

$$[N_2] = \begin{bmatrix} c' & 0 & s' \\ 0 & 1 & 0 \\ -s' & 0 & c' \end{bmatrix} \tag{12.71}$$

where c' and s' are determined from $[N_1][H]$. Similarly for $[N_3]$:

$$[N_3] = \begin{bmatrix} 1 & 0 & 0 \\ 0 & c'' & s'' \\ 0 & -s'' & c'' \end{bmatrix} \qquad (12.72)$$

For our zero injection example, we start with the $[H]$ and $[R]$ matrices as shown before:

$$[H] = \begin{bmatrix} 5.0 & -5.0 \\ 0 & -4.0 \\ 7.5 & -5.0 \end{bmatrix}$$

and

$$[R] = \begin{bmatrix} 10^{-4} & & \\ & 10^{-4} & \\ & & 10^{-20} \end{bmatrix}$$

Then, the $[H']$ matrix is

$$[H'] = \begin{bmatrix} 5.0 \times 10^2 & -5.0 \times 10^2 \\ 0 & -4.0 \times 10^2 \\ 7.5 \times 10^{10} & -5.0 \times 10^{10} \end{bmatrix}$$

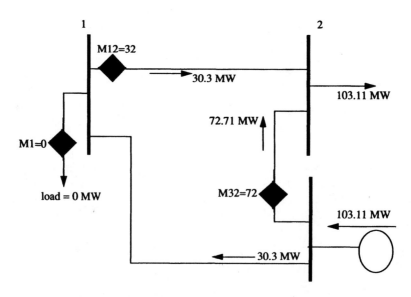

FIG. 12.14 State estimate resulting from orthogonal decomposition algorithm.

and the measurement vector is

$$\hat{\mathbf{z}} = \begin{bmatrix} 32 \\ 72 \\ 0 \end{bmatrix}$$

The resulting state estimate is shown in Figure 12.14. Note particularly that the injection at bus 1 is estimated to be zero, as we desired.

The orthogonal decomposition algorithm has the advantage that measurement weights can be adjusted to extreme values as demonstrated by the numerical example shown. As such, its robust numerical advantages have made it a useful algorithm for power system state estimators.

12.6   AN INTRODUCTION TO ADVANCED TOPICS IN STATE ESTIMATION

12.6.1   Detection and Identification of Bad Measurements

The ability to detect and identify bad measurements is extremely valuable to a power system's operations department. Transducers may have been wired incorrectly or the transducer itself may be malfunctioning so that it simply no longer gives accurate readings. The statistical theory required to understand and anlayze bad measurement detection and identification is straightforward but lengthy. We are going to open the door to the subject in this chapter. The serious student who wishes to pursue this subject should start with the chapter references. For the rest, we present results of these theories and indicate application areas.

To detect the presence of bad measurements, we will rely on the intuitive notion that for a given configuration, the residual, $J(\mathbf{x})$, calculated after the state estimator algorithm converges, will be smallest if there are no bad measurements. When $J(\mathbf{x})$ is small, a vector $\mathbf{x}$ (i.e., voltages and phase angles) has been found that causes all calculated flows, loads, generations, and so forth to closely match all the measurements. Generally, the presence of a bad measurement value will cause the converged value of $J(\mathbf{x})$ to be larger than expected with $\mathbf{x} = \mathbf{x}^{est}$.

What magnitude of $J(\mathbf{x})$ indicates the presence of bad measurements?

The measurement errors are random numbers so that the value of $J(\mathbf{x})$ is also a random number. If we assume that all the errors are described by their respective normal probability density functions, then we can show that $J(\mathbf{x})$

has a probability density function known as a *chi-squared distribution*, which is written as $\chi^2(K)$. The parameter K is called the degrees of freedom of the chi-squared distribution. This parameter is defined as follows:

$$K = N_m - N_s$$

where

N_m = number of measurements (note that a $P + jQ$ measurement counts as two measurements)

N_s = number of states = $(2n - 1)$

n = number of buses in the network

It can be shown that when $\mathbf{x} = \mathbf{x}^{est}$, the mean value of $J(\mathbf{x})$ equals K and the standard deviation, $\sigma_{J(\mathbf{x})}$, equals $\sqrt{2K}$.

When one or more measurements are bad, their errors are frequently much larger than the assumed $\pm 3\sigma$ error bound for the measurement. However, even under normal circumstances (i.e., all errors within $\pm 3\sigma$), $J(\mathbf{x})$ can get to be large—although the chance of this happening is small. If we simply set up a threshold for $J(\mathbf{x})$, which we will call t_J, we could declare that bad measurements are present when $J(\mathbf{x}) > t_J$. This threshold test might be wrong in one of two ways. If we set t_J to a small value, we would get many "false alarms." That is, the test would indicate the presence of bad measurements when, in fact, there were none. If we set t_J to be a large value, the test would often indicate that "all is well" when, in fact, bad measurements were present. This can be put on a formal basis by writing the following equation:

$$\text{prob}(J(\mathbf{x}) > t_J | J(\mathbf{x}) \text{ is a chi-squared}) = \alpha \qquad (12.73)$$
$$\text{with } K \text{ degrees of}$$
$$\text{freedom}$$

This equation says that the probability that $J(\mathbf{x})$ is greater than t_J is equal to α, given that the probability density for $J(\mathbf{x})$ is chi-squared with K degrees of freedom.

This type of testing procedure is formally known as *hypothesis testing*, and the parameter α is called the *significance level* of the test. By choosing a value for the significance level α, we automatically know what threshold t_J to use in our test. When using a t_J derived in this manner, the probability of a "false alarm" is equal to α. By setting α to a small number, for example $\alpha = 0.01$, we would say that false alarms would occur in only 1% of the tests made. A plot of the probability function in Eq. 12.73 is shown in Figure 12.15.

In Table 12.3, we saw that the minimum value for $J(\mathbf{x})$ was 40.33. Looking at Figure 12.12 and counting all $P + jQ$ measurements as two measurements,

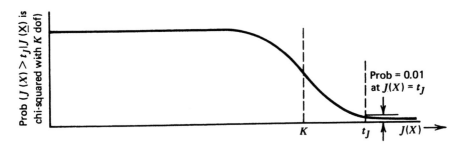

FIG. 12.15   Threshold test probability function.

we see that N_m is equal to 62. Therefore, the degrees of freedom for the chi-square distribution of $J(\mathbf{x})$ in our six-bus sample system is

$$K = N_m - N_s = N_m - (2n - 1) = 51$$

where

$$N_m = 62 \quad \text{and} \quad n = 6$$

If we set our significance level for this test to 0.01 (i.e., $\alpha = 0.01$ in Eq. 12.73), we get a t_J of 76.6.* Therefore, with a $J(\mathbf{x}) = 40.33$, it seems reasonable to assume that there are no "bad" measurements present.

Now let us assume that one of the measurements is truly bad. To simulate this situation, the state estimation algorithm was rerun with the M_{12} measurement reversed. Instead of $P = 31.5$ and $Q = -13.2$, it was set to $P = -31.5$ and $Q = 13.2$. The value of $J(\mathbf{x})$ and the maximum $\Delta|E|$ and $\Delta\theta$ for each iteration for this case are given in Table 12.5. The presence of bad data does not prevent the estimator from converging, but it will increase the value of the residual, $J(\mathbf{x})$.

The calculated flows and voltages for this situation are shown in Table 12.6. Note that the number of degrees of freedom is still 51 but $J(\mathbf{x})$ is now 207.94 at the end of our calculation. Since t_J is 76.6, we would immediately expect bad

TABLE 12.5   Iterative Results with Bad Measurement

Iteration	$J(\mathbf{x})$ at Beginning of Iteration (pu)	Largest $\Delta|E|$ at End of Iteration (pu V)	Largest $\Delta\theta$ at End of Iteration (rad)
1	3701.06	0.09851	0.06416
2	211.13	0.004674	0.001481
3	207.94	0.00002598	0.00004848

* Standard tables of $\chi^2(K)$ usually only go up to $K = 30$. For $K > 30$, a very close approximation to $\chi^2(K)$ using the normal distribution can be used. The student should consult any standard reference on probability and statistics to see how this is done.

TABLE 12.6 State Estimation Solution with Measurement M_{12} Reversed

	Base-Case Value			Measured Value			Estimated Value		
Measurement	kV	MW	MVAR	kV	MW	MVAR	kV	MW	MVAR
M_{V1}	241.5			238.4			240.6		
M_{G1}		107.9	16.0		113.1	20.2		99.3	21.9
M_{12}		28.7	−15.4		−31.5	+13.2		25.0	−12.2
M_{14}		43.6	20.1		38.9	21.2		40.6	21.9
M_{15}		35.6	11.3		35.7	9.4		33.7	12.3
M_{V2}	241.5			237.8			239.9		
M_{G2}		50.0	74.4		48.4	71.9		54.4	67.0
M_{21}		−27.8	12.8		−34.9	9.7		−24.4	9.2
M_{24}		33.1	46.1		32.8	38.3		35.0	44.1
M_{25}		15.5	15.4		17.4	22.0		16.3	14.7
M_{26}		26.2	12.4		22.3	15.0		25.1	11.3
M_{23}		2.9	−12.3		8.6	−11.9		2.3	−12.2
M_{V3}	246.1			250.7			244.6		
M_{G3}		60.0	89.6		55.1	90.6		61.4	86.3
M_{32}		−2.9	5.7		−2.1	10.2		−2.3	5.8
M_{35}		19.1	23.2		17.7	23.9		−20.5	22.2
M_{36}		43.8	60.7		43.3	58.3		43.2	58.2
M_{V4}	227.6			225.7			226.1		
M_{L4}		70.0	70.0		71.8	71.9		69.0	70.0
M_{41}		−42.5	−19.9		−40.1	−14.3		−39.6	−21.9
M_{42}		−31.6	−45.1		−29.8	−44.3		−33.5	−43.1
M_{45}		4.1	−4.9		0.7	−17.4		4.1	−5.0
M_{V5}	226.7			225.2			225.3		
M_{L5}		70.0	70.0		72.0	67.7		71.8	69.3
M_{54}		−4.0	−2.8		−2.1	−1.5		−4.1	−2.6
M_{51}		−34.5	−13.5		−36.6	−17.5		−32.7	−14.7
M_{52}		−15.0	−18.0		−11.7	−22.2		−15.8	−17.2
M_{53}		−18.0	−26.1		−25.1	−29.9		−19.3	−25.1
M_{56}		1.6	−9.7		−2.1	−0.8		0.1	−9.6
M_{V6}	231.0			228.9			230.0		
M_{L6}		70.0	70.0		72.3	60.9		66.9	66.7
M_{65}		−1.6	3.9		1.0	2.9		−0.1	3.9
M_{62}		−25.7	−16.0		−19.6	−22.3		−24.6	−15.0
M_{63}		−42.8	−57.9		−46.8	−51.1		−42.3	−55.6

measurements at our 0.01 significance level. If we had not known ahead of running the estimation algorithm that a bad measurement was present, we would certainly have had good reason to suspect its presence when so large a $J(\mathbf{x})$ resulted.

So far, we can say that by looking at $J(\mathbf{x})$, we can detect the presence of bad measurements. But if bad measurements are present, how can one tell which measurements are bad? Without going into the statistical theory, we give the following explanation of how this is accomplished.

Suppose we are interested in the measurement of megawatt flow on a particular line. Call this measured value z_i. In Figure 12.16(a) we have a plot of the normal probability density function of z_i. Since we assume that the error in z_i is normally distributed with zero mean value, the probability density function is centered on the true value of z_i. Since the errors on all the measurements are assumed normal, we will assume that the estimate, $\mathbf{x}^{est}$ is approximately normally distributed and that any quantity that is a function of $\mathbf{x}^{est}$ is also an approximately normally distributed quantity. In Figure 12.16(b), we show the probability density function for the calculated megawatt flow, f_i, which is a function of the estimated state, $\mathbf{x}^{est}$. We have drawn the density function of f_i as having a smaller deviation from its mean than the measurement z_i to indicate that, due to redundancy in measurements, the estimate is more accurate.

The difference between the estimate, f_i, and the measurement, z_i, is called the *measurement residual* and is designated y_i. The probability density function for y_i is also normal and is shown in Figure 12.16(c) as having a zero mean and a standard deviation of σ_{y_i}. If we divide the difference between the estimate f_i and the measurement z_i by σ_{y_i}, we obtain what is called a *normalized measurement residual*. The normalized measurement residual is designated y_i^{norm} and is shown in Figure 12.16(d) along with its probability density function, which is normal and has a standard deviation of unity. If the absolute value of y_i^{norm} is greater than 3, we have good reason to suspect that z_i is a bad measurement value. The usual procedure in identifying bad measurements is to calculate all f_i values for the N_m measurements once $\mathbf{x}^{est}$ is available from the state estimator. Using the z_i values that were used in the estimator and the f_i values, a measurement residual y_i can be calculated for each measurement. Also, using information from the state estimator, we can calculate σ_{y_i} (see references for details of this calculation). Using y_i and σ_{y_i}, we can calculate a normalized residual for each measurement. Measurements having the largest absolute normalized residual are labeled as prime suspects. These prime suspects are removed from the state estimator calculation one at a time, starting with the measurement having the largest normalized residual. After a measurement has been removed, the state estimation calculation (see Figure 12.11) is rerun. This results in a different $\mathbf{x}^{est}$ and therefore a different $J(\mathbf{x})$. The chi-squared probability density function for $J(\mathbf{x})$ will have to be recalculated, assuming that we use the same significance level for our test. If the new $J(\mathbf{x})$ is now less than the new value for t_J, we can say that the measurement that

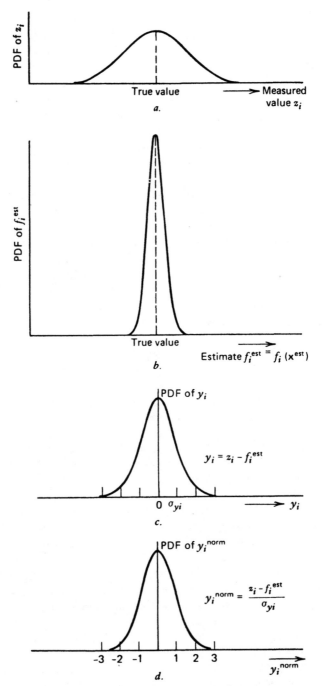

FIG. 12.16 Probability density function of the normalized measurement residual.

was removed has been identified as bad. If, however, the new $J(\mathbf{x})$ is greater than the new t_J, we must proceed to recalculate $f_i(\mathbf{x}^{est})$, σ_{y_i}, and then y_i^{norm} for each of the remaining measurements. The measurement with the largest absolute y_i^{norm} is then again removed and the entire procedure repeated successively until $J(\mathbf{x})$ is less than t_J. The references at the end of this chapter discuss a problem that the identification process may encounter, wherein several measurements may need to be removed to eliminate one "bad" measurement. That is, the identification procedure often cannot pinpoint a single bad measurement but instead identifies a group of measurements, one of which is bad. In such cases, the groups must be eliminated to eliminate the bad measurement.

The ability to detect (using the chi-squared statistic) and identify (using normalized residuals) are extremely useful features of a state estimator. Without the state estimator calculation using the system measurement data, those measurements whose values are not obviously wrong have little chance of being detected and identified. With the state estimator, the operations personnel have a greater assurance that quantities being displayed are not grossly in error.

12.6.2 Estimation of Quantities Not Being Measured

The other useful feature of a state estimator calculation is the ability to calculate (or estimate) quantities not being telemetered. This is most useful in cases of failure of communication channels connecting operations centers to remote data-gathering equipment or when the remote data-gathering equipment fails. Often data from some network substations are simply unavailable because no transducers or data-gathering equipment were ever installed.

An example of this might be the failure of all telemetry from buses 3, 4, 5, and 6 in our six-bus system. Even with the loss of these measurements, we can run the state estimation algorithm on the remaining measurements at buses 1 and 2, calculate the bus voltage magnitudes and phase angles at all six buses, and then calculate all network generations, loads, and flows. The results of such a calculation are given in Table 12.7. Notice that the estimate of quantities at the untelemetered buses are not as close to the base case as when using the full set of measurements (i.e., compare Table 12.7 to Table 12.4).

12.6.3 Network Observability and Pseudo-measurements

What happens if we continue to lose telemetry so that fewer and fewer measurements are available? Eventually, the state estimation procedure breaks down completely. Mathematically, the matrix

$$[[H]^T[R^{-1}][H]]$$

in Eq. 12.46 becomes singular and cannot be inverted. There is also a very interesting engineering interpretation of this phenomenon that allows us to alter the situation so that the state estimation procedure is not completely disabled.

TABLE 12.7   State Estimation Solution with Measurement at Buses 1 and 2 Only

Measurement	Base-Case Value			Measured Value			Estimated Value		
	kV	MW	MVAR	kV	MW	MVAR	kV	MW	MVAR
M_{V1}	241.5			238.4			238.8		
M_{G1}		107.9	16.0		113.1	20.2		112.4	20.5
M_{12}		28.7	−15.4		31.5	−13.2		30.6	−13.4
M_{14}		43.6	20.1		38.9	21.2		44.7	19.4
M_{15}		35.6	11.3		35.7	9.4		37.1	14.6
M_{V2}	241.5			237.8			237.6		
M_{G2}		50.0	74.4		48.4	71.9		48.2	71.7
M_{21}		−27.8	12.8		−34.9	9.7		−29.6	11.1
M_{24}		33.1	46.1		32.8	38.3		30.5	40.2
M_{25}		15.5	15.4		17.4	22.0		16.1	16.8
M_{26}		26.2	12.4		22.3	15.0		22.4	15.2
M_{23}		2.9	−12.3		8.6	−11.9		8.8	−11.7
M_{V3}	246.1						241.4		
M_{G3}		60.0	89.6					27.2	94.9
M_{32}		−2.9	5.7					−8.7	5.5
M_{35}		19.1	23.2					15.1	25.3
M_{36}		43.8	60.7					20.9	64.0
M_{V4}	227.6						225.0		
M_{L4}		70.0	70.0					67.6	61.2
M_{41}		−42.5	−19.9					−43.6	−18.9
M_{42}		−31.6	−45.1					−29.3	−39.7
M_{45}		4.1	−4.9					5.3	−2.6
M_{V5}	226.7						221.4		
M_{L5}		70.0	70.0					71.9	76.7
M_{54}		−4.0	−2.8					−5.2	−4.8
M_{51}		−34.5	−13.5					−35.9	−15.9
M_{52}		−15.0	−18.0					−15.5	−19.0
M_{53}		−18.0	−26.1					−14.0	−28.0
M_{56}		1.6	−9.7					−1.4	−9.0
M_{V6}	231.0						226.2		
M_{L6}		70.0	70.0					40.5	77.2
M_{65}		−1.6	3.9					1.4	3.4
M_{62}		−25.7	−16.0					−21.9	−18.8
M_{63}		−42.8	−57.9					−20.0	−61.8

If we take the three-bus example used in the beginning of Section 12.2, we note that when all three measurements are used, we have a redundant set and we can use a least-squares fit to the measurement values. If one of the measurements is lost, we have just enough measurements to calculate the states. If, however, two measurements are lost, we are in trouble. For example, suppose M_{13} and M_{32} were lost leaving only M_{12}. If we now apply Eq. 12.23 in a straightforward manner, we get

$$M_{12} = f_{12} = \frac{1}{0.2}(\theta_1 - \theta_2) = 5\theta_1 - 5\theta_2$$

Then

$$[H] = [5 \quad -5]$$
$$[R] = [\sigma_{M12}^2] = [0.0001]$$

and

$$\begin{bmatrix} \theta_1^{est} \\ \theta_2^{est} \end{bmatrix} = \left[\begin{bmatrix} 5 \\ -5 \end{bmatrix} [0.0001]^{-1} [5 \quad -5] \right]^{-1} [5 \quad -5][0.0001]^{-1}(0.55)$$

$$= \begin{bmatrix} 2500 & -2500 \\ -2500 & 2500 \end{bmatrix}^{-1} [5 \quad -5][0.0001]^{-1}(0.55) \qquad (12.74)$$

The matrix to be inverted in Eq. 12.74 is clearly singular and, therefore, we have no way of solving for θ_1^{est} and θ_2^{est}. Why is this? The reasons become quite obvious when we look at the one-line diagram of this network as shown in Figure 12.17. With only M_{12} available, all we can say about the network is that

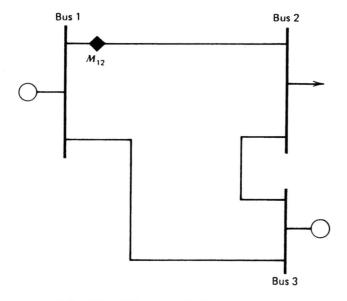

FIG. 12.17   "Unobservable" measurement set.

the phase angle across line 1-2 must be 0.11 rad, but with no other information available, we cannot tell what relationship θ_1 or θ_2 has to θ_3, which is assumed to be 0 rad. If we write down the equations for the net injected power at bus 1 and bus 2, we have

$$P_1 = 7.5\theta_1 - 5\theta_2$$
$$P_2 = -5\theta_1 + 9\theta_2$$

(12.75)

If measurement M_{12} is reading 55 MW (0.55 pu), we have

$$\theta_1 - \theta_2 = 0.11$$

(12.76)

and by substituting Eq. 12.75 into Eq. 12.76 and eliminating θ_1, we obtain

$$P_2 = 1.6P_1 - 1.87$$

(12.77)

Furthermore,

$$P_3 = -P_1 - P_2 = -0.6P_1 + 1.87$$

(12.78)

Equations 12.77 and 12.78 give a relationship between P_1, P_2, and P_3, but we still do not know their correct values. The technical term for this phenomenon is to say that the network is *unobservable*; that is, with only M_{12} available, we cannot observe (calculate) the state of the system.

It is very desirable to be able to circumvent this problem. Often a large power-system network will have missing data that render the network unobservable. Rather than just stop the calculations, a procedure is used that allows the estimator calculation to continue. The procedure involves the use of what are called *pseudo-measurements*. If we look at Eqs. 12.77 and 12.78, it is obvious that θ_1 and θ_2 could be estimated if the value of any one of the bus injections (i.e., P_1, P_2, or P_3) could be determined by some means other than direct measurement. This value, the pseudo-measurement, is used in the state estimator just as if it were an actual measured value.

To determine the value of an injection without measuring it, we must have some knowledge about the power system beyond the measurements currently being made. For example, it is customary to have access to the generated MW and MVAR values at generating stations through telemetry channels (i.e., the generated MW and MVAR would normally be measurements available to the state estimator). If these channels are out and we must have this measurement for observability, we can probably communicate with the operators in the plant control room by telephone and ask for the MW and MVAR values and enter them into the state estimator calculation manually. Similarly, if we needed a load-bus MW and MVAR for a pseudo-measurement, we could use historical records that show the relationship between an individual load and the total system load. We can estimate the total system load fairly accurately by knowing the total power being generated and estimating the network losses. Finally, if we have just experienced a telemetry failure, we could use the most recently

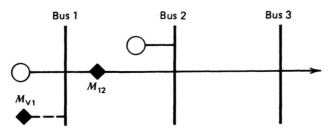

FIG. 12.18 Unobservable system showing importance of location of pseudo-measurements.

estimated values from the estimator (assuming that it is run periodically) as pseudo-measurements. Therefore, if needed, we can provide the state estimator with a reasonable value to use as a pseudo-measurement at any bus in the system.

The three-bus sample system in Figure 12.18 requires one pseudo-measurement. Measurement M_{12} allows us to estimate the voltage magnitude and phase angle at bus 2 (bus 1's voltage magnitude is measured and its phase angle is assumed to be zero). But without knowing the generation output at the generator unit on bus 2 or the load on bus 3, we cannot tell what voltage magnitude and phase angle to place on bus 3; hence, the network is unobservable. We can make this three-bus system observable by adding a pseudo-measurement of the net bus injected MW and MVAR at bus 2 or bus 3, but not at bus 1. That is, a pseudo-measurement at bus 1 will do no good at all because it tells nothing about the relationship of the phase angles between bus 2 and bus 3.

When adding a pseudo-measurement to a network, we simply write the equation for the pseudo-measurement injected power as a function of bus voltage magnitudes and phase angles as if it were actually measured. However, we do not wish to have the estimator treat the pseudo-measurement the same as a legitimate measurement, since it is often quite inaccurate and is little better than a guess. To circumvent this difficulty, we assign a large standard deviation to this measurement. The large standard deviation allows the estimator algorithm to treat the pseudo-measurement as if it were a measurement from a very poor-quality metering device.

To demonstrate the use of pseudo-measurements on our six-bus test system, all measurements were removed from buses 2, 3, 4, 5, and 6 so that bus 1 had all remaining measurements. This rendered the network unobservable and required adding pseudo-measurements at buses 2, 3, and 6. In the case, the pseudo-measurements were just taken from our base-case power flow. The results are shown in Table 12.8. Notice that the resulting estimates are quite close to the measured values for bus 1 but that the remaining buses have large measurement residuals. The net injections at buses 2, 3, and 6 do not closely match the pseudo-measurements since the pseudo-measurements were weighted much less than the legitimate measurements.

TABLE 12.8 State Estimation Solution with Measurements at Bus 1 and Pseudo-measurements at Buses 2, 3, and 6

Measurement	Base-Case Value			Measured Value			Estimated Value		
	kV	MW	MVAR	kV	MW	MVAR	kV	MW	MVAR
M_{V1}	241.5			238.4			238.4		
M_{G1}		107.9	16.0		113.1	20.2		111.4	19.5
M_{12}		28.7	−15.4		31.5	−13.2		33.3	−12.5
M_{14}		43.6	20.1		38.9	21.2		40.7	21.9
M_{15}		35.6	11.3		35.7	9.4		37.4	10.1
M_{V2}	241.5						236.2		
M_{G2}		50.0	74.4	Pseudo:	50.0	74.4		37.5	67.7
M_{21}		−27.8	12.8					−32.1	10.5
M_{24}		33.1	46.1					19.5	44.9
M_{25}		15.5	15.4					14.1	11.5
M_{26}		26.2	12.4					30.0	12.7
M_{23}		2.9	−12.3					6.0	−11.9
M_{V3}	246.1						240.5		
M_{G3}		60.0	89.6	Pseudo:	60.0	89.6		52.6	86.6
M_{32}		−2.9	5.7					−6.0	5.7
M_{35}		19.1	23.2					14.3	19.5
M_{36}		43.8	60.7					44.2	61.4
M_{V4}	227.6						223.8		
M_{L4}		70.0	70.0					51.9	73.3
M_{41}		−42.5	−19.9					−39.6	−21.8
M_{42}		−31.6	−45.1					−18.3	−44.6
M_{45}		4.1	−4.9					6.0	−6.9
M_{V5}	226.7						224.0		
M_{L5}		70.0	70.0					63.9	55.5
M_{54}		−4.0	−2.8					−5.9	−0.4
M_{51}		−34.5	−13.5					−36.3	−11.8
M_{52}		−15.0	−18.0					−13.7	−14.4
M_{53}		−18.0	−26.1					−13.6	−22.9
M_{56}		1.6	−9.7					5.5	−5.9
M_{V6}	231.0						224.9		
M_{L6}		70.0	70.0	Pseudo:	70.0	70.0		77.9	73.4
M_{65}		−1.6	3.9					−5.5	0.3
M_{62}		−25.7	−16.0					−29.3	−15.6
M_{63}		−42.8	−57.9					−43.2	−58.1

12.7 APPLICATION OF POWER SYSTEMS STATE ESTIMATION

In this last section, we will try to present the "big picture" showing how state estimation, contingency analysis, and generator corrective action fit together in a modern operations control center. Figure 12.19 is a schematic diagram showing the information flow between the various functions to be performed in an operations control center computer system. The system gets its information about the power system from remote terminal units that encode measurement transducer outputs and opened/closed status information into digital signals that are transmitted to the operations center over communications circuits. In addition, the control center can transmit control information such as raise/lower commands to generators and open/close commands to circuit breakers and switches. We have broken down the information coming into the control center as breaker/switch status indications and analog measurements. The analog measurements of generator output must be used directly by the AGC program (see Chapter 9), whereas all other data will be processed by the state estimator before being used by other programs.

In order to run the state estimator, we must know how the transmission lines are connected to the load and generation buses. We call this information the *network topology*. Since the breakers and switches in any substation can cause the network topology to change, a program must be provided that reads the telemetered breaker/switch status indications and restructures the electrical model of the system. An example of this is shown in Figure 12.20, where the opening of four breakers requires two electrical buses to represent the substation instead of one electrical bus. We have labeled the program that reconfigures the electrical model as the *network topology program*.* The network topology program must have a complete description of each substation and how the transmission lines are attached to the substation equipment. Bus sections that are connected to other bus sections through closed breakers or switches are designated as belonging to the same electrical bus. Thus, the number of electrical buses and the manner in which they are interconnected can be changed in the model to reflect breaker and switch status changes on the power system itself.

As seen in Figure 12.20, the electrical model of the power system's transmission system is sent to the state estimator program together with the analog measurements. The output of the state estimator consists of all bus voltage magnitudes and phase angles, transmission line MW and MVAR flows calculated from the bus voltage magnitude and phase angles, and bus loads and generations calculated from the line flows. These quantities, together with the electrical model developed by the network topology program, provide the basis for the economic dispatch program, congtingency analysis program, and generation corrective action program. Note that since the complete electrical model of the transmission system is available, we can directly calculate bus penalty factors as shown in Chapter 4.

---

* Alternative names that are often used for this program are "system status processor" and "network configurator."

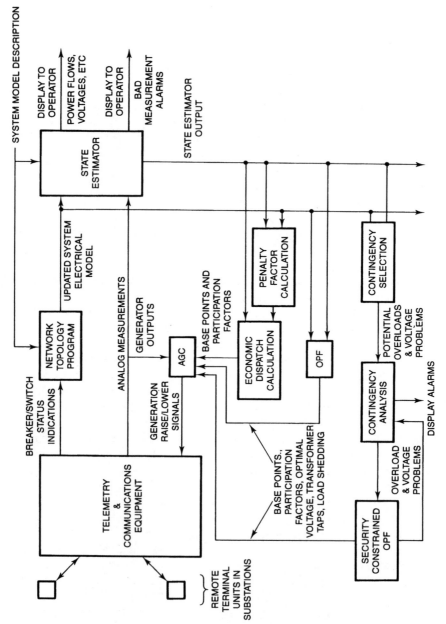

FIG. 12.19 Energy control center system security schematic.

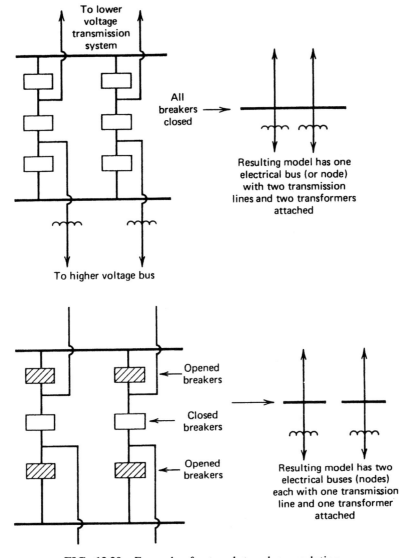

FIG. 12.20   Example of network topology updating.

APPENDIX
Derivation of Least-Squares Equations

One is often confronted with problems wherein data have been obtained by making measurements or taking samples on a process. Furthermore, the quantities being measured are themselves functions of other variables that we wish to estimate. These other variables will be called the state variables and designated $\mathbf{x}$, where the number of state variables is N_s. The measurement values will be called $\mathbf{z}$. We will assume here that the process we are interested in can be modeled using a linear model. Then we say that each measurement z_i is a linear function of the states x_i; that is,

$$z_i = h_i(\mathbf{x}) = h_{i1}x_1 + h_{i2}x_2 + \ldots + h_{iN_s}x_{N_s} \tag{12A.1}$$

We can also write this equation as a vector equation if we place the h_{ij} coefficients into a vector $\mathbf{h}$; that is,

$$\mathbf{h}_i = \begin{bmatrix} h_{i1} \\ h_{i2} \\ \vdots \\ h_{iN_s} \end{bmatrix} \tag{12A.2}$$

Then Eq. 12A.1 becomes

$$z_i = \mathbf{h}_i^T \mathbf{x} \tag{12A.3}$$

where

$$\mathbf{x} = \begin{bmatrix} x_1 \\ x_2 \\ \vdots \\ x_{N_s} \end{bmatrix}$$

Finally, we can write all the measurement equations in a compact form

$$\mathbf{z} = [H]\mathbf{x} \tag{12A.4}$$

where

$$\mathbf{z} = \begin{bmatrix} z_1 \\ z_2 \\ \vdots \\ z_{N_m} \end{bmatrix}$$

$$[H] = \begin{bmatrix} h_{11} & h_{12} & \cdots & h_{1N_s} \\ h_{21} & h_{22} & \cdots & \\ \vdots & & & \\ h_{N_m 1} & & \cdots & h_{N_m N_s} \end{bmatrix}$$

where row i of $[H]$ is equal to vector $\mathbf{h}_i^T$ (see Eq. 12A.2).

With N_m measurements we can have three possible cases to solve. That is,

N_s, the number of states, is either less than N_m, equal to N_m, or greater than N_m. We will deal with each case separately.

The Overdetermined Case ($N_m > N_s$)

In this case, we have more measurements or samples than state variables; therefore, we can write more equations, $h_i(x)$, than we have unknowns x_j. One way to estimate the x_i values is to minimize the sum of the squares of difference between the measurement values z_i and the estimate of z_i that is, in turn, a function of the estimates of x_i. That is, we wish to minimize

$$J(\mathbf{x}) = \sum_{i=1}^{N_m} [z_i - h_i(x_1, x_2, \ldots, x_{N_s})]^2 \tag{12A.5}$$

Equation 12A.5 can be written as

$$J(\mathbf{x}) = \sum_{i=1}^{N_m} (z_i - \mathbf{h}_i^T \mathbf{x})^2 \tag{12A.6}$$

and this can be written in a still more compact form as

$$J(\mathbf{x}) = (\mathbf{z} - [H]\mathbf{x})^T (\mathbf{z} - [H]\mathbf{x}) \tag{12A.7}$$

If we wish to find the value of $\mathbf{x}$ that minimizes $J(\mathbf{x})$, we can take the first derivative of $J(\mathbf{x})$ with respect to each x_j ($j = 1, \ldots, N_s$) and set these derivatives to zero. That is,

$$\frac{\partial J(\mathbf{x})}{\partial x_j} = 0 \quad \text{for } j = 1 \ldots N_s \tag{12A.8}$$

If we place these derivatives into a vector, we have what is called the gradient of $J(\mathbf{x})$, which is written $\nabla_x J(\mathbf{x})$. Then,

$$\nabla_x J(\mathbf{x}) = \begin{bmatrix} \dfrac{\partial J(\mathbf{x})}{\partial x_1} \\[2mm] \dfrac{\partial J(\mathbf{x})}{\partial x_2} \\[2mm] \vdots \end{bmatrix} \tag{12A.9}$$

Then the goal of forcing each derivative to zero can be written as

$$\nabla_x J(\mathbf{x}) = \mathbf{0} \tag{12A.10}$$

where $\mathbf{0}$ is a vector of N_s elements, each of which is zero. To solve this problem, we will first expand Eq. 12A.7:

$$J(\mathbf{x}) = (\mathbf{z} - [H]\mathbf{x})^T(\mathbf{z} - [H]\mathbf{x})$$
$$= \mathbf{z}^T\mathbf{z} - \mathbf{x}^T[H]^T\mathbf{z} - \mathbf{z}^T[H]\mathbf{x} + \mathbf{x}^T[H]^T[H]\mathbf{x} \qquad (12A.11)$$

The second and third term in Eq. 12.A.11 are identical, so that we can write

$$J(\mathbf{x}) = \mathbf{z}^T\mathbf{z} - 2\mathbf{z}^T[H]\mathbf{x} + \mathbf{x}^T[H]^T[H]\mathbf{x} \qquad (12A.12)$$

Before proceeding, we will derive a few simple relationships.

The gradient is always a vector of first derivatives of a scalar function that is itself a function of a vector. Thus, if we define $F(\mathbf{y})$ to be a scalar function, then its gradient $\nabla_y\mathbf{F}$ is:

$$\nabla_y\mathbf{F} = \begin{bmatrix} \dfrac{\partial F}{\partial y_1} \\[2mm] \dfrac{\partial F}{\partial y_2} \\[2mm] \vdots \\[2mm] \dfrac{\partial F}{\partial y_n} \end{bmatrix} \qquad (12A.13)$$

Then, if we define F as follows:

$$F = \mathbf{y}^T\mathbf{b} = \begin{bmatrix} y_1 & y_2 & \cdots \end{bmatrix}\begin{bmatrix} b_1 \\ b_2 \\ \vdots \end{bmatrix} \qquad (12A.14)$$

where $\mathbf{b}$ is a vector of constants b_i, $i = 1, \ldots, n$, then, F can be expanded as

$$F = y_1b_1 + y_2b_2 + y_3b_3 + \cdots \qquad (12A.15)$$

and the gradient of F is

$$\nabla_y\mathbf{F} = \begin{bmatrix} \dfrac{\partial F}{\partial y_1} \\[2mm] \dfrac{\partial F}{\partial y_2} \\[2mm] \vdots \\[2mm] \dfrac{\partial F}{\partial y_n} \end{bmatrix} = \begin{bmatrix} b_1 \\ b_2 \\ \vdots \\ b_n \end{bmatrix} = \mathbf{b} \qquad (12A.16)$$

It ought to be obvious that writing F with $\mathbf{y}$ and $\mathbf{b}$ reversed makes no difference. That is,

$$F = \mathbf{b}^T\mathbf{y} = \mathbf{y}^T\mathbf{b} \qquad (12A.17)$$

and, therefore, $\nabla_y(\mathbf{b}^T\mathbf{y}) = \mathbf{b}$.

Suppose we now write the vector $\mathbf{b}$ as the product of a matrix $[A]$ and a vector $\mathbf{u}$.

$$\mathbf{b} = [A]\mathbf{u} \qquad (12A.18)$$

Then, if we take F as shown in Eq. 12A.14,

$$F = \mathbf{y}^T\mathbf{b} = \mathbf{y}^T[A]\mathbf{u} \qquad (12A.19)$$

we can say that

$$\nabla_y F = [A]\mathbf{u} \qquad (12A.20)$$

Similarly, we can define

$$\mathbf{b}^T = \mathbf{u}^T[A] \qquad (12A.21)$$

If we can take F as shown in Eq. 12A.17,

$$F = \mathbf{b}^T\mathbf{y} = \mathbf{u}^T[A]\mathbf{y}$$

then

$$\nabla_y F = [A]^T\mathbf{u} \qquad (12A.22)$$

Finally, we will look at a scalar function F that is quadratic, namely,

$$F = \mathbf{y}^T[A]\mathbf{y}$$

$$= [y_1 \quad y_2 \quad \cdots \quad y_n]\begin{bmatrix} a_{11} & a_{12} & \cdots \\ a_{21} & a_{22} & \cdots \\ \vdots & \vdots & \end{bmatrix}\begin{bmatrix} y_1 \\ y_2 \\ \vdots \\ y_n \end{bmatrix}$$

$$= \sum_{i=1}^{n}\sum_{j=1}^{n} y_i a_{ij} y_j \qquad (12A.23)$$

Then

$$\nabla_y F = \begin{bmatrix} \dfrac{\partial F}{\partial y_1} \\[2mm] \dfrac{\partial F}{\partial y_2} \\[2mm] \vdots \\[2mm] \dfrac{\partial F}{\partial y_n} \end{bmatrix} = \begin{bmatrix} 2a_{11}y_1 + 2a_{12}y_2 + \cdots \\ 2a_{21}y_1 + 2a_{22}y_2 + \cdots \\ \vdots \end{bmatrix}$$

$$= 2[A]\mathbf{y} \qquad (12A.24)$$

Then, in summary:

$$
\begin{aligned}
&1. \ F = \mathbf{y}^T\mathbf{b} && \nabla_y F = \mathbf{b} \\
&2. \ F = \mathbf{b}^T\mathbf{y} && \nabla_y F = \mathbf{b} \\
&3. \ F = \mathbf{y}^T[A]\mathbf{u} && \nabla_y F = [A]\mathbf{u} \\
&4. \ F = \mathbf{u}^T[A]\mathbf{y} && \nabla_y F = [A]^T\mathbf{u} \\
&5. \ F = \mathbf{y}^T[A]\mathbf{y} && \nabla_y F = 2[A]\mathbf{y}
\end{aligned}
\qquad (12A.25)
$$

We will now use Eq. 12A.25 to derive the gradient of $J(\mathbf{x})$, that is $\nabla_x J$, where $J(\mathbf{x})$ is shown in Eq. 12A.12. The first term, $\mathbf{z}^T\mathbf{z}$ is not a function of $\mathbf{x}$, so we can discard it. The second term is of the same form as (4) in Eq. 12A.25, so that,

$$
\nabla_x(-2\mathbf{z}^T[H]\mathbf{x}) = -2[H]^T\mathbf{z} \qquad (12A.26)
$$

The third term is the same as (5) in Eq. 12A.25 with $[H]^T[H]$ replacing $[A]$; then,

$$
\nabla_x(\mathbf{x}^T[H]^T[H]\mathbf{x}) = 2[H]^T[H]\mathbf{x} \qquad (12A.27)
$$

Then from Eqs. 12A.26 and 12A.27 we have

$$
\nabla_x J = -2[H]^T\mathbf{z} + 2[H]^T[H]\mathbf{x} \qquad (12A.28)
$$

But, as stated in Eq. A.10, we wish to force $\nabla_x J$ to zero. Then

$$
-2[H]^T\mathbf{z} + 2[H]^T[H]\mathbf{x} = 0
$$

or

$$
\mathbf{x} = [[H]^T[H]]^{-1}[H]^T\mathbf{z} \qquad (12A.29)
$$

If we had wanted to put a different weight, w_i, on each measurement, we could have written Eq. 12A.6 as

$$
J(\mathbf{x}) = \sum_{i=1}^{N_m} w_i(z_i - \mathbf{h}_i^T\mathbf{x})^2 \qquad (12A.30)
$$

which can be written as

$$
J(\mathbf{x}) = (\mathbf{z} - [H]\mathbf{x})^T[W](\mathbf{z} - [H]\mathbf{x})
$$

where $[W]$ is a diagonal matrix. Then

$$
J(\mathbf{x}) = \mathbf{z}^T[W]\mathbf{z} - \mathbf{x}^T[H]^T[W]\mathbf{z} - \mathbf{z}^T[W][H]\mathbf{x} + \mathbf{x}^T[H]^T[W][H]\mathbf{x}
$$

If we once again use Eq. 12A.25, we would obtain

$$\nabla_x J = -2[H]^T[W]z + 2[H]^T[W][H]x$$

and

$$\nabla_x J = 0$$

gives

$$x = ([H]^T[W][H])^{-1}[H]^T[W]z \qquad (12A.31)$$

The Fully-Determined Case ($N_m = N_s$)

In this case, the number of measurements is equal to the number of state variables and we can solve for x directly by inverting $[H]$.

$$x = [H]^{-1}z \qquad (12A.32)$$

The Underdetermined Case ($N_m < N_s$)

In this case, we have fewer measurements than state variables. In such a case, it is possible to solve for many solutions x^{est} that cause $J(x)$ to equal zero. The usual solution technique is to find x^{est} that minimizes the sum of the squares of the solution values. That is, we find a solution such that

$$\sum_{j=1}^{N_s} x_j^2 \qquad (12A.33)$$

is minimized while meeting the condition that the measurements will be solved for exactly. To do this, we treat the problem as a constrained minimization problem and use Lagrange multipliers as shown in Appendix 3A.

We formulate the problem as

Minimize: $$\sum_{j=1}^{N_s} x_j^2$$

$$(12A.34)$$

Subject to: $$z_i = \sum_{j=1}^{N_s} h_{ij}x_j \quad \text{for } i = 1, \ldots, N_m$$

This optimization problem can be written in vector–matrix form as

$$\begin{array}{ll} \min & x^T x \\ \text{subject to} & z = [H]x \end{array} \qquad (12A.35)$$

The Lagrangian for this problem is

$$\mathcal{L} = x^T x + \lambda^T(z - [H]x) \qquad (12A.36)$$

Following the rules set down in Appendix 3A we must find the gradient of $\mathscr{L}$ with respect to $\mathbf{x}$ and with respect to λ. Using the identities found in Eq. 12A.25 we get

$$\nabla_x \mathscr{L} = 2\mathbf{x} - [H]^T \lambda = 0$$

which gives

$$\mathbf{x} = \frac{1}{2}[H]^T \lambda$$

and

$$\nabla_\lambda \mathscr{L} = \mathbf{z} - [H]\mathbf{x} = 0$$

which gives

$$\mathbf{z} = [H]\mathbf{x}$$

Then

$$\mathbf{z} = \frac{1}{2}[H][H]^T \lambda$$

or

$$\lambda = 2[[H][H]^T]^{-1}\mathbf{z}$$

and finally,

$$\mathbf{x} = [H]^T[[H][H]^T]^{-1}\mathbf{z} \qquad (12A.37)$$

The reader should be aware that the matrix inversion shown in Eqs. 12A.29, 12A.32, and 12A.37 may not be possible. That is, the $[[H]^T[H]]$ matrix in Eq. 12A.29 may be singular, or $[H]$ may be singular in Eq. 12A.32, or $[[H][H]^T]$ may be singular in Eq. 12A.37. In the overdetermined case $(N_m > N_s)$ whose solution is Eq. 12A.29, and the fully determined case $(N_m = N_s)$ whose solution is Eq. 12A.32, the singularity implies what is known as an "unobservable" system. By unobservable we mean that the measurements do not provide sufficient information to allow a determination of the states of the system. In the case of the underdetermined case $(N_m < N_s)$ whose solution is Eq. 12A.37, the singularity simply implies that there is no unique solution to the problem.

PROBLEMS

12.1 Using the three-bus sample system shown in Figure 12.1, assume that the three meters have the following characteristics.

Meter	Full Scale (MW)	Accuracy (MW)	σ (pu)
M_{12}	100	± 6	0.02
M_{13}	100	± 3	0.01
M_{32}	100	± 0.6	0.002

a. Calculate the best estimate for the phase angles θ_1 and θ_2 given the following measurements.

Meter	Measured Value (MW)
M_{12}	60.0
M_{13}	4.0
M_{32}	40.5

b. Calculate the residual $J(x)$. For a significance level, α, of 0.01, does $J(x)$ indicate the presence of bad data? Explain.

12.2 Given a single transmission line with a generator at one end and a load at the other, two measurements are available as shown in Figure 12.21. Assume that we can model this circuit with a DC load flow using the line reactance shown. Also, assume that the phase angle at bus 1 is 0 rad. Given the meter characteristics and meter readings telemetered from the meters, calculate the best estimate of the power flowing through the transmission line.

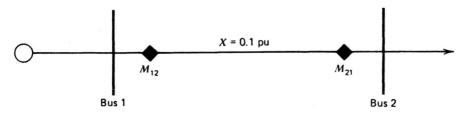

FIG. 12.21   Measurement configuration for Problem 12.2.

Meter	Full Scale (MW)	Meter Standard Deviation (σ) in Full Scale	Meter Reading (MW)
M_{12}	200	1	62
M_{21}	200	5	-52

Note: M_{12} measures power flowing from bus 1 to bus 2; M_{21} measures power flowing from bus 2 to bus 1.

Use 100 MVA as base.

12.3 You are given in the following network with meters at locations as shown in Figure 12.22.

Branch Impedances (pu)

$$X_{12} = 0.25$$
$$X_{13} = 0.50$$
$$X_{24} = 0.40$$
$$X_{34} = 0.10$$

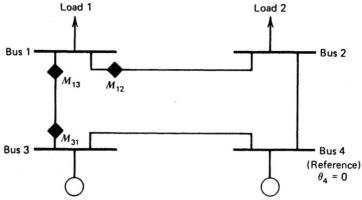

FIG. 12.22   Four-bus system with measurements for Problem 12.3.

Measurement Values	Measurement Errors
$M_{13} = -70.5$	$\sigma_{13} = 0.01$
$M_{31} = 72.1$	$\sigma_{31} = 0.01$
$M_{12} = 21.2$	$\sigma_{12} = 0.02$

a. Is this network observable? Set up the least-squares equations and try to invert $[H^T R^{-1} H]$.

b. Suppose we had a measurement of generation output at bus 3 and included it in our measurement set. Let this measurement be the following:

$$M_{3\,\text{gen}} = 92 \text{ MW} \quad \text{with } \sigma = 0.015$$

Repeat part a including this measurement.

12.4 Given the network shown in Figure 12.23, the network is to be modeled with a DC power flow with line reactances as follows (assume 100-MVA base):

$$x_{12} = 0.1 \text{ pu}$$
$$x_{23} = 0.25 \text{ pu}$$

The meters are all of the same type with a standard deviation of $\sigma = 0.01$ pu for each. The measured values are:

$$M_3 = 105 \text{ MW}$$
$$M_{32} = 98 \text{ MW}$$
$$M_{23} = -135 \text{ MW}$$
$$M_2 = 49 \text{ MW}$$
$$M_{21} = 148 \text{ MW}$$

a. Find the phase angles which result in a best fit to the measured values.

b. Find the value of the residual function J.

c. Calculate estimated generator output of each generator and the estimated power flow on each line.

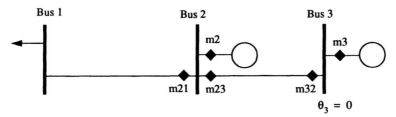

FIG. 12.23 Network for Problem 12.4.

d. Are there any errors in the measurements? If you think so, explain which meters are apt to be in error and why. Remove the suspected bad measurement and try to resolve the state estimator.

12.5 You are to purchase and install a set of programs that are to act as a monitor for the system security of a major utility company. You have solicited bids from major manufacturers of computer systems and are responsible for reviewing the technical contents of each bid. One of the manufacturers proposes to install a system with the flowchart and description given in Figure 12.24.

Bidders design:

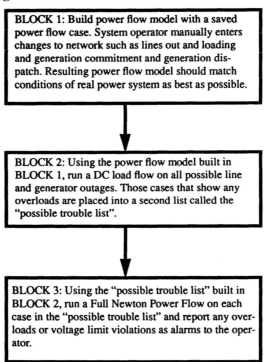

FIG. 12.24 Diagram for Problem 12.5.

  a. Write down as many of the design flaws that you can find in this bidder's design.

  b. Create a new design that you think will be a state-of-the-art system.

FURTHER READING

State estimation originated in the aerospace industry and only came to be of interest to power systems engineers in the late 1960s. Since then, state estimators have been installed on a regular basis in new energy control centers and have proved quite useful. References 1–4 provide a good introduction to this topic. Reference 4, in particular, is a carefully written overview with a good bibliography of literature up to 1974. References 5 and 6 show the variety of algorithms used to solve the state-estimation problem.

  The remaining references cover some of the subtopics of state estimation. The use of the state estimator to detect bad measurements and model parameter errors is covered in references 7–10. Network observability determination is covered in references 11 and 12. Methods of automatically updating the network model topology to match switching status are covered in references 13 and 14. Finally, orthogonal decomposition methods are covered in references 15 and 16.

  1. Schweppe, F. C., Wildes, J., "Power System Static State Estimation, Part I: Exact Model," *IEEE Transactions on Power Apparatus and Systems*, Vol. PAS-89, January 1970, pp. 120–125.
  2. Larson, R. E., Tinney, W. F., Peschon, J., "State Estimation in Power Systems, Part I: Theory and Feasibility," *IEEE Transactions on Power Apparatus and Systems*, Vol. PAS-89, March 1970, 345–352.
  3. Larson, R. E., Tinney, W. F., Hajdu, L. P., Piercy, P. S., "State Estimation in Power Systems, Part II: Implementation and Applications," *IEEE Transactions on Power Apparatus and Systems*, Vol. PAS-89, March 1970, pp. 353–359.
  4. Schweppe, F. C., Handschin, E., "Static State Estimation in Power Systems," *IEEE Proceedings*, Vol. 62, July 1974, pp. 972–982.
  5. Dopazo, J. F., Klitin, O. A., Stagg, G. W., VanSlyck, L. S., "State Calculation of Power Systems from Line Flow Measurement," *IEEE Transactions on Power Apparatus and Systems*, Vol. PAS-89, September/October 1970, pp. 1698–1708.
  6. Dopazo, J. F., Klitin, O. A., VanSlyck, L. S., "State Calculation of Power Systems from Line Flow Measurements, Part II," *IEEE Transactions on Power Apparatus and Systems*, Vol. PAS-91, January/February 1972, pp. 145–151.
  7. Dopazo, J. F., Klitin, O. A., Sasson, A. M., "State Estimation for Power Systems: Detection and Identification of Gross Measurement Errors," Proceedings 8th PICA Conference, Minneapolis, June 1973.
  8. Merrill, H. M., Schweppe, F. C., "On-Line System Model Error Correction," IEEE Winter Power Meeting, 1973, Paper C73-106-2.
  9. Debs, A. S., "Estimation of Steady-State Power System Model Parameters," *IEEE Transactions on Power Apparatus and Systems*, Vol. PAS-93, No. 5, 1974.
  10. Handschin, E., Schweppe, F. C., Kohlar, J., Fiechter, A., "Bad Data Analysis for Power System State Estimation," *IEEE Transactions on Power Apparatus and Systems*, Vol. PAS-94, March/April 1975, pp. 329–337.

11. Clements, K. A., Wollenberg, B. F., "An Algorithm for Observability Determination in Power System State Estimation," IEEE Summer Power Meeting, 1975, Paper A75-447-3.

12. Krumpholz, G. R., Clements, K. A., Davis, P. W., "Power System Observability: A Practical Algorithm using Network Topology," *IEEE Transactions on Power Apparatus and Systems*, Vol. 99, July/August 1980, pp. 1534–1542.

13. Sasson, A. M., Ehrman, S. T., Lynch, P., VanSlyck, L. S., "Automatic Power System Network Topology Determination," *IEEE Transactions on Power Apparatus and Systems*, Vol. PAS-92, March/April 1973, pp. 610–618.

14. DyLiacco, T. E., Ramarao, K., Weiner, A., "Network Status Analysis for Real Time Systems," Proceedings 8th PICA Conference. Minneapolis, June 1973.

15. Simoes-Costa, A., Quintana, V., "An Orthogonal Row Processing Algorithm for Power System Sequential Estimation," *IEEE Transactions on Power Apparatus and Systems*, Vol. PAS-100, No. 8, August 1981, pp. 3791–3800.

16. Vempati, N., Slutsker, I. W., Tinney, W. F., "Enhancements to Givens Rotation for Power System State Estimation," *IEEE Transactions on Power Systems*, Vol. 6, No. 2, May 1991, pp. 842–849.

13 Optimal Power Flow

13.1 INTRODUCTION

The optimal power flow of OPF has had a long history in its development. It was first discussed by Carpentier in 1962 (reference 1) and took a long time to become a successful algorithm that could be applied in everyday use. Current interest in the OPF centers around its ability to solve for the optimal solution that takes account of the security of the system.

In Chapter 3, we introduced the concept of economic dispatch. In the economic dispatch we had a single constraint which held the total generation to equal the total load plus losses. Thus, the statement of the economic dispatch problem results in a Lagrangian with just one constraint:

$$L = \sum F(P_i) + \lambda(P_{\text{load}} + P_{\text{losses}} - \sum P_i) \qquad (13.1)$$

If we think about the single "generation equals load plus losses" constraint:

$$P_{\text{load}} + P_{\text{losses}} - \sum P_i = 0 \qquad (13.2)$$

we realize that what it is actually saying is that the generation must obey the same conditions as expressed in a power flow—with the condition that the entire power flow is reduced to one simple equality constraint. There is good reason, as we shall see shortly, to state the economic dispatch calculation in terms of the generation costs, and the entire set of equations needed for the power flow itself as constraints. The power flow equations were introduced in Chapter 4. This formulation is called an optimal power flow.

We can solve the OPF for the minimum generation cost (as in Chapter 3) and require that the optimization calculation also balance the entire power flow—at the same time. Note also that the objective function can take different forms other than minimizing the generation cost. It is common to express the OPF as a minimization of the electrical losses in the transmission system, or to express it as the minimum shift of generation and other controls from an optimum operating point. We could even allow the adjustment of loads in order to determine the minimum load shedding schedule under emergency conditions. Regardless of the objective function, however, an OPF must solve so that the entire set of power constraints are present and satisfied at the solution.

Why set up the generation dispatch calculation as an OPF?.

1. If the entire set of power flow equations are solved simultaneously with the generation cost minimization, the representation of incremental losses is exact. Further, with an objective function that minimizes the losses themselves, the power flow equations are quite necessary.

2. The economic dispatch solutions in Chapter 3 only observed the generation limits $P_i^- \le P_i \le P_i^+$. With all of the power flow constraints included in the formulation, many more of the power system limits can be included. These include limits on the generator reactive power, $Q_i^- \le Q_i \le Q_i^+$, limits on the voltage magnitude at generation and load buses, $|E_i|^- \le |E_i| \le |E_i|^+$, and flows on transmission lines or transformers expressed in either MW, amperes or MVA (e.g. $\text{MVA}_{ij}^- \le \text{MVA}_{ij} \le \text{MVA}_{ij}^+$). This set of operating constraints now allows the user to guarantee that the dispatch of generation does not, in fact, force the transmission system into violating a limit, which might put it in danger of being damaged.

3. The OPF can also include constraints that represent operation of the system after contingency outages. These "security constraints" allow the OPF to dispatch the system in a defensive manner. That is, the OPF now forces the system to be operated so that if a contingency happened, the resulting voltages and flows would still be within limit. Thus, constraints such as the following might be incorporated:

$$|E_k|^- \le |E_k| \text{ (with line } nm \text{ out)} \le |E_k|^+ \qquad (13.3)$$

$$\text{MVA}_{ij}^- \le \text{MVA}_{ij} \text{ (with line } nm \text{ out)} \le \text{MVA}_{ij}^+ \qquad (13.4)$$

which implies that the OPF would prevent the post-contingency voltage on bus k or the post-contingency flow on line ij from exceeding their limits for an outage of line nm. This special type of OPF is called a "security-constrained OPF," or SCOPF.

4. In the dispatch calculation developed in Chapter 3, the only adjustable variables were the generator MW outputs themselves. In the OPF, there are many more adjustable or "control" variables that be be specified. A partial list of such variables would include:

- Generator voltage.
- LTC transformer tap position.
- Phase shift transformer tap position.
- Switched capacitor settings.
- Reactive injection for a static VAR compensator.
- Load shedding.
- DC line flow.

Thus, the OPF gives us a framework to have many control variables adjusted in the effort to optimize the operation of the transmission system.

5. The ability to use different objective functions provides a very flexible analytical tool.

Given this flexibility, the OPF has many applications including:

1. The calculation of the optimum generation pattern, as well as all control variables, to achieve the minimum cost of generation together with meeting the transmission system limitations.
2. Using either the current state of the power system or a short-term load forecast, the OPF can be set up to provide a "*preventative dispatch*" if security constraints are incorporated.
3. In an emergency, that is when some component of the system is overloaded or a bus is experiencing a voltage violation, the OPF can provide a "corrective dispatch" which tells the operators of the system what adjustments to make to relieve the overload or voltage violation.
4. The OPF can be used periodically to find the optimum setting for generation voltages, transformer taps and switched capacitors or static VAR compensators (sometimes called "voltage–VAR" optimization).
5. The OPF is routinely used in planning studies to determine the maximum stress that a planned transmission system can withstand. For example, the OPF can calculate the maximum power that can safely be transferred from one area of the network to another.
6. The OPF can be used in economic analyses of the power system by providing "*bus incremental costs*" (BICs). The BICs are useful to determine the marginal cost of power at any bus in the system. Similarly, the OPF can be used to calculate the incremental or marginal cost of transmitting power from one outside company—through its system—to another outside company.

13.2 SOLUTION OF THE OPTIMAL POWER FLOW

The optimal power flow is a very large and very difficult mathematical programming problem. Almost every mathematical programming approach that can be applied to this problem has been attempted and it has taken developers many decades to develop computer codes that will solve the OPF problem reliably.

Chapter 3 introduced the concept of the lambda-iteration methods, the gradient method and Newton's method. We shall review all of these here and introduce two new techniques, the linear programming (LP) method and the interior point method. The attributes of these methods are summarized next.

- **Lambda iteration method:** Losses may be represented by a $[B]$ matrix, or the penalty factors may be calculated outside by a power flow. This forms the basis of many standard on-line economic dispatch programs.
- **Gradient methods:** Gradient methods are slow in convergence and are difficult to solve in the presence of inequality constraints.
- **Newton's method:** Very fast convergence, but may give problems with inequality constraints.
- **Linear programming method (LPOPF):** One of the fully developed methods now in common use. Easily handles inequality constraints. Nonlinear objective functions and constraints handled by linearization.
- **Interior point method:** Another of the fully developed and widely used methods for OPF. Easily handles inequality constraints.

We introduced and analyzed the lambda-iteration method in Chapter 3. This method forms the basis of standard on-line economic dispatch codes. The technique works well and can be made to run very fast. It overlooks any constraints on the transmission system and does not produce a dispatch of the generation that will avoid overloads, voltage limit violations, or security constraint violations.

We shall derive the gradient method using the same mathematics used in Chapter 3, only with various advanced models of the transmission system instead of the "load plus losses equals generation" constraint used in Chapter 3. It is then a simple step to go on to develop the Newton's method applied with these same constraints. Finally, the LPOPF and interior point methods are presented.

The objective function in the OPF is usually minimized. In some cases, such as power transfers, it may be maximized. We shall designate the objective function as f. The equations that guarantee that the power flow constraints are met will be designated as

$$\mathbf{g(z)} = 0 \qquad (13.5)$$

Note that here we shall only be concerned with a variable vector **z**. This vector contains the adjustable controls, the bus voltage magnitudes, and phase angles, as well as the fixed parameters of the system, Later, we shall break the variables up into sets of state variables, control variables, and fixed parameters.

The OPF can also solve for an optimal solution with inequality constraints on dependent variables, such as line MVA flows. These will be designated

$$\mathbf{h}^- \le \mathbf{h(z)} \le \mathbf{h}^+ \qquad (13.6)$$

In addition, limits may be placed directly on state variables or control variables:

$$\mathbf{z}^- \le \mathbf{z} \le \mathbf{z}^+ \qquad (13.7)$$

The OPF problem then consists of minimizing (or maximizing) the objective function, subject to the equality constraints, the inequality constraints, and the state and control variable limits.

The developments and illustrative examples in this chapter concentrate (but not exclusively) on the LPOPF technique. The method is widely used and only requires an AC or DC power flow program, plus a suitable LP package for solving illustrative examples and (homework) problems.

13.2.1   The Gradient Method

In this section, we shall consider the objective function to be total cost of generation (later examples will demonstrate how other objectives can be used). The objective function to be minimized is:

$$\sum_{\text{all gen.}} F_i(P_i)$$

where the sum extends to all generators on the power system, including the generator at the reference bus.

We shall start out defining the unknown or state vector $\mathbf{x}$ as:

$$\mathbf{x} = \begin{bmatrix} \left. \begin{array}{c} \theta_i \\ |E_i| \end{array} \right\} \text{on each } PQ \text{ bus} \\ \\ \theta_i \quad \text{on each } PV \text{ bus} \end{bmatrix} \tag{13.8}$$

another vector, $\mathbf{y}$, is defined as:

$$\mathbf{y} = \begin{bmatrix} \left. \begin{array}{c} \theta_k \\ |E_k| \end{array} \right\} \text{on the reference bus} \\ \\ \left. \begin{array}{c} P_k^{\text{net}} \\ Q_k^{\text{net}} \end{array} \right\} \text{on each } PQ \text{ bus} \\ \\ \left. \begin{array}{c} P_k^{\text{net}} \\ |E_k|^{\text{sch}} \end{array} \right\} \text{on each } QV \text{ bus} \end{bmatrix} \tag{13.9}$$

Note that the vector $\mathbf{y}$ is made up of all of the parameters that must be specified. Some of these parameters are adjustable (for example, the generator output, P_k^{net}, and the generator bus voltage). Some of the parameters are fixed, as far as the OPF calculation is concerned, such as the P and Q at each load bus. To make this distinction, we shall divide the $\mathbf{y}$ vector up into two parts, $\mathbf{u}$ and $\mathbf{p}$:

$$\mathbf{y} = \begin{bmatrix} \mathbf{u} \\ \mathbf{p} \end{bmatrix} \tag{13.10}$$

where **u** represents the vector of control or adjustable variables, and **p** represents the fixed or constant variables. Note also that we are only representing equality constraints at this point.

Finally, we shall define a set of m equations that govern the power flow:

$$\mathbf{g(x, y)} = \begin{bmatrix} P_i(|\mathbf{E}|, \boldsymbol{\theta}) - P_i^{\text{net}} \\ Q_i(|\mathbf{E}|, \boldsymbol{\theta}) - Q_i^{\text{net}} \\ P_k(|\mathbf{E}|, \boldsymbol{\theta}) - P_k^{\text{net}} \end{bmatrix} \begin{matrix} \Big\} \text{for each } PQ \text{ (load) bus } i \\ \\ \text{for each } PV \text{ (gen.) bus } k, \text{ not} \\ \text{including the reference bus} \end{matrix} \qquad (13.11)$$

Note that these equations are the same bus equations as shown in Chapter 4 for the Newton power flow (Eq. 4.18).

We must recognize that the reference-bus power generation is not an independent variable. That is, the reference-bus generation always changes to balance the power flow; we cannot specify it at the beginning of the calculation. We wish to express the cost or objective function as a function of the control variables and of the state variables. We do this by dividing the cost function as follows:

$$\text{cost} = \sum_{\text{gen}} F_i(P_i) + F_{\text{ref}}(P_{\text{ref}}) \qquad (13.12)$$

where the first summation does not include the reference bus, The P_i are all independent, controlled variables whereas P_{ref} is a dependent variable. We say that the P_i are in the vector **u** and the P_{ref} is a function of the network voltages and angles:

$$P_{\text{ref}} = P_{\text{ref}}(|\mathbf{E}|, \boldsymbol{\theta}) \qquad (13.13)$$

then the cost function becomes:

$$\sum_{\text{gen}} F_i(P_i) + F_{\text{ref}}[P_{\text{ref}}(|\mathbf{E}|, \boldsymbol{\theta})] = \mathbf{f(x, u)} \qquad (13.14)$$

We can now set up the Lagrange equation for the OPF as follows:

$$\mathscr{L}(\mathbf{x, u, p}) = \mathbf{f(x, u)} + \boldsymbol{\lambda}^t \mathbf{g(x, u, p)} \qquad (13.15)$$

where

$\mathbf{x}$ = vector of state variables

$\mathbf{u}$ = vector of control variables

$\mathbf{p}$ = vector of fixed parameters

$\boldsymbol{\lambda}$ = vector of Lagrange multipliers

$\mathbf{g}$ = set of equality constraints representing the power flow equations

$\mathbf{f}$ = the objective function

This Lagrange equation is perhaps better seen when written as:

$$\mathscr{L}(\mathbf{x}, \mathbf{u}, \mathbf{p}) = \sum_{gen} F_i(P_i) + F_{ref}[P_{ref}(|\mathbf{E}|, \boldsymbol{\theta})] + [\lambda_1 \lambda_2, \ldots, \lambda_m] \begin{bmatrix} P_i(|\mathbf{E}|, \boldsymbol{\theta}) - P_i^{net} \\ Q_i(|\mathbf{E}|, \boldsymbol{\theta}) - Q_i^{net} \\ P_k(|\mathbf{E}|, \boldsymbol{\theta}) - P_k^{net} \\ \vdots \end{bmatrix}$$

(13.16)

We now have a Lagrange function that has a single objective function and m Lagrange multipliers, one for each of the m power flow equations.

To minimize the cost function, subject to the constraints, we set the gradient of the Lagrange function to zero:

$$\nabla \mathscr{L} = 0 \qquad (13.17)$$

To do this, we break up the gradient vector into three parts corresponding to the variables $\mathbf{x}$, $\mathbf{u}$, and λ:

$$\nabla \mathscr{L}_x = \frac{\partial \mathscr{L}}{\partial \mathbf{x}} = \frac{\partial \mathbf{f}}{\partial \mathbf{x}} + \left[\frac{\partial \mathbf{g}}{\partial \mathbf{x}}\right]^T \lambda \qquad (13.18)$$

$$\nabla \mathscr{L}_u = \frac{\partial \mathscr{L}}{\partial \mathbf{u}} = \frac{\partial \mathbf{f}}{\partial \mathbf{u}} + \left[\frac{\partial \mathbf{g}}{\partial \mathbf{u}}\right]^T \lambda \qquad (13.19)$$

$$\nabla \mathscr{L}_\lambda = \frac{\partial \mathscr{L}}{\partial \lambda} = \mathbf{g}(\mathbf{x}, \mathbf{u}, \mathbf{p}) \qquad (13.20)$$

Some discussion of the three gradient equations above is in order. First, Eq. 13.18 consists of a vector of derivatives of the objective function with respect to the state variables, $\mathbf{x}$. Since the objective function itself is not a function of the state variable *except for the reference bus*, this becomes:

$$\frac{\partial \mathbf{f}}{\partial \mathbf{x}} = \begin{bmatrix} \dfrac{\partial}{\partial P_{ref}} F_{ref}(P_{ref}) \dfrac{\partial P_{ref}}{\partial \theta_1} \\ \dfrac{\partial}{\partial P_{ref}} F_{ref}(P_{ref}) \dfrac{\partial P_{ref}}{\partial |E_1|} \\ \vdots \end{bmatrix} \qquad (13.21)$$

The $[\partial g/\partial x]$ term in Eq. 13.18 actually is the Jacobian matrix for the Newton power flow, which was developed in Chapter 4. That is:

$$
\left[\frac{\partial g}{\partial x}\right] =
\begin{bmatrix}
\dfrac{\partial P_1}{\partial \theta_1} & \dfrac{\partial P_1}{\partial |E_1|} & \dfrac{\partial P_1}{\partial \theta_2} & \dfrac{\partial P_1}{\partial |E_2|} & \cdots \\[2ex]
\dfrac{\partial Q_1}{\partial \theta_1} & \dfrac{\partial Q_1}{\partial |E_1|} & \dfrac{\partial Q_1}{\partial \theta_2} & \dfrac{\partial Q_1}{\partial |E_2|} & \cdots \\[2ex]
\dfrac{\partial P_2}{\partial \theta_1} & \dfrac{\partial P_2}{\partial |E_1|} & & & \cdots \\[2ex]
\dfrac{\partial Q_2}{\partial \theta_1} & \dfrac{\partial Q_2}{\partial |E_1|} & & & \cdots \\[2ex]
\vdots & & & \vdots
\end{bmatrix}
\tag{13.22}
$$

Note that this matrix must be transposed for use in Eq. 13.18.

Equation 13.19 is the gradient of the Lagrange function with respect to the control variables. Here the vector $\partial f/\partial u$ is a vector of derivatives of the objective function with respect to the control variables:

$$
\frac{\partial f}{\partial u} =
\begin{bmatrix}
\dfrac{\partial}{\partial P_1} F_1(P_1) \\[2ex]
\dfrac{\partial}{\partial P_2} F_2(P_2) \\[2ex]
\vdots
\end{bmatrix}
\tag{13.23}
$$

The other term in Eq. 13.19, $[\partial g/\partial u]$, actually consists of a matrix of all zeros with some -1 terms on the diagonals, which correspond to equations in $g(x, u, p)$ where a control variable is present. Finally, Eq. 13.20 consists simply of the power flow equations themselves.

The solution of the gradient method of OPF is as follows:

1. Given a set of fixed parameters p, assume a starting set of control variables u.
2. Solve a power flow. This guarantees that Eq. 13.20 is satisfied.
3. Solve Eq. 13.19 for lambda;

$$
\lambda = -\left[\frac{\partial g}{\partial x}\right]^{T-1} \frac{\partial f}{\partial x}
\tag{13.24}
$$

4. Substitute λ into Eq. 13.18 to get the gradient of $\mathscr{L}$ with respect to the control variables.

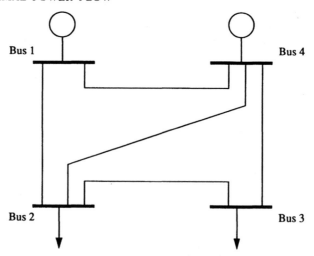

FIG. 13.1   Four-bus system for Example 13A.

The gradient will give the direction of maximum increase in the cost function as a function of the adjustments in each of the **u** variables. Since we wish to decrease the objective function, we shall move in the direction of the negative of the gradient. The gradient method gives no indication how far along the negative gradient direction we should move. Assuming that a distance is picked that reduces the objective, one must start at step 2 above, and repeat steps 2, 3, and 4 over and over until the gradient itself becomes sufficiently close to the zero vector, indicating that all conditions for the optimum have been reached.

EXAMPLE 13A

The following is a very simple example presented to show the meaning of each of the elements in the gradient equations. Example 13B will be a more practical example of the gradient method.

The four-bus system in Figure 13.1 will be modeled with a DC power flow. The following are known:

$$P_2, P_3, \text{ and } \theta_4 = 0$$

Line reactances: $x_{12}, x_{14}, x_{24}, x_{23}, \text{ and } x_{34}$

Cost functions: $F_1(P_1)$ and $F_4(P_4)$

All $|E|$ values are fixed at 1.0 per unit volts

The only independent control variable in this problem is the generator output P_1, or:

$$u = P_1 \tag{13.25}$$

The state variables are θ_1, θ_2, and θ_3, or:

$$\mathbf{x} = \begin{bmatrix} \theta_1 \\ \theta_2 \\ \theta_3 \end{bmatrix} \tag{13.26}$$

We wish to minimize the total generation cost while maintaining a solved DC power flow for the network. To do this with the gradient method we form the Lagrangian:

$$\mathscr{L} = F_1(P_1) + F_4[P_4(\theta_1 \ldots \theta_4)] + \begin{bmatrix} \lambda_1 & \lambda_2 & \lambda_3 \end{bmatrix} \begin{bmatrix} P_1(\theta_1 \ldots \theta_4) - P_1 \\ P_2(\theta_1 \ldots \theta_4) - P_2 \\ P_3(\theta_1 \ldots \theta_4) - P_3 \end{bmatrix} \tag{13.27}$$

In terms of the equations presented earlier:

$$\mathbf{f}(\mathbf{x}, \mathbf{u}) = F_1(P_1) + F_4[P_4(\theta_1 \ldots \theta_4)] \tag{13.28}$$

$$g(\mathbf{x}, \mathbf{u}) = \begin{bmatrix} P_1(\theta_1 \ldots \theta_4) - P_1 \\ P_2(\theta_1 \ldots \theta_4) - P_2 \\ P_3(\theta_1 \ldots \theta_4) - P_3 \end{bmatrix} \tag{13.29}$$

Note that in $g(\mathbf{x}, \mathbf{u})$, the P_1 is the control variable and P_2 and P_3 are fixed.
  We shall now expand $g(\mathbf{x}, \mathbf{u})$ as follows:

$$\mathbf{g}(\mathbf{x}, \mathbf{u}) = \begin{bmatrix} P_1(\theta_1 \ldots \theta_4) - P_1 \\ P_2(\theta_1 \ldots \theta_4) - P_2 \\ P_3(\theta_1 \ldots \theta_4) - P_3 \end{bmatrix} = \begin{bmatrix} \dfrac{1}{x_{12}}(\theta_1 - \theta_2) + \dfrac{1}{x_{14}}(\theta_1 - \theta_{24}) - P_1 \\ \vdots \end{bmatrix} \tag{13.30}$$

The result is:

$$\mathbf{g}(\mathbf{x}, \mathbf{u}) = [B'] \begin{bmatrix} \theta_1 \\ \theta_2 \\ \theta_3 \end{bmatrix} - \begin{bmatrix} P_1 \\ P_2 \\ P_3 \end{bmatrix} \tag{13.31}$$

and the Lagrange function becomes:

$$F_1(P_1) + F_4[P_4(\theta_1 \ldots \theta_4)] + \begin{bmatrix} \lambda_1 & \lambda_2 & \lambda_3 \end{bmatrix} \left([B'] \begin{bmatrix} \theta_1 \\ \theta_2 \\ \theta_3 \end{bmatrix} - \begin{bmatrix} P_1 \\ P_2 \\ P_3 \end{bmatrix} \right) \tag{13.32}$$

We now proceed to develop the three gradient components:

$$\nabla \mathscr{L}_\lambda = \mathbf{g}(\mathbf{x}, \mathbf{u}) = 0 \tag{13.33}$$

which simply says that we need to start by always maintaining the DC power flow:

$$\begin{bmatrix} \theta_1 \\ \theta_2 \\ \theta_3 \end{bmatrix} = [B']^{-1} \begin{bmatrix} P_1 \\ P_2 \\ P_3 \end{bmatrix} \tag{13.34}$$

The next component:

$$\Delta \mathscr{L}_x = \begin{bmatrix} \dfrac{\partial \mathscr{L}}{\partial \theta_1} \\[2ex] \dfrac{\partial \mathscr{L}}{\partial \theta_2} \\[2ex] \dfrac{\partial \mathscr{L}}{\partial \theta_3} \end{bmatrix} = \begin{bmatrix} \dfrac{\partial F_4}{\partial P_4} & \dfrac{\partial P_4}{\partial \theta_1} \\[2ex] \dfrac{\partial F_4}{\partial P_4} & \dfrac{\partial P_4}{\partial \theta_2} \\[2ex] \dfrac{\partial F_4}{\partial P_4} & \dfrac{\partial P_4}{\partial \theta_3} \end{bmatrix} + [B']^T \begin{bmatrix} \lambda_1 \\ \lambda_2 \\ \lambda_3 \end{bmatrix} = 0 \tag{13.35}$$

This can be used to solve the vector of Lagrange multipliers:

$$\begin{bmatrix} \lambda_1 \\ \lambda_2 \\ \lambda_3 \end{bmatrix} = (-1)[B']^{T-1} \begin{bmatrix} \dfrac{\partial P_4}{\partial \theta_1} \\[2ex] \dfrac{\partial P_4}{\partial \theta_2} \\[2ex] \dfrac{\partial P_4}{\partial \theta_3} \end{bmatrix} \dfrac{\partial F_4}{\partial P_4} \tag{13.36}$$

where

$$\begin{bmatrix} \dfrac{\partial P_4}{\partial \theta_1} \\[2ex] \dfrac{\partial P_4}{\partial \theta_2} \\[2ex] \dfrac{\partial P_4}{\partial \theta_3} \end{bmatrix} = \begin{bmatrix} -\dfrac{1}{x_{14}} \\[2ex] -\dfrac{1}{x_{24}} \\[2ex] -\dfrac{1}{x_{34}} \end{bmatrix} \tag{13.37}$$

It can be easily demonstrated that:

$$[B']^{T-1} \begin{bmatrix} \dfrac{\partial P_4}{\partial \theta_1} \\[2ex] \dfrac{\partial P_4}{\partial \theta_2} \\[2ex] \dfrac{\partial P_4}{\partial \theta_3} \end{bmatrix} = \begin{bmatrix} -1 \\ -1 \\ -1 \end{bmatrix} \tag{13.38}$$

so that

$$\begin{bmatrix} \lambda_1 \\ \lambda_2 \\ \lambda_3 \end{bmatrix} = \begin{bmatrix} 1 \\ 1 \\ 1 \end{bmatrix} \frac{\partial F_4}{\partial P_4} \tag{13.39}$$

Finally,

$$\frac{\partial \mathbf{g}}{\partial u} = \begin{bmatrix} -1 \\ 0 \\ 0 \end{bmatrix} \tag{13.40}$$

and

$$\nabla \mathcal{L}_u = \frac{\partial F_1}{\partial P_1} + \frac{\partial \mathbf{g}^T}{\partial u} \begin{bmatrix} \lambda_1 \\ \lambda_2 \\ \lambda_3 \end{bmatrix} = \frac{\partial F_1}{\partial P_1} + \begin{bmatrix} -1 & 0 & 0 \end{bmatrix} \left(\begin{bmatrix} 1 \\ 1 \\ 1 \end{bmatrix} \frac{\partial F_4}{\partial P_4} \right) = \frac{\partial F_1}{\partial P_1} - \frac{\partial F_4}{\partial P_4} \tag{13.41}$$

the gradient with respect to the control variable is zero when the two incremental costs are equal, which is the common economic dispatch criterion (assuming neither generator is at a limit). Since the DC power flow represents a linear lossless system, the result simply confirms that the gradient method will produce a result that is the same as economic dispatch.

EXAMPLE 13B

In this example, we shall minimize the real power losses (MW losses) on the three-bus AC system in Figure 13.2. To work this example, the student must be able to run an AC power flow on the three-bus system. (This example is taken from reference 4.)

Given the three-bus network shown in Figure 13.2, where

$$P_3 + jQ_3 = 2.0 + j1.0 \text{ per unit}$$

and

$$P_2 = 1.7 \text{ per unit}$$

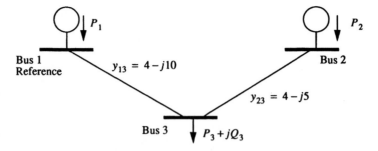

FIG. 13.2   Three-bus example for Example 13B.

In this problem, the generation at bus 2 will be fixed, *the only control variables will be the voltage magnitude at buses 1 and 2.* That is

$$\mathbf{u} = \begin{bmatrix} |E_1| \\ |E_2| \end{bmatrix} \tag{13.42}$$

The state variables will be the phase angles at buses 2 and 3 and the voltage at bus 3:

$$\mathbf{x} = \begin{bmatrix} \theta_2 \\ \theta_3 \\ |E_3| \end{bmatrix} \tag{13.43}$$

The fixed parameters are

$$\mathbf{p} = \begin{bmatrix} \theta_1 \\ P_2 \\ P_3 \\ Q_3 \end{bmatrix} \tag{13.44}$$

We shall solve for the minimum losses using the gradient method. This requires that we solve, repeatedly, the following:

$$\nabla \mathscr{L}_u = \begin{bmatrix} \dfrac{\partial P_{\text{losses}}}{\partial |E_1|} \\ \dfrac{\partial P_{\text{losses}}}{\partial |E_2|} \end{bmatrix} \tag{13.45}$$

Starting at an initial set of voltages:

$$\begin{bmatrix} |E_1| \\ |E_2| \end{bmatrix}^0 = \begin{bmatrix} 1.1 \\ 0.9 \end{bmatrix} \tag{13.46}$$

we proceed using

$$\begin{bmatrix} |E_1| \\ |E_2| \end{bmatrix}^1 = \begin{bmatrix} |E_1| \\ |E_2| \end{bmatrix}^0 + (-1)(\nabla \mathscr{L}_u)\alpha \tag{13.47}$$

where α was set to 0.03 after several trials.

As previously:

$$\mathbf{g(x, u)} = \begin{bmatrix} P_2(\mathbf{E}|, \boldsymbol{\theta}) - P_2 \\ P_3(|\mathbf{E}|, \boldsymbol{\theta}) - P_3 \\ Q_3(|\mathbf{E}|, \boldsymbol{\theta}) - Q_3 \end{bmatrix} \tag{13.48}$$

where the above represents the AC power flow equations as shown in Chapter 4. When we take the derivative,

$$\frac{\partial \mathbf{g}}{\partial \mathbf{x}} = \begin{bmatrix} \dfrac{\partial P_2}{\partial \theta_2} & \dfrac{\partial P_2}{\partial \theta_3} & \dfrac{\partial P_2}{\partial |E_3|} \\[2ex] \dfrac{\partial P_3}{\partial \theta_2} & \dfrac{\partial P_3}{\partial \theta_3} & \dfrac{\partial P_3}{\partial |E_3|} \\[2ex] \dfrac{\partial Q_3}{\partial \theta_2} & \dfrac{\partial Q_3}{\partial \theta_3} & \dfrac{\partial Q_3}{\partial |E_3|} \end{bmatrix} \tag{13.49}$$

these derivatives are calculated as shown in Chapter 4, Eq. 4.22 and the above represents the Jacobian matrix that would be used in the Newton power flow solution to this network. Similarly:

$$\frac{\partial \mathbf{g}}{\partial \mathbf{u}} = \begin{bmatrix} \dfrac{\partial P_2}{\partial |E_1|} & \dfrac{\partial P_2}{\partial |E_2|} \\[2ex] \dfrac{\partial P_3}{\partial |E_1|} & \dfrac{\partial P_3}{\partial |E_2|} \\[2ex] \dfrac{\partial Q_3}{\partial |E_1|} & \dfrac{\partial Q_3}{\partial |E_2|} \end{bmatrix} \tag{13.50}$$

One special note, the objective function, P_{losses} can be expressed in two different ways. The first is simply to write out the losses as:

$$P_{losses} = \text{Re}\left(\sum_{both \ lines} I^2 R \right) \tag{13.51}$$

or one can use the simple observation that since P_2 and P_3 are fixed, any change in the losses due to adjustments of V_1 and V_2 will be directly reflected in changes in P_1. That is, $\Delta P_{\text{losses}} = \Delta P_1$ and

$$\frac{\partial P_{\text{losses}}}{\partial \mathbf{x}} = \frac{\partial P_1}{\partial \mathbf{x}} \tag{13.52}$$

We shall use the second form of the objective so that

$$\mathbf{f} = P_1 \tag{13.53}$$

and then:

$$\frac{\partial \mathbf{f}}{\partial \mathbf{x}} = \begin{bmatrix} \dfrac{\partial P_1}{\partial \theta_2} \\[2ex] \dfrac{\partial P_1}{\partial \theta_3} \\[2ex] \dfrac{\partial P_1}{\partial |E_3|} \end{bmatrix} \tag{13.54}$$

The solution to the first AC power flow, with

$$\begin{bmatrix} |E_1| \\ |E_2| \end{bmatrix}^0 = \begin{bmatrix} 1.1 \\ 0.9 \end{bmatrix} \tag{13.55}$$

gives per unit losses of 0.3906 (39.06 MW losses on 100-MVA base). The reference-bus power, P_1, is 0.6906 per unit MW. Taking this solved power flow as the starting point, we have:

$$\frac{\partial \mathbf{g}}{\partial \mathbf{x}} = \begin{bmatrix} 8.14 & 8.14 & 1.54 \\ 6.96 & 12.0 & 3.85 \\ -4.5 & -7.85 & 10.0 \end{bmatrix} \tag{13.56}$$

$$\frac{\partial \mathbf{f}}{\partial \mathbf{x}} = \begin{bmatrix} 0 \\ 4.36 \\ 4.14 \end{bmatrix} \tag{13.57}$$

$$\begin{bmatrix} \lambda_1 \\ \lambda_2 \\ \lambda_3 \end{bmatrix} = (-1) \left[\frac{\partial \mathbf{g}}{\partial \mathbf{x}} \right]^{T-1} \frac{\partial \mathbf{f}}{\partial \mathbf{x}} = \begin{bmatrix} 0.743 \\ -0.98 \\ -0.154 \end{bmatrix} \tag{13.58}$$

$$\frac{\partial f}{\partial \mathbf{u}} = \begin{bmatrix} \dfrac{\partial P_1}{\partial V_1} \\[2ex] \dfrac{\partial P_1}{\partial V_2} \end{bmatrix} = \begin{bmatrix} 5.533 \\ 0 \end{bmatrix} \tag{13.59}$$

$$\left[\frac{\partial \mathbf{g}}{\partial \mathbf{u}}\right]^T = \begin{bmatrix} 0 & 3.354 & 5.0 \\ 4.94 & 4.5 & 6.96 \end{bmatrix} \tag{13.60}$$

Then,

$$\Delta \mathcal{L}_u = \frac{\partial f}{\partial \mathbf{u}} + \left[\frac{\partial \mathbf{g}}{\partial \mathbf{u}}\right]^T \begin{bmatrix} \lambda_1 \\ \lambda_2 \\ \lambda_3 \end{bmatrix} = \begin{bmatrix} 2.25 \\ -1.78 \end{bmatrix} \tag{13.61}$$

and, with $\alpha = 0.03$, we obtain a new set of voltages:

$$\begin{bmatrix} |E_1| \\ |E_2| \end{bmatrix}^1 = \begin{bmatrix} 1.1 \\ 0.9 \end{bmatrix} - \begin{bmatrix} 2.25 \\ -1.787 \end{bmatrix} 0.03 = \begin{bmatrix} 0.95 \\ 1.03 \end{bmatrix} \tag{13.62}$$

This represents the new control variable settings that must be fed back to the AC power flow.

The new AC power flow, with the above new voltages, results in $P_{\text{losses}} = 0.2380$ per unit and the generation at the reference bus of $P_1 = 0.5380$. Another iteration of the gradient calculation yields $P_{\text{losses}} = 0.2680$ per unit for a controls setting of:

$$\begin{bmatrix} V_1 \\ V_2 \end{bmatrix} = \begin{bmatrix} 0.86 \\ 0.86 \end{bmatrix} \tag{13.63}$$

Note that for this simple problem, the gradient is able to find a reduction in losses after the first iteration, but the next iteration caused the losses to increase. Eventually, it will need tuning, in the form of additional adjustments to the value of α, so that it will not simply oscillate around a minimum. Further, we never specified any voltage limits for V_1 and V_2. As we reduce losses, we may very well run into voltage limits on buses 1 or 2, or both. Here, the gradient method loses whatever simplicity it has and tends to become unmanageable. This would further be the case if we were to place a limit on V_3, which would be a functional inequality and would be very difficult to express in the gradient formulation we have used.

13.2.2 Newton's Method

The problems with the gradient method lie mainly in the fact that the direction of the gradient must be changed quite often and this leads to a very slow convergence. To speed up this convergence, we can use Newton's method, where we take the derivative of the gradient with respect to $\mathbf{x}$, $\mathbf{u}$, and λ. Then, the optimal solution becomes:

$$
\begin{bmatrix} \Delta\mathbf{x} \\ \Delta\mathbf{u} \\ \Delta\lambda \end{bmatrix} = -\begin{bmatrix} \dfrac{\partial}{\partial\mathbf{x}}\nabla\mathscr{L}_x & \dfrac{\partial}{\partial\mathbf{u}}\nabla\mathscr{L}_x & \dfrac{\partial}{\partial\lambda}\nabla\mathscr{L}_x \\[2mm] \dfrac{\partial}{\partial\mathbf{x}}\nabla\mathscr{L}_u & \dfrac{\partial}{\partial\mathbf{u}}\nabla\mathscr{L}_u & \dfrac{\partial}{\partial\lambda}\nabla\mathscr{L}_u \\[2mm] \dfrac{\partial}{\partial\mathbf{x}}\nabla\mathscr{L}_\lambda & \dfrac{\partial}{\partial\mathbf{u}}\nabla\mathscr{L}_\lambda & \dfrac{\partial}{\partial\lambda}\nabla\mathscr{L}_\lambda \end{bmatrix}^{-1}\begin{bmatrix} \nabla\mathscr{L}_x \\ \nabla\mathscr{L}_u \\ \nabla\mathscr{L}_\lambda \end{bmatrix} \tag{13.64}
$$

The form of Eq. 13.22 is essentially the same as that derived in Section 3.5 on Newton's method. This matrix equation is a very formidable undertaking to compute and manipulate. It is extremely sparse and requires special sparsity logic.

Handling inequality constraints is very difficult in either gradient or Newton approaches. The usual method is to form a constraint "penalty" function as follows. Suppose the voltage at a bus must meet limits:

$$
|E_i|^{min} \leq |E_i| \leq |E_i|^{max} \tag{13.65}
$$

It is possible to enforce this constraint by inventing the following exterior penalty functions:

$$
h(|E_i|) = \begin{bmatrix} K(|E_i| - |E_i|^{min})^2 \\ 0 \\ K(|E_i|^{max} - |E_i|)^2 \end{bmatrix} \quad \begin{matrix} \text{for } |E_i| < |E_i|^{min} \\ \text{for } E \text{ within limits} \\ \text{for } |E_i| > |E_i|^{max} \end{matrix} \tag{13.66}
$$

This penalty function is shown in Figure 13.3.

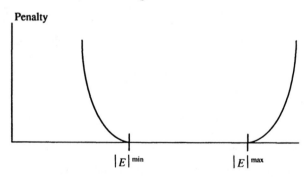

FIG. 13.3 Exterior penalty functions for voltage magnitude violations.

To solve the OPF with the voltage inequality constraint, we add the penalty function to the objective function, f. The resulting function will be large if the voltage is outside its limit, and thus the OPF will try to force it within its limits as it minimizes the objective.

Since Newton's method has the second derivative information built into it, it does not have great difficulty in converging and it can handle the inequality constraints as well. The difficulty with Newton's method arises in the fact that near the limit the penalty is small, so that the optimal solution will tend to allow the variable, a voltage in the example above, to float over its limit. The seemingly simple tuning procedure of raising the value of K may eventually cause the matrices to become ill-conditioned and the method fails. When there are few limits to be concerned with and the objective function is "shallow," that is, the variability of f with adjustments in the control variables is very low, Newton's method is the best method to use.

References 5–7 give examples of the development of Newton's method to solve the full AC OPF.

13.3   LINEAR SENSITIVITY ANALYSIS

Before continuing with the discussion of the linear programming and interior OPF methods, we shall develop the concept of linear sensitivity analysis. Linear sensitivity coefficients give an indication of the change in one system quantity (e.g., MW flow, MVA flow, bus voltage, etc.) as another quantity is varied (e.g., generator MW output, transformer tap position, etc.) These linear relationships are essential for the application of linear programming. Note that as the adjustable variable is changed, we assume that the power system reacts so as to keep all of the power flow equations solved. As such, linear sensitivity coefficients can be expressed as partial derivatives for example:

$$\frac{\partial \text{MVA flow}_{ij}}{\partial \text{MW gen}_k}$$

shows the sensitivity of the flow (MVA) on line (i to j) with respect to the power generated at bus k.

Some sensitivity coefficients may change rapidly as the adjustment is made and the power flow conditions are updated. This is because some system quantities vary in a nonlinear relationship with the adjustment and resolution of the power flow equations. This is especially true for quantities that have to do with voltage and MVAR flows. Sensitivities such as the variation of MW flow with respect to a change in generator MW output are rather linear across a wide range of adjustments and lead to the usefulness of the DC power flow equations and the "a" and "d" factors introduced in Chapter 11.

For this reason, the value represented by a sensitivty coefficient is only good for small adjustments and the sensitivities must be recalculated often.

13.3.1   Sensitivity Coefficients of an AC Network Model

The following procedure is used to linearize the AC transmission system model for a power system. To start, we shall define two general equations giving the power injection at a bus. That is, the net power flowing into a transmission system from the bus. This function represents the power flowing into transmission lines and shunts at the bus:

$$
P_i(|\mathbf{E}|, \boldsymbol{\theta}) = Re\left[\left(\sum_j E_i[(E_i - t_{ij}E_j)y_{ij}]^* \right) + E_i\left(E_i \sum_i y_{\text{shunt}_i} \right)^* \right]
$$

$$
Q_i(|\mathbf{E}|, \boldsymbol{\theta}) = Im\left[\left(\sum_j E_i[(E_i - t_{ij}E_j)y_{ij}]^* \right) + E_i\left(E_i \sum_i y_{\text{shunt}_i} \right)^* \right]
$$

(13.67)

where

$E_i = |E_i| \angle \theta_i$

t_{ij} = the transformer tap in branch ij

y_{ij} = the branch admittance

y_{shunt_i} = the sum of the branch and bus shunt admittances at bus i

Then, at each bus:

$$
P_i(|\mathbf{E}|, \boldsymbol{\theta}) = P_{\text{gen}_i} - P_{\text{load}_i}
$$
$$
Q_i(|\mathbf{E}|, \boldsymbol{\theta}) = Q_{\text{gen}_i} - Q_{\text{load}_i}
$$

(13.68)

The set of equations that represents the first-order approximation of the AC network around the initial point is the same as generally used in the Newton power flow algorithm. That is:

$$
\sum \frac{\partial P_i}{\partial |E_j|} \Delta |E_j| + \sum \frac{\partial P_i}{\partial \theta_j} \Delta\theta_j + \sum \frac{\partial P_i}{\partial j_{ij}} \Delta t_{ij} = \Delta P_{\text{gen}_i}
$$

$$
\sum \frac{\partial Q_i}{\partial |E_j|} \Delta |E_j| + \sum \frac{\partial Q_i}{\partial \theta_j} \Delta\theta_j + \sum \frac{\partial Q_i}{\partial t_{ij}} \Delta t_{ij} = \Delta Q_{\text{gen}_i}
$$

(13.69)

This can be placed in matrix form for easier manipulation:

$$
\begin{bmatrix} \dfrac{\partial P_1}{\partial |E_1|} & \dfrac{\partial P_1}{\partial \theta_1} & \cdots \\[2ex] \dfrac{\partial Q_1}{\partial |E_1|} & \dfrac{\partial Q_1}{\partial \theta_1} & \cdots \\[2ex] & \vdots & \end{bmatrix} \begin{bmatrix} \Delta|E_1| \\[1ex] \Delta\theta_1 \\[1ex] \vdots \end{bmatrix} = \begin{bmatrix} -\dfrac{\partial}{\partial t_{ij}}(P_i) & 1 & 0 \\[2ex] -\dfrac{\partial}{\partial t_{ij}}(Q_i) & 0 & 1 \end{bmatrix} \begin{bmatrix} \Delta t_{ij} \\[1ex] \Delta P_{\text{gen}_i} \\[1ex] \Delta Q_{\text{gen}_i} \end{bmatrix}
$$

(13.70)

This equation will be placed into a more compact format that uses the vectors **x** and **u**, where **x** is the state vector of voltages and phase angles, and **u** is the vector of control variables. The control variables are the generator MW, transformer taps, and generator voltage magnitudes (or generator MVAR). Note that at any given generator bus we can control a voltage magnitude only within the limits of the unit VAR capacity. Therefore, there are times when the role of the state and control are reversed. Note that other controls can easily be added to this formulation. The compact form of Eq. 12.30 then is written:

$$[J_{px}]\Delta x = [J_{pu}]\Delta u \tag{13.71}$$

Now, we will assume that there are several transmission system dependent variables, **h**, that represent, for example, MVA flows, load bus voltages, line amperes, etc., and we wish to find their sensitivity with respect to changes in the control variables. Each of these quantities can be expressed as a function of the state and control variables; that is, for example:

$$\mathbf{h} = \begin{bmatrix} \text{MVA flow}_{nm}(|E|, \theta) \\ |E_k| \end{bmatrix} \tag{13.72}$$

where $|E_k|$ represents only load bus voltage magnitude.

As before, we can write a linear version of these variables around the operating point

$$\Delta \mathbf{h} = \begin{bmatrix} \dfrac{\partial h_1}{\partial |E_1|} & \dfrac{\partial h_1}{\partial \theta_1} & \cdots \\[2mm] \dfrac{\partial h_2}{\partial |E_1|} & \dfrac{\partial h_2}{\partial \theta_1} & \cdots \\[2mm] \vdots & & \cdots \end{bmatrix} \begin{bmatrix} \Delta |E_1| \\ \Delta \theta_1 \\ \vdots \end{bmatrix} + \begin{bmatrix} \dfrac{\partial h_1}{\partial u_1} & \dfrac{\partial h_1}{\partial u_2} & \cdots \\[2mm] \dfrac{\partial h_2}{\partial u_1} & \dfrac{\partial h_2}{\partial u_1} & \cdots \\[2mm] \vdots & & \cdots \end{bmatrix} \begin{bmatrix} \Delta u_1 \\ \Delta u_2 \\ \vdots \end{bmatrix} \tag{13.73}$$

where

$$h_1 = \text{the line } nm \text{ MVA flow}$$

$$h_2 = \text{the bus } k \text{ voltage magnitude}$$

Again, we can put this into a compact format using the vectors **x** and **u** as before:

$$\Delta \mathbf{h} = [J_{hx}]\Delta x + [J_{hu}]\Delta u \tag{13.74}$$

We will now eliminate the Δx variables; that is:

$$\Delta x = [J_{px}]^{-1}[J_{pu}]\Delta u \tag{13.75}$$

Then, substituting:

$$\Delta \mathbf{h} = [J_{hx}][J_{px}]^{-1}[J_{pu}]\Delta u + [J_{hu}]\Delta u \tag{13.76}$$

This last equation gives the linear sensitivity coefficients between the transmission system quantities, **h**, and the control variables, **u**.

13.4 LINEAR PROGRAMMING METHODS

The gradient and Newton methods of solving an OPF suffer from the difficulty in handling inequality constraints. Linear programming, however, is very adept at handling inequality constraints, as long as the problem to be solved is such that it can be linearized without loss of accuracy.

Figure 13.4 shows the type of strategy used to create an OPF using linear programming. The power flow equations could be for the DC representation, the decoupled set of AC equations, or the full AC power flow equations. The choice will affect the difficulty of obtaining the linearized sensitivity coefficients and the convergence test used.

In the formulation below, we show how the OPF can be structured as an LP. First, we tackle the problem of expressing the nonlinear input–output or cost functions as a set of linear functions. This is similar to the treatment in Section 7.9 for hydro-units. Let the cost function be $F_i(P_i)$ as shown in Figure 13.5.

We can approximate this nonlinear function as a series of straight-line segments as shown in Figure 13.6. The three segments shown will be represented as P_{i1}, P_{i2}, P_{i3}, and each segment will have a slope designated:

$$s_{i1}, s_{i2}, s_{i3}$$

then the cost function itself is

$$F_i(P_i) = F_i(P_i^{\min}) + s_{i1}P_{i1} + s_{i2}P_{i2} + s_{i3}P_{i3} \tag{13.77}$$

and

$$0 \le P_{ik} \le P_{ik}^+ \quad \text{for } k = 1, 2, 3 \tag{13.78}$$

and finally

$$P_i = P_i^{\min} + P_{i1} + P_{i2} + P_{i3} \tag{13.79}$$

The cost function is now made up of a linear expression in the P_{ik} values.

In the formulation of the OPF using linear programming, we only have the control variables in the problem. We do not attempt to place the state variables into the LP, nor all the power flow equations. Rather, constraints are set up in the LP that reflect the influence of changes in the control variables only. In the examples we present here, the control variables will be limited to generator real power, generator voltage magnitude, and transformer taps. The control variables will be designated as the u variables (see earlier in this chapter).

The next constraint to consider in an LPOPF are the constraints that

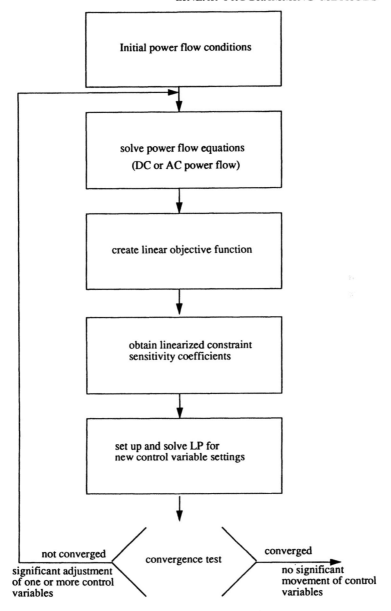

FIG. 13.4   Strategy for solution of the LPOPF.

represent the power balance between real and reactive power generated, and that consumed in the loads and losses. The real power balance equation is:

$$P_{\text{gen}} - P_{\text{load}} - P_{\text{loss}} = 0 \qquad (13.80)$$

The loss term here represents the I^2R losses in the transmission lines and

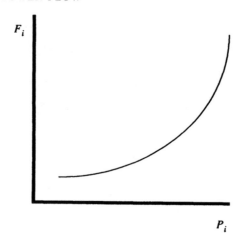

FIG. 13.5   A nonlinear cost function characteristic.

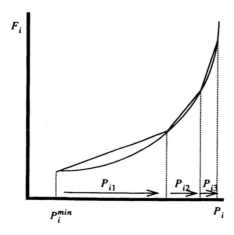

FIG. 13.6   A linearized cost function.

transformers. We can take derivatives with respect to the control variables, u, and this results in:

$$\sum_u \left(\frac{\partial P_{gen}}{\partial u} \right) \Delta u - \sum_u \left(\frac{\partial P_{load}}{\partial u} \right) \Delta u - \sum_u \left(\frac{\partial P_{loss}}{\partial u} \right) \Delta u = 0 \qquad (13.81)$$

If we make the following substitution:

$$\Delta u = u - u^0 \qquad (13.82)$$

then, the power balance equation becomes

$$\sum_u \left(\frac{\partial P_{gen}}{\partial u} \right) u - \sum_u \left(\frac{\partial P_{load}}{\partial u} \right) u - \sum_u \left(\frac{\partial P_{loss}}{\partial u} \right) u = K_p \qquad (13.83)$$

where

$$K_p = \sum_u \frac{\partial P_{\text{gen}}}{\partial u} u^0 - \sum_u \frac{\partial P_{\text{load}}}{\partial u} u^0 - \sum_u \frac{\partial P_{\text{loss}}}{\partial u} u^0 \tag{13.84}$$

A similar equation can be written for the reactive power balance:

$$\sum_u \left(\frac{\partial Q_{\text{gen}}}{\partial u}\right)\Delta u - \sum_u \left(\frac{\partial Q_{\text{load}}}{\partial u}\right)\Delta u - \sum_u \left(\frac{\partial Q_{\text{loss}}}{\partial u}\right)\Delta u = 0 \tag{13.85}$$

where the loss term is understood to include $I^2 X$ as well as the charging from line capacitors and shunt reactors. A substitution using $\Delta u = u - u^0$, as above, can also be done here.

The LP formulation, so far, would need to restrict control variables to move only within their respective limits, but it does not yet constrain the OPF to optimize cost within the limits of transmission flows and load bus voltages. To add the latter type constraints, we must add a new constraint to the LP. For example, say we wish to constrain the MVA flow on line nm to fall within an upper limit:

$$\text{MVA flow}_{nm} \leq \text{MVA flow}_{nm}^{\text{max}} \tag{13.86}$$

We model this constraint by forming a Taylor's series expansion of this flow and only retaining the linear terms:

$$\text{MVA flow}_{nm} = \text{MVA flow}_{nm}^0 + \sum_u \left(\frac{\partial}{\partial u} \text{MVA flow}_{nm}\right)\Delta u \leq \text{MVA flow}_{nm}^{\text{max}} \tag{13.87}$$

Again, we can substitute $\Delta u = u - u^0$ so we get:

$$\sum_u \left(\frac{\partial}{\partial u} \text{MVA flow}_{nm}\right)u \leq \text{MVA flow}_{nm}^{\text{max}} - K_f \tag{13.88}$$

where

$$K_f = \text{MVA flow}_{nm}^0 + \sum_u \frac{\partial}{\partial u} \text{MVA flow}_{nm} u^0 \tag{13.89}$$

Other constraints such as voltage magnitude limits, branch MW limits, etc., can be added in a similar manner. We add as many constraints as necessary to constrain the power system to remain within its prescribed limits. Note, of course, that the derivatives of P_{loss} and MVA flow$_{nm}$ are obtained from the linear sensitivity coefficient calculations presented in the previous section.

13.4.1   Linear Programming Method with Only Real Power Variables

As an introduction to the LPOPF, we will set up and solve a power system example which only has generator real powers as control variables. Further, the model for the power system power balance constraint will assume that load is constant and that the losses are constant. Finally, since the entire model used in the LP is based on a MW-only formulation, we shall use the "a" and "d" factors derived in Chapter 11 to model the effect of changes in controls on the constraints. As indicated in Figure 13.4, we shall solve the LP and then make the adjustments to the control variables and solve a power flow in each main iteration. This guarantees that the total generation equals load plus losses, and that the MW flows are updated properly. The cost functions can be treated as before using multiple segmented "piecewise linear" approximations.

The "power balance" equation for this case is as follows:

$$P_1 + P_2 + \ldots + P_{\text{ref}} = P_{\text{load}} + P_{\text{losses}} = \text{constant} \tag{13.90}$$

To constrain the power system, we need the expansion of the constraints, such as MW flows, bus voltages, etc., as linear functions of the control variables. In this case, the linear control variables will be represented as a vector $\mathbf{u}$:

$$u = \begin{bmatrix} P_1 \\ P_2 \\ \vdots \\ P_{\text{ref}} \end{bmatrix} \tag{13.91}$$

This is done with the linear sensitivity approach, as derived in the previous section. The result is a set of constraints:

$$\mathbf{h(u)} \leq \mathbf{h}^+ \tag{13.92}$$

which is written as

$$\mathbf{h(u)} = h(\mathbf{u}^0) + \frac{\partial h}{\partial \mathbf{u}} (\mathbf{u} - \mathbf{u}^0) \leq \mathbf{h}^+ \tag{13.93}$$

However, we shall observe that the derivatives $\partial h/\partial \mathbf{u}$ can be replaced with the "a" sensitivity coefficients developed in Chapter 11.

Thus, for a MW flow constraint on line rs we have:

$$\text{MW}_{rs} = \text{MW}_{rs}^0 + \sum_u a_{rs-u}(u - u^0) \leq \text{MW}_{rs}^{\max} \tag{13.94}$$

or

$$\text{MW}_{rs} = \sum_u a_{rs-u} u \leq \text{MW}_{rs}^{\max} - \left(\text{MW}_{rs}^0 - \sum_u a_{rs-u} u^0 \right) \tag{13.95}$$

TABLE 13.1   Line Flows: Power Flow 0

Line	Limit	MW Flow
1-2	30	28.69
1-4	50	43.58
1-5	40	35.60
2-3	20	2.93
2-4	40	33.09
2-5	20	15.51
2-6	30	26.25
3-5	20	19.12
3-6	60	43.77
4-5	20	4.08
5-6	20	1.61

Similar constraints are added for any power system network quantity that is to held within its limit.

EXAMPLE 13C

We shall use the LPOPF reduced model method to solve an OPF problem. An LP and an AC power flow will be used to solve a series of dispatch problems. The transmission system will be the six bus system introduced in Chapter 4, the MW limits on the transmission lines will be those introduced in Example 11B and shown in Table 13.1. The generator cost functions are those found in Example 4E and linearized as shown below.

We shall solve a series of LP–AC power flow calculations as follows.

Step 0

Run a base AC power flow (this will be the AC power flow shown in Figure 4.8 and it will be designated as POWER Flow 0 in numbering the various power flow calculations in this example). Looking at Figure 4.8 and the limit set we are using from Example 11B, also shown below, we note that there are no overloads.

The generation values for this power flow are:

$$P_1 = 107.87 \text{ MW}, \ P_2 = 50 \text{ MW}, \quad \text{and} \quad P_3 = 69 \text{ MW} \ \textit{power flow 0: result}$$

The total cost for this initial dispatch is 3189.4 R/h.

Step 1

We now set up the LP to solve for the optimum cost with only the power balance equation in the LP constraint set. By the nature of the cost curve

TABLE 13.2   Generator Unit Break Point MWs

Unit	Break Point 1 (unit min)	Break Point 2	Break Point 3	Break Point 4 (unit max)
1	50	100	160	200
2	37.5	70	130	150
3	45	90	140	180

TABLE 13.3   Generator Cost Curve Segment Slope

Generator	s_{i1}	s_{i2}	s_{i3}
1	12.4685	13.0548	13.5875
2	11.2887	12.1110	12.8222
3	11.8333	12.5373	13.2042

segments, we also incorporate the limits on the generators. The generator cost functions are as follows:

Generator on bus 1: $F_1(P_1) = 213.1 + 11.669P_1 + 0.00533P_1^2$ R/h
with limits of: 50.0 MW $\leq P_1 \leq$ 200.0 MW

Generator on bus 2: $F_2(P_2) = 200.0 + 10.333P_2 + 0.00889P_2^2$ Rh
with limits of: 37.5 MW $\leq P_2 \leq$ 150.0 MW

Generator on bus 3: $F_3(P_3) = 240.0 + 10.833P_3 + 0.00741P_3^2$ R/h
with limits of: 45.0 MW $\leq P_3 \leq$ 180.0 MW

The LP will be run with the unit cost functions broken into three straight-line segments such that the break points are located as shown in Table 13.2. The generator cost function segment slopes are computed as follows:

$$s_{if} = \frac{F_i(P_{ij}^+) - F_i(P_{ij}^-)}{P_{ij}^+ - P_{ij}^-} \tag{13.96}$$

where P_{ij}^+ and P_{ij}^- are the values of P_i at the end of the jth cost curve segment. The values are shown in Table 13.3. The segment limits are shown in Table 13.4. The LP cost function is:

$$[F_1(P_1^{min}) + 12.4685P_{11} + 13.0548P_{12} + 13.5878P_{13}]$$
$$+ [F_2(P_2^{min}) + 11.2887P_{21} + 12.1110P_{22} + 12.8222P_{23}] \tag{13.97}$$
$$+ [F_3(P_3^{min}) + 11.8333P_{31} + 12.5373P_{32} + 13.2042P_{33}]$$

TABLE 13.4 Segment Limits

Segment	Min MW	Max MW
P_{11}	0	50
P_{12}	0	60
P_{13}	0	40
P_{21}	0	32.5
P_{22}	0	60
P_{23}	0	20
P_{31}	0	45
P_{32}	0	50
P_{33}	0	40

Since the $F_i(P_i^{min})$ terms are constant, we can drop them in the LP. Then, the cost function becomes:

$$12.4685P_{11} + 13.0548P_{12} + 13.5878P_{13} + 11.2887P_{21}$$
$$+ 12.1110P_{22} + 12.8222P_{23} + 11.8333P_{31} \qquad (13.98)$$
$$+ 12.5373P_{32} + 13.2042P_{33}$$

The generation, load, and losses equality constraint is

$$P_1 + P_2 + P_3 = P_{load} + P_{losses} \qquad (13.99)$$

The load is 210 MW and the losses from the initial power flow are 7.87 MW. Substituting the equivalent expression for each generator's output in terms of its three linear segments, we obtain:

$$P_1^{min} + P_{11} + P_{12} + P_{13} + P_2^{min} + P_{21} + P_{22} + P_{23} + P_3^{min}$$
$$+ P_{31} + P_{32} + P_{33} = P_{load} + P_{losses} \qquad (13.100)$$

This results in the following after the P_i^{min}, P_{load}, and P_{loss} values are substituted:

$$P_{11} + P_{12} + P_{13} + P_{21} + P_{23} + P_{33} + P_{31} + P_{32} + P_{33}$$
$$= 210 + 7.87 - 50 - 37.5 - 45 = 85.37 \qquad (13.101)$$

We now solve the LP with the cost function and equality constraint given above, and with the six variables representing the generator outputs. The solution to this LP is shown in Table 13.5.

TABLE 13.5   First LP Solution

Variable	Min MW	Solution MW	Max MW
P_{11}	0	0.0	50
P_{12}	0	0.0	60
P_{13}	0	0.0	40
P_{21}	0	32.5	32.5
P_{22}	0	7.87	60
P_{23}	0	0.0	20
P_{31}	0	45.0	45
P_{32}	0	0.0	50
P_{33}	0	0.0	40

The total generation on each generator is:

$$P_i = P_i^{\min} + P_{i1} + P_{i2} + P_{i3} \tag{13.102}$$

then the generator optimal outputs are

$P_1 = 50$ MW, $P_2 = 77.87$ MW,   and $P_3 = 90$ MW LP 1: *result*

Note that this solution of necessity will have only one of the variables not at a break point while the others will be at a break point. Note also that the output on bus 1 is at its low limit. When we substitute these values for the generation at buses 1, 2, and 3, and run the power flow, we get the following:

$P_1 = 48.83$ MW, $P_2 = 77.87$ MW,   and $P_3 = 90$ MW *power flow 1: result*

The total cost for this dispatch is 3129.1 ₽/h. This illustrates the fact that the LP uses a linear model of the power system and when we put its results into a nonlinear model, such as the power flow, there are bound to be differences. Since the losses have changed (to 6.70 MW), the power output of the reference bus must decrease to balance the power flow. However, the solution to the optimal LPOPF has the reference-bus power output below its minimum of 50 MW. To correct this condition we set up another LP solution with the same cost function but with a slightly different equality constraint that reflects the new value of losses. The result of this LP is:

$P_1 = 50$ MW, $P_2 = 76.7$ MW,   and $P_3 = 90$ MW LP 1.1: *result*

Once again, we enter these results into the power flow and obtain:

$P_1 = 49.99$ MW, $P_2 = 76.7$ MW   and $P_3 = 90$ MW *power flow 1.1 result*

TABLE 13.6   Line Flows: Power Flow 1.1

Line	Limit	MW Flow
1-2	30	4.28
1-4	50	25.60
1-5	40	20.11
2-3	20	−6.42
2-4	40	48.75[a]
2-5	20	17.75
2-6	30	20.88
3-5	20	28.91[a]
3-6	60	54.63
4-5	20	1.84
5-6	20	3.87

[a] Overloaded line.

The total cost for this dispatch is 3129.6 ₽/h and the losses are 6.7 MW. This represents the least cost dispatch that we shall obtain in this example. As constraints are added later to meet the flow limits, the cost will increase.

Note also that we have two overloads on the optimum cost dispatch as shown in Table 13.6.

Step 2

The LP and power flow executions in step 1 resulted in a less-costly dispatch than the original power flow, but in doing so we have overloaded two transmission lines. We shall refer to these overloads as $(n - 0)$ overloads. This notation means that there are n lines minus zero outages in the network at the time of the overload. [Later we shall use the notation $(n - 1)$ to indicate that there are n lines minus one line (that is, a single-line outage) in the network at the time of the overloads. This notation can be used for further levels of overload such as $(n - 2)$, $(n - 3)$, etc. However, many electric utility transmission operations departments only go as far as $(n - 1)$ in dispatching their systems.]

We must redispatch the power system at this point to remove the $(n - 0)$ overloads. To do this, we add two constraints to the LP, one for each overloaded line. The power flow constraint on line 2-4 is modeled as:

$$f_{2\text{-}4} = f_{2\text{-}4}^0 + a_{2\text{-}4,1}(P_1 - P_1^0) + a_{2\text{-}4,2}(P_2 - P_2^0) + a_{2\text{-}4,3}(P_3 - P_3^0) \le 40 \tag{13.103}$$

Substituting 48.75 for $f_{2\text{-}4}^0$, 76.7 for P_2^0, and 90 for P_3^0, we get the following for the constraint for line 2-4 (note that $a_{2-4,1} = 0$) and, finally, we expand P_2 and P_3 in terms of the segments:

$$48.75 + 0.31(37.5 + P_{21} + P_{22} + P_{23} - 76.7)$$
$$+ 0.22(45 + P_{31} + P_{32} + P_{33} - 90) \leq 40 \qquad (13.104)$$

or

$$0.31P_{21} + 0.31P_{22} + 0.31P_{23} + 0.22P_{31} + 0.22P_{32} + 0.22P_{33} \leq 13.302$$
$$(13.105)$$

The constraint for line 3-5 is built similarly and results in:

$$0.06P_{21} + 0.06P_{22} + 0.06P_{23} + 0.29P_{31} + 0.29P_{32} + 0.29P_{33} \leq 6.492$$
$$(13.106)$$

The solution to the LP gives:

$$P_1 = 87.02 \text{ MW}, \ P_2 = 70.0 \text{ MW} \quad \text{and} \quad P_3 = 59.66 \text{ MW } \textit{LP 2: result}$$

Also note that only the first transmission line constraint is binding in the LP, the remaining constraint is "slack," that is, it is not being forced up against its limit. When these values are put into the power flow we obtain:

$$P_1 = 87.54 \text{ MW}, \ P_2 = 70.0 \text{ MW} \quad \text{and} \quad P_3 = 59.66 \text{ MW } \textit{power flow 2: result}$$

The flows on the two constrained lines are:

$$f_{2-4} = 39.40 \text{ MW} \quad \text{and} \quad f_{3-5} = 20.36 \text{ MW}$$

The total operating cost has now increased to 3155.0 ₽/h.

We now run another complete LP–power flow iteration to account for changes in losses and to bring the constraints closer to their limits. The solution to the second-iteration LP gives:

$$P_1 = 86.16 \text{ MW}, \ P_2 = 73.3 \text{ MW} \quad \text{and} \quad P_3 = 57.73 \text{ MW } \textit{LP 2.1: result}$$

Both transmission line constraints are binding in the second LP. When these values are put into the power flow we obtain:

$$P_1 = 86.16 \text{ MW}, \ P_2 = 73.3 \text{ MW} \quad \text{and} \quad P_3 = 57.73 \text{ MW } \textit{power flow 2.1: result}$$

The flows on the two constrained lines are:

$$f_{2-4} = 39.99 \text{ MW} \quad \text{and} \quad f_{3-5} = 20.06 \text{ MW}$$

The total operating cost has now decreased slightly to 3153.3 ₽/h. There are no more $(n - 0)$ line overloads.

TABLE 13.7   Line Flows: Power Flow 2.1 (with Line 2-3 Out)

Line	Limit	MW Flow
1-2	30	18.1
1-4	50	36.37
1-5	40	31.74
2-3	20	—
2-4	40	40.73[a]
2-5	20	19.19
2-6	30	31.11[a]
3-5	20	18.26
3-6	60	39.47
4-5	20	4.59
5-6	20	1.17

[a] Overloaded line.

Step 3

We have now achieved an optimal dispatch with all $(n - 0)$ overloads met. This dispatch will satisfy generation and all line flow limits; however, if we have a transmission line outage contingency, we may have overloads. By modeling the first contingency overloads, or the so-called $(n - 1)$ overloads, we can guarantee that should the contingency outage take place, there would be no resulting overloads. This is the scheme involved in security-constrained OPF, or SCOPF, and is the subject of Section 13.5.

In this example, to make matters simple we shall only study the result of one contingency outage. In our sample system, we shall start from the result of power flow 2.1 and take out line 2-3. The flows that result from this contingency power flow are shown in Table 13.7.

We now must form a new LP that has the generation, load, losses equality constraint and the original two $(n - 0)$ line flow constraints done in step 2, and two new constraints for each of the $(n - 1)$ overloads (i.e., on line 2-4 and line 2-6). To model line 2-4 with line 2-3 removed, we use the following constraint, as derived in Appendix 11A of Chapter 11.

$$\Delta f_l^k = \sum_i (a_{li} + d_{l,k} a_{ki}) \Delta P_i \qquad (13.107)$$

This now becomes:

$$f_l^k = \sum_i (a_{li} + d_{l,k} a_{ki})(P_i - P_i^0) + f_l^0 \le f_i^{\max} \qquad (13.108)$$

The new LP has five constraints. The first result of this LP gives:

$$P_1 = 91.39 \text{ MW}, P_2 = 66.96 \text{ MW}, \quad \text{and} \quad P_3 = 58.84 \text{ MW} \; LP \; 3: \; result$$

The $(n = 0)$ constraint on line 3-5 is binding and the $(n - 1)$ constraint on line 2-6 is binding. When these values are put into the power flow, we obtain (note that this power flow has all lines in):

$$P_1 = 91.52 \text{ MW}, \ P_2 = 66.96 \text{ MW}, \quad \text{and} \quad P_3 = 58.8 \text{ MW} \ \textit{power flow 3: result}$$

The flows on the two $(n - 0)$ constrained lines are:

$$f_{2\text{-}4} = 38.23 \text{ MW} \quad \text{and} \quad f_{3\text{-}5} = 19.94 \text{ MW}$$

A second power flow with line 2-3 out is also run with the same generation values. The results of this power flow show that the two $(n - 1)$ flow constraints are:

$$f_{2\text{-}4}^{\text{contingency}} = 38.86 \text{ MW} \quad \text{and} \quad f_{2\text{-}6}^{\text{contingency}} = 30.00 \text{ MW}$$

The total operating cost has now increased to 3160.5 ₨/h. A complete second iteration of the LP and power flows is run and results in the following power flows:

$$P_1 = 90.53 \text{ MW}, \ P_2 = 67.92 \text{ MW}, \quad \text{and} \quad P_3 = 58.84 \text{ MW} \ \textit{power flow 3.1: result}$$

The flows on the two $(n - 0)$ constrained lines are:

$$f_{2\text{-}4} = 38.54 \text{ MW} \quad \text{and} \quad f_{3\text{-}5} = 20.00 \text{ MW}$$

A second power flow with line 2-3 out is also run with the same generation values. The results of this power flow show that the two $(n - 1)$ flow constraints are:

$$f_{2\text{-}4}^{\text{contingency}} = 39.18 \text{ MW} \quad \text{and} \quad f_{2\text{-}6}^{\text{contingency}} = 30.09 \text{ MW}$$

The total operating cost has now increased to 3159.1 ₨/h.

13.4.2 Linear Programming with AC Power Flow Variables and Detailed Cost Functions

OPF programs that optimize the AC power flow of a power system go beyond the LPOPF introduced in the last section, in several respects.

First, they do not usually use fixed break points. Rather, the break points are added as needed as the solution progresses and can become close enough so that no error is perceptible between the piecewise linear approximation and the true nonlinear input–output curve of the generators. "Second, the AC quantities of voltage magnitude and perhaps phase angle become variables in the LP and the constraints are set up as linear functions using the sensitivity coefficients methods shown in Section 13.3. Usually, however, the nonlinear representations of the bus power and reactive power injections and the line or transformer MVA flows are not well represented as linear functions. To cope with the nonlinear nature of these constraints involves restricting the movement of each variable and then relinearlizing the equality and inequality constraints quite often. The result is an LP that "converges" on the optimal AC power flow, meeting all the power flow equality constraints and inequality constraints.

Reference 9 is an example of such an OPF code built around an LP.

13.5 SECURITY-CONSTRAINED OPTIMAL POWER FLOW

In Chapter 11, we introduced the concept of security analysis and the idea that a power system could be constrained to operate in a secure manner. Programs which can make control adjustments to the base or pre-contingency operation to prevent violations in the post-contingency conditions are called "security-constrained optimal power flows," or SCOPF.

We have seen previously that an OPF is distinguished from an economic dispatch by the fact that it constantly updates a power flow of the transmission system as it progresses toward the minimum of the objective function. One advantage of having the power flow updated is the fact that constraints can be added to the OPF that reflect the limits which must be respected in the transmission system. Thus, the OPF allows us to reach an optimum with limits on network components recognized.

An extension to this procedure is to add constraints that model the limits on components during contingency conditions. That is, these new "security constraints" or "contingency constraints" allow the OPF to meet precontingency limits as well as post-contingency limits. There is a price to pay, however, and that is the fact as we iterate the OPF with an AC power flow, we must also run power flows for all the contingency cases being observed. This is illustrated in Figure 13.7.

The SCOPF shown in Figure 13.7 starts by solving an OPF with $(n - 0)$ constraints only. Only when it has solved for the optimal, constrained conditions is the contingency analysis executed. In Figure 13.7, the contingency analysis starts by screening the power system and identifying the potential worst-contingency cases. As was pointed out in Chapter 11, not all of these cases are going to result in a post-contingency violation and it is important to limit the number of full power flows that are executed. This is especially important in the SCOPF, where each contingency power flow may result in new contingency constraints being added to the OPF. We assume here that only the M worst cases screened by the screening algorithm are added. It is possible to make $M = 1$, in which case only the worst potential contingency is added.

Next, all the $(n - 1)$ contingency cases that are under consideration must be solved by running a power flow with that contingency reflected in alterations to the power flow model. When the power flow results in a security violation,

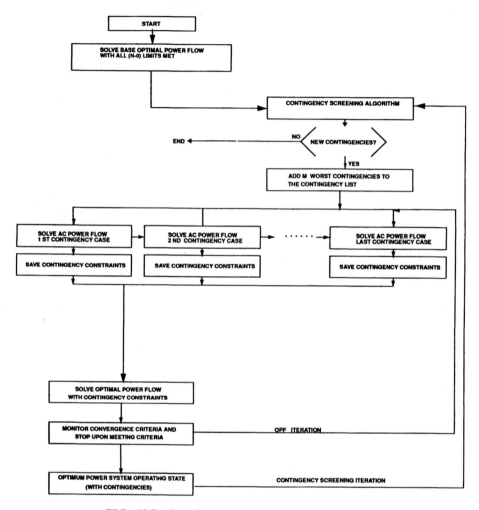

FIG. 13.7   Security-constrained optimal power flow.

the power system model is used to create a contingency constraint. In fact, what is done is to run a network sensitivity calculation (See Section 13.3) on the model with the contingency outage and save the resulting constraint sensitivities. When all contingency power flows are complete, all the contingency constraints are added to the OPF model and it is solved.

Note, in Figure 13.7, there are two main loops to be executed. The loop labeled "OPF Iteration" requires the OPF and each of the contingency power flows to be re-executed until the OPF has solved with all contingency constraints met. Next, the outer loop labeled "Contingency Screening Iteration" is tried. If the contingency screening algorithm does not pick up any new contingencies the SCOPF can end; if new contingencies are found, it must add them to the list and continue.

Why is all this necessary? The optimum operation conditions for a power system will often result in violation of system security. This is especially true when a large amount of interchange power is available at a favorable price. In this instance, the selling power system can be modeled in the OPF with its price of production set accordingly, and the OPF will then raise the interchange up to the point where transmission system components are limiting. Now, when the contingency analysis is run, there may be many cases which result in contingency violations and the OPF, with contingency constraints added, will have to back off the interchange power in order to meet the contingency limits.

It should also be noted that when some contingency constraints are added to the OPF, it will redispatch generation, and adjust voltages and transformers to meet these constraints. The process of adjustments may result in many new contingency violations when the screening algorithm and the power flows are run. The need to iterate between the OPF and the contingency screening represents an effort to find the "most constraining" contingencies.

SCOPF was introduced as step 3 in Example 13C and will also be illustrated in Example 13D, which follows.

EXAMPLE 13D

This example shows the results of running the same six-bus case used in Example 13C, with the same generator cost functions. However, we now are using a full AC OPF so that we will use line MVA limits and bus voltage limits as well. The MVA limits are shown in Table 13.8. The bus voltages are also limited, with bus 5 being the only one to hit its upper limit of 1.0 pu voltage magnitude.

The full AC OPF has six control variables: three generator outputs and three generator voltage magnitude schedules. In addition, the AC OPF can be used to minimize either MW losses, or to minimize operating cost. Table 13.9 summarizes these results.

TABLE 13.8   Line MVA Flows: Power Flow 0

Line	MVA Limit	MVA Flow
1-2	40	32.57
1-4	60	48
1-5	40	37.34
2-3	40	12.61
2-4	60	56.71
2-5	30	21.83
2-6	90	29.03
3-5	70	30.04
3-6	80	74.86
4-5	20	6.41
5-6	40	9.80

TABLE 13.9 Full AC OPF Results

Case	P_1	P_2	P_3	$\|E_1^{sched}\|$	$\|E_2^{sched}\|$	$\|E_3^{sched}\|$	MW Losses	Cost	Binding Constraints
Base case	107.9	50.0	60.0	1.05	1.0499	1.0429	7.87	3189.4	Generators 2 and 3 at max VAR limit
Min cost, adjust generator MW only	86.9	59.3	71.0	1.05	1.05	1.0458	7.14	3157.9	Line 2-4 MVA limit, generator 3 at max VAR limit
Min losses, adjust generator voltage only	107.1	50.0	60.0	1.05	1.0429	1.0499	7.1	3179.5	Bus 5 max voltage, line 1-5 MVA limit, generator 3 at max VAR limit
Min cost with generator MW adjustment, then min losses with adjustment of gen voltages	86.3	59.3	71.0	1.05	1.0429	1.0499	6.54	3150.3	Bus 5 max voltage, generator 3 at max VAR limit
Min cost with both generator MW and voltage adjustment	52.0	87.5	77.0	1.05	1.0429	1.0499	6.41	3124.6	Bus 5 max voltage, line 2-4 MVA limit, generator 3 at max VAR limit
Min cost, adjust generator MW with line 3-6 out	70.0	91.8	62.0	1.05	1.031	1.07	15.05	3219.7	Before, OPF line 1-5 has 12% MVA overload; after, line 2-4 at MVA limit, generator at max VAR limit

Note the variety of ways that a power system can be optimized using an OPF. For example, some power system operators may wish to simply reduce system losses through the adjustment of generator voltage schedules—this is often done with hydrosystems where the generator MW outputs must be kept on a fixed time schedule to meet hydro-requirements.

13.6   INTERIOR POINT ALGORITHM

In 1984, Karmarkar (refeence 10) presented a new solution algorithm for linear programming problems that did not solve for the optimal solution by following a series of points that were on the "constraint boundary" but, rather, followed a path through the interior of the constraints directly toward the optimal solution on the constraint boundary. This solution was much faster than conventional LP algorithms.

In 1986, Gill *et al.* (reference 11) showed the relationship between Karmarkar's algorithm and the so-called "logarithmic barrier function algorithm." This algorithm has become the basis for many OPF solution algorithms and is explained in reference 12.

In this derivation, no distinction is made between the control variables and the state variables; rather, all variables are considered in the $\mathbf{x}$ vector. The objective function will be $f(\mathbf{x})$. The constraints will be brokem into equality constraints and inequality constraints. The equality constraints are $\mathbf{g}(\mathbf{x}) = 0$ and the inequality constraints are

$$\mathbf{h}^- \le h(\mathbf{x}) \le \mathbf{h}^+ \tag{13.109}$$

where the $\mathbf{h}^-$ and $\mathbf{h}^+$ vectors are the lower and upper limits on the inequality constraints, respectively. Finally, we restrict the variables themselves to be within lower and upper bounds

$$\mathbf{x}^- \le \mathbf{x} \le \mathbf{x}^+ \tag{13.110}$$

The first step in transforming this problem is to add slack variables so that all the equations become equality constraints. We then obtain the following set of equations:

$$\min f(\mathbf{x})$$

$$g(\mathbf{x}) = 0$$

$$h(\mathbf{x}) + \mathbf{s}_h = \mathbf{h}^+$$

$$\mathbf{s}_h + \mathbf{s}_{sh} = \mathbf{h}^+ - \mathbf{h}^- \tag{13.111}$$

$$\mathbf{x} + \mathbf{s}_x = \mathbf{x}^+$$

$$\mathbf{x} - \mathbf{x}^- \ge 0, \mathbf{s}_x, \mathbf{s}_h, \mathbf{s}_{sh} \ge 0$$

Note that we now have a set of equations with all equality constraints except the final consisting of nonnegativity conditions on $\mathbf{x} - \mathbf{x}^-$ and the slack variables. These nonnegativity conditions are handled by adding what is called a "logarithmic barrier function" to the objective. Basically, this is a form of penalty function which becomes very large as the function or variable gets close to zero. The new objective function then looks like:

$$f_\mu = f(\mathbf{x}) - \mu \sum_j \ln(x - x^-)_j - \mu \sum_j \ln(s_x)_j - \mu \sum_i \ln(s_h)_i - \mu \sum_i \ln(s_{sh})_i$$

$$(13.112)$$

The parameter, μ, is called the "barrier parameter" and is a positive number that is forced to go to zero as the algorithm converges to the optimum. This then presents us with the Lagrange equation:

$$\begin{aligned}
\mathcal{L}_\mu = {}& f(\mathbf{x}) - \boldsymbol{\lambda}^T g(\mathbf{x}) - \boldsymbol{\lambda}_h^T[\mathbf{h}^+ - \mathbf{s}_h - h(\mathbf{x})] \\
& - \boldsymbol{\lambda}_{sh}^T(\mathbf{h}^+ - \mathbf{h}^- - \mathbf{s}_h - \mathbf{s}_{sh}) - \boldsymbol{\lambda}_x^T(\mathbf{x}^+ - \mathbf{x} - \mathbf{s}_x) \\
& - \mu \sum_j \ln(x - x^-)_j - \mu \sum_j \ln(s_x)_j - \mu \sum_i \ln(s_h)_i - \mu \sum_i \ln(s_{sh})_i
\end{aligned}$$

$$(13.113)$$

The solution to this Lagrangian equation is obtained by setting its gradient to zero:

$$\begin{aligned}
\nabla_x \mathcal{L}_\mu &= \nabla f(\mathbf{x}) - \nabla_g(\mathbf{x})^T \boldsymbol{\lambda} + \nabla h(\mathbf{x})^T \boldsymbol{\lambda}_h + \boldsymbol{\lambda}_x - \mu(\mathbf{x} - \mathbf{x}^-)^{-1} e = 0 \\
\nabla_{s_h} \mathcal{L}_\mu &= \boldsymbol{\lambda}_h + \boldsymbol{\lambda}_{sh} - \mu_h^{-1} e = 0 \\
\nabla_{sh} \mathcal{L}_\mu &= \boldsymbol{\lambda}_{sh} - \mu s_{sh}^{-1} e = 0 \\
\nabla_{s_x} \mathcal{L}_\mu &= \boldsymbol{\lambda}_x - \mu s_s^{-1} e = 0 \\
\nabla_\lambda \mathcal{L}_\mu &= -g(\mathbf{x}) \\
\nabla_{\lambda_{sh}} \mathcal{L}_\mu &= h(\mathbf{x}) + s_h - h^+ \\
\nabla_{\lambda_x} \mathcal{L}_\mu &= \mathbf{x} + s_x - x^+ \\
\nabla_{\lambda_h} \mathcal{L}_\mu &= s_h + s_{sh} - h^+ + h^-
\end{aligned}$$

$$(13.114)$$

These nonlinear equations are then solved iteratively by Newton's method, and the value of μ is adjusted toward zero.

The solution produces the values of the dual variables, some of which are the marginal costs for the real and reactive power at the buses. These bus incremental costs, BICs are the subject of the next section. Note that in Chapter 10, the BICs were calculated using an interior point OPF.

13.7 BUS INCREMENTAL COSTS

If we take the classical Lagrange equation for an optimal power flow:

$$\mathscr{L}(\mathbf{x}, \mathbf{u}, \mathbf{p}) = f(\mathbf{x}, \mathbf{u}) + \lambda^t g(\mathbf{x}, \mathbf{u}, \mathbf{p}) \tag{13.115}$$

and we asume that we have an optimal solution to this equation, then we can ask an interesting question: "What is the change in the optimal operating cost if we change one of the parameters $\mathbf{p}$?" More specifically: 'What is the change in optimal operating cost if we change the power produced or consumed at a bus in the network?" Thus, what we want is the following derivative:

$$\frac{\partial \mathscr{L}}{\partial P_i}$$

If we expand the Lagrange equation as follows:

$$\mathscr{L}(\mathbf{x}, \mathbf{u}, \mathbf{p}) = \sum_{gen} F_i(P_i) + F_{ref}[P_{ref}(/E|, \theta)] + [\lambda_1 \lambda_2, \ldots, \lambda_m] \begin{bmatrix} P_i^{net} - P_i(|\mathbf{E}|, \theta) \\ Q_i^{net} - Q_i(|\mathbf{E}|, \theta) \\ \vdots \end{bmatrix}$$

$$\tag{13.116}$$

The derivative of $\mathscr{L}$ with respect to P_i is simple, since the parameters only appear in the second part of the Lagrange equation. The resulting derivative for bus i is:

$$\frac{\partial \mathscr{L}}{\partial P_i} = \lambda_i \tag{13.117}$$

We see that the interpretation of the vector of Lagrange multipliers is that they indicate the increment in optimal cost with respect to small changes in the parameters of the network. In the case of small change in power, either consumed or produced at a bus, the Lagrange multiplier for that bus then indicates the incremental cost that will be incurred as a result of this change. This cost has been given the name "bus incremental cost" or BIC and is the same incremental cost we dealt with in the beginning of the text, where we derived the incremental cost of delivery of power from a generator. A power system is in economic dispatch when the BIC for each generator matches the generator's own incremental cost for the power it is producing.

The BIC is a useful concept for nondispatched generator buses and for evaluating the marginal cost of wheeling. In some proposed schemes, this bus incremental cost is used to establish the spot market price for energy.

One point is worth noting before we leave this topic. The above discussion assumed that one has the vector of Lagrange multipliers for an optimal solution. However, depending on the method used to solve the OPF, this may not be the case. Certainly, in the case of the OPF that is based on linear programming, the λ values are not available unless a special formulation is used—yet we need the BICs for the buses.

The Lagrange equation at the optimal solution can be used to solve for the Lambda vector, even through it was not used in the OPF alogorithm. This is because, at the optimal solution to the OPF, the Lagrange equation is assumed to satisfy,

$$\nabla \mathcal{L} = 0 \qquad (13.118)$$

or, for the state variable, $\mathbf{x}$, we have:

$$\nabla \mathcal{L}_x = \frac{\partial \mathcal{L}}{\partial \mathbf{x}} = \frac{\partial f}{\partial \mathbf{x}} + \left[\frac{\partial \mathbf{g}}{\partial \mathbf{x}} \right]^T \lambda = 0 \qquad (13.119)$$

which can be used to solve for λ as follows:

$$\left[\frac{\partial \mathbf{g}}{\partial \mathbf{x}} \right]^T \lambda = - \frac{\partial f}{\partial \mathbf{x}} \qquad (13.120)$$

The problem here is that the matrix

$$\left[\frac{\partial \mathbf{g}}{\partial \mathbf{x}} \right]^T \qquad (13.121)$$

has N rows where N equals the number of state variables, and M columns corresponding to M binding constraints. We shall assume that $N \leq M$. The vector

$$\frac{\partial f}{\partial \mathbf{x}}$$

has N elements and the lambda vector, λ, has M elements. Thus, the equation which can be used to solve for the lambda vector is overdetermined; that is, there are more elements in the lambda vector than rows in the matrix or the right-hand side. This type of equation has many solutions for the lambda vector. The correct one is found by applying a least-squares technique, as explained in Chapter 12 on state estimation. Further, the usual method of solving for the lambda vector is to apply the QR algorithm (also explained in Chapter 12). Thus, we can use *any method* to solve for the optimal state vector for an OPF and then develop the matrix and right-hand side shown above and solve for the BIC vector.

EXAMPLE 13E

This example gives the bus incremental costs for the same six-bus sample used in Examples 13C and 13D. For the case where both generation MW and

PROBLEMS **555**

TABLE 13.10 **Bus Incremental Costs**

Bus	₹/MWh	₹/MVARh
1	12.22	0
2	11.89	0
3	11.97	0.1
4	12.98	0.81
5	12.59	0.51
6	12.29	0.38

scheduled voltages are adjusted to obtain minimum cost, the bus incremental costs are given in Table 13.10.

There is a cost for increasing the MW delivered, as well as the MVAR delivered from or to any bus in the network. In Table 13.10, the bus incremental costs for delivering MW at buses 1, 2, and 3 are equal to the incremental costs of the generator cost functions at the optimal dispatch. The bus incremental cost to deliver MVAR at buses 1 and 2 is zero since these generators are not at their maximum VAR limit and can generate incremental MVAR for "free." The incremental cost to deliver more MVARs at bus 3 is nonzero since generator 3 is at maximum VAR limit and one would have to generate the extra VARs at buses 1 and 2. Finally, the delivery points have higher bus incremental costs since they require that all MW and MVAR consumed at these buses must be delivered via the transmission system, which will cost the system in MW and MVAR losses.

In addition to the bus incremental costs, the procedure outlined above can also be used to generate the cost of changing the limit at any binding constraint. In the case of the dispatch used in Table 13.10, line 2-4 is at an MVA linit and bus 5 is at maximum voltage. The incremental cost with respect to changing the MVA limit on line 2-4 is −1.01 ₹/MVAh, indicating that if the limit were increased the system operating cost would decrease. Last of all, the incremental cost of changing the bus 5 upper voltage limit −88.4 ₹/pu volt.

PROBLEMS

13.1 You are going to use a linear program and a power flow to solve an OPF. The linear program will be used to solve constrained dispatch problems and the power flow will confirm that you have done the correct thing. For each of the problems, you should use the power flow data for the six-bus problem found in Chapter 4.

The following data on unit cost functions applies to this problem:

Unit 1 (bus 1): $F(P) = 600.0 + 6.0P + 0.002P^2$

$P_{min} = 70$ MW

$P_{max} = 250.0$ MW

Unit 2 (bus 2): $F(P) = 220.0 + 7.3P + 0.003P^2$

$P_{min} = 55$ MW

$P_{max} = 135$ MW

Unit 3 (bus 3): $F(P) = 100.0 + 8.0P + 0.004P^2$

$P_{min} = 60$ MW

$P_{max} = 160$ MW

When setting up the LP you should use three straight-line segments with break points as below:

Unit 1, break points at: 70, 130, 180, and 250 MW

Unit 2, break points at: 55, 75, 95, and 135 MW

Unit 3, break points at: 60, 80, 120, and 160 MW

When using the LP for dispatching you should ignore losses.
Set up the power flow as follows:

Load = 300 MW

Generation on bus 2 = 100 MW

Generation on bus 3 = 100 MW

This should lead to a flow of about 67 MW on line 3-6.

Using the linear program, set up a minimum cost LP for the three units using the break points above and the generation shift (or "*a*") factors from Figure 11.7. You are to constrain the system so that the flow on line 3-6 is no greater than 50 MW.

When you obtain an answer from the LP, enter the values for P_2 and P_3 found in the LP into the load flow and see if, indeed, the flow on line 3-6 is close to the 30 MW desired. (Be sure the load is still set to 300 MW.)

13.2   Using the six-bus power flow example from Chapter 4 with load at 240 MW, try to adjust the MW generated on the three generators and the voltage on each generator to minimize transmission losses. Keep the generators within their economic limits and the voltages at the generators within 0.90 to 1.07 pu volts. Use the following as MVAR limits:

Bus 2 generator: 100 MVAR max

Bus 3 generator: 60 MVAR max

13.3 Using the six-bus power flow example from Chapter 4, set up the base case as in Problem 13.1 (300 MW load, 100 MW on generator buses 2 and 3). Solve the base conditions and note that the load voltages on buses 4, 5, and 6 are quite low. Now, drop the line from bus 2 to bus 3 and resolve the power flow. (Note that the VAR limits on buses 2 and 3 should be the same as in Problem 13.2.)

This results in a severe voltage drop at bus 6. Can you correct this voltage so it comes back into normal range (e.g., 0.90 per unit to 1.07 per unit)? Suggested options: Add fixed capacitance to ground at bus 6, raise the voltage at one or more of the generators, reduce the load MW and MVAR at bus 6, etc.

13.4 You are going to solve the following optimal power flow in two different ways. Given a power system with two generators, P_1 and P_2, with their corresponding cost functions $F_1(P_1)$ and $F_2(P_2)$. In addition, the voltage magnitudes on the generator buses are also to be scheduled.

The balance between load and generation will be assumed to be governed by a linear constraint:

$$\sum_i \beta_i P_i = \sum_i \beta_i P_i^0$$

In addition, two constraints have been identified and their sensitivities calculated. The first is a flow constraint where:

$$\Delta \text{flow}_{nm} = \sum_i af_i \Delta P_i + \sum_i av_i \Delta V_i$$

The second constraint involves a voltage magnitude at bus k which is assumed to be sensitive only to the generator voltages:

$$\Delta V_k = \sum_i \gamma_i \Delta V_i$$

a. Assume that the initial generator outputs are P_1^0 and P_2^0 and that the initial voltage magnitudes are V_1^0 and V_2^0 and that you have obtained the initial flow, flow_{nm}^0, and the initial voltage, V_k^0, from a power flow program.

Further assume that there are limits to be constrained flow and voltage: flow_{nm}^+ and flow_{nm}^- and for the voltage V_k^+ and V_k^-.

Express the flow on line *nm* and the voltage on bus *k* as linear functions of the four control variables: P_1, P_2, V_1, V_2.

b. Show how to obtain the minimum cost with the gradient method. In this case, you may assume that the flow constraint and the voltage constraint are equality constraints where we desire the constraints to be scheduled to the upper limit. Any matrices in this formulation should be shown with all terms; if the inverse is needed, just express it as an inverse matrix—do not try to show all the terms in the inverse itself.

c. Show the same minimun cost dispatch solution with an LP where we break each cost function into two segments.

FURTHER READING

Reference 1 is considered the classic paper that first introduced the concept of an optimal power flow. References 2 and 3 give a good overview of the techniques and methods of OPFs. Reference 4 is a good introduction to the basic mathematics of the gradient method, and references 5–7 cover the Newton OPF method.

Reference 8 shows how the bus incremental costs are calculated using a least-squares approach. Reference 9 is an excellent paper dealing with the application of linear programming to the OPF solution. References 10 and 11 introduce the concept of the interior point algorithm. References 12 and 13 deal with the application of the interior point algorithm to the OPF solution. References 14 and 15 talk extensively about how to incorporate security constraints into the OPF, while reference 16 shows some of the special AGC logic needed when an OPF is holding a line flow constraint.

1. Carpienter, J., "Contribution e l'étude do Dispatching Economique," *Bulletin Society Française Electriciens*, Vol. 3, August 1962.
2. Carpentier, J., "Optimal Power Flows," *Int. J. Electric Power and Energy Systems*, Vol 1, April 1979, pp. 3–15.
3. Huneault, M., Galiana, F. D., "A Survey of the Optimal Power Flow Literature," *IEEE Transactions on Power Systems*, Vol. 6, No. 2, May 1991, pp. 762–770.
4. Dommel, H. W., Tinney, W. F., "Optimal Power Flow Solutions," *IEEE Transactions on Power Apparatus and Systems*, Vol. PAS-87, October 1968, pp. 1866–1876.
5. Burchett, R. C., Happ, H. H., Vierath, D. R., "Quadratically Convergent Optimal Power Flow," *IEEE Transactions on Power Apparatus and Systems*, Vol. PAS-103, November 1984, pp. 3267–3275.
6. Sun, D. I., Ashley, B., Brewer, B., Hughes, A., Tinney, W. F., "Optimal Power Flow by Newton Approach," *IEEE Transactions on Power Apparatus and Systems*, Vol. PAS-103, October 1984, pp. 2864–2880.
7. Bjelogrlic, M., Calovic, M. S., Ristanovic, P., Babic, B. S., "Application of Newton's Optimal Power Flow in Voltage/Reative Power Control," *IEEE Transactions on Power Systems*, Vol. 5, No. 4, November 1990, pp. 1447–1454.
8. Venkatesh, S. V., Liu, W. E., Papalexopoulos, A. D., "A Least Squares Solution for Optimal Power Flow Sensitivity Calculation," *IEEE Transactions on Power Systems*, Vol. 7, No. 3, August 1992, pp. 1394–1401.

9. Alsac, O., Bright, J., Prais, M., Stott, B., "Further Developments in LP-Based Optimal Power Flow," *IEEE Transactions on Power Systems*, Vol. 5, No. 3, Augsut 1990, pp. 697–711.

10. Karmarkar, N., "A New Polynomial-Time Algorithm for Linear Programming," *Combinatorica*, Vol. 4, No. 4, 1984, pp. 373–395.

11. Gill, P. E., Murray, W., Saunders, A., Tomlin, J. A., Wright, M. H., "On Projected Newton Barrier Methods for Linear Programming and an Equivalence to Karmarkar's Projective Method," *Mathematical Programming*, Vol. 36, pp. 183–209.

12. Wu, Y., Debs, A. S., Marsten, R. E., "Direct Nonlinear Predictor–Corrector Primal–Dual Interior Point Algorithm for Optimal Power Flows," 1993 *IEEE Power Industry Computer Applications Conference*, pp. 138–145.

13. Vargas, L. S., Quintana, V. H., Vannelli, A., "Tutorial Description of an Interior Point Method and Its Applications to Security–Constrained Econimic Dispatch," *IEEE Transactions on Power Systems*, Vol. 8, No. 3, August 1993, pp. 1315–1325.

14. Alsac, O., Stott, B., "Optimal Power Flow with Steady-State Security," *IEEE Transactions on Power Apparatus and Systems*, Vol. PAS-93, May/June 1974, pp. 745–751.

15. Burchett, R. C., Happ, H. H., "Large-Scale Security Dispatching: An Exact Model," *IEEE Transactions on Power Apparatus and Systems*, Vol. PAS-102, September 1983, pp. 2995–2999.

16. Bacher, R., Van Meeteren, H. P., "Real-Time Optimal Power Flow in Automatic Generation Control," *IEEE Transactions on Power Systems*, Vol. 3, No. 4, November 1988, pp. 1518–1529.

Appendix: About the Software

The accompanying diskette contains an install application that loads on your computer seven power generation analysis programs useful in teaching a course centered on this book. The programs are intended for educational use and are not meant for commercial use. They are provided free of charge and may be copied and distributed to students. Copies of the program source code and documentation on the disks may also be copied and distributed.

A.1 INSTALLING THE DISKETTE FILES

The files require about 1 MB of hard drive space. To install the files, perform the following steps:

1. Assuming you will be using the drive A as the floppy drive for your diskette, at the A:> prompt type INSTALL. You may also type A:INSTALL at the C:> prompt.
2. Follow the instructions displayed by the installation program. The default choice for the installation directory is POWERGEN and the default drive is C.

A.2 DESCRIPTION OF PROGRAM FILES

The programs, which are written in Turbo Pascal 6.0, have also been tested under Borland Pascal 7.0 and can be run under DOS on a standard PC. The programs can be run separately from the DOS command line or through the point-and-click POWERGEN interface as shown in Figure A.1. Each program has a documentation file, a source file of the program, an executable file, several data files that contain cases to be run by the program, and in some cases a file describing some exercises you may wish to give students.

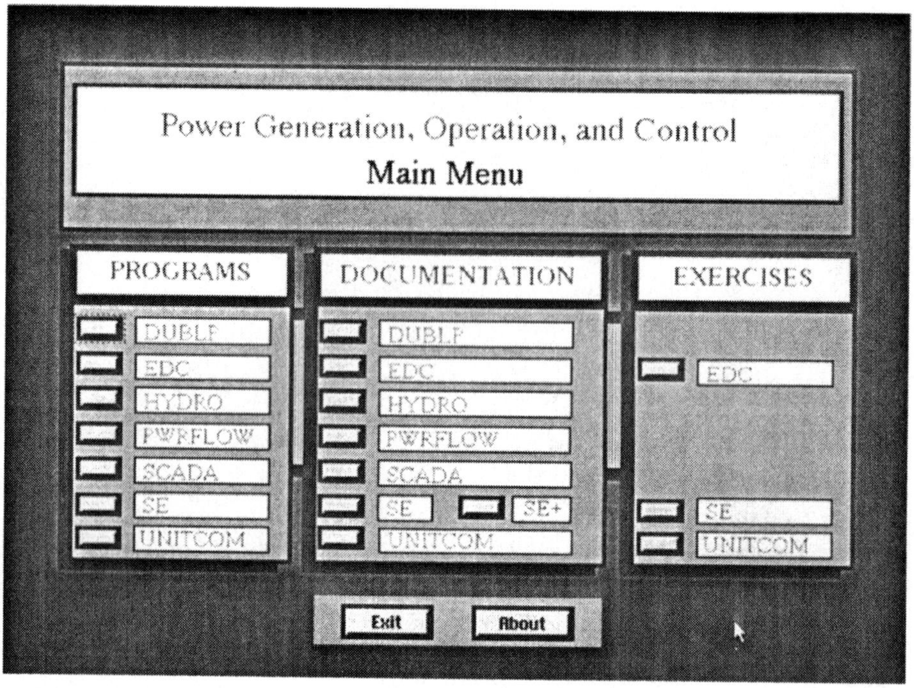

FIG. A.1

For example for program XYZ we might have:

XYZ.DOC	Documentation file
XYZ.PAS	PASCAL Source File
XYZ.EXE	PASCAL Executable File
XYZ1.DAT	First data file
XYZ2.DAT	Second data file
XYZ.EXR	Student exercises etc.

The documentation file should be read before you attempt to run the program. You can view and print these ASCII files with any standard text editor and you can also read the files by pressing the appropriate DOCUMENTATION button within the POWERGEN interface. The documentation explains how the program works and how to set up data files for the homework problems. Three of the programs also include exercise files that include suggestions for setting up experiments. The exercise files can be viewed with a text editor or by pressing a button in the EXERCISES section of the POWERGEN interface.

The source files for the PASCAL programs are also ASCII files. They can be printed in the same way as the documentation file if you care to see the programs.

A.3   USING THE PROGRAMS

The disk you received has the following programs:

Economic Dispatch (EDC)
Unit Commitment (UNITCOM)
Linear Programming (DUBLP)
Hydro Scheduling via Dynamic Programming (HYDRO)
Power Flow (PWRFLOW)
State Estimation (SE)
Supervisory Control and Data Acquisition (SCADA)
Power Generation interface program (POWERGEN)

To run the programs in the POWERGEN directory individually from the DOS command line, enter XYZ, where XYZ is the name of the PASCAL executable file for the program. Again, please read the documentation before running the programs to determine which data files are appropriate to use as input files.

You may also use the POWERGEN program to run the programs from a pushbutton interface. To run this program, type POWERGEN, press the ENTER key to continue past the title screen, then select the button of the desired application from under the PROGRAM section on the left.

If you have a printer available, it is advisable to run the programs with the printer option on so that all details of the program execution are printed.

The Pascal programs are written in Turbo Pascal 6.0 and if you wish to make any changes to them you must obtain that compiler or a later version from Borland. Please note that with the Pascal compilers from Borland, real variables must be formatted properly to be read directly into the program from a data file. By looking at any of the data files you can see that all real variables are formatted with a zero in front of the decimal in the case of a number less than one and a zero after the decimal for a number with no fractional part. Thus .001 is not acceptable, you must put in 0.001 and 100. is not acceptable you must put in 100.0 (also, 100 is not acceptable if the variable is a real number, it must be 100.0). Integers have no restrictions. Finally, separate all variables by spaces, not commas.

A.4   GETTING HELP

If you need help, call (612) 626-7192 (8 am–5 pm Central Time USA); leave a message if no answer. FAX (612) 625-4583, or email to wollenbe@ee.umn.edu. To replace defective disks or for inquiries about installing the software, call (212) 850-8717 for John Wiley technical support.

Any comments on the usefulness of these programs and/or program bugs found would be greatly appreciated and should be sent to:

Bruce Wollenberg
Electrical Engineering Department
University of Minnesota
200 Union Street S.E., Room 4-174
Minneapolis, MN 55455

INDEX